P9-CKY-037

VOLUME 2
700-1449

Science and Its Times

Understanding the Social Significance of Scientific Discovery

VOLUME 2
700-1449

Science and Its Times

Understanding the Social Significance of Scientific Discovery

Neil Schlager, Editor
Josh Lauer, Associate Editor

Produced by Schlager Information Group

Detroit
New York
San Francisco
London
Boston
Woodbridge, CT

Science and Its Times

VOLUME 2

700-1449

NEIL SCHLAGER, *Editor*
JOSH LAUER, *Associate Editor*

GALE GROUP STAFF

Amy Loerch Strumolo, *Project Coordinator*
Christine B. Jeryan, *Contributing Editor*

Mark Springer, *Editorial Technical Specialist*

Maria Franklin, *Permissions Manager*
Margaret A. Chamberlain, *Permissions Specialist*
Debra Freitas, *Permissions Associate*

Mary Beth Trimper, *Production Director*
Evi Seoud, *Assistant Production Manager*
Stacy L. Melson, *Buyer*

Cynthia D. Baldwin, *Product Design Manager*
Tracey Rowens, *Senior Art Director*
Barbara Yarrow, *Graphic Services Manager*
Randy Bassett, *Image Database Supervisor*
Mike Logusz, *Imaging Specialist*
Pamela A. Reed, *Photography Coordinator*
Leitha Etheridge-Sims *Image Cataloger*

ISBN: 0-7876-3934-6

Printed in the United States of America
10 9 8 7 6 5 4 3 2 1

Library of Congress Cataloging-in-Publication Data

Science and its times : understanding the social significance of scientific discovery / Neil Schlager, editor.
p.cm.
Includes bibliographical references and index.
ISBN 0-7876-3933-8 (vol. 1 : alk. paper) — ISBN 0-7876-3934-6 (vol. 2 : alk. paper) — ISBN 0-7876-3935-4 (vol. 3 : alk. paper) — ISBN 0-7876-3936-2 (vol. 4 : alk. paper) — ISBN 0-7876-3937-0 (vol. 5 : alk. paper) — ISBN 0-7876-3938-9 (vol. 6 : alk. paper) — ISBN 0-7876-3939-7 (vol. 7 : alk. paper) — ISBN 0-7876-3932-X (set : hardcover)
1. Science—Social aspects—History. I. Schlager, Neil, 1966-
Q175.46 .S35 2001
509—dc21 00-037542

Contents

Contents

700-1449

Mathematics

Physical Sciences

Technology and Invention

Preface

The interaction of science and society is increasingly a focal point of high school studies, and with good reason: by exploring the achievements of science within their historical context, students can better understand a given event, era, or culture. This cross-disciplinary approach to science is at the heart of *Science and Its Times*.

Readers of *Science and Its Times* will find a comprehensive treatment of the history of science, including specific events, issues, and trends through history as well as the scientists who set in motion—or who were influenced by—those events. From the ancient world's invention of the plowshare and development of seafaring vessels; to the Renaissance-era conflict between the Catholic Church and scientists advocating a sun-centered solar system; to the development of modern surgery in the nineteenth century; and to the mass migration of European scientists to the United States as a result of Adolf Hitler's Nazi regime in Germany during the 1930s and 1940s, science's involvement in human progress—and sometimes brutality—is indisputable.

While science has had an enormous impact on society, that impact has often worked in the opposite direction, with social norms greatly influencing the course of scientific achievement through the ages. In the same way, just as history can not be viewed as an unbroken line of ever-expanding progress, neither can science be seen as a string of ever-more amazing triumphs. *Science and Its Times* aims to present the history of science within its historical context—a context marked not only by genius and stunning invention but also by war, disease, bigotry, and persecution.

Format of the Series

Science and Its Times is divided into seven volumes, each covering a distinct time period:

Volume 1: 2000 B.C.-699 A.D.

Volume 2: 700-1449

Volume 3: 1450-1699

Volume 4: 1700-1799

Volume 5: 1800-1899

Volume 6: 1900-1949

Volume 7: 1950-present

Dividing the history of science according to such strict chronological subsets has its own drawbacks. Many scientific events—and scientists themselves—overlap two different time periods. Also, throughout history it has been common for the impact of a certain scientific advancement to fall much later than the advancement itself. Readers looking for information about a topic should begin their search by checking the index at the back of each volume. Readers perusing more than one volume may find the same scientist featured in two different volumes.

Readers should also be aware that many scientists worked in more than one discipline during their lives. In such cases, scientists may be featured in two different chapters in the same volume. To facilitate searches for a specific person or subject, main entries on a given person or subject are indicated by bold-faced page numbers in the index.

Within each volume, material is divided into chapters according to subject area. For volumes 5, 6, and 7, these areas are: Exploration and Discovery, Life Sciences, Mathematics, Medicine, Physical Sciences, and Technology and Invention. For volumes 1, 2, 3, and 4, readers will find that the Life Sciences and Medicine chapters have been combined into a single section, reflecting the historical union of these disciplines before 1800.

Arrangement of Volume 2: 700-1449

Volume 2 begins with two notable sections in the frontmatter: a general introduction to science and society during the period, and a general chronology that presents key scientific events during the period alongside key world historical events.

The volume is then organized into five chapters, corresponding to the five subject areas listed above in "Format of the Series." Within each chapter, readers will find the following entry types:

Chronology of Key Events: Notable events in the subject area during the period are featured in this section.

Overview: This essay provides an overview of important trends, issues, and scientists in the subject area during the period.

Topical Essays: Ranging between 1,500 and 2,000 words, these essays discuss notable events, issues, and trends in a given subject area. Each essay includes a Further Reading section that points users to additional sources of information on the topic, including books, articles, and web sites.

Biographical Sketches: Key scientists during the era are featured in entries ranging between 500 and 1,000 words in length.

Biographical Mentions: Additional brief biographical entries on notable scientists during the era.

Bibliography of Primary Source Documents: These annotated bibliographic listings feature key books and articles pertaining to the subject area.

Following the final chapter are two additional sections: a general bibliography of sources related to the history of science, and a general subject index. Readers are urged to make heavy use of the index, because many scientists and topics are discussed in several different entries.

A note should be made about the arrangement of individual entries within each chapter: while the long and short biographical sketches are arranged alphabetically according to the scientist's surname, the topical essays lend themselves to no such easy arrangement. Again, readers looking for a specific topic should consult the index. Readers wanting to browse the list of essays in a given subject area can refer to the table of contents in the book's frontmatter.

Additional Features

Throughout each volume readers will find sidebars whose purpose is to feature interesting events or issues that otherwise might be overlooked. These sidebars add an engaging element to the more straightforward presentation of science and its times in the rest of the entries. In addition, each volume contains photographs, illustrations, and maps scattered throughout the chapters.

Comments and Suggestions

Your comments on this series and suggestions for future editions are welcome. Please write: The Editor, *Science and Its Times,* Gale Group, 27500 Drake Road, Farmington Hills, MI 48331.

Advisory Board

Amir Alexander
Research Fellow
Center for 17th and 18th Century Studies
UCLA

Amy Sue Bix
Associate Professor of History
Iowa State University

Elizabeth Fee
Chief, History of Medicine Division
National Library of Medicine

Lois N. Magner
Professor Emerita
Purdue University

Henry Petroski
A.S. Vesic Professor of Civil Engineering and Professor of History
Duke University

F. Jamil Ragep
Associate Professor of the History of Science
University of Oklahoma

David L. Roberts
Post-Doctoral Fellow, National Academy of Education

Morton L. Schagrin
Emeritus Professor of Philosophy and History of Science
SUNY College at Fredonia

Hilda K. Weisburg
Library Media Specialist
Morristown High School, Morristown, NJ

Contributors

Kristy Wilson Bowers
University of Maryland

Sherri Chasin Calvo
Freelance Writer

Matt Dowd
Graduate Student
University of Notre Dame

Thomas Drucker
Graduate Student, Department of Philosophy
University of Wisconsin

H. J. Eisenman
Professor of History
University of Missouri-Rolla

Ellen Elghobashi
Freelance Writer

Loren Butler Feffer
Independent Scholar

Keith Ferrell
Freelance Writer

Randolph Fillmore
Freelance Science Writer

Richard Fitzgerald
Freelance Writer

Maura C. Flannery
Professor of Biology
St. John's University, New York

Donald R. Franceschetti
Distinguished Service Professor of Physics and Chemistry
The University of Memphis

Diane K. Hawkins
Head, Reference Services—Health Sciences Library
SUNY Upstate Medical University

Robert Hendrick
Professor of History
St. John's University, New York

James J. Hoffmann
Diablo Valley College

Leslie Hutchinson
Freelance Writer

P. Andrew Karam
Environmental Medicine Department
University of Rochester

Evelyn B. Kelly
Professor of Education
Saint Leo University, Florida

Rebecca Brookfield Kinraide
Freelance Writer

Judson Knight
Freelance Writer

Lyndall Landauer
Professor of History
Lake Tahoe Community College

Josh Lauer
Editor and Writer
President, Lauer InfoText Inc.

Adrienne Wilmoth Lerner
Department of History
Vanderbilt University

Brenda Wilmoth Lerner
Science Correspondent

Contributors

700-1449

K. Lee Lerner
Prof. Fellow (r), Science Research & Policy Institute
Advanced Physics, Chemistry and Mathematics, Shaw School

E. D. Lloyd-Kimbrel
Freelance Writer

Eric v. d. Luft
Curator of Historical Collections
SUNY Upstate Medical University

Lois N. Magner
Professor Emerita
Purdue University

Amy Lewis Marquis
Freelance Writer

Ann T. Marsden
Writer

William McPeak
Independent Scholar
Institute for Historical Study (San Francisco)

Leslie Mertz
Biologist and Freelance Science Writer

J. William Moncrief
Professor of Chemistry
Lyon College

Stacey R. Murray
Freelance Writer

Stephen D. Norton
Committee on the History & Philosophy of Science
University of Maryland, College Park

Neil Schlager
Editor and Writer
President, Schlager Information Group

Dean Swinford
Ph.D. Candidate
University of Florida

Lana Thompson
Freelance Writer

Todd Timmons
Mathematics Department
Westark College

Philippa Tucker
Post-graduate Student
Victoria University of Wellington, New Zealand

David Tulloch
Graduate Student
Victoria University of Wellington, New Zealand

Stephanie Watson
Freelance Writer

Michael T. Yancey
Freelance Writer

Introduction: 700–1449

Overview

The centuries between 700 and 1449 encompassed the bulk of the Middle Ages, the first glimmerings of the Renaissance, dramatic technical and cultural advances in Asia, the expansion and contraction of the Muslim Empire, and the pinnacle of the Mayan and Incan civilizations in the New World.

It was a time of ferment and chaos, but also a period of stasis. This was particularly true in Europe, where the collapse of the Roman Empire left a centuries-long void that no single nation or unifying body was able to fill. Lacking a central presence to focus culture, and without the economic resources generated by large, well-organized alliances, much of Europe descended into purely local governance, often centered around lords who ruled their immediate vicinities without the means to pursue any larger ambitions.

The often-desperate poverty that covered Europe was matched by a deepening ignorance. Intellectual pursuits like education, philosophy, and the study of science were luxuries that held little appeal when starvation and disease were rampant. In addition, local rulers were far more concerned with maintaining their own fragile power than with becoming patrons of the arts and sciences. Instead, monasteries became centers of learning and played an enormous role in keeping the spark of scholarship glowing through the darkest of these centuries, known as the Dark Ages. This loss of knowledge was perhaps the greatest risk Europe faced.

In the Islamic nations, however, learning was not only alive, but flourishing. Mathematics and chemistry benefited particularly from the Arabic preservation of ancient Greek manuscripts and treatises. Islam, the Muslim religion and the heart of Arabic culture, placed great importance on the works of scholars and artists. Islamic rulers endowed schools, and in doing so underwrote a body of knowledge that would flow readily westward as Europe, early in the next millennium, began to regain its strength and rebuild its culture. When European Crusaders ventured into Arab lands beginning in 1096, they returned with many of the Greek classics preserved by the Arabs. In addition, early Arabic explorers and traders were vital conduits for the transit of both preserved classical knowledge and imported Asian knowledge.

Asian cultures and civilizations grew greatly during this period, producing many technological and scientific accomplishments that would be copied by Western nations or discovered independently hundreds of years later. The Chinese discovered the magnetic compass, invented gunpowder, and invented printing. Indian mathematicians developed numeral system we use today, and gave the world the mathematical gift of the zero.

In the Americas, great civilizations including the Maya, the Pueblo, the Inca, and the Aztec flourished. Some, such as the Maya, would not last much past the 1440s. Others would not survive their encounters with Europeans, which came in the late 1400s and early 1500s.

Protecting and Transmitting Knowledge

By 700 the great civilizations of Rome and Greece were receding farther and farther into the past, leaving a chaotic tangle of European nations, states, and dominions with little interest in learning. The intellectual curiosity of the Arab world, though, salvaged much of classical culture, preserving it and, indeed, celebrating its accomplishments and building upon them. Perhaps most dramatically, the Baghdad Academy of Science, begun in 800 and sponsored by Harun al-Raschid, became one of the world's great centers of learning, and the source for much of the mathematical innovation that would flow outward over the next two centuries.

In China and Japan, the preservation of knowledge remained an innate aspect of culture—hardly surprising given that their civilizations had not collapsed into chaos as had Europe's. Written language was especially important to the large Asian civilizations, and its dissemination led to the development of three key technologies: ink, paper, and printing.

Paper and ink were both developed by the Chinese, although ink was also known to ancient Egyptians. Ink was in use by about 2500 B.C., and paper around A.D. 105. Block printing was developed in China around the sixth century. By carving an entire page of a document into a single block, multiple copies of the document could be duplicated rapidly and efficiently.

Adapted by the Chinese in the following century, block printing became the centerpiece of an entire cultural industry, with millions of books being printed and distributed. The collection and preservation of knowledge was further enhanced by the sheer portability of printed material. Chinese books traveled westward and with them many of the insights and findings of classical Chinese culture. During subsequent centuries the Chinese and Koreans also introduced early versions of movable type.

Worlds of Numbers

Mathematics lies at the heart of most scientific disciplines and technologies, and while the mathematical innovations introduced between 700 and 1449 do not equal in volume those that came in the century immediately afterward, they remain among the most important and indispensable of all mathematical tools.

By far the most important of these tools is the zero. First postulated in India as early as A.D. 500, the zero traveled with traders to the Arab lands, where it took root, as had other Indian mathematical innovations. ("Arabic" numerals themselves are an Indian invention.) In 810 Muhammad ibn al-Khwarizmi (780-850) wrote a book that gave the zero and its properties, which simultaneously simplified mathematics and increased their power, to the world. This book also contained the first use of the word *al-jabr,* which we know today as algebra.

Arabic numerals completed their journey from India to Europe through the work Leonardo Fibonacci (c. 1170-c. 1240), who wrote of the efficiency of the numerical system and of al-Khwarizmi's mathematical insights. Despite the clear advantages offered by Fibonacci, the unwieldy system of Roman numerals would persist throughout Europe for another 300 years.

China's commitment to the preservation and distribution of printed knowledge also made a tremendous contribution to the survival and growth of mathematics. The thirteenth-century volumes *Discussion of the Old Sources* and *Mathematical Treatise in Nine Sections* were essentially encyclopedias of the mathematical universe; their translation and ongoing publication ensured China's role in mathematical development even after Chinese mathematical innovation entered decline early in the fourteenth century.

Gunpowder and Weaponry

Much technological innovation is driven by military ends, but the most central of all military innovations—gunpowder—seems to have come into existence for lighter purposes. Gunpowder was developed by the Chinese during the 700s, and for several centuries its primary purpose seems to have been to brighten the night sky in the form of fireworks, although by the thirteenth century it was being used as weapon, albeit ineffectively, against Mongol invaders.

While gunpowder would not come into its own as a weapon until the arrival of high-quality cast iron in the mid-1400s, advances in metalworking brought other weapons to prominence during the Middle Ages. In 732 Charles Martel (c. 688-741) led a Frankish army to victory over invading Muslims largely by virtue of the heavy armor with which his cavalry was equipped. For three centuries afterward armor played an important and often decisive role in military conflict. In 1050, however, improved crossbow designs emerghed, using advances in mechanics (cranks that amplified muscle power, enabling greater potential energy to be stored in the bowstring) and metalworking (steel-shafted arrows) to build a weapon able to penetrate chain mail and other armors.

Two and a half centuries later the crossbow met its match in the Welsh longbow. Loaded and fired by hand, the longbow could dispatch more arrows—and thus more enemies—in less time than crossbows. More importantly, skilled longbowmen could fire accurately as far as 300 yards, much farther than crossbows. This tactical advantage was exploited with particular ferocity by the English.

But crossbows and longbows were powered by men, their effectiveness was limited by the strength of their archers. In the mid-1300s, gunpowder became the most devastating of military technologies in Europe. Cannons were used as early as 1346 by Edward III of England in the opening battles of the Hundred Years' War against France. Those cannons, however, proved less effective than longbows, primarily because their barrels were poorly made.

This disadvantage did not last long. Advances in ironworking, especially the ability by the early 1400s to cast molten iron into hard, seamless objects rather than hammering it into a more brittle shape, were immediately applied to cannon manufacture. (The Chinese had possessed such metallurgical skills, notably the blast furnace, a full millennia before Europe, but seem not to have applied the skill to cannon making.) In 1439 Charles VII of France commissioned the casting of large numbers of cannons that allowed the walls of castles, once impervious to longbows and crossbows alike, to be breached with ease. With the development of the Spanish harquebus (an early matchlock gun) in 1450, the cannon itself was reduced to a size that could be operated, albeit with some difficulty, by a single person. Another century would elapse before individual guns would become common and effective, but the arrival of the harquebus can be seen as the starting point for the development of the modern rifle—and all of the consequences of gunpowder-driven military technology that have shaped the years since.

Practical Innovation

Another great driver of technological advance is practical necessity. In medieval Europe, that necessity often centered around food, and large advances in agricultural technology helped Europe climb out of chaos and into the light of the Renaissance.

Around 900 in northern Europe, a major step forward came with the invention of the horse collar. Previous harnessing systems had fastened around the horse's neck and throat—these allowed the horse to pull only so hard before it began to choke. The collar, placed against the horse's shoulders, permitted the horse to pull with full force. This almost unimaginable increase in animal power, when joined with the moldboard plow invented 300 years earlier, and the iron horseshoe, which dated from about 770, marked the beginning of an agricultural revolution.

Animals were not the only source of power. Waterwheels had been in use for centuries, and around 700 Persian inventors turned the same principle to harnessing of the wind. By 1180 the idea, with some improvements, most notably a vertical orientation, had found its way to France, where windmills began to spring up like giant flowers. They soon came to be the most common method of powering grain mills and water pumps.

Heat is as central a pragmatic concern as nourishment, particularly in cool northern climates. By the early 1200s, faced with the difficulty of transporting firewood over increasingly long (and increasingly deforested) distances, the English began to burn more coal, which had been used throughout Europe and China as a minor fuel for millennia. More portable than wood, coal also could be used to create hotter fires, vital to the emerging and advancing science of metallurgy.

Even when they're fed and warmed, people must be clothed, and by 1290 another Indian innovation had made the journey to Europe. The spinning wheel removed the endless drudgery of hand-spinning fiber into thread, replacing it with a mechanism powered by a foot-pedal. In addition to being an advance in fiber-spinning, the first wheels were also the first known examples of belt-driven machines: transmitting energy across axles to produce a spinning reaction.

Shelter, too, benefited from technological innovation, as better glass admitted more light into buildings and homes—and revealed more dirt. A slow increase in hygiene was an unexpected consequence of windows.

Perhaps the greatest of European architectural innovations came in 1137 with the introduction of the flying buttress, which enabled walls to support far greater weight and height than ever before. Buttresses concentrated support for heavier roofs, which made it possible not only to build less massive walls, but also to punctuate them with windows. The advance made possible many of the great European cathedrals that stand to this day.

Even vaster architecture of a completely different type appeared throughout the Americas. In 1050 the Mexican city of Casas Grandes, for example, began the excavation of an immense underground water supply system. A century earlier the Maya ended construction of a religious edifice in Uxmal, Mexico, the greatest of their architectural accomplishments.

In what would become Illinois, Mississippian Indians spent two centuries (900-1100) building a terraced mound burial site over 14 acres. The mound rose as high as 100 feet (30.48 meters), and was topped by an earthen building another 50 feet (15.24 meters) tall. It is estimated that more than 50 million cubic feet (15 million cubic meters) of earth were moved by hand to accomplish this, one of the greatest of earth-engineering accomplishments.

Exploration and Trade

The exchange of ideas is an important, if often unplanned, consequence of exploration and foreign trade. During the centuries between 700 and 1450, hundreds of ideas—papermaking, gunpowder, mathematical concepts, early printing techniques, and countless others—traveled with explorers and merchants. The technology of travel itself, particularly the arts of navigation and shipbuilding, also improved markedly during this period, due especially to the development of two key naval technologies: magnetic compasses and rudders.

The property of magnetism had been recognized as far back as the sixth century B.C. (it was named for the qualities of a mineral found near the city of Magnesia in Asia Minor) and had been described by Greeks including Thales. Around A.D. 200 the Chinese first recorded the north–south orientation of magnetic materials. It was in England, however, in 1180 that magnetic materials were first used as direction-finding tools, with improvements following rapidly, primarily from French experimenters, who gave the name *compass* to devices that identified magnetic north and south. In 1269 Petrus Peregrinus de Maricourt, a French scholar, recorded much of his research into the scientific nature of magnetic poles. It was the practical application of the compass, though, that had the greatest effect. Free of visible landmarks, sailors were able to determine their direction and fare ever farther afield. The compass opened the widest expanses of the seas and distant lands to expeditions confident of the directions in which they traveled, if not of the destinations they would discover.

Simple steering devices—generally oars held out behind ships and boats—had of course been used since the earliest days of seafaring, but by 1241 shipbuilders in northern Europe were simplifying these devices and incorporating them into the fixed design and construction of ships. The advantages in maneuverability were obvious and immediate, and the rudder became a universal element in ship design—and an important tactical tool for military ships. More maneuverable ships held a decided advantage over less agile ones.

While the compass and the rudder would combine to give European sailors the tools needed to sail the world's uncharted oceans, it should be noted that the impulse to explore was not completely fettered without them. Between 870 and 1000 the Vikings, aboard well built but relatively unsophisticated sailing craft, discovered the Arctic Circle, Greenland, Iceland, and Vinland (Labrador and Newfoundland), feats of European exploration that wouldn't be equaled until the circumnavigations of the world that lay nearly five centuries in the future. In the Americas, by 1250 the Maya had expanded their knowledge of navigation routes, sailing as far south as what is now Nicaragua.

New Ways of Seeing

While the golden age of physics awaited the arrival of the Renaissance and the birth of higher mathematics, important discoveries were made during the Middle Ages, many of them focusing on the way people view the world.

Al-Haytham, an Arab physicist known in the West as Alhazen (965-1039), speculated in 1025 that vision was enabled by rays of light reaching the eyes; previous theorists had proposed that the eyes themselves transmitted the beams that comprised vision. After making this discovery, al-Haytham devoted much of the rest of his life to studying the properties of lenses of various dimensions and curvatures, determining clearly that their effect on light was determined by the lenses' shape, not in the rays of light reaching them. He is considered the father of the science of optics.

By 1249 in England, Roger Bacon (c.1220-1292) applied optical principles to the development of lenses to overcome defective vision. Eyeglasses made a contemporaneous appearance in China; it is not known whether the idea arose independently in the two cultures, or was transmitted between them. Bacon was able to produce only convex lenses, most useful for the farsighted. An equally spectacular discovery was made in 1451, with the introduction of concave lenses, which improved the vision of the nearsighted.

Forty years later, in 1291, improvements in glass production, primarily in Venice, resulted in more nearly transparent glass. This greater transparency made it possible for a pane of glass to be placed over a piece of polished metal, producing a superior reflecting device. The mirror had been invented.

Looking Skyward

Advances in optics would ultimately lead to the telescope (1608), but even without its assistance the skies exerted a large attraction on the curious—and the superstitious. The great comet of 1066 (probably Halley's Comet) was viewed by many as an ominous portent of the Battle of Hastings (1066). It is also an important astronomical phenomenon preserved in artwork of the time. An even more dramatic event had been observed (and recorded by the Chinese) a dozen years earlier when a new star blazed brightly for three weeks in the constellation Taurus.

Less periodic heavenly occurrences were studied as well. In 1252 under the guidance of Alfonso X of Castile, an astronomer as well as a ruler, the first wholly new catalog of the planets since the time of Ptolemy 100 years earlier was undertaken. Although hampered by the lack of telescopic devices and in need of mathematical tools yet to be invented, the Alfonsine tables, as they came to be known, were a large contribution to the advance of astronomy.

The Darkness Brightens

By 1449 the major elements of the Renaissance and its spectacular flowering of art, science, culture, and technology were in place. Many of those elements had, of course, existed in civilizations such as China for centuries, but it was the European nations that most aggressively exploited and exported them. Ironically, in the centuries following 1449, both China and Japan would basically withdraw from commerce and correspondence with the West, remaining insular until well after the Industrial Revolution.

The long period of European intellectual dormancy would prove fertile soil for science and technology: The period between 1450 and the present is far shorter than the period covered in this essay, yet it has seen us travel from tentative exploration of the oceans to a permanent presence in outer space, from the Scientific Revolution to the Industrial Revolution to the Information Age.

Even as 1449 drew to a close, the key to the future was taking shape in the hands of Johannes Gutenberg (c. 1390-1468), who began experiments with movable type in 1435. By 1454 his printing press would prove its worth. It was this technology, more than any other, which ensured that human culture would never again face the risk that whole bodies of knowledge would be lost to darkness.

KEITH FERRELL

Chronology: 700–1449

c. 700-c. 900 Arab doctors systematically translate and adapt ancient Greek medical texts, ensuring that the advanced learning of antiquity is preserved.

750-751 A pivotal two years: in 750, the Umayyad Caliphate gives way to the Abbasids in the Middle East; in the Americas, the Huari civilization and the city-state of Teotihuacán begin to die out. In 751, the Carolingian dynasty replaces the Merovingians in Western Europe, and China's T'ang Dynasty begins to decline after defeat by the Arabs at Talas.

800 Charlemagne is crowned "emperor of all the Romans" by Pope Leo III on Christmas Day; this is the origin of the Holy Roman Empire, and of a sometimes close—but more often tempestuous—relationship between Church and state during the Middle Ages.

c. 800-c. 1000 The Vikings burst out of Scandinavia to terrorize Western Europe, colonize lands far to the west, establish Russia, and in their later incarnation as Normans ("Northmen"), continue to influence history.

820 Al-Khwarizmi, an Arab mathematician, writes a mathematical text that introduces the word "algebra" (*al-jabr* in Arabic), as well as Indian numerals, including zero; henceforth these are mistakenly referred to as Arabic numerals.

c. 850-c. 950 First serious challenges to Ptolemy: Irish-born philosopher John Scotus Erigena suggests that the planets revolve around the Sun (c. 850); Albategnius clarifies a number of fine points in Ptolemy (900); and Abd al-Rahman al-Sufi revives Ptolemy's catalog of fixed stars, preparing an accurate map of the sky that will remain in use for centuries.

962 Having defeated the Magyars in 955, Otto the Great is crowned emperor and proceeds to reinvigorate the Holy Roman Empire, establish the power of the state over the Church, and revitalize a Europe that had sustained a series of defeats over the preceding centuries.

c. 1000 The magnetic compass is developed in China.

1054 After centuries of growing animosity between the Greek Orthodox and Roman Catholic churches, the two officially split over the issue of clerical marriage.

1066 William the Conqueror launches an invasion of England; his victory at the Battle of Hastings forever changes the course of English history, spawning centuries of dynastic tension between England and France, and, more permanently, introducing a Latin element to the English language.

1071 The Seljuk Turks, who replaced the Abbasid Caliphate as the dominant power in the Middle East, defeat Byzantine forces at the Battle of Manzikert in Armenia—a blow from which the Byzantine Empire will never fully recover.

1095-1099 Petitioned by the Byzantine emperor for military aid against the Turks, Pope Urban II instead launches the First Crusade, which ends with the capture of Jerusalem and the establishment of four crusader states in the Levant; though the crusades will continue for many centuries, this marks the high point of Western European success in the Holy Land.

1204 The Fourth Crusade ends with western Europeans' capture of a major eastern stronghold—Constantinople, home of their fellow Christians in the Byzantine Empire. They hold the city for 57 years, further weakening Byzantium, which had long shielded western Europe from invaders to the east.

1211-1279 The Mongols conquer most of the known world, destroying Seljuk power in the Middle East, nearly taking eastern Europe, seizing Russia (maintaining control there for the next two centuries), establishing the Yüan Dynasty in China, and later spawning the Mogul Empire of sixteenth-century India.

1224 Holy Roman Emperor Frederick II issues laws regulating the study of medicine, thus elevating the status of real physicians and diminishing the number of quacks; later, in 1241, he becomes the first major European ruler to permit dissection of cadavers, formerly prohibited by religious law.

1271-1295 With his father and uncle, Marco Polo undertakes one of the most celebrated journeys in history, through the Middle East and India to China and the court of Kublai Khan, and later to Southeast Asia and the East Indies; his subsequent account of his travels is perhaps the most important geographical work in western Europe during the Middle Ages.

1288 The first known guns are made in China; firearms are first mentioned in Western accounts 25 years later, in 1313.

1291 The Western presence in the Holy Land officially comes to an end with the Muslim reconquest of Acre; despite their brutality, the Crusades have given western Europeans exposure to the advanced cultures of Islam and Byzantium, and thus ultimately have laid the foundations for the Renaissance.

c. 1300-c. 1325 Two very different scholars both advance scientific thinking: John Duns Scotus distinguishes between causal laws and empirical generalizations, laying the groundwork for the scientific method; and William of Ockham formulates "Ockham's Razor," which states that when two theories equally fit all observed facts, the one requiring the fewest or simplest assumptions is preferable.

1309-1417 The power of the papacy is broken when the papal throne is moved to Avignon in 1309; later, in 1378, divisions between pro-Rome and pro-Avignon groups lead to the Great Schism, which debilitates the Church just as it has begun to face the first stirrings of the Reformation.

1347-1351 The Black Death ravages Europe, killing between 25% and 45% of its population, which had been about 100 million in 1300; not until about 1500 will figures return to their preplague levels.

c. 1400 Persian mathematician al-Kashi is the first to use decimal fractions.

c. 1420-1460 Prince Henry the Navigator of Portugal, who never undertook any voyages himself, operates an extremely influential school of navigation that virtually inaugurates the Age of Exploration, sending pupils on voyages of discovery to places such as the Madeiras (1420), Cape Verde (1445), and the mouth of the Gambia River (1446).

1429 Joan of Arc leads a tiny force against the English at Orléans, turning the tide of the Hundred Years' War, which will end with French victory in 1453.

Exploration and Discovery

Chronology

860 Viking mariners discover Iceland.

c. 915 The Arab journeyer al-Mas'udi travels through Persia and India, and later recounts his experiences in several written works.

982 Erik the Red discovers Greenland.

c. 1000 Leif Erikson, son of Erik the Red, explores a region he calls "Vinland"—the eastern coast of North America.

1154 Al-Idrisi, an Arab journeyer and geographer, produces one of the most important geographical works of the medieval period, translated as *The Would-Be Traveler's Stroll Across the Horizons of the Globe*, or *The Delight of Him Who Desires to Journey through the Climates*.

1200s Mongol conquests open up trade routes between the East and West.

1271-1295 Together with his father and uncle, Marco Polo undertakes one of the most celebrated journeys in history, through the Middle East and India to China and the court of Kublai Khan, and later to Southeast Asia and the East Indies; his subsequent account of his travels is perhaps the most important geographical work in Western Europe during the Middle Ages.

c. 1314-1330 Odoric of Pordenone, an Italian Franciscan missionary, journeys to a number of eastern lands, including Tibet, a place virtually unseen by Westerners; later, his account is plagiarized by the author of *The Travels of Sir John Mandeville, Knight*.

1325-1354 Moroccan journeyer Ibn Battuta explores the Islamic world and beyond, from Beijing to Timbuktu, providing one of the first written accounts of the latter in his *Rihlah*.

1405-1433 Admiral Cheng Ho (Zheng He), under the direction of Ming Dynasty emperor Yung-lo, makes a number of voyages to Southeast Asia, the Indian subcontinent, the Middle East, and sub-Saharan Africa.

1406 French navigators Jean de Béthencourt and Gadifer de La Salle conquer the Canary Islands for Spain.

c. 1420-1460 Prince Henry the Navigator of Portugal, who never undertook any voyages himself, operates an extremely influential school of navigation that virtually inaugurates the Age of Exploration, sending pupils on voyages of discovery to places such as the Madeiras (1420), Cape Verde (1445), and the mouth of the Gambia River (1446).

Overview: Exploration and Discovery 700-1449

Throughout the centuries human curiosity about the unknown has led individuals on adventures to the far reaches of the globe. Ancient exploration was largely in the context of military conquest. Perhaps the best early example is that of Alexander the Great (356-323 B.C.), whose exploration created an empire was so vast that it remained unmatched for more than a thousand years, until the Vikings set out across Europe and the Atlantic. The Roman Empire also expanded its borders—to the north as far as Britain (Albion) and to the south as far as the Atlas Mountains in northern Africa—but with a greater interest in colonization, not exploration. In addition to conquest and colonization, the search for new routes to commerce, especially for the luxury commodity of silk, and new opportunities for religious conversion prompted exploration. The Chinese ventured westward with silk, which was much desired by the Romans, and from the fourth century on Chinese monks journeyed long distances to the West to visit the birthplace of Buddha and to study Buddhist scriptures. Fa-Hsien (374?-462?) and Hsuan-tsang (602-664) were two of the most well-traveled Chinese monks, both journeying for many years throughout China and India.

In the Middle Ages, as the civilizations of the world developed and expanded, the desire to explore and conquer new lands and peoples intensified. Merchants, monks, and mariners (and combinations of all three) ventured forth on expeditions. The dominant sea power in Europe from 800 to 1150 was the Vikings, prime examples of this fundamental urge to discover and conquer. With their technologically advanced longships, skilled seamanship, and military raiding parties, the Vikings exerted their influence from Russia to Greenland, which was colonized by Vikings led by Erik the Red (950?-1001?) around 982, and established peripheral contact with the Byzantine Empire and the shores of North America. They established extensive trade routes, and their raiding hordes, which changed the political map of the medieval world, became the impetus for nation building in Europe.

Norwegian outlaws, exiles, and adventurers began colonizing Iceland around 874, after the 850 discovery of the island by Naddoddur and its circumnavigation several years later by another Swede, Gardar Svafarsson. The Vikings discovered and made landfall in North America over 500 years before Italian navigator Christopher Columbus (1451-1506) would receive credit for the same feat. In 986 Bjanri Herjolfsson was the first European to sight the eastern coast of North America. He was followed by Leif Erikson (980?-1020?), who explored the coastline from Baffin Island to Cape Cod, making several landfalls in 1001. In fact, the first European attempt to establish a permanent settlement in North America was led, in 1010, by Thorfinn Karlsefni (980?-?) in the region of Newfoundland.

Another nomadic military power during the Middle Ages was the great Mongol empire created by Genghis Khan (1162-1227), under whose leadership the ruthless and marauding Mongols expanded into northern China, Persia, and Russia. He was followed by other great khans who led the Mongols in the creation of a vast empire that stretched across Asia and signified the first extensive Asian exploration of the West, which, in turn, stimulated European exploration and trade with the Far East. From 1245-47 a Franciscan friar named Giovanni da Pian del Carpini (1180?-1252) traveled to Mongolia and Central Asia, met with Mongol leader Batu Khan (?-1255), and opened new routes to the Far East, providing important cultural and geographic descriptions of the Mongols and their territories in his *History of the Mongols,* the first accurate Western account of the Mongol Empire. In the 1280s Chinese ecclesiastic Rabban bar Sauma (1220?-1294) became the envoy of the Mongols, traveling from Beijing through Central Asia, Persia, and Asia Minor to Italy, where he met the newly elected pope, Nicholas IV, in Rome, and then to France, where he met King Philip IV in Paris. His diary gives an unusual outsider's view of medieval Europe. Another Franciscan friar, Odoric of Pordenone (1286?-1331) journeyed throughout Asia Minor, Persia, India, southeast Asia, and China from about 1316-31. He brought back an account of his journey, during which he is said to have baptized over 20,000 persons. Odoric's account appears to have been plagiarized in a fourteenth-century English work known as *The Voyage and Travels of Sir John Mandeville.*

Perhaps the best-known adventurer in Central Asia is Marco Polo (1254?-1324), whose extensive 24-year journey with his father Niccolò and uncle Maffeo included 17 years in Mongol-controlled China. Polo's account of his adventures was published in 1298 by Rusticiano as *Divisament dou Monde,* now generally known as *The Travels of Marco Polo,* and served to excite the nations of Europe about the riches in trade and culture that might be found in unfamiliar areas of the world, such as the Far East. Other explorers made significant geographical and cultural journeys through Asia, including Abu al-Hasan 'Ali al-Mas'udi (895?-957), who traveled through Persia and India and throughout other areas of the Middle East, recording his travels and observations in his book *Meadows of Gold.* Another was Niccolò de Conti (1395?-1469), who traveled for 30 years in southern Asia, from Persia to the eastern coast of China. French Catholic monk Jordanus of Séverac (1290-1354) traveled to India and wrote *Mirabilia* (translated as *Book of Marvels*), in which he described that region's geography and peoples. Another writer, al-Biruni (973-1048), a Persian scholar and scientist, authored *Ta'rikh al-Hind,* considered one of the greatest medieval works of travel and social analysis, in which he discussed and described the history, geography, and religion of India.

Before the Vikings began their raids and the Mongols began their expansion, religion was a rapidly spreading factor in unifying peoples. Christianity, which had become the official religion of the Roman Empire in the late 300s, was offset by Islam, founded by the prophet Muhammad in 610. By the eleventh century, pockets of Muslim believers extended from Spain to India and circled the Mediterranean Sea, especially concentrated in Arabic lands. The changes in the region prompted the travels of Ahmad ibn Fadlan (908?-932), sent on a diplomatic mission to Russia to explain Islamic law. His account of his journeys, the *Risala,* includes details of his visits and experiences among various Turkic peoples as well as the Vikings. Arab Muhammad ibn-Ahmad al-Maqdisi (945-1000) traveled throughout the Muslim world, from the Iberian peninsula to Africa, Syria, the Arabian desert, Persia, Central Asia, and Indonesia, and wrote of his travels and observations in *Best Division for Knowing the Provinces* (begun 985), considered among the most accurate geographic descriptions of the Islamic world during the Middle Ages.

Other important Arab explorers include geographer al-Idrisi (1100-1165?), who traveled extensively in Asia, Africa, and Europe and created a book, commonly known as the *Book of Roger,* which contained detailed maps and records, along with important geographic information about these regions derived from his own travel experiences and other eyewitness accounts, including information from Greek and Arabic sources. In addition, Ibn Battuta (1304-1368), a Muslim from North Africa, spent 25 years traveling to every civilized part of the known non-Western world—a journey of some 75,000 miles (120,701 km)—and wrote of his adventures in the highly informative travelogue, the *Rihla.*

The Crusades, which included eight expansive military expeditions from 1096 to 1270, brought Christian Europeans to the Holy Land, introduced Islamic culture (and the science of cartography, expertly refined by Islamic mapmakers) to the West, and resulted in Christian occupations of Palestine, Syria, Greece, and the Baltic. Also created was a heightened desire for adventure and an undeniable drive to visit distant places and peoples. Further encouraged by tales of wealth and culture in the East, related by Marco Polo and other adventurers, the nations of Europe were spurred into a period known as the Age of Discovery, during which the quest for new lands and trade routes by sea became a major objective.

The beginning of this period of European maritime discovery can be traced to the fifteenth-century Portuguese prince known as Henry the Navigator (1394-1460), who established a navigational school at Sagres, near Cabo de São, Portugal. Under his sponsorship in the early 1400s, expeditions explored and colonized the Madeiras, discovered by João Gonçalves Zarco in 1418. The Madeiras became an important foothold for Portuguese exploration in the following centuries. The Portuguese explorers under Prince Henry also journeyed along much of Africa's west coast, including voyages past Cape Blanc, Cape Bojador, and Cape Verde. Spain also began a period of maritime exploration, colonizing the Canary Islands in 1402 under the leadership of Frenchman Jean de Béthencourt (1360?-1422?).

In the fifteenth and sixteenth centuries Europe's expanding horizons were flung open by the pursuit of trade, especially luxury goods such as those found in the Far East. In the 1600s national exploration was challenged by commercial organizations such as the East India Company, which made extensive ocean voyages to Asia and the South Pacific. Flourishing trade routes led to permanent trading posts and eventually resulted in colonial occupations, including those founded in the New World, which Europeans initially established while seeking ocean routes to China and the Far East. By the end of the seventeenth century exploration was no longer limited to purely nationalistic or economic pursuits, but attracted men and women with personal motives, including missionaries, religious exiles, scientists, and adventurers who traveled to proselytize, escape oppression, study the world, and for the satisfaction of new experiences.

ANN T. MARSDEN

Al-Maqdisi Travels Throughout the Muslim World

Overview

The most important writings on geography and exploration during the period from the tenth to the twelfth centuries emerged from the Muslim world. There a series of journeyers and geographers chronicled their travels and categorized the towns and physical features of the Islamic realms, a vast network of empires that stretched from Spain to India, and from eastern Europe to the desert kingdoms of West Africa. Among the first of these writers was al-Maqdisi, who traveled throughout much of the Arab and Muslim world, and who in 985 began writing about his journeys in *Best Division for Knowing the Provinces.*

Background

It is understandable that both scientific geography and the art of travel writing would flourish in the realms controlled by the Muslims. Not only did those lands enjoy the greatest flowering of civilization in the Western world up to that time since the golden age of Rome, if not of Greece, the Muslim caliphates constituted by far the largest Western empires since Rome. Thus it became increasingly necessary to possess knowledge concerning the many towns, roads, and physical features of the lands where Allah was the acknowledged God, and Arabic the lingua franca.

When Muhammad (c. 570-632) began his ministry as prophet of Allah in 613, the Arabian peninsula was a remote, forgotten corner of the world. By the time he died in 632, Muslims controlled the western and southern portions of the peninsula, but the sophisticated urban centers to the north—Damascus, Jerusalem (the town of al-Maqdisi's birth), and Baghdad—remained beyond the reach of Islam. That situation changed rapidly in the decades immediately following Muhammad's death, however, as the four caliphs who succeeded him as spiritual and political leaders of Islam conquered Syria, Egypt, Iraq, and much of Persia.

Under the Umayyad caliphate (661-750), the boundaries of Islamic lands spread to the edges of India and China in the East, and to Spain and North Africa in the West. Islam's westward expansion halted with the defeat of Muslim troops by the Frankish majordomo Charles Martel (c. 688-741) at Tours in 732, and though the Muslims gained a victory over China's T'ang dynasty at Talas in 751, the momentum had gone out of Arab efforts to conquer the world. The Abbasid caliphate (750-1258) simply maintained the gains established under its predecessors until it lost its authority, first to the Turks and later to the non-Muslim Mongol invaders.

Yet it was during the period from the tenth to the twelfth centuries, when the caliphate's strength had not yet been fully dissipated, that the cultural centers of Baghdad, Damascus, Jerusalem, and other cities produced some of the Medieval era's greatest thinkers. These included the historian and geographer al-Mas'udi (d. 957), known as "the Herodotus of the Arabs" for his contributions to Middle Eastern historiography. Yet al-Mas'udi was not simply a scholar abstracted from the real world: like the Greek historian Herodotus (c. 484-c. 424 B.C.) to whom he was compared, he traveled widely, compiling notes for his writings.

The writings of Ahmad Ibn Fadlan (fl. 920s), most of which have been lost but which exerted considerable influence in medieval Islam, represented another strain in Muslim geographical writing: the nonscholarly work of a journeyer. Thus Ibn Fadlan wrote memorably about his experiences among the Varangians or Vikings of Russia.

Impact

In assessing the work of Muhammad ibn Ahmad al-Maqdisi (945-1000), sometimes known as al-Muqaddasi, it is useful to compare his career to that of his contemporary Ibn Hawkal (920-990). Both traveled throughout most of the Muslim world, though the journeys of Ibn Hawkal—which included forays to Spain, West Africa, India, and Sicily—were more extensive. Both wrote about their travels, Ibn Hawkal in *Of Ways and Provinces* and *On the Shape of the Earth,* and al-Maqdisi in *Ahsan al-taqasim fi ma'rifat al-aqalim,* whose title is translated in English as *The Best Division for Knowing the Provinces* or *The Best Divisions for Knowledge of the Regions.*

Most interesting of all, both were suspected as agents of the Fatimid regime in Egypt, whose leaders belonged to the Ismaili sect of Shi'a Islam. Though Shi'ites are most commonly associated in the modern mind with the grim fundamentalists who seized power in Iran in 1979, a number of Medieval Shi'ite groups were characterized by a much greater degree of tolerance. Such was the case with the Fatimids, named after Muhammad's daughter Fatima, who (in the beginning, at least) espoused a highly tolerant faith with elements of a universal religion.

Claiming leadership over all Islamic lands, the Fatimids began their conquests in northwestern Africa in 893. During the reign of the Fatimid caliph Moizz (953-975), they seized power over Egypt, and eventually their realms stretched from Sicily and Algeria to western Arabia and Palestine. The center of their empire, however, remained in Egypt, where in 973 they established their capital at Cairo.

Ibn Hawkal's trips took place during the period of Fatimid ascendancy in Egypt, and some scholars believe that he functioned as a spy for the Fatimids. In writing about the Spanish city of Cordoba, for instance, he speculated that the Umayyad remnant that controlled the Iberian peninsula might be vulnerable to foreign attack—an observation some have interpreted as a field report to the Fatimid leaders in North Africa. Similarly, Ibn Hawkal just happened to be in Egypt in 969, as the Fatimids were completing their decades-long effort to win control of that country.

Al-Maqdisi, on the other hand, has been regarded as a Fatimid propagandist but not as an outright spy, and in reading his work, his prejudices are clear. Not only did he favor Shi'ites over the mainstream Sunni Muslims, he was outspoken in his preference for Muslim lands over those of Christians. He did not bother to visit the latter, he indicated, because he did not consider Christians worthy of study, and in the places he visited, he judged the presence of Christians and Jews as a sign of religious impurity in the Muslim majority. In this he prefigured the intolerance that would come to characterize the Fatimids as their regime, which remained in power until 1171, began to go into decline.

Though he came from a famous family of architects and builders, al- Maqdisi chose to pursue a much more varied career that involved him in numerous professions. As for his travels, these began in 966, when he was 21 years old, and seem to have continued after the time he began writing *The Best Division* at age 40. The exact order of his journeys is not known, though it is clear he sailed all the way around the Arabian peninsula; journeyed deep into Central Asia as far north as Samarkand and Bukhara in modern-day Uzbekistan; spent a year in Yemen; and crossed deserts in Persia and Arabia numerous times.

The extent of al-Maqdisi's journeys may not have been as great as those of Ibn Hawkal, but like Ibn Hawkal he was careful to include copious details on each place he visited. Indeed, al-Maqdisi's painstaking attention to matters such as climate, local economies, ethnicities, cultures, and units of weight and measurement make *The Best Division* one of the great works of Islamic geography.

Furthermore, al-Maqdisi is one of the first examples of a traveler who became fully engaged in his environment, literally working his way from country to country. At various times he was employed as a teacher, scribe, courier, doctor, lawyer, papermaker, bookbinder, and even a *muezzin,* a Muslim temple crier who calls the faithful to prayer. He associated with all social classes, experiencing the hard knocks of life at the bottom of the social ladder, as well as the luxuries of the privileged few at the top. He was robbed several times and thrown into jail as a spy, but at other times he rode in sedan chairs alongside the wealthy, and interacted socially as an equal with nobles.

All these factors made *The Best Division* a highly readable and influential work, and one that provided particularly useful knowledge concerning Mesopotamia, Syria, and Central Asia. In his career, al-Maqdisi helped set the pattern for a number of traveler/geographers who followed, among them al-Idrisi (1100-c. 1165), who fled political troubles in the Arab world to work as geographer for the Norman ruler Roger II of Sicily, and the former slave Yaqut (1179-1229), whose work provides a lasting portrait of Central Asia just before the Mongols arrived and forever changed the character of the place.

But perhaps the most obvious link is with the medieval Arab journeyer/geographer who is least obscure in the Christian West: Ibn Battuta (1304-1368). As with al-Maqdisi, who made the *hajj* or pilgrimage to Mecca three times, Ibn Battuta's journeys centered around a series of pilgrimages—in his case, four. Certainly Ibn Battuta traveled much further than al-Maqdisi—indeed, Ibn Battuta probably traveled further than anyone in the premodern era, including Marco Polo (1254-1324)—but he followed the pattern of engagement in the local culture established by al-Maqdisi and others. And yet for all the breadth of his travels, even Ibn Battuta did not experience as wide a variety of social interactions, nor did he work in as varied a range of professions, as al-Maqdisi.

Thus it is unfortunate to note that al-Maqdisi is virtually unknown in the West, and hardly more recognized in the Arabic-speaking world. The bulk of scholarship in English on his travels and writings centers on the work of Basil Anthony Collins, who translated *The Best Division*. Thus in a group whose most prominent figure, Ibn Battuta, is hardly a household name to begin with, al-Maqdisi is even more shrouded in obscurity. What makes this doubly unfortunate is the fact that al-Maqdisi and other early writers kept the geographer's profession alive at a time when Europe was turned inward. By the time of Ibn Battuta, at least, Europeans had rediscovered the outside world, largely through their contacts with Arabs and Byzantines in the Crusades. Ironically, however, few Europeans then or now recognized the debt they owed al-Maqdisi and other Arab writers of his time.

JUDSON KNIGHT

Further Reading

Alavi, S. M. Ziauddin. *Arab Geography in the Ninth and Tenth Centuries.* Aligarh, India: Aligarh Muslim University, 1965.

Collins, Basil Anthony. *Al-Muqaddasi: The Man and His Work: With Selected Passages Translated from the Arabic.* Ann Arbor: University of Michigan, 1974.

Al-Muqaddasi. *The Best Divisions for Knowledge of the Regions: A Translation of Ahsan al-taqasim fi ma'rifat al-aqalim,* translated by Basil Anthony Collins. Reading, England: Centre for Muslim Contribution to Civilisation, 1994.

Ibn Battuta Explores the Non-Western World

Overview

Over the space of a quarter-century, the Moroccan journeyer Ibn Battuta (1304-1368) traveled to every civilized portion of the known non-Western world. From Morocco to China, from Russia to Mali, from Spain to Sumatra, he covered a staggering amount of ground: some 75,000 miles or 120,000 kilometers, not counting many detours. In so doing he gathered material for a highly informative travelogue, the *Rihla.* Yet in spite of the fact that he saw far more of the planet than did Marco Polo (1254-1324), he is much less well-known—even in Middle Eastern nations.

Background

Comparisons with Polo are virtually inevitable: not only were both men travelers of the medieval world, but they were contemporaries for 20 years. By 1304, when Ibn Battuta was born in the Moroccan city of Tangier, Polo had written his memoirs, a book that earned him a reputation as a teller of tall tales if not an outright liar. Ibn Battuta, who set out on his own journeys in 1325, a year after Polo's death, would one day publish his own book—and he, too, would be branded a fabricator of falsehoods.

The most significant difference between the two men, of course, is the fact that Polo came from Christian Europe and Ibn Battuta from Muslim North Africa, and this distinction bore heavily on the experiences each would encounter. Europe in Polo's time was rapidly awakening from the long period of isolation that had characterized the Early Middle Ages (c. 500-c. 1000), whereas the Muslim world of Ibn Battuta's era was on the decline from its former glory. In the seventh and eighth centuries, Arab warriors had greatly expanded territories under the control of the Umayyad (661-750) and the Abbasid (750-1258) caliphates, and Arabs had come to dominate the world from Morocco to the edge of China. The caliphs had imposed their own version of the Pax Romana, creating a world of peace and prosperity—a realm in which, from the tenth to the twelfth centuries, some of the medieval period's greatest thinkers had thrived.

But several factors had conspired to bring about the Abbasids' decline. One was the arrival of the Turks, a nomadic people from Central Asia who became the dominant political power in the Near East from the tenth century onward. Another was the Crusades (1095-1291), Western Europe's assault on the Holy Land and Byzantium, which led to massive slaughter on both sides and engendered religious tensions that remain alive today. And finally there were the internal contradictions within the caliphate itself, most of all the fact that its line of rulers had grown increasingly weak until in 1258 the last of them was killed by Hulagu Khan (c. 1217-1265).

Hulagu represented a new power, one that united the Near East, Central Asia, East Asia, and parts of Eastern Europe under a single system: the Mongol khanates. Mongol rule, in fact, had helped make Polo's journey possible, because for the first time since the Roman Empire had controlled the eastward routes, it was possible for a European to travel to India and lands beyond. By the time of Ibn Battuta, the Mongols too were in decline, but the trade routes remained open. And because Battuta was not a light-skinned Christian, much else was open to him as well, and he saw lands Marco Polo could never have visited.

Impact

Ibn Battuta's journeys began, in fact, with a pilgrimage to a city that is quite literally forbidden to non-Muslim visitors: Mecca. All Muslims are encouraged to make the *hajj* or pilgrimage to the holy city, located in what is now Saudi Arabia, at least once if they can afford to do so. Ibn Battuta would manage to make the hajj a total of four times. A member of a wealthy family in Tunis, he set out on his journey with the stated purpose of making the hajj, then returning home for a career as an Islamic judge. He would indeed return, but not for 24 years.

First he crossed North Africa, a 10-month journey, before arriving in the Egyptian port of Alexandria. In those days the Pharos Lighthouse, one of the Seven Wonders of the Ancient World, was still standing, and Ibn Battuta visited it. Only one other of the Seven Wonders, the Great Pyramid of Giza, was still standing (as it is today), and he soon saw that one too, when he took a boat ride up the Nile to Aswan. He then journeyed overland to the Red Sea port of Aidhab, where he hoped to board a ship for Jeddah in western Arabia. A local rebellion, however, forced him to return to Cairo, from whence he crossed the Sinai Peninsula to Jerusalem.

Jerusalem was also a pilgrimage site for Muslims, as it was of course for Christians and Jews, and after stopping in Jerusalem for a time, Ibn Battuta went on to one of the Islamic world's greatest cultural centers: Damascus. In the Syrian city he studied Islamic law and took a second wife. (Apparently he had married earlier, but in a fashion typical of his time and culture, Ibn Battuta made little mention of the women in his life. Over the course of his journey, he would take numerous wives.) By September 1326, Ibn Battuta was on his way to Mecca. Instead of turning homeward after completing his hajj, however, he joined a group of pilgrims returning to the Muslim world's other great cultural center, former capital of the fallen Abbasid caliphate: Baghdad. He then traveled throughout Iraq and Persia before returning to Baghdad and joining a caravan headed back to Mecca.

His second sojourn in the holy city lasted much longer: three years, from September 1327 to the fall of 1330. During this time, Ibn Battuta furthered his legal studies and became a Muslim legal scholar, or *qadi*. He finally left Mecca for Jeddah, then sailed down the Red Sea for Yemen, and after some time in southern Arabia, sailed on to Somalia. He later continued southward as far as Kilwa in what is now Tanzania. Kilwa, a trading city established by Arabs, Persians, and Africans, is some 600 miles (960 km) below the Equator, and this too was a place no Westerner would see for several centuries—particularly because Europeans at that time believed that anyone who crossed the Equator would burn to death.

Always on the move, Ibn Battuta sailed from East Africa to Oman on the eastern side of the Arabian peninsula, then made his third hajj. Intent on going to India, he took a roundabout route, sailing from Jeddah to Egypt, then skirting the Levantine coast to eastern Anatolia (modern-day Turkey). Crossing Anatolia, he boarded a ship across the Black Sea, and reached Kaffa, a port established by Genoese sailors on the Crimean Peninsula.

This was one of the few Christian cities he visited in his many long years of travel, but Ibn

Battuta was soon to see the capital of the Eastern Orthodox Church. At first he moved eastward, into the lands of the Mongol khan Özbeg, from whose name that of the Uzbek people is drawn. However, one of Özbeg's wives, a Greek, persuaded him to join her on a trip back to Constantinople, where Ibn Battuta was presented to Byzantine Emperor Andronicus III.

In time Ibn Battuta moved eastward again, through the khan's lands and on into Central Asia. He entered the Chagatai Khanate, another Mongol realm, then veered southward into Afghanistan. Finally his party crossed the Hindu Kush Mountains—Ibn Battuta was the first traveler to record their name—and reached the Indus River in September 1335.

Much of the subcontinent at that time was under the control of the Delhi Sultanate, and the capital at Delhi was a great center of culture. But Sultan Muhammad ibn Tughluq (r. 1325-51) was a cruel and ruthless leader, and it is a mark of Ibn Battuta's abilities as a diplomat that he not only made a place for himself at Tughluq's court, but even managed to prosper there. Tughluq even once paid off Ibn Battuta's debts, but when the latter consulted a local soothsayer, the emperor was angered and placed Ibn Battuta under house arrest.

After six months, Ibn Battuta was rehabilitated and sent on a diplomatic mission to the Mongols' "Great Khan" in China. His boat wrecked off the southern coast of India, a disaster in which one of Ibn Battuta's children was killed, and rather than return to face the vicious Tughluq, he sailed southward to the Maldive Islands. There he came under the protection of a Muslim queen, but was eventually forced to leave because of political pressures.

In Sri Lanka Ibn Battuta visited Adam's Peak, a high mountain sacred to Muslims, Buddhists, and Hindus alike. Sailing northward along India's eastern coast, his vessel was attacked by pirates, but he straggled into Bengal and boarded a Chinese junk for Sumatra. Sumatra, in what is now Indonesia, had a Muslim ruler who befriended Ibn Battuta and supplied him for a journey onward to China. This phase of his travels took him the furthest distance away from his homeland—it is rumored he traveled as far north as Peking (modern Beijing)—and Ibn Battuta recorded copious observations regarding Chinese culture and civilization.

Ibn Battuta's trip home was a varied one, involving stops in Sumatra, India, Arabia, Persia, and Syria, but though he witnessed the ravages of the Black Death (1347-51), it was a less eventful journey than his eastward travels had been. He made a final hajj in November 1348, then traveled overland to Egypt and then by boat along the North African coast to Morocco. On November 8, 1349, he returned to his hometown of Tangier. He was 45 years old, and had been away for 24 years.

One more great journey remained for Ibn Battuta, who took part in a military expedition to defend the city of Ceuta in northern Morocco against a Christian invading force. From Ceuta he crossed the Straits of Gibraltar to Spain, then still in Muslim hands, before returning to Africa for a journey with a caravan across the Atlas Mountains and the Sahara Desert. He visited the empire of Mali, and there became one of the first outsiders to write see Timbuktu, a city that would reach its peak about a century later.

After a visit to what is now Niger, Ibn Battuta ended his travels, finally settling in Morocco to practice law. The sultan assigned a young writer named Ibn Juzayy to assist him in recording his observations, and the result was the *Rihla,* whose title means simply "travel book." Completed in 1335, the work initially earned Ibn Battuta a number of detractors, as well as no small share of supporters, among Muslim readers. But in the years that followed, as the Arab world went further into decline and the torch of exploration passed to the West, the book all but disappeared.

Ironically, when the *Rihla* was finally resurrected in the nineteenth century, it was by Westerners, and the book was soon translated into French, German, and English. In time Ibn Battuta came to be accorded his just recognition as a man who had recorded many sights and facts that would simply have been beyond the reach of a Western traveler.

Aside from his visits to Mecca and other "forbidden" spots, Ibn Battuta gave valuable accounts of Muslim naval power, slavery, and marriage practices, as well as a uniquely Islamic view on tensions with Christianity and other religions. He also helped popularize the name of one of the world's great mountain ranges, and in turn a crater on the Moon has been named in honor of Ibn Battuta. The Tangier airport, as well as a ferry across the Straits of Gibraltar, are both named for him, a fitting tribute to a man who set off from Tangier to see the world.

JUDSON KNIGHT

Further Reading

Abercrombie, Thomas. "Ibn Battuta, Prince of Travelers." *National Geographic*. Photographs by James L. Stanfield. December 1991, pp. 2-49.

Dunn, Ross E. *The Adventures of Ibn Battuta: A Muslim Traveler of the 14th Century*. Berkeley: University of California Press, 1986.

Ibn Battuta's Trip. http://www.sfusd.k12.ca.us/schwww/sch618/islam/nbLinks/Ibn_Battuta_map_sites.html (August 30, 2000.)

Finding Mecca: Mapmaking in the Islamic World

Overview

The Islamic tradition of mapmaking dates almost to the very dawn of Islam, driven in part by the necessity for all Muslims to face Mecca during their daily calls to prayer, and by the need to properly orient mosques to also face Mecca. Over the centuries, Islamic mathematicians and cartographers brought mathematical cartography to new levels of sophistication, drawing on their own research as well as incorporating many tools from the Greek and Hindu cultures. The result of this mixture of science, mathematics, religion, and cultures resulted was unique collection of maps and tables.

Background

Islam, which literally translates as "submission," is a religion dating to the beginning of the seventh century A.D. Within 30 years, Islam had spread throughout the Arabian peninsula. Within three centuries, it had been established from the western coast of Africa through the Indian subcontinent. The rise of Islam corresponds roughly with the decline of the Roman Empire, and Arabs had extensive contact with both the Greek intellectual legacy from Alexandria, Egypt, and with the Hindu civilization of the East. It must be pointed out that *Arab* refers to the people living in a certain part of the world, *Muslim* refers to followers of the religion of Islam, *Islam* refers to the religion itself, and *Islamic* refers to the activities and properties of followers of Islam. Not all Arabs were (or are) Muslims, and not all Muslims are Arab. In fact, the majority of Muslims live in the Indian subcontinent and in Indonesia, and many Arabs are Jewish or Christian, as well as Muslim.

Islam requires certain obligations of all Muslims, including the requirement to pray several times daily while facing Mecca. It is also common for mosques to be constructed so that worshipers face Mecca while they pray. To fulfill both these obligations, it is essential to know in which direction Mecca lies.

As Islam spread across Africa and Asia, Muslim cartographers and mathematicians began working to develop accurate maps and tables that, for any location in the known world, could help the faithful know in which direction Mecca lay. These cartographers enlisted the help of mathematicians skilled in the use of spherical trigonometry and adapted the best mathematical tools they could find to their purposes. The result was a cartographic tradition different from those of Europe or Asia, since the main focus of Islamic maps was more to locate Mecca from any point in the world than it was to help describe the world.

However, there was more to Islamic mapmaking than just helping the faithful find Mecca. Arab traders ranged widely. Accurate information about coasts, topography, towns, and other features was important for them, and these features were included in many maps. Arab astronomers also devoted a great deal of time to mapping the locations of stars and the wanderings of the planets, and Arab mariners helped map coastlines as they sailed the oceans. In their cartographic efforts, Muslim mapmakers developed new ways of depicting the world, helped to better define some basic concepts (such as the length of a degree of latitude), and constructed incredibly lengthy and accurate mathematical tables of calculations and results.

Part of this latter effort was an outgrowth of the Arab penchant for "universalism." The mathematicians preferred to spend time constructing a universal table of directions and distances to Mecca that could be used from anywhere in the world, rather than constructing individual tables for each separate city or country in which Muslims found themselves. However, the universal approach was a long and tedious process. The longest single table we know of had nearly 500,000 entries, which were all calculated by hand and copied (also by hand!) for each new user.

Perhaps the most interesting artifacts from the Islamic cartographers of this time are two Persian maps that did not come to light until 1989 and 1995. These maps were quite advanced for their time. Although they probably date to the eighteenth or nineteenth century, many of their features almost certainly date to the eleventh century or earlier, suggesting the cartographers of that age had an even more sophisticated understanding of spherical trigonometry and other mapmaking techniques than was previously suspected.

In general, Islamic maps from this time are no more and no less accurate than comparable maps from other cultures. It must not be forgotten that knowledge of the world was still very much in its infancy, and ignorance of many facets of geography was nearly universal. One example of this is that, almost invariably, European and Islamic maps showed a long, eastward projection from the southern tip of Africa that enclosed a major portion of the Indian Ocean. Another example is the nearly universal practice of forcing known geographic features into a map that was either aesthetically pleasing to the cartographer or that fit certain preconceptions of the times. However, this should not take away from the genuine advances made by Islamic cartographers during the time period that was known in Europe as the "Dark Ages."

Impact

Islamic cartographers and their advances left a notable mark on their society and others. Arab and Muslim traders traveled to Europe, throughout Africa, and as far east as China. This helped them to spread their maps through most of the known world, and their position at the junction between East and West helped them to convey innovations from one sphere to the other. The primary impact of this mapmaking tradition was in the areas of geodesy (determining the size and shape of Earth), applying spherical trigonometry to cartographic problems, and developing new and useful coordinate systems for mapping Earth.

By saying "projection," a cartographer is referring to the method by which the spherical surface of Earth is portrayed on a flat piece of paper. Probably the best-known projection is the Mercator projection, in which the outline of Earth's surface is simply spread over a square, with lines of latitude and longitude drawn like the lines marking the squares on a chessboard. The major problem with the Mercator projection is that it is extremely inaccurate at high latitudes-objects farther north or south from the equator begin to look increasingly (and erroneously) large. Early Muslim cartographers did not face this problem, because over relatively small areas that are not far north or far south, a Mercator-like projection does not introduce many inaccuracies. However, as their religion spread, it became increasingly necessary for later cartographers to construct maps that would show believers living anywhere from Europe to China how to face Mecca. To do this, cartographers turned to mathematicians to help construct new map projections using spherical trigonometry.

These projection methods greatly improved the accuracy of Islamic maps, and the maps made using these techniques were widely used and copied for centuries. In addition, the same mathematical techniques used for mapping the outer surface of Earth could be used to map the "inner" surface of the night sky, and Muslim astronomers made some finely detailed star maps. Maps made to show the position of Mecca with respect to other parts of the globe could be used with equal facility to show foreign coasts, trade routes, and other geographic phenomena. As a result, Muslim maps became valuable tools for traders, sailors, and astronomers, in addition to the Islamic faithful.

Another problem addressed by Islamic mapmakers was that of defining the length of a degree of latitude or longitude. The answer to this important question proved surprisingly difficult to determine. In fact, centuries later, expeditions were dispatched to the mountains and jungles of Peru and to the northern wastes of the Scandinavian peninsula with the sole purpose of precisely measuring this quantity.

The driving force behind the cartographic advances of Islamic mapmakers was a need to help their religious brethren accurately locate Mecca from anywhere in the known world. With this starting point, Muslims used mathematical and geographic techniques learned from Hindu and Greek cultures to develop their own mapmaking techniques. These techniques led to maps that were generally superior to those that had preceded them, and the same techniques were later employed to construct maps that were also put to use

by Muslim traders and explorers. As well as proving valuable to the Muslim world, these accurate maps were also important to the development of other nations to which the maps were taken.

P. ANDREW KARAM

Further Reading

Harley, J. B., and Woodward, David, ed. *The History of Cartography,* Vol. 2, Book 1: *Cartography in the Traditional Islamic and South Asian Societies*. Chicago: University of Chicago Press, 1992.

Hourani, Albert. *A History of the Arab Peoples*. Cambridge, MA: Harvard University Press, 1991.

King, David. *World-maps for Finding the Direction and Distance of Mecca*. Koninklijk Brill NV, Leiden, The Netherlands, 1999.

The Crusades

Overview

The Crusades were a series of eight military campaigns between the years 1096 and 1270 in which Europeans attempted to wrest control of the Holy Land from the Muslims who ruled the Middle East. The Crusades failed to achieve their objective and cost untold lives. However, they did expose Western Europe to new ideas, and resulted in a heightened desire for adventure and an urge to see distant places. This curiosity was eventually channeled into the exploration of the New World.

Background

In the course of the fall of the vast Roman Empire during the fourth and fifth centuries A.D., it broke into two parts. The western part splintered further as German chieftains such as the rulers of the Goths, Vandals, Angles, Saxons, Jutes, and Franks carved out their individual kingdoms. The eastern part became the Byzantine Empire. It was ruled from the ancient city of Byzantium, now Istanbul, Turkey. At that time the city had been renamed Constantinople, after the Emperor Constantine.

Christianity, which had become the official religion of the Roman Empire in the late 300s, itself became divided as a result of the political schism. In the West, the bishops of Rome, successors to Saint Peter, gained great influence in the Church. This resulted in the development of the institution of the papacy. The patriarch of Constantinople was an important religious leader in the Byzantine Empire. Power struggles increased over the centuries, and theological differences widened. Finally, after Pope Nicholas I denounced Patriarch Photius in the eleventh century, the eastern cleric declared he was no longer under the pope's authority. The final break between what became known as the Roman Catholic and Eastern Orthodox churches occurred in 1054.

Islam was also spreading rapidly in the Middle Ages. Founded by the prophet Muhammad in A.D. 610, by the eleventh century it extended from Spain to India, and circled the Mediterranean Sea. Its dominions included the ancestral homeland of the Jewish people, where Jesus had been born, taught, and died. Before the Muslim takeover the eastern Mediterranean had been part of the Byzantine Empire, and the Roman Empire before that. The "Holy Land" was now sacred to three religions. At its center was Jerusalem: the ancient capital of the Jews where the last remaining wall of the Second Temple still stands; the scene of many events in Jesus' life and the site of his crucifixion; and the location from which, according to Muslim belief, Muhammad ascended to Heaven to talk with God.

Many Jews and Christians continued to live in the Holy Land, and for centuries were generally well treated by the Muslim authorities. They were allowed freedom of religion and access to their holy places, and thousands of Christian pilgrims streamed in every year unhindered. This changed in the eleventh century, when warlike Seljuk Turks from Central Asia, newly converted to Islam, began interfering with the pilgrims and upsetting the balance of power in the Middle East between the Muslims and the Byzantine Empire.

In 1071, in the Battle of Manzikert, the Seljuk Turks defeated the armies of the Byzantine Emperor Romanus IV Diogenes. The Turks eventually advanced to within 100 miles (about 161 km) of Constantinople, and wrested control

of both Jerusalem and Baghdad from their fellow Muslims. They adopted Arab customs and regarded themselves as the champions of Islam in opposition to the Christians.

Impact

In 1081, Alexius I Comnenus ascended to the Byzantine throne. Previously a military commander, he was determined to regain the provinces lost to the Seljuk Turks. He knew he could not do this without assistance from the West, but doubted, with good reason, that his territorial aspirations would prompt help from that direction. Instead, he borrowed a concept from his Muslim neighbors: the *jihad*, or holy war, by which they had extended their sway over such a large area.

Alexius, on behalf of Byzantine Christians, wrote to Pope Urban II appealing to him and to the Western princes under his religious authority for military aid. He brought up the necessity of defending Constantinople, site of many Christian shrines and relics. He stressed the vulnerability of Jerusalem, and warned that the tomb of Jesus, or Holy Sepulcher, might be destroyed. Leaving no stone unturned, he also mentioned the "treasures" and the "beautiful women" of the Orient.

Pope Urban II had his own problems to deal with. Conflicts between the European kings and nobles were threatening the papacy and, he believed, Christendom itself. He needed a cause behind which to unite them. On November 27, 1095, after a meeting of a church council in France, the pope made what the historian Will Durant called "the most influential speech in medieval history." In it, he called for knights to go to the Holy Land, free the Christians there from Muslim rule, and regain the Holy Sepulcher. Traveling preachers spread the word throughout Europe. This was the start of the eight Crusades that were to span the next 200 years.

Along with the mounted knights, many more foot soldiers were to "take the cross," or become Crusaders. These included archers, crossbowmen, spearmen and foragers. Under the feudal system, they owed their allegiance to landowning lords. Crusades were costly enterprises. The money for outfitting the soldiers on their expeditions to the East might be provided by the feudal lords or raised through taxes, sales of land or other property, or loans. Payment of the loans could be delayed until their return, and no doubt some had dreams of coming home with some of the treasure hinted at by the Byzantine emperor. Of course, a great many never came home at all.

In the First Crusade (1096-1099), the common people set out first. Poorly trained and equipped, many starved or were killed by Eastern Europeans. Those who survived the trek through Europe were slaughtered as they ventured into Asia Minor. Behind them came the knights, who fared much better. In 1099, they captured Jerusalem. They set up four states on the eastern shore of the Mediterranean, called the Latin States of the Crusaders. These were the County of Edessa, the Principality of Antioch, the County of Tripolis, and the Kingdom of Jerusalem.

The Second Crusade (1147-1149) was prompted by the Turkish conquest of the County of Edessa, after too few crusaders were left behind to defend it. King Louis VII of France and Emperor Conrad III of Germany heeded the preachings of Saint Bernard and led their armies into Asia Minor. However, they refused to cooperate with each other, leading to their defeat by the Turks.

DAMASCUS STEEL

When the Crusaders went to liberate the Holy Land, they were amazed at the quality of the steel blades wielded by their foes. Made of Damascus steel, these swords were hard, flexible, and beautifully patterned. This sort of metallurgy was unknown in Medieval Europe and, indeed, until very recently, the details remained largely unknown, a secret of the Islamic metallurgists. We have since learned that Damascus steel is wrought iron, hard in its own right, to which a high level of carbon has been added. The carbon, as in any other carbon steel, helps to harden the metal far beyond the iron and low-carbon steels used by Crusaders. Not only did the blades hold a shaper edge, but they retained their edge longer.

The other feature that made Damascus steel unique was its flame-like pattern, resulting from a combination of metalworking and the metal itself. The metal itself, because it was worked by hand, contained different carbon concentrations, giving different colors in the blade. Then, when the steel was worked, it was folded, hammered, folded again, and so on, sometimes dozens of times, to make the steel tougher yet. This, too, left a distinctive pattern in the metal, making it both beautiful and deadly.

P. ANDREW KARAM

A fifteenth-century illuminated manuscript depicting Louis IX's conquest of Damietta, Egypt, in 1250. *(Archivo Iconografico, S.A./Corbis. Reproduced with permission.)*

After the Muslims, under the leadership of the great warrior Saladin, recaptured Jerusalem and most of the rest of the Holy Land, the Third Crusade (1189-1192) was organized. Again the Christian campaign was hampered by lack of co-operation among its leaders. King Philip II of France went home early in order to scheme against the English King Richard I, called the Lion-Hearted, who remained in the East. After a two-year siege, Richard recaptured Acre (now Akko). In 1192, he and Saladin agreed upon a truce in which the Christians would keep the

Mediterranean coast. The Muslims would control the interior, but would allow Christian pilgrims to enter Jerusalem.

The Fourth Crusade (1201-1204) never reached the Holy Land at all. Its major impact was to allow a force of Venetians to seize Constantinople and rule the Byzantine Empire until 1261. The Byzantine treasure captured in the raids, and especially access to Byzantine trading markets, greatly increased the wealth and influence of Venice.

In the Fifth Crusade (1217-1221) the Christians captured Damietta, at the mouth of the Nile. However, they soon returned it in order to obtain a truce. The Holy Roman Emperor Frederick II led the Sixth Crusade (1228-1229). An expert negotiator, he managed to talk the Muslims into giving up Jerusalem without a fight.

Jerusalem remained under Christian control until the Muslims took it again in 1244. The conquest prompted the Seventh Crusade (1248-1254). King Louis IX of France (Saint Louis) and his noblemen were captured by the Turks and held until an enormous ransom was paid. Once freed, Louis organized the Eighth Crusade (1270) to seek revenge. He died in North Africa, and his army returned to Europe.

Historians generally agree that the Crusades were a tragic failure by almost any measure. All the territories the Christians had gained at enormous cost in lives on both sides were eventually recaptured. The ruling Muslims, hardened by conflict, were no longer as tolerant of other religions as they had been before. The Byzantine Empire, which had set the entire chain of events in motion in order to regain its lost provinces, never recovered. The prestige of Church leaders in the West was also weakened.

A few more attempts to organize crusades in the fourteenth and fifteenth centuries met with little enthusiasm. Europeans had been exposed to new ideas as a result of the Crusades, and had become more interested in exploration, but their sights were beginning to turn west to the Atlantic. With their mapmaking and shipbuilding skills improved by three centuries of long-distance military campaigns, they were preparing for the voyages that would lead them to the New World.

SHERRI CHASIN CALVO

Further Reading

Biel, Timothy Levi. *The Crusades.* San Diego: Lucent Books, 1995.

Billings, Malcolm. *The Crusades: Five Centuries of Holy Wars.* New York: Sterling, 1996.

Chazan, Robert. *In the Year 1096: The First Crusade and the Jews.* Philadelphia: Jewish Publication Society, 1996.

Francesco, Gabrieli, ed. and transl. *Arab Historians of the Crusades.* E.J. Costello, transl. from the Italian. New York: Barnes and Noble, 1993.

Hallam, Elizabeth, ed. *Chronicles of the Crusades.* Godalming, UK: Bramley Books, 1996.

Payne, Robert. *The Dream and the Tomb: A History of the Crusades.* Chelsea, MI: Scarborough House, 1991.

Tate, Georges. *The Crusaders: Warriors of God.* L. Frankel, transl. New York: Harry N. Abrams, 1996.

Treece, Henry. *The Crusades.* New York: Barnes and Noble, 1994.

Al-Idrisi and Representations of the Medieval Muslim World

Overview

Ash-Sharif al-Idrisi (1100-1165?) wrote one of the greatest works of medieval geography and produced the first world map to use a grid system of vertical and horizontal lines to designate geographic subdivisions and climatic zones. As a geographer and adviser to Roger II, the Norman king of Sicily, he also helped to bridge the distinct cultures of Europe and the Islamic world. While in Sicily, al-Idrisi constructed a silver planisphere that was covered with a map of the world. This map, which featured trade routes, major cities, and geographic details, was remarkably accurate for the time.

Furthermore, al-Idrisi composed the *Kitab Nuzhat al-Mushtaqfi Ikhtiraq al-Afaq,* or *The Delight of Him Who Desires to Journey Through the Climates.* This text, also known as the *Al-Kitab ar-Rujari,* or *The Book of Roger,* was intended to accompany the silver planisphere. It contains detailed maps and records important geographical information on Asia, Africa, and European countries. Al-Idrisi compiled material from personal experience and eyewitness reports along with information taken from Arabic and Greek maps and geographic texts.

Background

Al-Idrisi was born in Sabtah, a Spanish settlement in Morocco. He came from a long line of nobility, caliphs, and holy men. His closest ancestors were the Hammudids of a caliphate in Spain and North Africa that lasted from 1016 until 1058. Al-Idrisi spent his youth traveling through this area. He also traveled through Portugal, northern Spain, and the French Atlantic coast. He had even journeyed as far as Asia Minor by the age of 16.

There is some dispute regarding the importance of al-Idrisi's geographic works. The maps that Roger II commissioned him to make exhibit great detail, but are not particularly innovative or creative. While Roger II was displeased with Greek and Muslim maps, al-Idrisi's maps simply compiled Greek and Muslim information into a single form.

Impact

Al-Idrisi's *Kitab ar-Rujari* is often considered more influential than his maps. It represents a serious attempt to link descriptive and astronomical geography. However, al-Idrisi has been criticized for merely compiling information from previous sources. Likewise, some scholars have argued that al-Idrisi was unable to accurately master the mathematical skills necessary to record geographic details accurately. His *Kitab ar-Rujari* is significant because it was distributed widely and in Latin. Also, it was indispensable as a source of information for areas such as the Mediterranean basin and the Balkans.

His maps are also significant in that they include several unique features. Al-Idrisi developed a cartographic system that divided the world into seven distinct climatic zones. These zones move from north to south and predate lines of latitude. The seven climates, called aqalim, are divided into 10 sections that move from east to west. These sections are called ajza. The 70 sections that result from the intersection of aqalim and ajza are provided with their own maps in the *Kitab ar-Rujari*. This is the first instance in European cartography of purely geographic lines of distinction. Older maps marked political divisions, but were not organized into the geographic sections that al-Idrisi employed.

Of course, even political divisions were problematic for cartography at that time. Borders were constantly shifting or were not clearly defined. Also, political units were not the solid monoliths with which we are familiar. Even the dar al-Islam, the preeminent power of the time, was a confusing amalgamation of places, people, and cultures. Sea borders were easily recognized, but land borders were much more problematic. Al-Idrisi's maps, for instance, mark centers of power, but do not clearly delineate their boundaries. Such maps reflect a view of the dar al-Islam as a series of loosely connected points, and not as a single discrete unit.

In such a system, some of the points were especially isolated and surrounded by hostile powers. In order to understand al-Idrisi's geographic work, one must also consider the role of Spain, or al-Andalus, in the dar al-Islam. In addition, study of al-Idrisi's work requires that the cultural connotations of terms such as "near" and "far" be considered. The texts and maps that al-Idrisi produced in Sicily help to clarify the conceptual distances that separated or united areas and cultures.

The time al-Idrisi spent studying in Cordoba as a young man more than likely shaped his awareness of distance and cultural distinctions. Indeed, al-Idrisi's entrance into the service of Roger II of Sicily in about 1145 exemplifies such divisions in the medieval world. As-Safadi, a fourteenth-century Arab scholar, indicates that Roger II invited al-Idrisi to Sicily with these words:

> *You are a member of the caliphal family. For that reason, when you happen to be among Muslims, their kings will seek to kill you, whereas when you are with me you are assured of the safety of your person.*

Scholars are uncertain about al-Idrisi's reasons for relocating to Sicily. Some have surmised that he was viewed as a renegade by Muslims only after he began to serve a Christian king. Others, however, contend that al-Idrisi was in serious danger of assassination attempts before he even accepted Roger II's offer.

Regardless, his relocation to Sicily secured his fame, and is indicative of major developments in the medieval world. Muslim geographers had long produced accurate maps and documents of the world. However, by the twelfth century, the "center" of the world was shifting, for numerous political and cultural reasons, from the dar al-Islam (the political, cultural, and economic entity that extended from Spain, referred to as al-Andalus, to the Middle East) to Western Europe.

By the end of the twelfth century, the Dar al-Islam had controlled Mediterranean commerce, culture, and science for over three centuries. During this period, the Muslim world stretched from Spain to the Middle East. When the Umayads were defeated by the Abbasids in the middle of the ninth century, the Muslim political, cultural, and economic focus shifted from Damascus, a city near Jerusalem, to Baghdad, which is landlocked and further east.

Most of Europe was removed from the Muslim sphere of influence, and the nascent European nation-states were unable to generate enough power or political stability to successfully confront and overtake the Muslim world by military or economic means.

The one European exception to this situation was Spain, which existed at the western fringes of Muslim consciousness. In Cordoba, the intellectual heart of al-Andalus, citizens mimicked the fashions and manners of Baghdad and traveled eastward in search of cultural, spiritual, and intellectual fulfillment.

Al-Andalus, which constituted the southern half of the Iberian peninsula, became part of the Islamic world in 711, when a Muslim army crossed the Straits of Gibraltar and ousted the Visigoths. The period from the tenth to the twelfth century marked a time of considerable commercial stability for al-Andalus. The east-west axis that the Muslims had established across the Mediterranean helped to secure their power. Likewise, al-Andalus benefited considerably from the movement of goods between west and east. Andalusian markets eagerly consumed eastern goods, and could rely on the Middle East as a stable market for the exportation of Iberian products.

However, another component also aided medieval Spain. The proximity of al-Andalus to the Christian European world helped to boost its economic prestige. Indeed, during this period the Iberian peninsula was part of both the Muslim and Christian worlds. While the border between northern Christian Spain and southern Muslim Spain shifted towards the south over the centuries, the Iberian peninsula operated as a gateway between these two spheres of influence.

The Spain that al-Idrisi encountered during his education at Cordoba was a unique hybrid of European and Muslim influence. However, while Andalusian cities such as Cordoba embraced Muslim influence, Christian Europe fought against it. Indeed, Roger II's father, Roger de Hauteville, helped to cut the Muslim stranglehold on Mediterranean trade. Roger de Hauteville's conquest of Sicily assumed the trappings of a crusade; in 1063, Pope Alexander II presented him with a papal banner that was to be carried at the head of the army. Likewise, the Pope granted absolution to all soldiers who helped in the battle effort. The upheaval in Italy paralleled the violence brought on by the crusades. In fact, the crusaders stormed Jerusalem in 1099, only a few years after Roger de Hauteville secured Sicily. In 1072, Roger I crushed the Byzantine navy, and relied on his own naval power to establish a new kingdom on the Mediterranean.

His son, Roger II, continued this path. He took power in 1112 at the young age of 17. By 1122 he had attacked North Africa in an attempt to avenge an attack on Italy by Saracen and Spanish Muslim fleets. However, as his offer to al-Idrisi attests, he was also interested in learning the secrets the Muslims had employed to maintain control of the Mediterranean for such a long time.

Indeed, al-Idrisi's fame is linked to Roger's success in asserting the power of a European nation on the Mediterranean. Navigators and traders from the Mediterranean, as well as from the Atlantic and the North Sea, frequented Sicily after Roger's victory. Sicily quickly asserted itself as a new center of influence. Even after al-Idrisi's death, his books remained extremely popular among European audiences for several centuries.

DEAN SWINFORD

Further Reading

Ahmad, Nafis. *Muslim Contribution to Geography.* Lahore: Muhammed Ashraf, 1947.

Brauer, R.W. *Boundaries and Frontiers in Medieval Muslim Geography.* Philadelphia: American Philosophical Society, 1995.

Constable, Olivia Remie. *Trade and Traders in Muslim Spain: The Commercial Realignment of the Iberian Peninsula, 900-1500.* Cambridge: Cambridge University Press, 1994.

Fletcher, Richard. *Moorish Spain.* London: Weidenfeld and Nicolson, 1992.

Imamuddin, S.M. *Muslim Spain 711-1492 AD: A Sociological Study.* Leiden: E.J. Brill, 1981.

The Mongols Conquer an Empire, Opening Trade and Communication between East and West

Overview

The Middle Ages in Europe and the Middle East were marked by three invasions of Central Asian nomads: the Huns, the Turks, and finally the Mongols. The latter would conquer the largest empire of all and exert an enormous influence on history, paving the way for the Age of Exploration. Other than the Crusades, in fact, no single series of events had as much to do with Europe's reawakening from centuries of confusion following the fall of the Western Roman Empire in the fifth century A.D. The Mongols' impact is all the more impressive in light of the fact that they had no cities, no written language, and indeed no real history of any kind until they were united under one of the most dynamic leaders in all of history, Genghis Khan.

Background

Central Asia is a loosely defined region, a vast landlocked area bounded on the north by Siberia; on the east by China's densely populated eastern half; on the south by the Indian subcontinent and Iran; and on the west by the Caspian Sea and Ural Mountains. Within this great expanse, wider than the United States and almost as broad from north to south, is an ocean of grasslands and (in Mongolia and western China) deserts. It is not suitable for crops, only for herds, thus encouraging a nomadic way of life.

The existence of "barbarians" beyond its northern and western borders was a central factor in premodern Chinese history, spurring the Chinese to unite under their first emperor, Ch'in Shih-huang-ti (259-210 B.C.), in 221 B.C. Protection against the nomadic Hsiung-Nu, in fact, served as the basis for the building of the Great Wall, begun under his reign, and though the wall never fully kept out the nomads, it at least discouraged many among them.

A great number of Hsiung-Nu, in fact, began slowly making their way westward, looking for better grazing lands. By A.D. 372, they had reached the Volga River, where they became known by a different name: Huns. Crossing the Volga, they displaced the Ostrogoths, setting in motion a long domino-like chain of events that ultimately led to the destruction of the Western Roman Empire in 476.

In the wake of the Huns came various groups of Turkic-speaking Central Asian tribes: first the Avars, who introduced the horse stirrup to the West; then the Khazars, Bulgars, and by the tenth century the Oghuz Turks. From the latter came the Ghaznavids, who invaded what is now Afghanistan; and the Seljuks, who became the first Turks to conquer parts of the country that today bears their name. At that time, the land now known as Turkey was an integral part of the Byzantine Empire, and had been culturally linked with Greece since ancient times. Thus when the Seljuks defeated the Byzantines at the Battle of Manzikert in Armenia in 1071, it was a crippling blow to Byzantium, an event regarded as the beginning of the end of the Byzantine Empire.

In the late eleventh century, tensions between Christian Europe and the Muslim Seljuks would help to spawn a series of "holy wars" called the Crusades (1095-1291). The latter, for all their barbaric cruelty—and despite their failure as a military enterprise—awakened the West from its isolation and exposed Westerners to the advanced civilizations of the Muslim and Byzantine worlds. Before the Crusades had played themselves out, a second great awakening would come with the arrival of a third and final group of invaders from Central Asia: the Mongols.

Impact

Unlike the Huns or Turks, the Mongols did not begin their conquests primarily in reaction to attacks by outside forces; rather, they were welded into a mighty fighting force by a single man, a chieftain named Temujin. History, however, knows him better as Genghis Khan. The latter title, meaning "ruler of all men," was bestowed on him in 1206, after he became the first Mongol leader to unite all of that nation's tribes. Before long, the new Mongol khan received a visit from an official of China's Sung Dynasty, demanding an oath of loyalty. Genghis's response was to sweep into China in 1211 at the head of his army, a group of extraordinarily skilled and well-organized horsemen. By 1215 Genghis had conquered a city he called Khanbalik, which is today the Chinese capital of Beijing.

The westward thrust of Genghis's campaigns began in 1216, when he invaded the realms of a defiant Turkic khan in southwestern Asia. Another Mongol force, led by his son Juchi, moved deep into Russia in 1223. However, in 1226 Genghis himself turned back toward China to deal with a revolt. He died on the way, and in the aftermath Juchi relented from his attack on Russia, returning to the Mongol capital at Karakorum to participate in choosing a successor.

It was a pattern that would be repeated throughout the short but highly eventful period of Mongol conquests: just when they had the momentum on their side, they pulled back from the attack to spend months upon months in bickering over who should take leadership. In this case, nearly three years passed before Genghis's sons agreed that the youngest, Ogodai (r. 1229-41), should rule. Ogodai, however, lacked his father's vision as well as his ruthlessness, and though the Mongol realms would grow considerably in the years that followed, the driving force behind their expansion was gone.

In 1235 Ogodai sent Juchi's son Batu Khan to resume the attack on Russia with a combined force of Mongols and Tatars, another Central Asian nation that had once been the Mongols' rivals. By 1240 they had sacked Moscow and Kiev, and in the following year they devastated Poland and neighboring Silesia. They poured into Hungary, and by July 1241 were prepared to take Vienna. Then suddenly they were gone: Batu had received word that Ogodai was dead, and like Juchi before him, he hastened back to Karakorum to participate in choosing a successor.

The question of what would have happened if Batu had kept going is one of history's great "what ifs." As it was, the Mongol-Tatar force quickly turned from invaders to lazy administrators more intent on receiving tribute than on expanding their realms. In Russia they became known as the Golden Horde because of their wealth, and because their word for their tents, *yurtu,* sounded like *horde.* Mongol rule in Russia was not extraordinarily harsh, and the conquerors interfered little with the affairs of the locals, but they did expect huge payments of tribute, and more important, they kept Russia isolated from the changes taking place in Europe by that time. This in turn greatly influenced Russia's turn toward political authoritarianism combined with technological and economic backwardness.

Five years passed before the choosing of a new khan, Ogodai's son Kuyuk, in 1246, and when they finally renewed their efforts in the west, the Mongols shifted their focus from Europe to the Middle East. This, combined with the fact that Kuyuk had taken an interest in Nestorian Christianity, led to European rumors associating him with Prester John, a fabled Christian king in the East.

When Kuyuk died in 1248, it took the Mongols three more years to choose his cousin Mangu (r. 1251-59). The latter sent Hulagu (c. 1217-1265), yet another cousin, into Persia and Mesopotamia, where he destroyed the Assassins in 1256 before sweeping into Baghdad and killing the last Abbasid caliph in 1258. Upon Mangu's death, Hulagu gave himself the title Il-khan, and thenceforth all of southwestern Asia would be a separate khanate under his rule. Soon he invaded Syria, which inspired more European hope that the Mongols would defeat the principal Muslim threat, the Turkish Mamluks. But in a battle at Goliath Spring in Nazareth on September 3, 1260, it was the Mamluks who defeated the Mongols. This brought an end to Mongol conquests in the west.

Though historians often refer to the Mongol realms as though they were a single empire, in the period after Genghis's death these actually became four separate khanates: the Golden Horde in Russia; the realm of the Il-Khan in Southwest Asia; the Chagatai khanate in the western portion of Central Asia; and the realm of the Great Khan in Mongolia and China. As for China, though the Mongols had been making inroads there since Genghis's first thrust in the early thirteenth century, the conquest was completed by his grandson, the only successor who possessed anything approaching the leadership skills of Genghis: Kublai Khan (1215-1294).

Kublai, who became Great Khan in 1260, finally defeated the last Chinese forces in 1279. He had by then established the Yüan Dynasty (1264-1368), the first foreign ruling house to control China—and one of the shorter dynasties in Chinese history. Among the reasons for this lack of longevity was Chinese contempt for the Mongols, a serious problem given the fact that the less civilized Mongols depended on the Chinese to run the country for them.

It is interesting to note that whereas the Chinese were accustomed to looking down on all outsiders, the Mongols were some of the most open-minded people of the Middle Ages. Precisely because they lacked a sophisticated culture, they admired those of the peoples they ruled, and there were many of these: under the forceful Kublai, lands from Korea to Tibet to Vietnam bowed to the military power of the Mongols. Kublai did not annex these countries, but made them vassals, and the unity of the Mongol realm facilitated travel through areas formerly dominated by bandits and competing warlords.

This in turn made possible a degree of communication and trade between East and West that had not existed since the glory days of Rome. Among the first Westerners to travel eastward were the Polo brothers, Niccolò and Maffeo, who in 1271 brought with them a 17-year-old boy destined to become one of the most celebrated travelers of all time: Niccolò's son Marco (1254-1324).

Marco's writings about his experiences in the Orient became the most famous of their kind, but soon other Westerners were journeying to the East, bringing back with them such new ideas as gunpowder, paper money, the compass, kites, and even playing cards. From the Mongol realms, in turn, came a journeyer to the West: Rabban Bar Sauma (c. 1220-1294), a Turkish Nestorian monk who traveled to Europe and met Pope Nicholas IV (r. 1288-92), with whom he joined in an unsuccessful attempt to raise another crusade against the Muslims.

The Mongol period also saw a flowering in the arts of China, an unintended result of Chinese contempt for their rulers. Rather than serve the "barbarians," many talented Chinese opted to become artists and educators rather than civil servants. Of course the Mongols were most eager to absorb the refined culture of China, and this produced yet another unintended effect: in becoming more sophisticated, they lost the brutal toughness that had aided them in their conquests, and so become vulnerable to overthrow.

A series of failed invasions, both against Japan and Java, hastened the decline of Mongol power even under Kublai, and none of his successes proved to be anything like his equal as a leader. Furthermore, the Mongols lacked the sheer numbers to truly dominate China: not only were the Chinese older and wiser, in terms of their civilization, they were also more numerous. In 1368 a rebel named Chu Yüan-chang seized control of Khanbalik, and established China's last native-ruled dynasty, the Ming (1368-1644).

The Mongols had one last moment of glory under Timur Lenk (1336-1405), or "Timur the Lame," who became known to Europeans as Tamerlane. Though he was not related to Genghis Khan, Tamerlane saw himself as a successor to the great conqueror, and set out to build an empire of his own, conquering much of southwestern Asia. Tamerlane would be remembered for his cruelty as a conqueror, and for his establishment of Samarkand, in what is now Uzbekistan, as a great cultural center. In 1526 his descendant Babur (1483-1530) established the Mogul dynasty, which would rule India until the eighteenth century.

Even by the era of Tamerlane, however, the time of the Mongols' greatest importance to world history was long past. Their impact, however, spread across time and space. To cite a significant example, the Mongols' opening of trade routes can in part be blamed for the spread of the plague or Black Death. An epidemic that started in Asia, the plague soon moved westward, carried by rats on merchant ships, to kill a third of Europe's population in the years 1347-51.

Much more significant than the plague, however, was the fact that the Mongols opened Europeans to the idea of international travel, trade, and exploration. As Mongol power waned at the end of the fourteenth century, Europeans found themselves confronted by hostile Muslim rulers who blocked the eastward routes. Thus arose a need to find a sea route to Asia, and a new era in exploration was born. Central among the figures who brought about this era was Portugal's Prince Henry the Navigator (1394-1460), who had grown up reading Polo's work. So, too, had Christopher Columbus (1451-1506), who in 1492 set out to reach the East by sailing west—and instead discovered the New World.

JUDSON KNIGHT

Further Reading

Books

The Editors of Time-Life Books. *The Mongol Conquests: Time Frame A.D. 1200-1300.* Alexandria, Virginia: Time-Life Books, 1989.

Grosset, René. *The Empire of the Steppes: A History of Central Asia.* New Brunswick, New Jersey: Rutgers University Press, 1970.

Howorth, Henry H. *History of the Mongols.* New York: B. Franklin, 1965.

National Geographic Society (U.S.) Cartographic Division. *Mongol Khans and Their Legacy.* Washington, D.C.: National Geographic Society, 1996.

Prawdin, Michael and Michael Charol. *The Mongol Empire: Its Rise and Legacy.* New York: Macmillan, 1940.

Internet Sites

Oestmoen, Per Inge. "The Realm of the Mongols." http://home.powertech.no/pioe/index.html.

Giovanni da Pian del Carpini Travels to Mongolia

Overview

The Mongols are often remembered as ruthless and marauding nomads who would let nothing stand in their way during the height of their power. This characterization is due, at least in part, to hostile historical sources that exaggerated their cruelty in an attempt to discredit them. Some descriptions of this barbarian horde, however, reflect the true nature of these people. It is difficult to separate the historical facts from propaganda, but a Westerner named Giovanni da Pian del Carpini wrote an excellent firsthand account of the Mongols called *History of the Mongols Whom We Call the Tartars* (1247). This work has often been cited as the best reference on the subject from this time period.

The Mongols constitute one of the principal ethnic groups in Asia. Their traditional homeland is centered in Mongolia, which is divided into the two present-day regions of the People's Republic of China and Mongolia. Geographically, Mongolia lies within a traditional migration corridor between China and Hungary, which has influenced much of their history. The term Mongol is sometimes confusing because at one time it was erroneously used as a racial characterization. However, Mongols exhibit a vast range of physical characteristics and the term should be taken as a group of people bound together by a common language and history.

Western Europeans lived in great fear of the Mongols in the thirteenth century. At this time, the Mongols were at the height of their power and controlled much of Europe and Asia. The Mongol Empire stretched from the China Sea in the east to the Caspian Sea in the west. From north to south, it stretched from Siberia to central China. The Mongols were fearless warriors who utilized armies of mounted archers to their tactical advantage. Despite this, Pope Innocent IV dispatched the first formal delegation to meet the Mongols. This mission had multiple goals. First and foremost, the pope wanted to convert the Mongols to the Christian faith. Second, he wanted to gain reliable information regarding the size and condition of the Mongol armies in addition to finding out what they were planning in the future. Third, he hoped to form an alliance with the Mongols so that he could persuade them from invading Christian territory and to form a possible partnership against the Islamic people. Last, he had hoped that the meeting would help protect traders along the legendary "Silk Road" to and from China. Pope Innocent IV saw this as an important mission and selected Giovanni da Pian del Carpini, who was already more than 60 years of age, as its leader.

Giovanni da Pian del Carpini was a Franciscan friar who had been selected by the Pope largely based on his previous experience. In light of the hardships he had faced with his vow of poverty and his religious background, Carpini was well suited to the challenges that his journey would present. He had played a leading role in the establishment of the Franciscan order, and he had been a leading Franciscan teacher and held important offices in a variety of different countries. Carpini had also been in Spain at the time of the great Mongol invasion and witnessed the disastrous Battle of Liegnitz in 1241. Based on these experiences, the pope selected Carpini, despite his advancing age, to head the mission in 1245, and chose Willem van Ruysbroeck to direct a second mission in 1253. The pope gave them instructions to find out all they could about the Mongols and to persuade them to receive the Christian faith.

Background

Carpini embarked on his journey on Easter Sunday in 1245. Initially, another friar accompanied Carpini, but that friar was eventually left in Kiev. Carpini also recruited a Franciscan interpreter named Benedict the Pole along the route. The group made their way to the Mongol posts at Kanev and then continued on to the Volga River where they met Batu Kahn. Batu was the supreme commander on the western frontiers of the Mongol Empire and the conqueror of Eastern Europe. Carpini gained an audience with Batu only after he had submitted to a Mongol purification ceremony, which involved passing between two fires. He then met with Batu and presented him with gifts. Batu ordered them to travel to see the supreme Kahn in Mongolia. The group fittingly set out on the second leg of their journey on Easter Sunday 1246.

In order to withstand the rigors of travel, Carpini's body was tightly bound for the long ride through Central Asia. Their group journeyed

3,000 miles (4,800 km) in a little over three months and arrived at the imperial camp of Sira Ordu near Karakorum in mid-July. The Franciscans arrived at Sira Ordu just as one supreme ruler was dying, and were present when that ruler's eldest son, Kuyuk, was elected to the throne. On August 24 they were presented to the supreme Khan. They were detained for some time and then allowed to return to Europe with a letter addressed to the Pope. This letter, written in three different languages, outlined the supreme Kahn's assertion that he was the "scourge of God" and the pope must swear allegiance to the Kahn. During the long journey back, the friars suffered great hardships, especially in the winter months. Finally, on June 9, 1247, the group reached Kiev, which was a Slavic Christian outpost. They were welcomed with open arms and the letter was eventually hand delivered to the Pope. In his report, Carpini seemed confident that they could convert the Mongols to Christianity despite the contents of the letter.

Impact

Not long after his return, Carpini was appointed archbishop of Antivari in Dalmatia where he recorded his observations from his trip in a large volume of work. Carpini was an astute observer of the tradition and customs of the Mongols while he was in their presence. He recorded his impressions in a manuscript containing various types of style and content, which he called, *History of the Mongols Whom We Call the Tartars*. He also wrote a second manuscript titled, *Book of the Tartars*. He had written various chapters concerning the Mongols' character, history, foreign policy, and military tactics, including a section on the best way to defeat or resist the Mongols in case of attack. He also included a travelogue of his journeys, factual evidence of the groups of people who had been conquered by the Mongols, groups of people who had successfully resisted invasion, a list of the Mongol rulers, and finally, a record of people who could corroborate his assertions. His book was the first Western account of the Mongol Empire written by someone who was relatively unbiased.

Carpini's book discredited much of the folklore associated with the Mongols at that time. It gave a clear account of the everyday lives of this group and showed that they were human, not an inhumane band of marauding barbarians. Much of his book was summarized into a widely distributed encyclopedia that served as the primary body of knowledge regarding the Mongol Empire.

The book also served as a model for other adventurers in its rigorous and detailed account of the history and events concerning a group of people. It is probably the best treatment of a cultural study done by any Christian writer of that era. It was vastly superior in most ways to the chronicle of Ruysbroeck, who wrote of the similar mission he had undertaken in the Mongol Empire in 1253. Ruysbroeck chronicled his travels in *A Journey to the Eastern Parts of the World*. He provided a more personalized account of his travels while providing confirmation for many of the facts Carpini reported. Ruysbroeck's account also provided much insight into the Mongol culture. Ruysbroeck had conversations with people who had been to China and gave the first Western accounts of paper money and other aspects of Chinese culture.

A consequence of Carpini's journey is that he proved that one could travel east and return without much harm. He was the first European in over 300 years to travel that far east and return safely. Certainly the journey was a hardship and Carpini had, at one point, been stricken very ill. However, by returning, he helped to open the door for other diplomats and adventurers to attempt to meet and study other cultures and societies. Carpini was the first in a long wave of explorers and certainly influenced many who came after him, although he is rarely thought of as such. The knowledge of the Mongol Empire unlocked a new pathway between East and West and brought stability to two continents. Though merchants and traders long traveled the Silk Road, never had so many traveled so far as during the Mongol era. For the first time, many Europeans sought out the promise of wealth in the cities of Asia. Carpini's accounts and those of others such as Ruysbroeck and Marco Polo aroused the European imagination and inspired the quest for new passages to the East, long after the Mongol Empire fell.

JAMES J. HOFFMANN

Further Reading

Giovanni da Pian del Carpini. *Historia Mongalorum quos Nos Tartaros appellamus*. (The story of the Mongols whom we call the Tartars). Translated by Erik Hildinger. Wellesley, M.A.: Branden Publishing Company, 1996.

Marshall, Robert. *Storm from the East: From Genghis Khan to Khubilai Khan*. Berkeley: University of California Press, 1993.

Morgan, D. *The Mongols*. Peoples of Europe Series. Oxford: Blackwell Publishing, 1990.

Mongolia and Europe: Personal Accounts of Cultural Overlap and Collision

Overview

For European explorers, merchants, and adventurers, the Orient presented a considerable challenge and exerted a powerful draw. Many of the products that revolutionized late-medieval Europe were originally imported from Asia. Paper, stirrups, and gunpowder were all products that European merchants eagerly desired to distribute in their homelands. Indeed, acquisition of these products, combined with the mastery and control of trade routes necessary to secure them, played an important role in preparing Europe for the Renaissance.

However, despite the clear advantages gained by Europeans through the acquisition of Eastern products and technologies, early travelers and adventurers who traded with the East often encountered a world that seemed diametrically opposed to their own.

Background

Though Western Europe gained much in terms of technological knowledge from interaction with Asia, the European view of Asian culture was frequently negative. In large part Europeans were terrified of the Mongols, who, in the thirteenth and fourteenth centuries, were engaged in a wholesale invasion of Asia and Eastern Europe. Indeed, Western Europe learned of the Mongols, whom they named the Tartars, after the second Mongol expedition of 1238 unleashed widespread destruction across a significant portion of Eastern Europe.

The following excerpt from the *Chronica majora* of Matthew Paris (?-1259) indicates the attitude with which many Europeans regarded those from eastern Asia:

> *That the joys of mortal men be not enduring, nor worldly happiness long lasting without lamentations, in this same year [1240] a detestable nation of Satan, to wit, the countless army of the Tartars, broke loose from its mountain-environed home, and piercing the solid rocks [of the Caucasus], poured forth like devils from the Tartarus, so that they are rightly called Tartari or Tartarians. Swarming like locusts over the face of the earth, they have brought terrible devastation to the eastern parts [of Europe], laying it waste with fire and carnage. After having passed through the land of the Saracens, they have razed cities, cut down forests, killed townspeople and peasants . . . . They are without human laws, know no comforts, are more ferocious than lions or bears, . . . [and] are rather monsters than men, thirsting for and drinking blood, [and] tearing and devouring the flesh of dogs and men.*

Indeed, the Mongolians were both powerful and horrifying. Their disregard for the people whom they conquered threatened the future of a Europe that was, at that time, embroiled in wars and intrigues. The Crusades had by that point also weakened the stability of European nations. Quarrels between the pope and the princes of European countries prevented the unification of power necessary to defeat a foe already deeply entrenched in eastern Europe.

Impact

In 1241, even as the Mongolian hordes threatened to push into western Europe, the invasions suddenly stopped. Ogotay Khan, the son of Genghis Khan (1154?-1227), died. Upon his death, the leaders of the Mongol forces acted as they had after the deaths of all the Khans—they swept back across Europe and Asia. Once they had returned to Mongolia, the leaders met in council in order to elect a new leader.

In Europe, Innocent IV was elected pope in 1243. Unlike Pope Gregory IX, his predecessor, Innocent IV took action in order to prevent the further spread of Tartar dominance. Innocent believed that the invasion had halted because the Mongol menace was threatened by fear of Divine Wrath for threatening Christendom. In order to reinforce this point, and to learn the Mongol's true intentions, Innocent IV arranged for missionaries to be sent east to the Mongols.

Friar Giovanni da Pian del Carpini (1180?-1252) was sent on such a mission and, after a long and arduous journey, returned with a letter for the pope that includes the following passage:

> *By the power of the eternal heaven, the order of the oceanic Khan of the people of the great Mongols. The conquered people must respect it and fear them. This is an*

order sent to the great Pope so that he may know it and understand it . . . . The petition of submission that you sent to us we have received through your ambassadors. If you act according to your own words, you who are the great Pope, with the kings, will come all together in person to render us homage and we will then have you learn our orders.

In this letter, Kuyuk Khan, the son of Ogotay, denies Innocent's request that Kuyuk be baptized and discontinue the conquest of Europe.

However, the invasion did not spread further west. Indeed, Mongol chiefs in western Asia ruled without the aid of the Great Khan. Under Kublai Khan (1215-1294), the Mongol empire succeeded in conquering China and stretched to its greatest size. But, even though the empire was huge, it lacked a single unifying principle, such as that provided by religion.

The Franciscan Order was involved in a continuous evangelical mission during this time, which involved sending monks to Asia in order to exert a moralizing influence on the rulers of the Mongol empire. Odoric of Pordenone (1286?-1331) was one of these friars sent to the East. He journeyed through Asia roughly 20 years after Marco Polo (1254-1324) returned to Europe with his tales of the court of Kublai Khan. At that point the Mongol empire began to slip apart. The outlying regions were too far from Mongolia. The Mongol dynasty, on the other hand, was subject to the softening process engendered by easy living. Elaborate palaces of gold and jade filled the once-desolate plains.

Odoric traveled through an empire in decline and provided an intelligent, though sometimes fantastic, account of the people of Asia. His account provides many details that Marco Polo fails to mention. When Odoric returned to Italy in 1330, he recounted his travels to a fellow Franciscan. This account, combined with those of Carpini and Willem van Ruysbroeck (1215?-1295?), were combined in *The Travels of Sir John Mandeville,* a popular travel narrative in medieval Europe.

It may be surmised from Odoric's narrative that Pope Innocent IV's desire to inundate the Mongol court with missionaries was successful to some degree. When describing "The Glory and Magnificence of the Great Khan" at royal feasts, Odoric mentions his own role in the emperor's court:

I, Friar Odoric, was present in person for the space of three years, and was often at the banquets, for we Minor Friars have a place of abode appointed out for us in the Emperor's court, and are enjoined to go and bestow our blessing upon him.

Odoric's description of royal protocol indicates that he was both an observer and participant. Likewise, Odoric's elaborate depiction of objects, such as the Khan's two-wheeled chariot, "upon which a majestic throne is built of the wood of aloe, being adorned with gold and great pearls, and precious stones, and four elephants bravely furnished [t]o draw the chariot," suggests the majesty of the Mongol dynasty and reveals the extent to which outsiders had penetrated into the royal circle.

Odoric and his fellow friars were occasionally in contact with the Great Khan. Odoric recounts one occasion in which the Khan called him and several other friars to the Khan's chariot. When the friars approached, the Great Khan "veiled his hat, or bonnet, being of an inestimable price, doing reverence to the cross. And immediately I put incense into the censer, and our bishop, taking the censer, perfumed him, and gave him his benediction." Encounters such as this seem to have been fairly common.

Odoric's account in "Of the Honour and Reverence Done to the Great Khan" offers other details indicative of the influence of Odoric and his fellow friars in the Khan's court. In this chapter, Odoric and his friars present gifts to a group of Mongol barons, "which had been converted to the faith by friars of our order, being at the same time in the army." This passage suggests that at that time high-ranking officers of the Mongol army were Christian. The depiction of the Mongols presented by Matthew Paris in the *Chronica majora* was becoming increasingly inaccurate.

However, the fact that Odoric's journey was a religious quest cannot be forgotten. His narrative is filled with description of "idolatrous" customs perpetrated by unsavory infidels. Odoric is often considered the first European to enter the Tibetan city of Lhasa, the home of the Dalai Lama. His account highlights the social barriers that prevented cultural understanding between East and West. In one section, in which he discusses his journey to Lhasa, Odoric makes direct, but often inaccurate and biased, comparisons between Buddhism and Christianity.

In Lhasa, Odoric writes, "their Abassi, that is to say, their Pope, is resident, being the head and prince of all idolaters." Odoric presents even greater disdain in his description of a Tibetan burial ceremony. In order to reveal the true limitations of the flesh and the transcendence of the spirit, the Tibetans dismembered their dead and feed them to the vultures. While the transcendence of the spirit is also a component of the Christian religion, Odoric mocks the practice. He ends his discussion of Tibet by stating that "many other vile and abominable things does this nation commit, which I mean not to write, because men neither can nor will believe, except they should have sight of them." Odoric, a true citizen of the medieval world, was unable to view a different culture as equal, and provided an account of the East that emphasized its cultural distance from Europe.

Following centuries of relative isolation, encounters between East and West increased dramatically during the Middle Ages. The accounts provided by friars such as Odoric offer personal glimpses into the impacts and implications of this cultural collision.

DEAN SWINFORD

Further Reading

Bartlett, Robert. *The Making of Europe: Conquest, Colonization and Cultural Change: 950-1350.* Princeton: Princeton University Press, 1993.

Komroff, Manuel. *Contemporaries of Marco Polo.* New York: Boni & Liveright, 1928.

Mandeville, John. *The Travels of Sir John Mandeville.* New York: Dover, 1964.

Phillips, J.R.S. *The Medieval Expansion of Europe.* Oxford: Oxford University Press, 1988.

Walsh, James J. *These Splendid Priests.* Freeport: Books for Libraries Press, 1968.

The Journeys of Marco Polo and Their Impact

Overview

Marco Polo (c. 1254-1324) was a Venetian merchant and adventurer who made an extended, twenty-four year (1271-95), journey with his father Niccolò and his uncle Maffeo into central Asia, including seventeen years spent in Mongol-controlled China. He was among the first Europeans to visit this part of the world and was the first to record in detail the many things he observed there. He included information on the culture and religion as well as the geography and government of the regions he visited. His account of the trip was published in 1298 as *Divisament dou Monde* (Description of the world), now known generally as "The Travels of Marco Polo." Although read widely when it appeared, it was regarded by most readers as a work of fiction. Only later was it realized that most of its contents are quite accurate. In any case, it served to excite Europeans about the riches in trade and culture which might be found in unfamiliar areas of the world and to encourage them to venture out in search of them.

Background

Marco Polo lived at an auspicious time in history. The Dark Ages that had followed the collapse of the Roman Empire were ending. Governments were becoming more stable and trade was increasing. Under Genghis Khan, the Mongols had conquered China and most of the rest of eastern Asia. They had also subjugated Russia and threatened Europe as well. The grandson of Genghis Khan, Kublai Khan (1215-1294) became the Great Khan in 1257 and ruled the immense Mongol Empire. Although the Mongols made no great effort to change the culture of the nations they conquered, they did not trust the natives' participation in government and looked to foreigners, especially Europeans, for help in administering their empire.

For many years, Europe and central Asia had been engaged in trading. The Chinese Empire had been a strong power with a well developed culture since the time of the Roman Empire, and there was active trade between the two empires along a 4,000-mile (6,400-km) caravan route, known as the Silk Road. Although trade declined when the Roman Empire degenerated, trade was revived by the Mongols during the thirteenth and fourteenth centuries. Few traders traveled the entire route from Europe to China; merchandise changed hands a number of times before reaching its destination. A few Europeans, however,

Marco Polo arriving in China. *(Bettmann/Corbis. Reproduced with permission.)*

had preceded the Polos in making the trip to the court of the Great Khan in China. For instance, Pope Innocent IV sent friars there to attempt to convert the Mongols to Christianity. Among these were Giovanni da Pian del Carpini in 1245 and Willem van Ruysbroeck in 1253.

Although challenged by other nations, Venice controlled the Mediterranean and dominated European trade with the Middle East and with much of the rest of Asia. Marco Polo's family included wealthy merchants and traders—prominent members of Venetian society. His father, Niccolò, and his uncle, Maffeo, traded extensively in the Middle East. They left on a trading mission in 1253, leaving behind Niccolò's pregnant wife who gave birth to Marco in his absence. The brothers traded with the ruler of the western territories of the Mongol Empire and in 1260 left Constantinople on a trip through Afghanistan and Uzbekistan to Shang-tu (also known as Xanadu), the summer residence of the Great Khan of the Mongol Empire, Kublai Khan. They arrived in 1265 and remained in Kublai Khan's court until 1269 when they were sent back to Europe as emissaries to Pope Innocent IV, requesting one hundred men to instruct and convert the Mongols and asking for oil from the lamp in the Holy Sepulchre in Jerusalem.

When Niccolò and Maffeo returned to Venice, Marco, now fifteen years old, met his father for the first time. His mother had died while he was very young, and he had been reared by an uncle and aunt.

Pope Clement IV had died, and the Polos waited for a new pope to be elected so that they could deliver Kublai Khan's requests. After two years, the cardinals could still not agree on a pope, and the Polos decided to return to China. Seventeen-year-old Marco accompanied them when they left in 1271. They traveled first to Acre in Palestine where they learned that their friend Teobaldo had been elected pope as Gregory X. The new pope provided them with two monks (instead of the one hundred requested), oil from the lamp in the Holy Sepulchre in Jerusalem, and papal communications to the Khan. Soon after they set out from Acre, the two monks, afraid of the dangers ahead, turned back, leaving the Polos to proceed alone.

Their journey took them through Turkey, Iran (Persia), Afghanistan, and Pakistan. Marco became ill in the desert but recovered in the cool regions of Afghanistan. They visited Kashmir and crossed the mountains into China. Following the Silk Road, they crossed the Gobi Desert and arrived at the Mongol summer capital, Shang-tu, in 1275.

Marco was introduced to Kublai Khan and the Mongol court and was impressed with its splendor. The Polos remained in the Mongol court in Shang-tu and in Beijing, serving as advisors to the Khan, for seventeen years. Marco was good at languages, and he immediately became a favorite of Kublai Khan who sent him as his emissary to various parts of the Mongol Empire. Marco was, therefore, able to explore and observe the country and people of much of China and of parts of India. He was the first European to visit Burma, and he traveled to Ceylon on a mission from Kublai Khan to buy Buddha's tooth and begging bowl. He kept notes on all he observed so that he could make detailed reports to the Khan on the conditions in the various parts of his realm.

After a number of years, the Polos wished to return to their native Venice. Kublai Khan was getting old, and they were concerned that they would not be safe among the Mongols after his death. Kublai Khan valued their service and, for a number of years, would not let them leave. He finally granted them permission to return to Europe in 1292, provided they accompany the Mongol princess Kokachin to Persia where she would marry Arghun, the Mongol Khan of Persia. They set out with six hundred escorts and fourteen ships. Marco was able to add a great deal to his store of knowledge of Asia on their voyage, which took them to Vietnam, the Malay Peninsula, Sumatra, Ceylon, and the shore of Africa before landing at Hormuz on the Persian Gulf. Only eighteen of the six hundred escorts survived the long hazardous trip. From Hormuz, the Polos accompanied Kokachin on to Khorasan, where she married the son of the recently deceased Mongol ruler who had requested her as a wife. The Polos then traveled on to Tabriz and Constantinople, across Armenia, and finally returned to Venice in 1295. Marco, who had been seventeen when he left home, was now forty-one years old. The Polos had changed so much since leaving home that they were not recognized initially by their family and friends. They had been robbed on the trip back from China, losing much of the wealth they had accumulated in the service of the Mongol Empire, but they managed to bring back ivory, jade, jewels, porcelain, and silk as proof of their tales of China and the Far East.

The Polos settled back into influential positions in the Venetian trade community. Soon after his return, Marco served as the gentleman-commander of a warship in a trade war with Genoa and in 1296 was taken prisoner at the battle of Curzola. While imprisoned in Genoa, he met a prisoner from Pisa named Rustichello (also known as Rusticiano), an author of some renown. Marco told the story of his Asian travels to Rustichello. When he was freed, Marco returned to Venice, married and had three daughters. The dictated memories of his travels were published in 1298 as *Divisament dou Monde* (Description of the world), and they gained him immediate notoriety. Unfortunately, many readers regarded the book as fiction, a chivalric fable similar to the King Arthur legend. Its truth was not realized until after Marco Polo's death.

Impact

"The Travels of Marco Polo," as Marco Polo's story is now generally known, had a significant impact on subsequent exploration. When they were first published, the tales he told seemed fantastic to Europeans who had never been exposed to the details of central Asia. There are certainly fantastic elements in the book. He describes animals and customs that are clearly fictional, and it is evident that he relates many things that he did not witness himself. He seems rather gullible in accepting the tales of others as truth. It is also possible that Rustichello embellished the material to make the book more interesting. Since it was published before the printing press, each copy had to be hand-reproduced, and changes in the original content undoubtedly occurred during this process. In fact, there are well over one hundred different manuscript versions. Nevertheless, most of what he reported was true, and as time passed it came to be accepted as such. The book became an important source of knowledge on the geography, people, culture, government, religions, technology, plants, and animals of the vast area to the east. It provided a new view of the world and opened up new possibilities for trade and exploration. It served as a significant enticement for merchants, explorers, rulers, and churchmen to seek greater understanding of central Asia and to form closer relationships with its inhabitants. Maps of Asia were based on his descriptions until the sixteenth century, and many of the ventures of the Age of Exploration of the fifteenth and sixteenth centuries, including the voyages of Christopher Columbus, were inspired by the "Travels." Marco Polo's book was not replaced as the most important source of information on central Asia until the nineteenth century and may, indeed, be the most influential travelogue ever written.

J. WILLIAM MONCRIEF

Further Reading

Polo, Marco. *The Travels of Marco Polo the Venetian*. Translated and edited by William Marsden, re-edited by Thomas Wright. Garden City: International Collectors Library (Doubleday), 1948.

Polo, Marco. *The Travels of Marco Polo*. New York: Orion Press, 1958. (From a fourteenth century manuscript in the Bibliotheque Nationale, Paris)

Niccolò de' Conti Immerses Himself in the East

Overview

Between 1419 and 1444, Niccolò de' Conti undertook an odyssey that has been compared with that of Marco Polo some 150 years before. Like Polo, Conti was a Venetian merchant who realized in his youth that international trade offered boundless opportunities for adventure, and like Polo he spent a quarter-century traveling in the East. Both men wrote of their journeys—in fact, their accounts have been published together—and though Polo and his writing are much better known, Conti's work also exerted considerable influence on Europeans' growing interest in exploration. As for the degree to which the two men became immersed in the life of the East, Conti exceeded Polo by a wide margin: Polo may have served in the court of an Eastern monarch, Kublai Khan, but Conti married an Indian wife, raised a family, and renounced Christianity.

Background

Western European contact with the East—not just the Far East, later visited by both Polo (1254-1324) and Conti (c. 1396-1469), but even parts of the Near or Middle East—had all but ceased with the decline of the Western Roman Empire in the fourth century A.D. It is more than a little ironic, then, that the beginnings of this isolation more or less coincided with Westerners' adoption of what is, after all, a Middle Eastern religion: Christianity.

Christian missionaries such as Peter and Paul had come from Judea to Rome, and though these original messengers had met with persecution and death, the religion had taken hold, and for a time the new faith had unified the Mediterranean basin, with bishops at Rome, Constantinople, Alexandria, Jerusalem, and Antioch. The rise of Islam in the seventh century, however, placed the last three of these cities under Muslim control, and established a virtual iron curtain between West and East.

Other divisions served to further isolate Western Europeans. Disagreements over the nature of Jesus of Nazareth—specifically, his dual role as both human and divine—had separated splinter groups such as the Arians and Nestorians from mainstream Christianity, as they would ultimately divide the main Western (Roman Catholic) and Eastern (Greek Orthodox) churches in 1054. This, too, led to loss of contact with the East: had there not been a break between Rome and the Nestorians, who gravitated toward India and China, it is possible their shared faith might have helped keep communication open between Christians in Western Europe and East Asia.

Except for limited diplomatic exchanges between Charlemagne (742-814) and the Abbasid caliph Harun al-Rashid (766-809), then, European contact with the East was minimal for a period of about half a millennium. Then two series of events opened the way for Europe's contact with the outside world, and indeed for the continent's awakening from the Dark Ages. First, the Crusades (1095-1291) exposed Europeans to the advanced civilizations of Byzantium and, to a much greater extent, the Muslim Middle East. Then, during a period of about two centuries beginning around 1220, the Mongol empires of Genghis Khan (1162-1227) and his successors served to unify a vast region between eastern Europe and the Far East under a single group of rulers.

Mongol rule made possible the first European journeys to the Far East, beginning with that of the priest Giovanni da Pian del Carpini (1182-1252), sent to the court of Kuyuk Khan by Pope Innocent IV in 1245. Later, Kublai Khan (1215-1294) sent a traveler *westward,* Rabban Bar Sauma (c. 1220-1294), who visited Pope Nicholas IV in about 1288. Nicholas in turn sent Giovanni da Montecorvino (1246-1328), who established the first Catholic mission in China. In the meantime, however, the first European merchants had begun to go East, among them brothers Niccolò and Maffeo Polo. Later the Polos made another journey, this time with Niccolò's son Marco. It is likely that another Niccolò—another merchant of Venice, born more than 70 years after Marco Polo died—grew up listening to the tall tales Marco had written in his book.

Conti's family had built its considerable wealth by trading with Egypt, and though the young Niccolò was certainly interested in the world of international business, he had his eye on even more exotic trading locales than those of Cairo and Alexandria. In 1419 he traveled to Syria, where he learned Arabic, which along with Farsi (or Persian) was essential for trade in the Middle East. In time Conti moved on to Baghdad and later Persia (modern Iran), where he learned to speak Farsi as well.

While in Persia, Conti established a trading company in association with local merchants, and began to conduct a brisk business at various ports along the Indian Ocean. Having apparently decided that his destiny lay in the East, he married an Indian woman, and together they began raising a family. Around this time, he also renounced Christianity, presumably in favor of Islam. This act would later become a subject of dispute, and indeed it has never been fully clear whether he did so willingly or not.

Among the areas Conti visited in India—all the while taking extensive notes—were Cambay, a state in the northwestern part of the subcontinent; Vijayanagar, capital of a powerful Hindu kingdom in the Deccan Plateau of central India; and Maliapur, where he found the shrine of St. Thomas the Apostle. The latter had, according to legend at least, gone to India as a missionary; in any case, the existence of a native Christian community in India (members of whom Conti met) is indisputable.

Moving far to the east, Conti next visited the island of Sumatra in what is now Indonesia, a region known at the time as the East Indies. There he observed cannibalism, and made note of the area's riches in gold and pepper. (At that time, the two were almost equally precious, since spices were necessary for the preservation of meat.) He then went to the Malay Peninsula, where he engaged in profitable trade with Muslim merchants; to Champa, or modern Vietnam; and later to Burma, where he visited the wealthy city of Pegu along the Irrawaddy River.

Conti made at least one more venture eastward, to Java, before beginning his long, slow journey toward Venice. Along the way, he stopped at a number of locales, including Calicut on India's Malabar Coast, and returned to Cambay before sailing on to Aden at the southern tip of the Arabian Peninsula. He stopped for a time in what is now Somalia, then sailed up the Red Sea and disembarked at Jidda, the port for Mecca. From there he made his way via overland routes first to Egypt, then to Mount Sinai in southern Judea. By 1444, he had returned to the city of his birth.

Impact

Over the course of a quarter-century, Conti had mingled with many varieties of peoples and had observed numerous cultures and ways of life firsthand. He had spent time among Muslims, Hindus, Buddhists, and practitioners of other religions, including breakaway forms of Christianity. These facts made him a great curiosity among the Venetians, as did his Indian wife and Indo-Italian family. Wealthy from his many business dealings, he was a celebrated figure, exotic and intriguing—but he still had to account for his renunciation of Christianity.

For this Pope Eugenius IV (r. 1431-1445) devised a unique form of penance: in order to pay for his sin, Conti was required to write an account of his journeys. He therefore went to work with Vatican secretary Poggio Bracciolini (1380-1459), who acted as scribe. (It is interesting to note that Eugenius was a strong advocate of humanism in education, and favored a rapprochement between Rome and the Greek Orthodox Church, the Nestorians, and other Christian groups.) As for Conti's record of his journeys, the papacy kept parts of those writings a secret for some time, no doubt intending to protect knowledge regarding trade; nonetheless, Conti's descriptions of spices in the East Indies greatly spurred the pace of European exploration.

Only in 1723, with the publication of *Historiae de veritate fortunae,* was Bracciolini's record of Conti's travels published. In the meantime, however, a Spanish version written by Rodrigo de Santaelle had appeared, and in 1579 it was translated into English by John Frampton. This version, which combines an account of Conti's journeys with those of Polo, is the one most familiar to readers in the English-speaking world.

Unfortunately, however, "most familiar" is a relative description: Conti's name is hardly a household word on the level of Polo's. He did gain some exposure, however, with the 1996 publication of *The Venetian's Wife,* a novel by Nick Bantock in which Conti appears as a character. Bantock, author of the highly acclaimed *Griffin & Sabine* (1991) and its sequels, said he first heard of Conti "in passing, on a PBS [Public Broadcasting System] program. I was amazed that he could be so historically unrecognized considering his extraordinary travels." Impressed by what he learned when he researched Conti's life, Bantock chose to include a highly fictionalized version of Conti in his sensual thriller.

JUDSON KNIGHT

Further Reading

Books

Bantock, Nick. *The Venetian's Wife: A Strangely Sensual Tale of a Renaissance Explorer, a Computer, and a Metamorphosis.* San Francisco: Chronicle Books, 1996.

Conti, Niccolò de', and Marco Polo. *The Most Noble and Famous Travels of Marco Polo Together with the Travels of Niccolò de' Conti,* translated by John Frampton, edited by N. M. Penzer. London: Argonaut Press, 1937.

Major, R. H., ed. *India in the Fifteenth Century.* London: Hakluyt Society, 1857.

Al-Mas'udi, the "Herodotus of the Arabs," Travels Widely and Writes Influential Works of History

Overview

The ranks of Arab writers, foremost among the medieval world's geographers, are full of adventurers who wrote about their journeys in travel memoirs. Works such as the *Rihla* of Ibn Battuta combine travelogue with analysis of local characteristics and customs. Yet few of these writers would qualify as scientific geographers in the modern sense—few, that is, aside from al-Mas'udi, author of numerous works on history and geography, the most famous of which is known in the West as *The Meadows of Gold.* In this, a universal history covering the period from the world's creation to his own time, al-Mas'udi gained a reputation as a historian on the order of the greatest among his profession: hence his title as "Herodotus of the Arabs."

Background

In order to understand al-Mas'udi (895-956), it is useful to examine the career of Herodotus (c. 484-c. 424 B.C.) This is so not simply because al-Mas'udi has been compared with the Greek "Father of History," but also because *The Meadows of Gold* mirrors Herodotus's *History* in scope, character, and research methodology.

A native of Halicarnassus in Asia Minor (modern Turkey), Herodotus grew up inspired by Homer's *Iliad* and *Odyssey.* Looking around him at the Greece of his day, fresh from its victory in the Persian Wars (499-449 B.C.), he realized that the recent conflict was the Trojan War of his own day. Thus was born the *History,* a chronicle of the world up to the conclusion of the Greeks' war with Persia.

In setting down his *History,* Herodotus drew on all manner of geographical, social, and political details. He traveled throughout the known world, and this heightened exposure—he later said that he interviewed people from 30 foreign nations—gave great depth to his work. His research took him to Phoenicia, Egypt, Libya, Mesopotamia, Persia, and the Black Sea region. As he went, he interviewed people, made notes, and collected material.

Prior to Herodotus, historical writing had consisted either of mythology, which was interesting but useless from a scientific standpoint, or of dry annals and king lists, which were long on fact but short on analysis (or, for that matter, interest to the average reader.) Herodotus was the first to transcend the dichotomy between myths and annals, and indeed he pioneered the methodology of the modern historian: gathering facts, weighing those facts for truth and falsehood, finding an overall picture among the many details, and then writing a narrative based on this picture.

Yet though he was a pioneer of scientific history, Herodotus was not above recounting myths and outlandish stories if he believed that these could illustrate some larger truth. "I must tell what is said," he wrote, "but I am not bound to believe it, and this comment of mine holds [true] about my whole *History.*" Thus while rumors and superstitions might not be "true," they often contained some grain of truth about the human condition.

Indeed, humanity was the ultimate subject of the *History,* and Herodotus might well be described as the first social scientist. Certainly his approach was in many ways surprisingly modern. For instance, he displayed a remarkable degree of regard for the customs and cultures of other lands—a surprising trait for someone from a civilization that used the word *barbaroi* to describe anyone who was not Greek. Like many aspects of Herodotus's worldview, this was a quality that would be shared by al-Mas'udi.

Impact

Because the Muslim holy book, the Quran, called upon the faithful to make a pilgrimage or *hajj* to the city of Mecca at least once in their lives if possible, Islam naturally encouraged travel and exploration. It did not, however, necessarily encourage toleration of other faiths; that element in the writing of al-Mas'udi was a reflection of the author's natural curiosity regarding the breadth of human experience.

Born in Baghdad, Abu al-Husayn Ali ibn al-Husayn al-Mas'udi probably began his travels at about the age of 30, when his writings tell of a trip to Persia. He continued east as far as Khurasan, in modern Afghanistan, and went on to India, traveling deep into the Deccan Plateau at the central part of the subcontinent. These journeys took him through lands inhabited by Zoroastrians in Persia and Hindus in India, and al-Mas'udi made it a point to visit their temples and learn about their customs.

Along the way, al-Mas'udi recorded extensive notes regarding plant and animal life, providing descriptions of coconuts, oranges, elephants, and peacocks. He also took down information he learned from others regarding China and Ceylon or Sri Lanka, but apparently did not visit those regions. (He associated Ceylon with "Sarandib," a land described in a passage from the *Thousand and One Nights* from which the English word "serendipity" is derived.)

After a few years' journey in the East, al-Mas'udi moved westward in 916 or 917, sailing to Oman and thence to East Africa. There, on the coastline of what is now Kenya, Tanzania, and Mozambique, Persian, and Arab merchants had been conducting trade for many years, and al-Mas'udi's is among the first written accounts of sub-Saharan Africa.

For a decade beginning in 918, al-Mas'udi traveled around Iraq, Syria, the Arabian peninsula, and Palestine. In the latter region, he visited Christian churches, talked with Jewish and Christian scholars, and went to Nazareth, where Jesus had spent his early years. He also observed the Samaritans, Jews who had remained in the area during the Babylonian captivity in the sixth century B.C. Their separation from mainstream Jews had made them a pariah class in the time of Jesus, as evidenced by the shock with which the disciples met Jesus' kindness to Samaritans in the New Testament.

In 927 al-Mas'udi visited the ancient center of Palmyra, perched on a caravan route through the Syrian Desert. He then traveled southward, where he met members of the Sabaeans, a sect distinct both from Judaism and Islam. By February 928, he was in the city of Hit in western Iraq, and there witnessed the city's siege by Karmathians, a violent Islamic splinter group linked with the Assassins.

Continuing northward and eastward, between 932 and 941 al-Mas'udi roamed throughout the Caucasus and the southwestern edge of Central Asia. There he explored the area around two of the world's largest inland seas, the Caspian and the Aral. The former, at more than 143,000 square miles (370,370 square km, or about the size of Montana) is by far the world's largest lake, and the latter is fourth-largest, at approximately 25,000 square miles or 64,750 square kilometers. However, it would be many centuries before explorers from civilized lands discovered the second- and third-largest, Lake Superior in North America and Lake Victoria in Africa, respectively; so as far as al-Mas'udi was concerned, the Aral Sea was second only to the Caspian. He provided the first written description of the Aral, and became the first geographer to correctly note that the fresh-water Caspian is not connected to the Black Sea.

While in the region, al-Mas'udi collected valuable information about non-Muslim peoples, and his work includes some of the first written descriptions concerning Russians, Bulgars, and Khazars, a Turkic people who adopted the Jewish religion.

Apparently it was only in 941 that the well-traveled al-Mas'udi got around to making the hajj to Mecca, after which he went to what is now Yemen at the southern tip of the Arabian peninsula. He followed this with a visit to Egypt, where in 942 he had an opportunity to observe yet another variety of religious experience among the Coptic Christians celebrating Epiphany. He spent the last decade of his life traveling back and forth between Syria and Egypt.

In the course of his career, al-Mas'udi wrote some 20 books, many of which have been completely lost. Among his works was the 30-volume *Akhbar az-zaman* (History of time), which—perhaps because of its overwhelming scope and intimidating title—failed to capture the attention of scholars. This was also the case with a second historical work, so al-Mas'udi resolved to condense the two ponderous books in a single, more concise work. The latter, written during his final years in Egypt and Syria, became *Muruj adh-dhahab wa ma'adin al-jawahir,* or *The Meadows of Gold and Mines of Gems.*

Like Herodotus's *History, Meadows of Gold* examines the grand sweep of history from the creation of the world to the present time, and includes among its narrative detailed observations on culture, customs, flora and fauna, and ethnicity. Pearl-diving in the Persian Gulf, the great temples of the world, hazards posed to mariners by waterspouts, and the burial customs of the Hindus are but a few of the topics covered in the *Meadows of Gold.* The latter volume succeeded where the others had failed, and earned al-Mas'udi acclaim throughout the Western world. Thus no less an authority than Ibn Khaldun (1332-1406), the renowned Islamic philosopher of history, pronounced al-Mas'udi an *imam*—a leader or guide—for all historians.

A map of the world drawn by al-Mas'udi illustrates his penetrating knowledge of geography. Among the features he depicted, some for the first time, was the meeting of the Indian and Atlantic oceans at the southern tip of Africa; the correct position of the Nile valley; the locations of the Indus and Ganges rivers of India, with Sri Lanka at the subcontinent's southern tip; and the outlines of the Caspian and Aral seas. Yet it was his writing that earned al-Mas'udi the reputation as "Herodotus of the Arabs." Indeed, he deserves to be called the inheritor of Herodotus's mantle not only among Arabs, but among all peoples. Certainly he continued his forebear's work as a pioneer of historical writing and the social sciences, in the process becoming one of the first writers to combine scientific geography, multifaceted history, and a detailed discussion of the peoples of the known world.

JUDSON KNIGHT

Further Reading

Books

Ahmad, S. Maqbul, and A. Rahman, eds. *Al-Mas'udi: Millennary Commemoration Volume.* Aligarh, India: Indian Society for the History of Science, 1960.

Horne, Charles F., ed. *The Sacred Books and Early Literature of the East,* Volume VI: *Medieval Arabia.* New York: Parke, Austin, and Lipscomb, 1917.

Al-Mas'udi. *The Meadows of Gold: the Abbasids.* Translated by Paul Lunde and Caroline Stone. London: Kegan Paul, 1989.

Shboul, Ahmad M. H. *Al-Mas'udi and His World: A Muslim Humanist and His Interest in Non-Muslims.* London: Ithaca Press, 1979.

Ibn Fadlan: An Arab Among the Vikings of Russia

Overview

In 921, the Arab traveler Ahmad ibn Fadlan (fl. 920s) went on a diplomatic mission to what is now Russia. There he encountered numerous Turkic peoples, among them the Khazars, one of the few groups in history outside of Israel to adopt Judaism. But perhaps the most memorable passages in the *Risala,* his account of his journeys, concern the Varangians, a group of Vikings known by a term that would eventually become the name of the surrounding country itself: *Rus*.

Background

Ibn Fadlan traveled on orders from al-Muqtadir (r. 908- 932), ruler of the Abbasid caliphate. Though by Ibn Fadlan's time the influence of the caliphs—imperial leaders who possessed religious as well as political authority—had declined somewhat, the Abbasid dynasty still remained the single most powerful force east of the Byzantine Empire and west of China. Through military might, combined with their fervent belief in Islam, the soldiers of the Umayyad caliphate (661-750) had extended Arab influence from Spain to India; and though the Abbasids (750-1258) proved less aggressive militarily, they were nonetheless eager missionaries for the Muslim faith.

Hence the purpose of Ibn Fadlan's mission: to explain Islamic law to the recently converted Volga Bulgars. The Volga Bulgars had moved into Eastern Europe from the frontiers of China during the sixth and seventh centuries, part of a great wave of migration that brought various Turkish peoples westward. ("Turks," a term that encompasses a number of nations, all speak languages of the Turkic family, and share origins in the far eastern reaches of Central Asia.) One group of Bulgars had continued moving to the western shores of the Black Sea, becoming Christianized and settling in the land that today bears their name. By contrast, the Volga Bulgars, as their name implied, had settled along the eastern shores of the Volga River in what is now Russia.

Another notable Turkic nation in the region were the Khazars, who in the eighth century had converted to Judaism. The Khazars lived on the southern end of the Volga, and as early as 568 had sent an ambassador to Byzantium. In time their realm came to be known as the Khazar Khanate, or simply Khazaria. It was one of the few lands in history, other than ancient or modern Israel—and a few Semitic states at the southern tip of the Arabian peninsula during the early centuries A.D.—where Judaism was the majority religion. Khazaria would flourish until its destruction by Kievan Rus or Russia in the eleventh century.

Russia in Ibn Fadlan's time had not yet earned its present name, though the people who came to be known as "Rus" had already arrived. They were Vikings, part of the vast group of nations that had burst onto the pages of history late in the eighth century. In the years that followed, they had fanned out across Europe, heading as far west as Iceland, Greenland, and later North America. Some even moved into southwestern Europe, where they came to be known as Normans in France and Sicily.

Then there were the Varangians, a group of Vikings who in 862 sailed out of their homeland in Sweden, eastward across the Baltic Sea and up the rivers of Eastern Europe, working their way further inland. Drawn by myths of a rich city of gleaming gold—no doubt Constantinople—they founded a great city of their own, Novgorod. They also established their control over Kiev, founded earlier by the Slavs, the people who first gave the Varangians the name "Rus." (In the context of early medieval Russia, the terms *Viking, Varangian,* and *Rus* are interchangeable.) Eventually one of the Varangians, a chieftain named Rurik (d. c. 879) emerged as their leader, founding a dynasty that would remain influential in Russian affairs until 1598.

Impact

Other than the fact that he was a theologian who served in the court of al-Muqtadir, little is known about Ahmad ibn Fadlan. From certain aspects of his writing style, scholars have guessed that he may not have been an Arab, but there is no certainty on this point. As for the purpose of his journey, it was a diplomatic mission: the actual leader of the group was a eunuch named Susan al-Rassi, and Ibn Fadlan went as a religious advisor charged with educating the Volga Bulgars on Islamic law.

The group set out from the Abbasid capital at Baghdad (now in Iraq) on June 21, 921, and followed established caravan routes eastward toward Bukhara, now part of Uzbekistan. Bukhara lay well to the east and south of their destination, however, and at Gurgan near the Caspian Sea in what is now northeastern Iran, they began heading northward on March 4, 922. The group moved along the Caspian's eastern shore, all the way around the sea, which is actually an inland lake. The Caspian is by far the world's largest inland lake, at 143,000 square miles (370,370 square km), or about the size of Montana. They skirted the Caspian's northern shore, finally reaching the delta of the Volga River.

Among the tribes Ibn Fadlan encountered along the way were the Oghuz Turks on the eastern shore of the Caspian, ancestors of the people who inhabit modern-day Turkmenistan. On the Ural River at the northern tip of the Caspian (today the Ural and the mountains of the same name are recognized as the boundary between Europe and Asia), the Arabs met the Pechenegs, another Turkish tribe. At the southern end of the Volga were the Khazars, and in what is now central Russia the party found yet another Turkish group, the Bashkirs. Finally, on May 12, 922, the group arrived at the Volga Bulgars' capital on the shores of the great river. There they were presented to the Bulgar khan, and Ibn Fadlan read aloud a letter from the caliph before presenting the khan with presents from the ruler in Baghdad.

But the Volga Bulgars, though they provided much of the purpose behind Ibn Fadlan's mission, do not occupy the most memorable passages in Ibn Fadlan's narrative. Rather, his focus is on a group of people he met during his time among the Bulgars, a strange blond-haired and blue-eyed race who had settled in the region, and with whom the Bulgars traded. These were the Varangians, who he described thus: "They are the filthiest of God's creatures. They have no modesty in defecation and urination.... they are like wild asses." Yet as he wrote elsewhere, "I have never seen more perfect physical specimens, tall as date palms, blond and ruddy...."

The ways of the Varangians—for instance, their practices with regard to sex, hygiene, and religion—seem to have filled Ibn Fadlan, a highly educated member of what was perhaps the world's most advanced civilization at the time, with a weird fascination. Particularly noteworthy was Ibn Fadlan's description of the ritual surrounding the burial of a Viking chieftain. The dead man's belongings were divided into three parts, one part for the wife and daughters of the deceased, one part to buy clothing for the corpse, and one part to pay for the vast amounts of alcohol that would be consumed by the men taking part in the 10-day-long funeral.

The female slaves of the deceased were asked, "Who among you will die with him?" and one eventually came forward and said "I." Presumably the volunteer was compelled by fear, combined with only a vague notion of what awaited her. She participated in gruesome rituals and then was killed. Her corpse was then placed alongside that of the deceased in a boat, which would be launched on the river and set ablaze. "If in this moment a wind blows and the fire is strengthened," Ibn Fadlan wrote, "...the man is accordingly one who belongs in Paradise; otherwise they take the dead to be one unwelcome at the threshold of bliss or even to be condemned."

These and other recollections of the Vikings appear in the *Risala,* Ibn Fadlan's account of his journeys, which he wrote upon his return to Baghdad. In some places, he appears to have exaggerated. For instance, he wrote that the Vikings all washed from a common bowl, and that it was typical for them to blow their noses and spit into the same basin in which they would wash their hands and faces. This was probably hyperbole occasioned by the shock with which a medieval Arab, to whom cleanliness was a necessity of religious faith as well as of health, would have viewed the habits of unhygienic Europeans.

In other places, however, Ibn Fadlan exhibited a cool-headedness that served him well as a travel writer. For instance, he adopted a somewhat skeptical tone when discussing reputed sightings of Gog and Magog, beastly creatures mentioned in the biblical book of Revelation and associated with the end of the world. Later in the Middle Ages, European scholars would write confidently that the monsters had been located somewhere in Central Asia; Ibn Fadlan, at least, reported such rumors merely as legends he had heard from others. All in all, his book constitutes an invaluable travelogue and a priceless source of ethnographic information concerning the peoples of Central Asia and Eastern Europe.

The final portion of the *Risala*, which presumably would have told about Ibn Fadlan's journey back and his later life, has been lost. The version known in the West today comes from the work of a Russian scholar, C. M. Fraehn, who in 1823 translated the text from Arabic to German. Though fragments have made their way into English, the book as a whole remains untranslated for English speakers.

While still in college during the 1960s, future best-selling author Michael Crichton (1942-) read the translated fragments, and was fascinated by the story of Ibn Fadlan. The result was his novel *Eaters of the Dead,* a highly fictionalized account of Ibn Fadlan's adventures. Published in 1976, the book was reissued in 1993, and in 1999 became the basis for the film *The 13th Warrior,* which starred Antonio Banderas as Ibn Fadlan.

JUDSON KNIGHT

Further Reading

Crichton, Michael. *Eaters of the Dead: The Manuscript of Ibn Fadlan, Relating His Experiences with the Northmen in A.D. 922.* New York: Ballantine Books, 1993.

Dunlop, D. M. *The History of the Jewish Khazars.* Princeton, NJ: Princeton University Press, 1954.

"*Risala:* Ibn Fadlan's Account of the Rus." http://www.realtime.net/~gunnora/ibn_fdln.htm.

The 13th Warrior. Touchstone Pictures, 1999.

The Viking Raids, A.D. 800-1150

Overview

The Vikings, or Norsemen, of Scandinavia, were the dominant sea power in Europe from about A.D. 800 to 1150, exploring the coastlines of Europe, the British Isles, and North Africa. Their technologically advanced longships, skilled seamanship, and military-like raiding parties exerted Viking influence from Russia to Greenland, and established peripheral contact with the Byzantine Empire and the shores of North America. The presence of Norse raiders had a profound impact on medieval Europe. Trade routes established by the Vikings promoted the flow of coins, sliver, and limited goods from the Middle East to Northern Europe. Norse settlements changed the political map of the Middle Ages, not only by expanding its physical bounds, but by encouraging the rise of strong local leaders and armies to defend populations from Viking marauders. The Norse established direct rule in some places, and simply plundered other locations. Regardless of the permanence of Viking rule, many historians of the period credit the Viking raids as the impetus for nation building in Europe, and note the proliferation of Norse law in the earliest codes of some modern European nations.

Background

No one reason can be identified as the primary catalyst for the beginning of the Viking raids. Historians have proposed numerous hypotheses—from domestic unrest in Scandinavia to overpopulation. Though the events of Scandinavian exploration, trade, and military expansion are collectively referred to as the Viking Invasion, or raids, the various Nordic groups actually developed different regions of interest. The Danes concentrated their efforts southward in England, France, and Frisia (present-day Netherlands). The Swedes were primarily interested in the eastern Baltic region, Russia, and trade relations with the Near East. The other major group, the Norwegians, sailed in the North Atlantic to the northern British Isles.

Western Europe certainly bore the brunt of the Viking Invasion, perhaps no peoples more than those of the British Isles. The first Anglo-Saxon Chronicle reported that Norseman had made small raids in northern England beginning in 750. By 793, the raids had escalated and the Vikings attacked and burned the great Irish monastery at Lindisfarne, perhaps the first great ecclesiastical center of the British Isles. The Vikings cemented their presence in the region when they founded the city of Dublin and fanned out in smaller settlements throughout Ireland and Wales. These early raids were followed in the ninth century by a more forceful Danish invasion. The persistent threats by Viking rulers over the course of the next century facilitated the unification of the various kingdoms in England into one nation. However, the Norse continued to have influence in England, occasionally through direct control of the Anglo-Saxon throne. Canute the Great, a Dane, ruled England in the eleventh century, and had a profound effect upon government and legal institutions.

The Norsemen, however, did not begin their famous raids in the west. Rather, the Viking search for trade goods and precious metals began with a series of raids in the region surrounding the Baltic Sea and on the various Slavic peoples in present-day Russia and Ukraine. During the eighth century, the Norsemen, predominantly Swedes and Danes, pushed east and south along major rivers toward the Black Sea. Along these routes, they established trade networks with the Arabs and the Byzantine Empire. This eastern trade route was one of the major sources of silver for the Vikings. In the ninth century, Turkish peoples (most likely Petchnegs and Bulgars) began migrating into the region, perhaps disrupting the stability of the Norse trade routes. Other histories claim that the invading Vikings demanded tribute from the Slavs who later drove the Norsemen out of the region. The diminishing success of these late-tenth century trade routes through Russia created the impetus for the Vikings to concentrate on other sources of wealth.

The end of Viking raids was the result of several factors. The domestic politics of the Scandinavian countries had stabilized to the point that the constant raids waged by various leaders to gain wealth, tribute, and men-at-arms were no longer as necessary. As in other areas of Europe, the outlines of modern nations emerged in Scandinavia during this era. Denmark, Sweden, and Norway were all united under separate monarchies. Viking colonization of islands in the North Sea and North Atlantic also declined as the colonies themselves either became more self-sufficient or were assumed into other nations. Lastly, by A.D. 1000, the Vikings had converted to Christianity, and began to normalize their relations with the other Christian nations of western Europe. The trade systems that were established during even the earliest decades of the Viking raids remained in use—disrupted only by the Crusades.

Impact

At the forefront of Viking technological innovation was the longship. The curved and highly decorated prow designs, and long, slender bodies of the boats became the aesthetic signature of Viking warships. However, the combination and modification of several marine technologies in the engineering of the Norse warship is what facilitated the Viking dominance of the sea and Northern Europe for nearly three centuries. The Viking shipbuilders expanded upon classic designs of rowboats by elongating the body, adding more rowers, and narrowing the girth of the ship. Historians believe that the Vikings added sail power to the craft by the late eighth century, however the earliest archaeological evidence of a mast on a Viking longship dates to around 810. The use of combination sail and row-powered small craft allowed for versatile navigation along coastlines, and shallow draft permitted easy landing. Military craft usually were longer, with more room for oarsmen, while commercial craft were smaller and had fewer oarsmen and a square-rigged sail. Though the Norsemen made substantial innovations to maritime craft in northern Europe, their feats of navigation and exploration are nonetheless amazing. The longships were largely open boats—only scant evidence suggests some sort of material cover—and the mariners and raiding parties traversed great distances with no navigational aid other than the various coastlines.

Longshipmen were also equipped for the more military aspects of the Viking raiding party, and were efficient warriors. The Vikings made use of weaponry that was largely similar to that of the rest of Western Europe. They were renowned for being ferocious and skilled warriors, but the greatest tactical advantage of the Vikings was the element of surprise that their fleets were able to employ in their raids. Some communities endured not one but tens of different raids during the 300 years of Viking dominance.

The notion that the Vikings were solely battle-hungry, ruthless, plunderers is mistaken. The introduction of Viking wealth, namely coin hordes, into western Europe redefined currency and political power. The Vikings were also a great source of cultural diffusion in the early medieval period. The Norsemen had very evolved myths and cosmologies, many of which were later chronicled in the twelfth century. Viking, mostly Danish, ideas of property ownership and legal structures were introduced into other societies, especially Anglo-Saxon England. Linguistic traces of old Norse dialects are found from Greenland to Wales.

There are few historical records of the Vikings and their accomplishments. Most of the narrative works and rune stones (carved stones with ancient Norse writing) that tell of the Viking raids date to several generations after the events took place. The most direct evidence scholars have of the extent of Norse influence in Europe and the technological advancements of their society comes from archaeological research that has been conducted over the past century. The most famous Viking sites are ship burials (the interment of important persons and their possessions in longships) and coin hordes (large deposits of silver coinage buried for safe keeping). Ship burials provide insight into the daily material culture of the upper echelon of Norse society and their burial ritual. Coin hordes, some containing silver currency from Byzantium and Arabia (the easternmost ends of the Viking trade routes), make it possible to evaluate the extent and vigor of the Viking trade routes and raids of conquest. These sites, as well as a newer interest in the excavation of Viking settlements and colonial outposts, have yielded the majority of our current accepted knowledge of the Norsemen in the early Middle Ages.

Not all historical and archaeological knowledge of the Viking raids, and the role of the Vikings in the shaping of Europe, is without a history of dispute. Viking antiquities and remains were some of the earliest sites to be explored in northern Europe using modern archaeological techniques in the nineteenth century. As the number of discovered Viking sites grew across Europe, the then-prominent theories of the evolution of the modern European states were challenged. Scholars in some countries were reluctant to accept the notion of a former Viking presence. These debates were short-lived for the most part. However, in Russia, for a variety of ideological reasons, the archaeological evidence and historical theories of Viking dominance of trade routes through Russia challenged the notion of Russian Slavic nationalism (the idea that Russia was created by the grouping together of various Slavic peoples, without coercion or help from others) and were hotly disputed. Only in the past 25 years has the role of Viking raids in Russian and Ukrainian history been more closely examined. The legacy of the Vikings even provides challenges for North American (pre-) history. Archaeological discoveries of temporary Norse settlements in Newfoundland in the 1970s called into question the widely accepted theory that Europeans did not have contact with North America until the 1492 voyage of Christopher Columbus.

ADRIENNE WILMOTH LERNER

Further Reading

Jones, Gwyn. *A History of the Vikings*, 2nd ed. Oxford: Oxford University Press, 1984

Lindberg, David C. *The Beginnings of Western Science*. Chicago: University of Chicago Press, 1992.

Rabban Bar Sauma, the "Reverse Marco Polo," Travels from Beijing to Bourdeaux

Overview

One could call Rabban Bar Sauma a "reverse Marco Polo": whereas Polo traveled from West to East, Bar Sauma's trek took him from what is now Beijing to the Bourdeaux region in France; and whereas Polo went on business, the priest Bar Sauma was on a religious mission. Of course, Polo and his journey are much better known, because they exposed technologically backward Europeans to the sophistication of Asia; but Bar Sauma, too, helped open the way for greater contact between continents and cultures.

Background

Bar Sauma (c. 1220-1294) belonged to the Nestorians, a sect named after the Persian priest Nestorius (d. 451). The latter, who became bishop of Constantinople, taught that Christ had two separate identities, one human and one divine. At the Council of Ephesus in 431, the Church declared Nestorianism a heresy, and soon this doctrinal separation led to physical separation, as the Nestorians began making their way eastward.

Initially they settled in Mesopotamia and Persia, or modern Iraq and Iran, but the conquest of those lands by Muslims in the mid-600s forced them eastward. As early as 635, a Nestorian community existed in China, and though that group was later suppressed, the Turkic-speaking Uighur peoples of Sinkiang province maintained the faith. The Nestorians benefited from the Mongol conquest of China under Genghis Khan (1162-1227) in the early thirteenth century, and soon they were allowed to resume their missionary work in China. Among the converts gained in those early years was Genghis's daughter-in-law, mother of the future Great Khan Kublai (1215-1294).

This was the setting in which Bar Sauma ("Rabban," similar to *rabbi,* is a title) was born. Though he came from Khanbalik, or modern Beijing, he was a member of the Uighur peoples—in other words, a Turk rather than Chinese. Nonetheless, he grew up speaking the language of the majority population, and though his Nestorian Christianity would certainly have posed a barrier to his absorption in the larger culture, it is likely that in many respects he was Chinese.

At the age of 20, Bar Sauma left his wealthy family to become a Nestorian monk and live in a cave. He spent several decades this way, preaching the Gospel and attracting followers, among them a youth named Markos. Some time between 1275 and 1280, Bar Sauma and Markos together went on a pilgrimage to Jerusalem, bringing with them a letter of recommendation from the bishop of Khanbalik, as well as a travel permit from Kublai Khan himself.

Impact

After passing through an area in Central Asia that was then in the grip of an anti-Mongol rebellion, they finally reached Khorasan, or present-day Afghanistan. From there they made their way to Maragheh, now in Azerbaijan. The latter was the capital of Kublai's brother Hulagu (c. 1217-1265), who had established a separate realm, the Il-khanate, that governed much of southwestern Asia. His widow Dokuz Khatun was also a Nestorian.

Traveling on to Baghdad, they met the catholicos, head of the Nestorian church, who asked them to go back to China as his messengers. The catholicos made Markos a bishop, and designated him metropolitan (a sort of archbishop in eastern churches) of northern China. Continued fighting in areas to the east prevented the two men from departing, however, and while they were waiting to do so, the catholicos died. The upshot of this was that a convocation of Nestorian bishops chose Markos as the new catholicos, with the title Mar Yaballaha III.

Markos and Bar Sauma then traveled to Maragheh to have the selection confirmed by the Il-khan Abaga. In 1282, however, Abaga died and his son Arghun (c. 1258-1291) took the throne. Desirous of gaining a victory over the Syrians and winning control of Palestine, Arghun told Markos that he wanted to make an alliance with the Christians of Europe. In response, Markos sent his old teacher and friend Bar Sauma as an emissary to the pope.

In 1287 Bar Sauma left Baghdad and headed north through Armenia to the Black Sea port of Trebizond, a Byzantine stronghold. From there he sailed to the capital at Constantinople, where he received an audience with Emperor Andronicus II Palaeologus (1260-1332) and visited the Hagia Sophia, Byzantium's magnificent cathedral. From Constantinople he went on to Italy by ship, passing Sicily along the way and witnessing an eruption of Mount Etna.

Finally arriving in Rome, Bar Sauma learned that Pope Honorius IV had recently died, and the College of Cardinals had yet to choose a successor. The cardinals began to quiz him regarding Nestorian tenets, but Bar Sauma did not dare become embroiled in a theological discussion with representatives of the church had declared his own faith heretical. Therefore he departed Rome for a trip through Italy and France while he waited for the cardinals to elect a new pope.

From Rome Bar Sauma traveled north to what he called Thuzkan—more familiar to Westerners as Tuscany—and thence to Genoa. He then made his way through *Frangestan,* or France, to the capital at Paris, where he spent a month at the court of Philip IV (1268-1314). After that he traveled to Gascony in southern France, which was then in English hands, and on to Bourdeaux, where he met England's king Edward I (1239-1307). Edward received Bar Sauma apparently with enthusiasm, and listened to his pleas for help—yet like his counterparts in Constantinople and Paris, he offered no concrete assistance.

Had Bar Sauma come to Europe 150 years before, or even a century before, he might have gotten exactly what he needed. But now the Crusades (1095- 1291), an effort to wrest the Holy Land from the Muslims, were drawing to a close, with Europe having lost all it had gained in the victories of the First Crusade (1095-99). Thus Europe was exhausted from crusading, and though the monarchs sympathized with the idea of a joint effort against Islam, they could and would give it little more than lip service. This was particularly the case in Byzantium, which had suffered the worst from the Crusades, during which its lands had been ravaged and even captured by the Greeks' alleged allies from Western Europe.

Of course Bar Sauma did not realize that his efforts were doomed, and he was thrilled when in 1288 the new pope, Nicholas IV (r. 1288-92), expressed an interest in joining the Il-khan in a war against the Muslims. He received communion from Nicholas on Palm Sunday, and this at least was something concrete: a clear sign that the eight centuries of hostility between Rome and the Nestorians had come to an end.

Confident that help was on its way, Bar Sauma headed eastward again and reported back to Arghun. He spent the remainder of his days in Baghdad, where he died in 1294. It was probably during those last years of his life that he wrote down the record of his travels, which in the twentieth century appeared in English as *The Monks of Kublai Khan,* translated by Sir E. A. Wallis Budge.

Though neither the pope nor the European rulers sent military assistance, Rome did dispatch a number of missionaries, most notably Giovanni da Montecorvino (1246-1328). Unfortunately for China's Christians, however, the era of toleration was about to come to a close with the ouster of the Mongols by the founders of the Ming Dynasty (1368-1644), China's last native-born ruling house.

The Nestorians of southwestern Asia suffered even worse hardships under Timur the Lame or Tamerlane (1336-1405), who though he claimed to be related to Genghis Khan was a Muslim and a harsh opponent of Christianity. Nonetheless, in the brief window of time that missionaries from Europe found it relatively easy to travel to China, they forged important links between East and West.

JUDSON KNIGHT

Further Reading

Books

Bar Sauma, Rabban. *The Monks of Kublai Khan, Emperor of China; or, The History of the Life and Travels of Rabban Sawma, Envoy and Plenipotentiary of the Mongol Khans to the Kings of Europe, and Markos Who as Mar Yahbh-Allaha III Became Patriarch of the Nestorian Church in Asia,* translated and edited by E. A. Wallis Budge. London: Religious Tract Society, 1928.

Mirsky, Jeanette. *The Great Chinese Travelers: An Anthology.* New York City: Pantheon Books, 1964.

Montgomery, J. A., translator and editor. *The History of Yaballaha III and His Vicar, Bar-Sauma, Mongol Ambassador to the Frankish Courts at the End of the Thirteenth Century.* New York City: Columbia University Press, 1927.

Newton, Arthur Percival. *Travel and Travellers of the Middle Ages.* Freeport, New York: Books for Libraries, 1962.

Internet Sites

Traveling to Jerusalem. "Bar Sauma." http://www.uscolo.edu/history/seminar/sauma.htm

Young, John L. *By Foot to China: Mission of The Church of the East, to 1400.* http://www.aina.org/byfoot.htm

Jean de Béthencourt and Gadifer de La Salle Colonize the Canary Islands for Spain

Overview

Located off the coast of western Africa, the Canary Islands are characterized by such stunning variations in geography and climate that visitors sometimes describe them as "a continent in miniature." Overall temperatures are pleasant, and annual rainfall is low, creating a dry climate unusual for a region where nothing is very far from the sea. Thus tourism is a thriving industry in the Canaries, which have belonged to Spain since their conquest in the early fifteenth century. The outside world has known about the Canaries since ancient times, but the islands remained in the possession of their Berber inhabitants for centuries before the 1402 arrival of Jean de Béthencourt and Gadifer de La Salle—an event that, as it turned out, marked the beginning of European colonialism.

Background

Consisting of numerous islands that together comprise 2,807 square miles (7,270 square km) of land surface, the Canaries are located off the coast of southern Morocco. The easternmost of the islands is just 67 miles (108 km) from the African mainland. The islands' volcanic origins are evident even today: at Fire Mountain on the eastern isle of Lanzarote, the underground heat is so intense that in some places it is possible to grill meat simply by placing it over a hole in the ground.

La Palma in the west, where a volcanic crater formed as recently as 1971, has the greatest altitude-to-area ratio of any place on Earth: the island is a veritable skyscraper in the middle of the Atlantic. Indeed, the five islands that make up the western group of the Canaries—

Tenerife, Gran Canaria, La Palma, Gomera, and Ferro or Hierro—are simply mountains rising from the ocean floor, and Teide Peak on Tenerife is, at 12,198 feet (3,718 m), the tallest mountain on Spanish soil. The eastern group consists of Lanzarote, Fuerteventura, and six tiny islets, all of which straddle the Canary Ridge plateau along the ocean floor.

The climate of the Canaries is highly variable across space but equable over time. The enormous height differences between the mountain peaks and the nearby beaches create extremes of warmth and coldness, and patterns of wind and tide influence variations in humidity. The northern or windward side of the islands benefits from the action of breezes against the high mountains, creating condensation and resultant moisture. Thus while much of the north is green and lush, the southern islands have deserts; in fact, on La Gomera trees tend to be damp on the windward side and dry on the leeward.

Despite these wide differences in climate on different parts of the islands, temperatures in the inhabited areas are surprisingly uniform. On a typical afternoon in August at the port of Las Palmas on Gran Canaria, for instance, the temperature is 79° F (21° C), whereas in January it is only 9° F (5° C) cooler. Rainfall, most of which occurs in November and December, averages only about 10 inches or 250 millimeters annually, though on the windward side of the islands it may be as great as 30 inches (750 mm) a year.

Its pleasant climate, combined with the variations in climatological regions and their resulting variation in plant life—the Canaries have been called a "botanist's paradise"—certainly explain the islands' appeal to modern tourists. These factors may also, however, explain how it was that the original inhabitants, who apparently came from the African mainland, forgot the art of sailing: few people want to leave the Canaries once they have arrived.

The original Canarians came to be known as Guanches, though that term more properly applies to the people on the western part of the islands, while those on the east were called Canarios. According to research conducted by Earnest A. Hooton (1887-1954) and others, it appears that the Guanches had their origins among the Cro-Magnon of southern or possibly central Europe, from whence they migrated to northern Africa. Linguistically the Guanches were linked to the Berber peoples of western North Africa, which they left probably as a result of the once-lush Sahara region's drying between c. 2000 and c. 600 B.C. Their physical appearance, however, demonstrated their European origins: though their skin was brown, their eyes were blue or gray, and their hair blondish. Many of the present-day Canaries share these physical characteristics, though the Spanish colonists rendered the Guanches themselves extinct as an ethnic group.

At the time of the Europeans' arrival in 1402, the Guanches were still at a Neolithic, or New Stone Age, level of technological development. Herding and rudimentary fruit-growing provided their means of subsistence, and they clothed themselves in tunics of leather or plaited reeds. Their knowledge of sophisticated embalming techniques, however, suggests exposure to an advanced civilization, presumably that of the Egyptians; unlike the Egyptians, and certainly unlike most uncivilized peoples, they practiced a monotheistic religion. Furthermore, they utilized a form of writing, with alphabet-like characters that have yet to be translated.

European knowledge of the Canaries dates at least to the time of Juba II (c. 50 B.C.-c. A.D. 24), king of Mauritania in North Africa, whose expedition to the islands first alerted Romans to their existence. It appears that the Greeks, however, also knew of the Canaries, which they associated in various legends with the sunken continent of Atlantis; the Elysian Fields of eternal bliss; or the Gardens of Hesperides, where Heracles (or Hercules) was sent to retrieve a golden apple from a tree guarded by a dragon. Ancient geographers regarded the island of Hierro (Ferro) as the western edge of the world, and reckoned all longitude from it. Historian and biographer Plutarch (A.D. 46?- c. 120) dubbed the Canaries "the Fortunate Isles," a name by which the ancients knew them. As for the name "Canary," it comes from the writings of Pliny the Elder (c. 23-79 A.D.), and has nothing to do with birds: rather, Pliny noted "the multitude of dogs"—*canes* in Latin—"of great size" who lived on the islands.

The medieval period largely wiped away European knowledge of the Canaries. An Arab expedition landed on Gran Canaria in 999, but only with the arrival of Genoese mariners in 1325 did Europeans become reacquainted with the islands. Even then, knowledge of them was not widespread, and thus the Canaries were still a land waiting—at least, in the view of the Europeans who seized it—to be conquered.

Impact

It is one of the great ironies of history that many of the most famous explorers did not come from the countries under whose flag they sailed. This was true both of Spain's most famous explorer, the Italian Christopher Columbus (1451-1506), and of its first, the Frenchmen Jean de Béthencourt (c. 1360-1422) and Gadifer de La Salle (fl. c. 1340-1415). Indeed, Béthencourt was not even, strictly speaking, French: a Norman, he was a descendant of the Vikings who gave their name to the Normandy region of France after they invaded it late in the ninth century. Béthencourt and Gadifer were adventurers—the latter had won acclaim as a soldier during the Hundred Years' War (1337-1453) against the English—and met on a crusade against the Muslim stronghold at Tunis in North Africa in 1390. The two struck up a friendship, and in time hatched a plan of making an expedition to the Canaries, an undertaking Béthencourt promised to finance.

They and their crew set sail from La Rochelle, France, on May 1, 1402, and arrived in the Canaries the following month. Not long afterward, Béthencourt, his own resources exhausted, returned to Europe for financing. He sought royal help, though not from his own king—France would not take an interest in exploration for some two centuries—but from Henry III (r. 1390-1406) in Spain.

The latter ruled Castile, most powerful of the Christian kingdoms that had long been engaged in a war to wrest the Iberian peninsula from the Muslims who had controlled it for nearly seven centuries. By 1402 the Christians were well on their way to complete victory, and as Spanish confidence grew, so did Spain's interest in the outside world. Henry had sent ambassadors as far away as the court of Tamerlane or Timur the Lame (1336-1405) in Persia, and thus when Béthencourt came to Henry with a proposal for the financing of an expedition to the Canaries, he found a ready audience.

Henry, in fact, declared Béthencourt "king of the Canary Islands," and the antipope Benedict XIII issued a papal bull bestowing his blessings on the Spanish conquest of the Canaries. The elevation of Béthencourt to a king, albeit a vassal, understandably angered Gadifer, who learned of the fact upon Béthencourt's return to the Canaries some 18 months after his departure. By then Gadifer had undertaken much of the difficult work of subduing the islands from his base on Gomera, overpowering the technologically primitive Guanches and putting Norman peasants to work as colonists. Gadifer demanded that Béthencourt go with him to seek arbitration from Henry III, who—again, not surprisingly—decided in Béthencourt's favor.

Gadifer went home to France, and presumably the two conquerors never saw one another again. Béthencourt himself did not remain long in the Canaries: after adopting the use of Norman colonists as pioneered by Gadifer, he placed his nephew, Maciot de Béthencourt, in charge of the islands, and in 1406 returned to France. There he died 16 years later.

By the time of Béthencourt's death, the Canaries had become the site of a colonial struggle—in fact, the first in a series of such conflicts among modern European nations, engagements that would culminate in World War I five centuries later. In this case the combatants were Spain and the second emerging European colonial power, Portugal. The latter invaded the islands in 1420, and in 1425 began half a century of dominance over the Canaries. Meanwhile Spain in 1469 united under the dual monarchy of Aragon's Ferdinand II (1452-1516) and Castile's Isabella I (1451—1504), who fought a four-year war with Portugal beginning in 1475. Included among the terms of a 1479 treaty was Portuguese recognition of Spanish sovereignty over the Canaries. Seventeen years later, in 1496, the Spanish destroyed the last Guanche stronghold.

When Columbus set out on his first voyage to the New World, he stopped at the Canaries, where his ships underwent repairs and took on provisions. Columbus would use the Canaries as a launchpad for his other three voyages, and the islands became an important staging point for other explorers. In 1584, for instance, English settlers passed by them on their way to the founding of the Roanoke Island colony in Virginia. In 1936 the islands became a staging-point of quite a different kind, when Generalissimo Francisco Franco used them as the initial base for his Nationalist revolt in the Spanish Civil War.

JUDSON KNIGHT

Further Reading

Books

Abreu de Galindo, Juan de, and Captain George Glas. *The History of the Discovery and Conquest of the Canary Islands: Translated from a Spanish Manuscript, Lately Found in the Island of Palma, With an Enquiry into the Origin of the Ancient Inhabitants, to Which Is Added a Description of the Canary Islands, Including the Modern History of the Inhabitants, and an Account of Their Manners, Customs, and Trade,* translated by George Glas. Dublin: D. Chamberlaine, 1767.

Bontier, Pierre and Jean Le Verrier. *The Canarian; or, Book of the Conquest and Conversion of the Canarians in the Year 1402, by Messire Jean de Béthencourt, Kt., Composed by Pierre Bontier and Jean Le Verrier.* Translated and edited by Richard Henry Major. New York: B. Franklin, 1969.

Hooton, Earnest Albert. *The Ancient Inhabitants of the Canary Islands.* Cambridge, Massachusetts: Peabody Museum of Harvard University, 1925.

Internet Sites

"The Canary Islands: The Fortunate Islands." http://www.ships-yachts.com/canaryislands.htm.

João Gonçalves Zarco Inaugurates the Era of Portuguese Exploration with the Rediscovery of the Madeira Islands, 1418-20

Overview

Located some 560 miles (about 900 km) west of Morocco, the Madeiras—also known as the Funchal Islands—today consist of two inhabited islands, Madeira and Porto Santo, and two uninhabited groups, the Desertas and the Selvagens. Together they comprise about 306 square miles (794 sq km) and they have remained a possession of Portugal since João Gonçalves Zarco discovered them in the period 1418-20. In fact Gonçalves was actually *re*discovering the islands, which may have been known since ancient times. In any case, history has assigned him a relatively small role in an event that marked the beginning of Portugal's overseas empire.

Background

Some scholars believe that Phoenician mariners of ancient times had known about the Madeiras, which lie north of the Canary Islands on a latitude just south of Casablanca, Morocco. Certainly it appears that sailors from Genoa, a great maritime power in the Middle Ages, knew of the island group: the Laurentian Portolano, a Genoese map dating to 1351, clearly depicts the Madeiras.

Yet somehow the islands were forgotten. It may be that Genoa, which had been hit hard by the Black Death (1347-51), scaled back its furthest westward sea routes; and no doubt the Genoese were not inclined (the map notwithstanding) to share their knowledge of the islands with others. Thus when Zarco caught sight of them in 1418, it was as though he were finding the island group for the first time.

Zarco's experience can hardly be separated from its context as part of the exploration efforts put into place by Portugal's remarkable Prince Henry the Navigator (1394-1460). The latter was destined never to reign, a fortunate thing for the history of exploration because it is hard to imagine how he could have made the great strides that he did as a patron of exploration if he had been distracted by affairs of state.

Henry participated in a successful crusade for Ceuta, a city in northern Morocco in 1415, and as a result was widely honored. The experience whetted his appetite to learn more about the African continent and outlying islands, as had his earlier reading of travelers' tales. It was Henry's desire to find the source of the gold in West African empires such as the Mali of the fabled Mansa Musa (c. 1280-c. 1337), and to locate the realm of the legendary Christian king Prester John. Therefore he began to gather around himself explorers, mariners, and cartographers in an informal "school" established in Sagres, a city in southwestern Portugal. Zarco's discovery of the Madeiras would be the first great achievement of Henry's school.

Impact

Of Zarco himself, little is known. It appears that his family was Jewish, an interesting fact in light of the fierce anti-Semitism that characterized the Christian kingdoms of the Iberian peninsula. This hatred of Jews had become particularly fierce as Portugal, Aragon, Castile, and other kingdoms had fought to reconquer the region from the Muslims who had held it since 711; yet the Zarco family, who came from the Portuguese city of Tomar, seemed to flourish in Catholic Portugal. João Gonçalves himself became a noble in the house of Prince Henry, and thus it is not surprising that the latter would have chosen Zarco to command one of his school's first voyages. (Henry himself never actually took part in any of the expeditions he organized.)

Though Zarco first saw the Madeiras in 1418, it was not until 1419 that he landed on the island of Porto Santo. With him was Tristão Vaz Teixeira, and the two men accurately calculated their position so as to be able to return shortly afterward with colonists. This time they were joined by Bartolemeu Perestrello, destined to become father-in-law to Christopher Columbus (1451- 1506). It would be another year before they discovered Madeira, the largest of the islands, though some scholars maintain that the Portuguese actually made this discovery during the reign of Afonso IV (1325-57).

The islands were completely bereft of human habitation, or indeed of land mammals, and the explorers must have felt as though they had reached the edges of the known world—as in fact they had. These events took place, after all, nearly 75 years before Columbus's epochal voyage to the New World, after which monumental discoveries became relatively common for a century. As it was, the claiming of the Madeiras was a project that seemed almost magical.

When the first men ventured away from Porto Santo and caught their earliest glimpses of Madeira (the two islands are about 26 miles or 42 kilometers apart), they reported that they had a seen a mass of dark clouds along the horizon to the south. A number of medieval superstitions plagued mariners of that time, as they would for many years: for instance, before Gil Eannes (fl. 1433-1445) sailed past Cape Bojador on the West African coast some 15 years later, European sailors believed the cape to be a point of no return. Thus it took enormous courage just to venture to the southwest from Porto Santo, where they found Madeira.

Madeira looked like a dark cloud precisely because, like Porto Santo, it was covered in dense vegetation. Even more mysterious than these two, the only inhabited parts of the Madeiras today, were the barren portions of the island group. There were the Desertas, some 11 miles (18 km) southeast of Madeira. Though in time the Portuguese would inhabit these four tiny islands, the land was inhospitable, good only for rabbits and wild goats. (The latter are in fact the only full-time residents of the aptly named Desertas today.) Even more forbidding were the Selvagens, or Salvage Islands, on which there was never any question of human habitation. The Selvagens are a group of three rocks, the largest with a circumference of some 3 miles (5 km), located about 156 miles (251 km) south of Madeira.

Concentrating their attention on Porto Santo and Madeira, the men set about colonizing the islands according to Henry's orders. In this they practiced a form of slash-and-burn agriculture, setting the forest alight and letting the flames clear the island. It was said that the fires raged for seven years.

They planted grapes from the eastern Mediterranean, though scholars are unsure whether the original plants came from Cyprus or Crete. By the seventeenth century, Madeira wine would become the island's most famous export, and even today the dark-brown wine, with its strong aftertaste, is known the world over. Supposedly the taste of Madeira, which varies from dry to sweet, resulted in part from the action of the turbulent Atlantic waters on casks shipped abroad during the era of seafaring; today makers of the wine produce a similar effect by artificially agitating the casks.

In the meantime, sugarcane had been brought to the Madeiras from Sicily, probably in about 1452. Madeira became home to a sugar plantation, reputedly the world's first, and its economy became increasingly dependent on the labor of African slaves. Thus the islands went through something of a crisis in 1775, when the reformist Portuguese Prime Minister Carvalho e Mello (1699-1782) ended slavery on the Madeiras.

Though sugar and wine production have remained mainstays of the local economy, a few other crops, most notably bananas, sweet potatoes, and taro root, have been added over the years. Handicrafts such as embroidery, introduced c. 1850 by an Englishwoman named "Mrs. Phelps," is also a major industry, as are woodworking and the making of wicker furniture. Fishing is significant, along with tourism—an industry in which the gorgeous Madeiras, with their breathtaking landscapes, pleasant climate, and exotic-yet-familiar charm, enjoy a distinct advantage.

During the nineteenth century, Britain, then the dominant world power, briefly took possession of the Madeiras. No doubt the British acted with the idea in mind that the islands' proximity to Africa and Europe, and their position at the gateway to the Atlantic, made them strategically significant. A number of factors conspired to reduce their military importance, however, most notably Britain's loss of its possessions in Africa, and Portugal soon regained control of the islands.

Yet in the modern world of supersonic transport and electronic communication, where the battlegrounds are financial and commercial rather than military, the Madeiras, no longer remote, are as strategically located as any other place. This was a point made by U.S. Ambassador to Portugal Gerald McGowan in a March 16, 1999, speech in Funchal, Madeira. "Standing here in Madeira today," McGowan said in the course of his speech, "I can't help thinking what must have been running through the mind of João Gonçalves Zarco when he landed on this beautiful island almost 600 years ago."

In fact McGowan was one of the few people outside the Madeiras who remembered Zarco and his great achievement in the period 1418-20. A statue of the explorer stands in downtown Funchal, before the Bank of Portugal building, but Zarco's name is hardly a household word anywhere else. One of the few books about his adventures in a language other than Portuguese is *Relation historique de la découverte de l'isle de Madère* (1671) by Francisco Alcoforado. Even in this case, the author was Portuguese; the book is only partially about Zarco himself; and like all other works on the explorer, this one has yet to see translation into English.

Despite this lack of attention to Zarco's life, it is fair to say that he struck the first blow in a campaign of exploration that would build one of the world's first and greatest overseas empires. Henry used the Madeiras as the point of departure for expeditions to new lands, gradually adding the Azores and parts of West Africa to Portuguese possessions in the 1420s. In time, Bartolomeu Dias (c. 1450-1500) and Vasco da Gama (c. 1460-1524) would round the southern tip of Africa, opening up eastward routes to Asia and in the process inaugurating the European colonization of the East.

JUDSON KNIGHT

Further Reading

Madeira Island's Regional Tourism Service. http://www.madeira-web.com/

University of Calgary. *European Voyages of Exploration.* "Africa: Ceuta—The First Step." http://www.acs.ucalgary.ca/HIST/tutor/eurvoya/africa.html

University of Calgary. *European Voyages of Exploration.* "Prince Henry the Navigator." http://www.acs.ucalgary.ca/HIST/tutor/eurvoya/henry1.html

Washington File. United States Information Service, Bucharest, Romania. "Ambassador to Portugal McGowan Speech in Madeira ." http://www.usembassy.ro/USIS/Washington-File/200/99-03-16/eur210.htm

Nuño Tristão: Early Portuguese Explorer

Overview

During the 1440s, Nuño Tristão explored a large portion of Africa's west coast in a series of expeditions commissioned by Portugal's Prince Henry the Navigator. Tristão, one of Prince Henry's most trusted sea captains, is credited with the discovery of Cape Blanc, Arguin Island, and the Gambia River. Tristão was one of the first Europeans to engage in the slave trade, and under his command Arguin Island became the staging area for slaves bound for Portugal. This fertile period of exploration of Africa's western coast established Portugal as a leader in the emerging colonial world.

Background

In 1415 Portugal began its era of geographic discovery and the establishment of colonies. At the time it had a relatively stable monarchy whose kings encouraged maritime trade. Portugal's natural geographic position on the Iberian peninsula encouraged exploration in Africa. In 1415 the reigning royal family of John I took part in the assault on the port of Ceuta in northern Morocco. Ceuta, highly prized, was the terminus of the trans-Sahara trade route. After the successful invasion of Ceuta, the small, unrecognized country of Portugal began the domination of Europe over Africa and Asia that was to last for the next 400 years.

The riddle of the unknown Atlantic had fascinated many generations of seamen. By the middle of the fourteenth century many cartographers felt that there was a sea route to India; however, the official doctrine of the Church was based on the *mappa-mundi*, the wheel map that

put Jerusalem in the center of the inhabited world with the continents placed around it.

Before Prince Henry, no one had attempted to venture out to the Sea of Obscurity or the Ocean of Darkness, as it was called, because of existing superstitions and frightening myths. According to accounts, monsters swam in the seas, and the overhead sun made the seawater boil and the pitch holding the ship's wooden beams together melt. The people inhabiting the African shore were known to be burned black by the sun and there was fear of being roasted alive. Magnetic rocks encouraged compasses to fail and if a ship escaped these hazards, it was only a matter of time before they would be caught in the steady flow of water that poured night and day over the edge of the flat world. Today it is known that the constant north-south flow or the Canary Current along the Saharan coast made it possible for ships to sail southward as far as West Africa but prevented a return voyage. Cape Bojador, south of the Canary Islands represented the point of no return.

The coast of Africa was only known as far as Cape Nao, because of rumors that those who ventured beyond it would never return. In 1419 or 1420, less than five years after the fall of Ceuta, a Portuguese expedition under the command of João Gonçalves Zarco rediscovered the island of Madeira. In 1427 and 1432, two expeditions succeeded in reaching the Azore Islands where large quantities of timber were used for shipbuilding. For nearly twelve years Prince Henry sent out fourteen expeditions with the aim of rounding Cape Bojador, on the coast of what is know the Western Sahara. In 1432, Gil Eannes successfully rounded Cape Bojador.

In 1441,Prince Henry outfitted two of the new caravel ships for separate journeys down the west coast of Africa with the instructions to travel south and capture native people for interrogation. Nuño Tristão, one of Henry's most trusted and experienced captains was given command of one ship and Antão Gonçalves the other. The two ships met at Rio de Oro in Western Sahara. They came across a market run by black Muslims dressed in robes and turbans. Although they had instructions to approach the local people peaceably, a skirmish ensued and ten prisoners were taken. Gonçalves sailed back to Portugal with the captives. Tristão continued an additional 320 miles south and named it Cape Blanc for its white sand.

In 1442, Prince Henry sent Nuño Tristão on another expedition. On this voyage he traveled past Cape Blanc and below Cape Anna and discovered the Arguin archipelago. This discovery played an important role in Portuguese history, for it was here on the main island that Prince Henry authorized a fort and trading post that marked the beginning of colonialism in Africa. In addition, Tristão captured fourteen natives on Arguin Island and returned to Portugal a hero. The island became a busy slave-trading station, and for the next few years, Tristão made a series of slave raids on the coast of West Africa.

In 1446, Tristão set out on his third and last voyage. Instructed by Prince Henry to push southward, he sailed beyond the Sénégal River, passed Cape Verde, the westernmost point in Africa, and sailed beyond the mouth of the Gambia River to an unnamed river. They launched the ship's boats and traveled upstream, where they were met by a barrage of poisoned arrows, hurled by hostile natives. Only five crew members survived the attack. The bodies of Nuño Tristão and eighteen of his men were buried at sea.

Impact

The Portuguese navigations during Prince Henry's influence opened a new and unprecedented chapter in human history. Prince Henry combined his political and organizational talents to catalyze the great era of Portuguese exploration and discovery of which Nuño Tristão was a part. Prince Henry established himself at Sagres and although he never went to sea, he was influential in guiding the direction of Portuguese exploration for generations to come. He set in motion a program emphasizing geographical exploration, instead of focusing on commercial voyaging. He surrounded himself with scientists, geographers, astronomers and cartographers that led to the design of the caravel, the compilation of sea charts and sailing rudders and improved methods of navigation.

Nuño Tristão was entrusted with sailing the first caravel, when he succeeded in reaching Cape Blanco in 1441. Twelve years of experimentation went into the evolution of this innovative design. The difficulty ships experienced when returning to Portugal against the prevailing north winds led to the evolution of the caravel. A hybrid, the ship's hull was modeled on the Atlantic fishing boat and the mast structure was borrowed from ships used in the Mediterranean. With its triangular sails, and small crews of thirty or less, the caravel was well adapted to maneuvering in and out of estuaries and shoals along the western coast of Africa.

Slave trade on the coast of West Africa. *(Bettmann/Corbis. Reproduced with permission.)*

The success of Portuguese navigation depended, in great measure, on the work of Prince Henry and the captains he inspired and supported. The sailing conditions in the southern Atlantic were considerably different than navigating the familiar waters and landmarks in the Mediterranean and the northern European coast. Rarely out of site of land, European sailors had reasonably accurate charts of the main landforms and they mainly traveled in north to south directions. However, when Prince Henry's ships began exploring the south Atlantic they had no charts or familiar landmarks. Expert cartographers and astronomers assisted Prince Henry in correlating information brought by returning sea captains and detailed charts and maps slowly became more accurate and useful.

The navigator's tools were the compass, sun, and stars, and the astrolabe. The astrolabe was an efficient instrument for obtaining a position by sun or star altitudes. Portuguese navigators used the compass, a small magnetized metal bar mounted over cards with wind roses painted on them. With these tools, the captains and crew would note the changing altitudes at various points along the route. As new capes, headlands, and bays were discovered, they were fixed for latitude and added to Prince Henry's master chart. Gradually the men became familiar with deep-sea voyaging, and as the tools of the trade became more sophisticated, the voyages became more routine. The boiling ocean, the flood of waters at the world's end, and the magnetic rocks that destroyed ships were forgotten as the scientific approach to navigation took hold.

As for Tristão's slave raids, they continued until 1445, when Prince Henry refused to allow his crews to kidnap Africans. Prince Henry, like many Europeans, saw Africans as potential Christian converts, and when first slaves came to Portugal, they were permitted to intermarry with the Christian Portuguese and absorbed into society. They were also permitted to buy their freedom. Prince Henry's dictum against *abducting* slaves, however, did little to stop the lucrative *trade* in human beings, which continued for the next 300 years.

In the fifteenth century, it must be remembered, the idea of slavery was not a new one. It was as old as civilization itself, appearing in ancient Egypt, Greece, and Rome. In addition, slavery was a recognized hazard for every European seaman who sailed the Mediterranean, since Moorish pirates sought both human and inanimate cargo. Men who were taken prisoner during warfare knew that they would probably become slaves. In addition, slavery had always been an integral part of African culture and society, and black slaves had been traded to the Islamic world since the middle of the seventh century.

The widespread nature of slavery made it an easy and profitable business for the Portuguese. Beginning about 1450, Portuguese merchants bought slaves from the northward-bound caravans from Arguim, tapping into a longstanding trans-Saharan trade. By the end of the fifteenth century, 1,200-2,500 slaves per year were being exported. This complex preexisting system contributed to the massive exportation of Africans during the next two centuries.

LESLIE HUTCHINSON

Further Reading

Bell, Christopher. *Portugal and the Quest for the Indies*. London: Constable and Company,1974.

Bradford, Ernle. *Southward the Caravels: The Story of Henry the Navigator*. London: Hutchinson of London, 1961.

Duffy, James. *Portugal in Africa*. Cambridge: Harvard University, 1962.

Outhwaite, Leanord. *Unrolling the Map, The Story of Exploration*. New York: John Day Company, 1972.

Thorton, John. *Africa and Africans in the Making of the Atlantic World, 1400-1800*. United Kingdom: Cambridge University Press, 1998.

Gil Eannes Passes the Point of No Return at Cape Bojador—And Inaugurates a New Era in Exploration

Overview

The name of Gil Eannes is hardly a household word; nor is that of the place associated with the Portuguese explorer, Cape Bojador. Nor indeed did Eannes *discover* the cape: the place had been known for many years. To journeyers of Eannes's time, Bojador represented an unbreachable barrier, a point of no return, and it was the achievement of this reluctant hero to pass that invisible boundary in 1434. In so doing, he opened new territory not only on land but in the mind, and thus made possible the golden age of Portuguese exploration, with all its glories and horrors.

Background

Located at 26°08' N, 14°30' W, or about 100 miles (160 km) south of the Canary Islands, Cape Bojador—"the Bulging Cape"— juts into the Atlantic from the west coast of Africa. Today it is part of the territory of Western Sahara, claimed by Morocco, in which it constitutes a central geographic mark, dividing the region between its northern third, Saguia el Hamra, and its southern two-thirds, Río de Oro.

The region's first exposure to the outside world probably occurred in about 600 B.C., when according to Herodotus (c.484-c.420 B.C.), Egypt's pharaoh Necho II (r. 610-595 B.C.) commissioned a group of Carthaginian mariners to circumnavigate the African continent. If indeed this voyage occurred, the sailors would most certainly have passed by the cape, which juts some 25 miles (40 km) from the African mainland. Further contact may have occurred during a voyage of Hanno (fifth cent. B.C.) down the west coast of Africa.

During medieval times, Sanhajah Berber tribes established their dominance in the area, only to be overtaken by Bedouins from further east. There was little other competition, however: squeezed as it was between the ocean and the desert, the area around Cape Bojador was hardly worth the trouble of conquering it. Added to this was the fearsome reputation the cape had acquired in the eyes of mariners.

The Arabs called the place *Abu khatar,* meaning "father of danger," and indeed Cape Bojador became the site of many a shipwreck. The reason for this was a network of reefs surrounding the cape, which created a sort of net for catching ships. As distant as a league (5 kilometers) from shore, the sea was only a fathom (about 2 meters) deep, and as though the shallows were not forbidding enough, the northern side of the cape was subject to violent waves and currents, while fogs and mists often covered the region as a whole. Strong prevailing winds made it almost impossible for a ship to return north of the cape once it had passed it, rendering the spot truly a point of no return—or, in the parlance of European sailors in the fourteenth and fifteenth centuries, the "Green Sea of Darkness."

Impact

The story of Eannes (fl. 1433-1445), and his feat in passing Cape Bojador, is inextricably tied with that of Prince Henry the Navigator (1394- 1460). A man who, more than any other individual, deserves the credit for initiating the Age of Exploration, Henry himself never traveled further from home than Tunis, and certainly never took part in any of the expeditions and voyages that he helped organize. Yet from his informal "school" in Sagres, on the extreme southwestern tip of Portugal, ships went out to what were then the edges of the known world: the Azores, the Madeiras, the Canaries—and beyond.

The first stop beyond the Canary Islands was Cape Bojador, which, in addition to posing a profound physical barrier, constituted an even more formidable psychological one. At that time conventional wisdom maintained that the Sun was boiling hot at the Equator. Thus even if a ship could get past Cape Bojador—which in fact is just above the Tropic of Cancer, some 2,000 miles or 3,200 kilometers north of the Equator—the equatorial Sun would eventually burn it to powder. Furthermore, should a vessel somehow make it past all other hazards, its crew would most surely meet unspeakable monsters in the subequatorial region known as the Antipodes.

The latter supposition was the result of impeccable medieval logic: since all men had descended from Adam and Eve, and since it was impossible to cross the Equator, all creatures in the Antipodes must necessarily be something other than human. Indeed, to live in the fearsome waste dubbed *Terra Incognita,* a being must surely be monstrous. Finally, there was the possibility that Africa itself blended into *Terra Incognita,* meaning that even if a ship got past all other obstacles, its crew would find themselves unable to sail further.

A fascinating figure who lived on the cusp between the medieval and modern worlds, Henry was driven as much by crusading fervor—a desire to win the souls of unknown lands to Christ—as by an urge to breach physical boundaries. Beginning in 1418, when João Gonçalves Zarco and Tristão Vaz Teixeira first caught sight of Porto Santo in the Madeiras, sailors from Henry's school had been edging further southward; yet Cape Bojador remained a forbidding wall against further exploration. By the early 1430s, Henry had determined to cross the barrier represented by the cape, and for the job he chose Eannes.

The latter had grown up in Lagos, near Sagres, and had served Henry from boyhood as a squire. Thus the prince was certain of Eannes's loyalty when in 1433 he ordered him to sail past the cape. In today's terms, this would be equivalent to a deep-space mission, and Eannes was no doubt terrified as he embarked. In any case, he did not get far on his first voyage: after reaching the Canaries, he returned to Portugal. The prince, however, was not willing to accept defeat: in 1434 he sent Eannes out again, this time with the charge, "Make the voyage from which, by the grace of God, you cannot fail to derive honor and profit."

Setting out in a small fishing boat—the larger Portuguese *barcas* would not come into use until later in the decade— Eannes managed to round the cape by sailing far to the west. Instead of the world's edge, he found himself on a calm sea. After skirting the cape, he landed on the desert coast and collected one of the few living specimens available, a plant that came to be known among the Portuguese as "St. Mary's roses." This he brought back to Henry's court as proof that he had landed.

Eannes returned a hero, and no doubt Henry and the Portuguese people showered him with riches and acclaim. Certainly he would have deserved such, given the fact that he had opened the way to the world beyond Cape Bojador. In 1435 he and Alfonso Gonçalves Baldaya sailed past Bojador, where they saw human and camel tracks, at that time the most southerly signs of life known to Europeans. Baldaya continued further south, and traded for sealskins, the first commercial cargo brought back to Europe from West Africa.

Eannes last appears on the pages of history on August 10, 1445, when he left Portugal with an armada of *caravels*—a new, sturdier sailing vessel that replaced *barcas*—bound for the island of Tidra off the coast of what is now Mauritania. There the Portuguese forces did battle with the Muslim inhabitants, a conflict in which Eannes may have died. In any case, he had opened the way for Portuguese vessels to move ever southward, and by 1456 Alvise Cadamosto (1432-1488) had reached the Gambia River, far down the coastline from Bojador. Just 32 years later, in 1488, Bartolomeu Dias (c. 1450-1500) rounded the Cape of Good Hope at the southern tip of Africa, and in 1497 Vasco da Gama (c. 1460-1524) sailed around the cape to India.

Stirring as these victories were, however, the exploration of West Africa is mingled with the taint of African slavery. The Portuguese captured their first natives in 1441, and it appears that in 1444, Eannes himself took part in an expedition to seize some 200 slaves near Cape Blanco, first reached in the preceding year. During the half-century up to 1500, some 300,000 people were captured in Africa, and as the modern age dawned, so did the era of modern chattel slavery. Though outlawed throughout Europe during the 1700s, the enslavement of Africans in the New World would continue until Cuba and Brazil finally outlawed the institution in the 1880s.

Also in the 1880s, the area around Cape Bojador became a Spanish colony. Despite its barrenness, the region had become the cause for a territorial dispute between Spain and Portugal from about 1450, and this rivalry continued even as the two Iberian countries' influence as world powers of exploration waned. While Britain, France, and other nations helped themselves to more desirable spots in Africa, Spain in 1860 signed the treaty of Tetuan with Morocco, which gave it rights to the region that came to be known as the Spanish Sahara.

Though it officially annexed the area in 1884, Spain did not fully occupy it until three decades later, when it took advantage of the fact that other nations' attentions were diverted by World War I. In 1957 Morocco began to press greater claims over the Spanish Sahara, and the discovery of phosphate deposits increased the seriousness of the dispute. Finally in 1975 Spain relinquished it, and thus Cape Bojador may justly be called the place where European exploration and colonization in Africa both began and ended. In the decades that followed, the territory would see sustained conflict between new claimants: Morocco, Mauritania, and native Polisario rebels. Today the Western Sahara is one of the most sparsely populated regions on Earth, with some 208,000 people on more than 100,000 square miles (259,000 square km).

JUDSON KNIGHT

Further Reading

Books

Bradford, Ernie. *A Wind from the North: The Life of Prince Henry the Navigator.* New York: Harcourt, Brace, 1960.

Prestage, Edgar. *The Portuguese Pioneers.* London: Adam & Charles Black, 1933.

Zurara, Gomes Eanes de. *Conquests and Discoveries of Henry the Navigator; Being the Chronicles of Azurara.* Edited by Virginia de Castro e Almeida, translated by Bernard Miall. London: Allen & Unwin, 1936.

Internet Sites

Granger, David A. "The Atlantic Slave Trade: Crime of the Millennium." *Stabroek News* [Guyana], October 17, 1999. http://topcities.com/Groove/awane.ns910174.htm.

Viking Settlers in Greenland

Overview

About 1,000 years ago, the North Atlantic island of Greenland, the largest island in the world, was colonized by a group of Vikings ruled by the famous Erik the Red. The colonies died out after about 400 years, their inhabitants perhaps victims of changing climate or conflicts with native peoples. During their existence, however, the Norse settlements in Greenland provided a springboard for further exploration westward, including the first journeys by Europeans to North America.

Background

The Vikings were a seafaring Nordic people of the Middle Ages, ancestors of today's Danes, Norwegians, Swedes and Icelanders. During the height of their influence, from about 750-1050, they ranged across northern Europe and as far south as the Mediterranean. They established a territory in northern England called the Danelaw. The Norse occupation of northern France is reflected in the name of the region of Normandy.

The Vikings have the reputation of being marauders, and this is not undeserved. They increased their wealth by ferocious raids in which they burned and looted villages and towns, killed their inhabitants, and headed back out to sea before any opposing force could be mustered. However, they were also skilled farmers, craftsmen and explorers. Their culture and ideas had a profound effect on the development of European civilization.

A major factor in their dominance at sea was their efficient shipbuilding methods. Their swift, low-slung longboats were propelled by a few dozen oarsmen, often assisted by a single square sail. The hulls of Viking ships were *clinker*-built, with overlapping planks for extra strength and to reduce leaks. Vikings had no compass for navigation, or method to compute their longitude. However, they did have an instrument called a bearing dial, which could be used to track the position of the North Star and maintain a steady east-west course.

Eventually the success of the Vikings led to an increase in their population beyond what their Scandinavian homelands could comfortably support. Piracy and conquest had been their typical response to such a situation in the early days. But more integrated into the civilization of medieval Europe by this time, they were less inclined to fall upon their neighbors as *berserkers*, their name for warriors, and destroy everything in their paths. Concentrated along the seashore, and with their sturdy seagoing vessels at the ready, they naturally saw the uninhabited Atlantic islands to their west as offering opportunities for expansion. Iceland in particular became a thriving settlement.

Impact

Iceland was short on land suitable for growing crops. Grain, timber and metals had to be imported from Norway. But, as they had on the mainland, Viking settlers fished and tended pigs, sheep, cattle and goats. They also hunted sea mammals for their skins, ivory, and oil. Falcons they caught and trained were in demand all over Europe. Icelanders set up a constitution and a general assembly, called the Althing, with a chief "law-speaker" or president. The Icelandic Althing continues to exist today, and is the oldest parliamentary body in the world.

By the tenth century, many Vikings had abandoned their Norse gods, such as Odin and Thor, and become Christians. Combining their own tradition of oral poetry and storytelling with Christian and Irish written narratives, they developed the *saga*, a literary form unique to Iceland written in an alphabet called *runic*. From these sagas, such as the *Eiriks* saga (Saga of Erik) and the *Groenlendinga* saga (Saga of the Greenlanders), we get much of our knowledge of Viking history and explorations. However it is a challenge to interpret and confirm the information gleaned from the sagas, as the stories include fanciful elements and sometimes contradict one another.

The Icelandic Vikings are believed to have first learned about Greenland from a man named Gunnbjorn Ulf-Krakuson, who reported islands to the west after his ship had blown off course in about the year 930. It was an explorer named Erik the Red who followed up on this knowledge. Erik was a violent man from a violent family. He was born in Norway, but his family emigrated to Iceland when he was a child, after his father Thorvald was banished for what the *Eiriks* saga calls "some killings." Erik in turn was banished from Iceland for three years in 982, after killing a total of four men in two family feuds.

With the help of his supporters, Erik outfitted and manned a ship and headed west. Within a few days, he found land, but it didn't look terribly promising. It was covered in a huge icecap, with glaciers and rocks guarding its inhospitable coast. Still, Erik continued around the island to the south, and up its western shore. Here he found a land of greater potential, with grassy meadows and natural harbors. During his three years of exile, Erik and his crew explored the island in the summer, scouting about 1,000 miles (about 1,600 km) of its coast and marking off their future land claims. They wintered in sod shelters, hunting and fishing for their food.

By the time he could return to Iceland, Erik had plans of mounting a larger expedition to settle the island. Like any enthusiastic land promoter, he gave his destination an attractive and rather optimistic name, Greenland, and touted its harbors, wildlife and pastures. Many were motivated to join him by a thirst for adventure, compounded by depleted pastures in Iceland and the rigors of a recent famine. Erik sailed from Iceland in 986 leading about 25 ships. These were not longboats, but larger, broad-beamed ships called *knarr*, designed for carrying cargo and passengers.

Eleven of the ships, according to the sagas, were either forced back or sank. The other 14, carrying about 400 settlers, landed safely. The colonists spread out among the fjords of Greenland's southwestern coast in three settlements. The largest one, to the south near Ericsfjord, was called the Eastern Settlement. At its peak, it consisted of several hundred farms. A smaller colony further up the coast, near the modern capital of Godthab, was called the Western Settlement. There was also a hamlet called the Middle Settlement, with about 20 farms, near today's Ivigtut.

Greenland had no timber, so the settlers built their homes and barns of sod, stone and driftwood. Although the land wasn't good for crops, it was productive as pasture, and the colonists devoted large farms to raising cattle and sheep. They also fished and hunted reindeer, bear, foxes, birds, whales, seals, and walrus. They exported fur, hides, wool, oils, whalebone, walrus ivory, falcons and polar bears to the European mainland. In return, they obtained lumber, grain, beer and wine, metal items like tools and weapons, and luxury goods such as finished garments.

It was only a few months after Greenland was settled that the seagoing trader Bjarni Herjolfsson was blown off course and noticed land even farther west. This accidental sighting prompted Erik's son Leif to mount the first European expedition to North America in the year 1000, establishing a camp in an area he called Vinland. Most scholars believe this was probably on the northern tip of Newfoundland. Leif's expedition was followed by several other journeys to the northeastern corner of North America.

When Leif returned to Greenland, having converted to Christianity himself, he likewise converted his mother, Tjodhild. Near the family farm in Brattahild, located in the most fertile region of the Eastern Settlement, she built the first Christian church in Greenland. This small sod structure, which was described in the sagas, has recently been excavated by archaeologists. Eventually there would be 12 parish churches, a monastery, and a cathedral with its own bishop in Greenland. The Greenlanders also established a government with a constitution and a code of laws. In 1261 they voted to unite with Norway. Norway united with Denmark in 1380, thus bringing Greenland under Danish rule.

The Viking settlements in Greenland, with a peak population of about 3,000, lasted until the fifteenth century. The exact reason for their demise is not known, although a drop in the already harsh temperatures in the 1400s probably contributed. There may also have been conflicts with Eskimos making their way down from the north. In 1712, with contact lost for centuries, the king of Denmark and Norway sent out the pastor Hans Egede to minister to any Viking descendants of the Christian faith who remained on Greenland. He found none. The island was by that time inhabited solely by Eskimos.

When the union between Denmark and Norway was dissolved in 1814, Greenland remained a colony of Denmark. Norway disputed the claim for over a century, but Denmark finally prevailed in the World Court in 1933. The new Danish constitution of 1953 elevated Greenland's status from a colony to a province. Under home rule, established in 1979, it governs its own internal affairs. Greenland's residents are of mixed Eskimo and Danish heritage. Most speak an Eskimo language called Greenlandic, and many speak Danish as well. The principal religion is the official Lutheran church of Denmark.

Greenland's location makes it an important base for forecasting North Atlantic storms. During World War II, when Denmark was occupied by Germany, the United States agreed to take over the defense of Greenland. Under NATO agreements, U.S. military bases in Greenland continue to form an important part of the North American "early warning" defense system.

SHERRI CHASIN CALVO

Further Reading

Brent, P. *The Viking Saga.* New York: G.P. Putnam's Sons, 1975.

Graham-Campbell, J. *Cultural Atlas of the Viking World.* New York: Facts on File, 1994.

Jones, G. *A History of the Vikings.* Oxford: Oxford University Press, 1968.

Logan, F. D.*The Vikings in History.* New York: Routledge, 1991.

Magnusson, M. *Vikings!* New York: Elsevier-Dutton, 1980.

The Vikings Explore North America

Overview

Five hundred years before Columbus, Vikings led by Leif Eriksson became the first Europeans known to have set foot in North America. Norse sagas and archaeological finds record their explorations and their contacts with the native peoples. Unfortunately for the Vikings, their relations with the indigenous inhabitants were not friendly, and they soon abandoned their encampments.

Background

In the tenth century, the Vikings had expanded from their Scandinavian homeland and settled the islands to their west, including the Faeroe Islands and Iceland. Vikings were accomplished sailors, well equipped to handle the rough waters of the North Atlantic. Their ships were sturdy, fast and flexible. They had overlapping *clinker* hull planking, to increase their strength and help prevent leaks. With a framework held together by animal sinews or spruce roots, they could flex rather than break apart under the stress of ocean swells. Deck planks were removable for arranging cargo and bailing out bilge water. The ships rode low enough to sail on the open ocean, but were shallow enough to beach and to navigate rivers. In addition to being propelled by their single square sail, they could also be rowed.

The famous Viking Erik the Red, who had been banished for three years from the established colony in Iceland after killing several men, spent the time exploring a large island to the west. When his exile was over, he returned to Iceland and recruited colonists to settle on the island, which he gave the enticing name of Greenland. In 986, he led an expedition to Greenland with hundreds of settlers. They established their farms along Greenland's southwestern coast, raising livestock on the rich pasture.

Neither Greenland nor Iceland had much timber, arable land, or metal for tools and weapons. They did, however, have abundant wildlife on both land and sea. Settlers obtained furs and hides, whalebone and walrus ivory to trade for the goods they needed. Soon traders in merchant ships were plying the waters between Greenland, Iceland, and the Norwegian mainland.

Viking sailors gave storms a wide berth, sometimes veering well out of their way to avoid them. As a result it was not uncommon to get lost or to be blown off course. In fact, it was one such lost sailor, Gunnbjorn, whose reports of land sighted to the west had led Erik on his explorations. When the Greenland settlement was only a few months old, another sailor, Bjarni Herjolfsson, went out from Norway in his trading ship to visit his father in Iceland. When he got there, however, he discovered that the older man had gone with Erik's expedition.

Bjarni determined to sail on to Greenland and winter with his father as he had planned. He and his crew became lost in a heavy fog, and by the time it lifted, Bjarni realized he had sailed much too far. There was land to the west, which we now call Labrador, Canada. Bjarni didn't know what it was, but he knew it wasn't Greenland, and he didn't go ashore there. He headed east, back into the open ocean, and luckily found Greenland in four days. He carried with him the tale of the new lands.

A replica of a Viking longship. *(Hulton Deutsch Collection/Corbis. Reproduced with permission.)*

Impact

Bjarni's adventures were told in the Saga of the Greenlanders. The *saga* is an Icelandic literary form, written in a *runic* alphabet. Sagas combine the oral poetry and storytelling tradition of the Vikings with the written narrative forms adopted from medieval Christian civilization. Sagas must be used with caution as historical evidence because they incorporate fantasy and are often contradictory. However, they are often the only documentation available about incidents in Viking history. Together with archaeological evidence, they allow scholars to piece together an account of the Viking journeys to North America.

Among the Greenlanders who heard Bjarni's story was Leif Eriksson, son of Erik the Red. Leif bought Bjarni's ship and, in the year 1000, sailed west with a crew of 35 men. They first made landfall on a mountainous, glacier-covered coast Leif called Helluland, or "Flat Stone Land." This was probably Baffin Island. Turning away from its barren coast, they proceeded south until they came upon a forested region they named Markland, or "Woodland," believed to be Labrador. Continuing still further south, they arrived at a place they called Vinland, where they set up their encampment of Leifsbudir. After wintering there, they returned to Greenland.

The location of Vinland has long been debated. One translation of Vinland is "Grassland." If this is its meaning, it suggests the coast of Newfoundland. On the other hand, Vinland may also mean "Wineland" or "Vineland." This presents a problem, as grapes do not grow that far north. Berries do, though, including lingonberries, which are traditionally used to make wine in Scandinavia. Cranberries, currants, and gooseberries are other candidates. The climate is believed to have been slightly warmer in Leif's day, so perhaps grapes could grow in Newfoundland at that time. There is also the possibility that Leif, like his father, simply gave the land he found an attractive name.

The discovery of a Viking site at L'Anse aux Meadows on the northern coast of Newfoundland has led most scholars to believe that this was Leif Eriksson's Vinland. The site at L'Anse aux Meadows was discovered and excavated in the early 1960s by the husband-and-wife team of archaeologists Helge and Anne Ingstad. While there are many dubious claims for "Viking" sites and artifacts in North America, the authenticity of L'Anse aux Meadows is unmistakable. Archaeologists have found structures of typical Scandinavian design and a Viking-style bronze pin. The buildings and artifacts have been dated to the early eleventh century. The absence of signs of long-term habitation, such as burial grounds, large garbage dumps, and rebuilt houses, fits in with Leifsbudir's use as a seasonal camp.

When Erik the Red died, Leif became head of the family, and was prevented from returning to Vinland by his new responsibilities. His younger brother Thorvald took Leif's ship and went instead. He and his crew of 30 men spent the summer exploring west of Leifsbudir, and then camped there for the winter. Thorvald's visit is notable in that it included the first recorded contact between Europeans and the indigenous people of North America, either Native Americans or Eskimos, whom the Vikings called *skraelings*, or barbarians.

Sadly, this first meeting set the tone for many that were to follow between Europeans and native North Americans. Thorvald and his crew sailed into a fjord and encountered nine native men in skin boats. A battle ensued in which eight of the nine natives were killed. The last man escaped and returned with reinforcements. The Vikings retreated, but not before Thorvald had been mortally wounded by an arrow. His men returned to Greenland, as the dying Thorvald had ordered. Another son of Erik the Red's, Thorstein, took the same ship out yet again, hoping to bring back his brother's body. But caught up in a storm, he found himself back on Greenland's coast, and died of an illness before he could set out once more.

According to the *Saga of the Greenlanders,* another expedition went out to Vinland shortly after the return of Thorvald's crew. It consisted of about 150 would-be colonists led by a wealthy Icelandic trader named Thorfinn Karlsefni. They set up an encampment not far from Leifsbudir, and surrounded it with a protective stockade. At first, they traded with the indigenous people, but relations soon turned hostile. When one native was killed while attempting to steal weapons, his companions returned with more men and boats. Although the Vikings defeated them, Thorfinn feared that the fighting would continue until all the colonists were killed. After the next winter, the settlement was abandoned. However, it had achieved one milestone. Thorfinn's wife Gudrid had given birth to a son, Snorri, the first European child born in North America.

Although some scholars doubt its veracity, there is another tale recorded in the sagas about an expedition led by Leif Eriksson's sister Freydis. This was apparently a commercial timber-cutting trip in partnership with two men from Iceland. Freydis is depicted as a ruthless and treacherous woman. Wanting to claim her partners' ship, she tricked her husband Thorvard into killing the two Icelanders and their crew. When Thorvard demurred at killing the five women among them, Freydis took on that task herself. Then she sailed back to Greenland with both ships and all the timber.

After Thorfinn, there were no more Viking attempts to colonize North America, although a few more journeys were recorded. In particular, Labrador was the destination for occasional timber-gathering expeditions. The Greenland settlement itself died out in the 1400s, as the climate cooled and supply ships no longer ventured through the icy seas. Little evidence remains of contact between the Vikings and the native North American peoples.

SHERRI CHASIN CALVO

Further Reading

Brent, P. *The Viking Saga.* New York: G.P. Putnam's Sons, 1975.

Graham-Campbell, J. *Cultural Atlas of the Viking World.* New York: Facts on File, 1994.

Logan, F.D. *The Vikings in History.* New York: Routledge, 1991.

Ingstad, H. *Westward to Vinland.* E.J. Friis, transl. New York: St. Martin's Press, 1969.

Ingstad, A. and H. Ingstad. *The Norse Discovery of America.* 2 vols. Oslo: Norwegian University Press, 1986.

Jones, G. *A History of the Vikings.* Oxford: Oxford University Press, 1968.

Magnusson, M. *Vikings!* New York: Elsevier-Dutton, 1980.

Magnusson, M. and Paulsson, H. *The Vinland Sagas.* New York: Penguin, 1965.

The Discovery and Settlement of Iceland

Overview

Iceland is the only European country whose history has a definite beginning. Norwegian outlaws, exiles, and adventurers began to settle this previously uninhabited land about 874. In 930 they established what is at the dawn of the twenty-first century the oldest parliamentary democracy in the world.

Background

The first visitors to Iceland may have been Romans but were probably Irish. The *kayaks* and *umiaks* of the Inuit could not have traveled as far as Iceland from Greenland or North America. Roman and early British records refer to a place called "Thule" or "Ultima Thule," which must have been Iceland. A very few Irish monks lived in Iceland in the eighth and ninth centuries, as the Irish monk Dicuil stated in 825 in *Liber de mensura orbis terrae* (Book of measuring the circle of the world), but they had either abandoned this refuge or been driven out by the time the Norse settlement began in the 870s.

About 850, a Swedish Viking named Naddoddur was blown off course west of the Faroe Islands and landed in the east fjords of Iceland, which he named "Snowland." Supposedly he was the first Scandinavian to see Iceland. A few years later, another Swede, Gardar Svafarsson, circumnavigated Iceland and stayed on its northern coast over the winter. Since he proved that the land comprised an island, it was renamed "Gardar's Island."

Hrafna-Flóki Vilgerdarson ("Raven Floki") sailed from Norway via the Shetland Islands to find Gardar's Island about 860. Floki got his nickname because he navigated with ravens. In search of land, he would release a raven from his ship, then follow its path if it flew straight. His first sight of Iceland was the peak Vesturhorn in the southeast, near the present town of Höfn. From there he sailed west along the south coast, rounded the peninsula of Reykjanes, crossed Faxaflói, the great bay named after one of his companions, Faxi, passed Snaefellsnes, continued north across Breidafjord, and finally landed at Vatnsfjord near Bardastrand in the west fjords.

Floki spent a severe winter and an unusually cold spring at Vatnsfjord. Fish sustained him and his men, but their livestock died. In late spring, Floki, disgusted at the sight of drift ice still in the fjord, gave the land its present name, "Iceland." He tried to sail back to Norway that summer, but could not tack around Reykjanes. He was forced to spend a second winter in Iceland, this time at Borgarfjord, an inlet of Faxaflói. When he finally was able to return to Norway the following summer, he had nothing good to say about Iceland, although several of his shipmates praised it. A town near Vatnsfjord is named Flókalundur in Floki's honor.

Impact

About 874, the settling of Iceland began in earnest when Ingólfur Arnarson, Hjorleif Hrodmarsson, and their party arrived. News of Ingólfur's success led about 30,000 more Norse to move to Iceland over the next 60 years. The period of Icelandic history from Ingólfur's landing until the establishment of the national assembly, the Althing, in 930, is called the "Age of Settlement." The lives of most of the leading settlers are chronicled in *Landnámabók* (The book of settlements) compiled in the thirteenth century. Most of today's Icelanders can trace their genealogies back to the pioneers mentioned in *Landnámabók.*

The Age of Settlement was concurrent with the long reign of the first king of Norway, Harald. That concurrence was not coincidental. Harald's tyranny drove many Norwegians into exile, and Iceland was a natural place for many of them to find new homes.

Harald was the son of the warlord Halfdan the Black. Having sworn as a young boy never to cut or comb his hair until he had conquered Norway, he was known as Harald Lúfa (Harald the Shaggy). He achieved his goal while still a teenager, sometime in the early 860s, and thereafter was called Harald Haarfager (Harald Fine-Hair). He ruled southwestern Norway and several tributary lands as a dictator until 930, when he abdicated in favor of his son, Eirik Blood-Axe. Harald died in 933 in his mid-eighties.

Among the Norwegian Viking chieftains who fell afoul of King Harald was Ketil Flatnose, son of Bjorn Buna, from whom a great many of the most prominent early settlers of Iceland were descended. Ketil conquered the Hebrides at Harald's request, but then refused to pay tribute to

Harald. In retaliation, Harald confiscated all of Ketil's Norwegian lands and outlawed Bjorn the Easterner, the only one of Ketil's five children who had remained in Norway. Ketil and his daughter, Jorunn Wisdom-Slope, stayed in the Hebrides, but Bjorn and his other three children, son Helgi Bjolan and daughters Aud the Deep-Minded and Thorunn Hyrna, all went to Iceland.

Helgi Bjolan received a land grant from Ingólfur and settled at Hof, near Esjuberg, across the bay from the present capital, Reykjavík. Jorunn's son, Ketil the Foolish, was the first settler at Kirkjubaer in the southeast. Thorunn and her husband, Helgi the Lean, took possession of the land around Eyjafjord in the north. One of the first Christians in Iceland, Helgi the Lean called his home "Kristness."

Another early Icelandic Christian was Helgi the Lean's sister-in-law Aud (sometimes known as Unn). As survivor of both her husband Olaf the White and her son Thorstein the Red, she led her own expedition to Iceland about 915 and became the matriarch of the fertile dales region at the head of Hvammsfjord in the west. Women in early Iceland enjoyed more rights than in any other medieval culture, and nearly full equality with men in legal matters such as divorce and inheritance. Among Aud's descendants were Hoskuld Dala-Kollsson, Thord Gellir, and Snorri Godi, all powerful chieftains frequently mentioned in the sagas about Iceland's early years as a settlement.

Thorolf Mostur-Beard hid Bjorn the Easterner from Harald in Norway until both men decided to emigrate. Thorolf went directly to Iceland about 882 and settled at Helgafell on the Snaefellsnes peninsula, where Hvammsfjord runs into Breidafjord. Bjorn went to the Hebrides first, then to Iceland two years later, where Thorolf granted him land on Snaefellsnes between Hraunsvík and Hraunsfjord. *Eyrbyggja Saga* and *Laxdaela Saga* concern, respectively, Bjorn's and Aud's descendants.

Kveld-Ulf Bjalfason was a Norwegian chieftain who did not oppose King Harald, but refused to become his retainer. Harald, enraged, conspired for years against Kveld-Ulf's powerful and influential family. On trumped-up charges, he had Kveld-Ulf's son Thorolf executed about 890. Kveld-Ulf and his surviving son, Skalla-Grim, fled to Iceland, but not before they had recovered a ship that Harald had stolen from Thorolf. Kveld-Ulf died during the voyage. Skalla-Grim threw his father's coffin overboard, buried it where it washed ashore at Myrar, north of Borgarfjord, and built his home there. Skalla-Grim's son, Egil, is the eponymous hero of *Egil's Saga*.

Ketil Trout, another Norwegian chieftain, defended Thorolf Kveld-Ulfsson against Harald and killed the two slanderers who had engineered Thorolf's downfall. Ketil then decided that Norway had become unsafe and brought his entire household to Iceland. He took possession of a large stretch of land between the Markar River and the Thjórsá Estuary and became the dominant chieftain in southern Iceland. Among those to whom he granted land was Sighvat the Red, many of whose descendants had roles in the most important Icelandic epic, *Njal's Saga*.

Laxdaela, Eyrbyggja, Egil's, and *Njal's* are generally recognized as the four greatest Icelandic sagas. They are works of historical fiction, written two or three centuries after the incidents they describe, yet they seem more history than fiction. They tell the story of early Iceland's unique political, social, and religious development.

In Viking society, power determined wealth. Raiders, sailors, and farmers who put their trust in their own skill, might, and ingenuity had no need of monarchy. Nearly all of the first Icelanders were there precisely to escape monarchy. Very early in the Age of Settlement, the new settlers formed regional democratic assemblies to adjudicate local disputes. Soon, as the population rapidly increased and as more disputes crossed local boundaries, the need grew for a national assembly on the same model. The chieftains chose a natural amphitheater at Thingvellir in southeastern Iceland for this annual national legislative and judicial meeting, the Althing, which convened for two weeks each June beginning in 930. The Althing decreed in 1000 that thenceforth Christianity would be the universal religion of Iceland, replacing the worship of the Norse gods. The Althing remains the world's oldest parliament, and now meets in a building in Reykjavík.

After about 930, Icelanders considered the land fully settled, and thenceforth few new immigrants were made welcome. The golden age of Iceland, the "Saga Age," lasted from the founding of the Althing until about the middle of the twelfth century. During the subsequent "Sturlung Age," corruption, clan rivalries, and blood vengeance weakened the social structure sufficiently for Norway to annex Iceland easily in 1262. When Denmark united with Norway and Sweden in 1397, Iceland became a Danish possession and remained so until the Danes granted it limited home rule in 1874 and full independence in 1944.

Until about the fourteenth century, Iceland was warmer than it is now. The sagas of the twelfth and thirteenth centuries refer matter-of-factly to trees and meadows in Iceland that could nowadays not survive. As the climate deteriorated, severe volcanic eruptions, ensuing famines, and other natural disasters also contributed to about six centuries of general Icelandic misery from 1262 to 1874. Since regaining its independence, Iceland has enjoyed a steady renaissance. It is now one of the most highly educated countries in the world, with nearly 100% literacy.

ERIC V.D. LUFT

Further Reading

The Book of Settlements: Landnámabók. Trans. Hermann Pálsson and Paul Edwards. Winnipeg, Canada: University of Manitoba Press, 1972.

Egil's Saga. Trans. Hermann Pálsson and Paul Edwards. Harmondsworth, England: Penguin, 1976.

Eyrbyggja Saga. Trans. Hermann Pálsson and Paul Edwards. Harmondsworth, England: Penguin, 1989.

Laxdaela Saga. Trans. Magnus Magnusson and Hermann Pálsson. Harmondsworth, England: Penguin, 1969.

Magnusson, Magnus. *Iceland Saga*. London: Bodley Head, 1987.

Njal's Saga. Trans. Magnus Magnusson and Hermann Pálsson. Harmondsworth, England: Penguin, 1960.

Swaney, Deanna. *Iceland, Greenland, and the Faroe Islands*. Hawthorn, Australia: Lonely Planet, 1994.

The Vinland Sagas: The Norse Discovery of North America: Graenlendinga Saga and Eirik's Saga. Trans. Magnus Magnusson and Hermann Pálsson. Harmondsworth, England: Penguin, 1965.

The Legend of Prester John Spurs European Exploration

Overview

Few among even the most educated modern people recognize "Prester John," a mythical Eastern Christian king whose existence Europeans widely believed during the late Middle Ages. Fewer still appreciate the enormous impact this strange myth had on the history of the West, particularly inasmuch as it inspired the Portuguese expeditions that inaugurated the Age of Exploration. The story of Prester John's legend encompasses a vast panorama, stretching over a period of some 500 years beginning in the mid-twelfth century; taking place in lands from western China to Italy to Ethiopia; and involving figures as varied as Genghis Khan and Henry the Navigator.

Background

To understand the background of the Prester John story, one must start at the middle, when the tale first made its appearance, then move forward to examine its impact. Only when this is done can one properly delve into the elements from the distant past that spawned it.

In Palestine in 1144, the Christian stronghold at Edessa fell, ending a period of crusader dominance in the Holy Land that had lasted for nearly a half-century, since the First Crusade. Early in 1145, Raymond of Antioch, grandson of the conquering crusader by the same name, sent Bishop Hugh of Jabala to seek help from Pope Eugenius II. In time, Europe would send forces that would wage the disastrous Second Crusade (1147-49), a failed attempt to win back the gains of the first; but the most significant outgrowth of Hugh's audience with the pope was a bizarre tale concerning a king named Presbyter Johannes, or John the Priest.

Present at the meeting between Hugh and Eugenius was one of the medieval world's great historians, Bishop Otto of Freising (c. 1111-1158). In his *History of the Two Cities,* Otto later wrote that Hugh spoke of "a certain John, a king and priest who lives in the extreme Orient, beyond Persia and Armenia, and who like his people is a Christian...." This Prester John, as he came to be known, subscribed to the Nestorian faith, which maintained that Christ had two separate identities, one human and one divine. The Church had declared Nestorianism a heresy in 431, and in subsequent centuries, the Nestorians had gravitated eastward.

Prester John, however, was on his way west, according to Hugh. He wished to aid the crusaders in Jerusalem, and had won a great victory over a Muslim army in Persia, but had been unable to cross the frozen Tigris River in Mesopotamia. Therefore he and his armies had turned north, hoping to find a place where the river was frozen in winter; but after years of trying, they had finally given up. Near the end of his account, Otto noted that John was "said to be a direct descendant of the Magi, who are mentioned in the Gospel, and to rule over the same peoples they governed, enjoying such glory and prosperity that he uses no scepter but one of emerald."

In *The Realm of Prester John*—a particularly notable study among dozens on the subject—Robert Silverberg noted that the quest for Prester John's land would last for half a millennium, in the process becoming "one of the great romantic enterprises of the Middle Ages." It was a journey across thousands of miles, first into Asia, and later to Africa; yet, "Tracing the origin of the legend of Prester John leads the scholar on a quest nearly as exhausting and difficult as those undertaken by the medieval explorers."

It seems clear that Hugh did not invent the story he told, Silverberg observed, not least because belief in Prester John would have hurt his cause: as long as there was a powerful Eastern king intent on coming to the crusaders' assistance, there was no need for Rome to help. Instead, it seems that Hugh was actually trying to counteract rumors about Prester John's invincibility that were already beginning to spread throughout Europe. Therefore one must look beyond Hugh, Silverberg wrote, "But to uncover the sources of the tale Hugh told requires a lengthy voyage on a sea of conjecture."

Impact

Twenty years after Hugh's meeting with the pope, in 1165, Byzantine emperor Manuel I Comnenus (r. 1143-80) received a letter purportedly from Prester John. In the lengthy missive, the putative monarch discusses the wealth and power of his kingdom, and does so in terms both boastful and condescending: "Our Majesty has been informed that you hold our Excellency in esteem, and that knowledge of our greatness has reached you." The epistle goes on to discuss the wealth of John's land, where all the people are pious and the king himself is served by vassal kings and bishops. Peppering the letter are also descriptions of numerous fantastic elements in John's kingdom: swirling oceans of sand, salamanders that live in fire, and a stone with a cavity holding water that cures all diseases.

As for the location of this fabled land, John identifies it as "the Three Indias." The latter was a vague, confusing expression. Two of the "Indias" were, at least, in or close to what is today known as India, Nearer or Lesser India being the area of mountains from the Caucasus to the Himalayas, and Farther or Greater India the main part of the Indian subcontinent. But "Middle India" was thousands of miles away in Ethiopia, a reflection of the fact that medieval Europeans believed everything east of the Nile to be part of Asia.

Scholars have long recognized the letter as a forgery, probably written by an imaginative monk. There are a number of reasons for this conclusion, not least the fact that the author followed none of the established conventions for diplomatic correspondence in the Middle Ages: for example, the letter includes no date or reference to its place of origin, and more important, its tone is hardly that of one who hopes to maintain the goodwill of his recipient.

Forgery or not, however, the letter began to make the rounds in Europe, appearing in some 100 versions—many with added passages—in a dozen languages. It quickly gained wide acceptance, such that in 1177 Pope Alexander III (c. 1105-1181) apparently attempted to answer the letter. Certainly it is known that the pope wrote a message to a Christian ruler in the East, cautioning him against boastfulness and inviting him to accept the Roman Church as the true one. It is also certain that Alexander gave this letter to his physician, Philippus—but as to where Philippus went with it, or even what became of the messenger himself, historians do not know.

Over the next two centuries, the Prester John legend went through numerous permutations, disappearing periodically, then reappearing in new forms. The next stage in the legend's development came in 1222, after yet another failed crusade, the fifth (1217-21). It was then that Jacques de Vitry (c. 1170-1240), bishop of Acre in the Holy Land, reported rumors that a certain King David, son or grandson of Prester John, was about to come to the crusaders' aid from the East. This was true, at least in part: a new, powerful king *was* on his way, and his host would indeed smash the power of Muslim rulers in the Middle East. But he was no Christian, and his name was not David; it was Genghis Khan (1162-1227).

As they, like their Muslim foes, faced the Mongol onslaught, Europeans soon realized that Genghis Khan was not their promised savior. Thus the early thirteenth-century version of the legend depicted Genghis as a usurper who had seized the throne from the real Prester John. For example, Marco Polo (1254- 1324), in his account of his travels among Mongol lands, stated matter-of-factly that Toghrul or Unc Khan, a chieftain overthrown by the young Genghis, was in fact Prester John.

By the mid-thirteenth century, the legend had again become attached to a Mongol leader, Genghis's grandson Kuyuk, who had taken an interest in Nestorian Christianity. Other travelers among the Mongols, including Giovanni da Pian del Carpini (1182-1252), Giovanni da Montecorvino (1246-1328), Odoric of Pordenone (c. 1265-1331), and John de' Marignolli (c. 1290-c. 1357), would add new elements to the tale of Prester John, as would the scholar Johannes of Hildesheim (d. 1375). So too would "Sir John Mandeville," alleged author of a fictionalized—but nonetheless highly popular—travelogue in c. 1360.

As Mongol power waned, however, new accounts depicted John as coming from Armenia. Then, in the early 1330s, the *Description of Marvels,* a travel memoir by Jordanus of Séverac (1290- 1354), depicted a new location for John's kingdom: Ethiopia. Thus was born the final version of the legend, one that would open the curtain on the modern era of exploration.

Born a century after Jordanus, Prince Henry the Navigator (1394-1460) grew up intrigued by the idea of searching for Prester John in Africa, and his desire to find the legendary Christian king spurred him to send explorers along the African coast. In time this aim became tied with a quest for a sea route to India. Thus in 1487, the same year Portugal's King John II (r. 1481- 95) sent Bartholomeu Dias (c. 1450-1500) and Pero da Covilhã (c. 1460- c. 1526) on expeditions to find routes to India, he ordered Afonso da Paiva to seek out Prester John in Ethiopia.

Paiva died in Cairo, so Covilhã, having completed his own mission along the Horn of Africa, was sent in his place to Ethiopia. There he did indeed meet up with a Christian monarch, the emperor of Ethiopia, who refused to let him leave; thus when a group of Portuguese explorers arrived in Ethiopia in 1520, they were met by an aging Covilhã. By then another Portuguese adventurer, Vasco da Gama (c. 1460-1524), had long since completed his successful 1497 voyage to India around the Cape of Good Hope—an expedition in which he carried with him a letter from his king to Prester John.

The myth would persist in one form or another for at least a century more, during which it continued to inspire European exploration, a great (if little-known) influence on the course of history. But where did it have its origins? What inspired the legend Hugh of Jabala passed on to Otto of Freising in 1145? A number of historical, semihistorical, and literary elements appear to have informed the tale.

There were, first of all, legends concerning St. Thomas, or "Doubting Thomas," the apostle who had supposedly gone to India as a missionary. There is indeed a native Indian Christian community in India, as well as a grave that supposedly is Thomas's; as for whether the apostle is truly buried there, or whether the ancient community of Indian Christians owes its existence to Thomas's missionary work, these matters are not known.

Legends of Thomas were bound up with those of his erstwhile colleague, the original "Priest John"—that is, St. John the disciple. Combined with the established fact of the Nestorian community, these elements all contributed to the idea of a Christian king from the East. But when Prester John failed to materialize in that direction, it was only logical that Europeans should look for him in Ethiopia, a land that had been Christian since the fourth century A.D.

Fantastic elements in the Prester John legend owe much to the considerable body of fiction and mythology concerning Alexander the Great (356-323 B.C.), writings collectively described as "the Romance of Alexander." This literature, an example of which is *Christian Legend Concerning Alexander* (c. 514), arose long after Alexander's time, and usually depicts the Macedonian conqueror alongside fixtures of Christian apocalyptic writing, such as the fierce giants Gog and Magog from the Book of Revelation.

Finally, there is the most clearly historical and concrete element of the Prester John legend, the defeat of a Muslim prince by forces from the East in 1141. These forces were the Kara-Khitai, a nomadic tribe from northern China under the leadership of Yeh-lü Ta-shih (1098-1135). On September 9 near Samarkand in what is now Uzbekistan, they won a victory over an army led by the Seljuk leader Sanjar from Persia. The story soon made its way westward, and as it passed from one reteller to another, it gained new elements and ultimately became an entirely different, and truly fantastic, tale.

JUDSON KNIGHT

Further Reading

Books

Alvares, Francisco. *Narrative of the Portuguese Embassy to Abyssinia During the Years 1520- 1527,* translated and edited by Lord Stanley of Alderley. London: Hakluyt Society, 1881.

Buchan, John. *Prester John.* Edited and with an introduction by David Daniell. New York City: Oxford University Press, 1994.

Gumilev, L. N. *Searches for an Imaginary Kingdom: The Legend of the Kingdom of Prester John.* Translated by R. E. F. Smith. New York City: Cambridge University Press, 1987.

Rachewiltz, Igor de. *Prester John and Europe's Discovery of East Asia.* Canberra: Australia National University Press, 1972.

Sanceau, Elaine. *The Land of Prester John: A Chronicle of Portuguese Exploration.* New York City: Knopf, 1944.

Silverberg, Robert. *The Realm of Prester John.* Athens: University of Ohio Press, 1972.

Internet Sites

Philadelphia Print Shop. "Mythical Geography: The Kingdom of Prester John." http://www.philaprintshop.com/presjohn.html.

The Arab-Persian Trading Cities of East Africa

Overview

Beginning with the arrival of Arab and Persian merchants in the period from the ninth to the twelfth centuries, trade flourished on the East African coast, and reached its peak between 1200 and 1500. During those centuries, ships plied the Indian Ocean, the Red Sea, and the Persian Gulf, bringing import and export goods between the East Indies, China, India, the Arab lands, Persia, and East Africa. It was a time of great contact between cultures, an age that saw the establishment of some 37 city-states in what is now Somalia, Kenya, Tanzania, and Mozambique, as well as Madagascar. Most notable among these were Mogadishu, Malindi, Kilwa, Mombasa, and Zanzibar.

Background

Starting in about 1200 B.C., groups of peoples began migrating southward from the region of modern-day Nigeria. Though they constituted a loose collection of tribes and nations, they were united by language: in each of their tongues, the word for "people" was and is the same—*bantu.* In Africa, one of the most ethnically diverse regions on Earth, the average language today has only half a million speakers, whereas the significant Bantu tongue, Swahili, has some 49 million speakers in Kenya, Tanzania, the Congo, and Uganda. Furthermore, Swahili serves as the lingua franca—"common tongue"—of southern Africa, a common language much as Arabic became in the Middle East and as Latin was among educated Europeans of the Middle Ages.

Outside of Ethiopian civilizations such as those in Kush and Aksum, the Bantu were the first civilized peoples of sub-Saharan Africa. (Although they lacked several ingredients of civilization, including a written language, they were skilled iron workers and practiced a sophisticated form of agriculture.) As they migrated southward and eastward, they displaced the Pygmies and the Khoisan (i.e. Bushmen), forcing them to vacate to the rainforests and deserts, respectively. By about A.D. 500, Bantu peoples controlled the choicest spots in southern Africa.

The flourishing East African commercial city-states of the medieval period were a product of Bantu contact with Arab and Persian traders. North of Bantu-speaking East Africa, Aksum (modern Eritrea) had been connected— commercially and even, for several centuries, politically—with several thriving principalities on the pre-Muslim Arabian peninsula. Likewise the area that is now Somalia had seen regular contact with Arab trading powers for some time; but from about 700, Arab and Persian commercial interests turned southward.

Arabs used the name Azania, from the root word *Zanj*—their term for Africans—to describe all of East Africa south of Somalia. From the beginning, their interest was purely in the coastal areas, and as the thriving cities of Kilwa, Mombasa, Malindi, and others took shape, they became islands of civilization cut off from the forbidding interior of Africa. (Zimbabwe, Kongo, and other empires of central southern Africa did not emerge until near the beginning of the modern era.) Most of these city-states were in fact islands, protected by heavy fortifications both from the less civilized peoples of the interior, and from seaborne invaders.

By the ninth century, Arab geographers identified four major areas along the East African coast: Berber territories in Somalia; the Zanj city-states; Sofala, a land in what is now Mozambique; and below Sofala a vaguely defined region known as Waqwaq. The first significant settlement in Azania was on the island of Qanbalu, which may have been Pemba Island off the coast of modern Tanzania. Zanzibar, too, was inhabited by Arabs as early as 1100, though the days of its greatest influence lay in the future.

Kilwa, now in southeastern Tanzania, had perhaps the best harbor of all the city-states. It had existed for several centuries before the first traders began arriving from the Persian Gulf in about 900, but the most significant ruins—city walls made of coral masonry, as well as stone mosques—date from the period that followed. Cowries, a type of seashell, constituted the principal form of currency. (These shells had also been used for the same purpose in ancient China.) Another significant city-state of the early period in this area was Manda in what is now Kenya, established as early as the ninth century. There, seawalls made of giant coral blocks weighing as much as a ton (more than 900 kilograms) protected the city against the Indian Ocean waves.

Among the products exported from East Africa were gold, much of it transported overland a great distance, as well as iron tools, ivory, tortoiseshell, and rhinoceros horn. These went to ports in Arabia, India, southeast China, and the East Indies. In turn the African city-states imported cotton and glass beads from India; silk and porcelain from China; pottery from Arabia, and other items.

A powerful force in the area from the twelfth century onward were the Shirazi tribe from the Persian Gulf, who established their influence in Kilwa and other regions—including the far-off Indian Ocean island of the Comoros. Under the leadership of Abu al-Mawahib, the Shirazi built the palace of Husuni Kubwa at Kilwa. With more than 100 rooms, it was the largest single structure in sub-Saharan Africa for many centuries.

During the early fifteenth century, Mombasa—still an important port in Kenya today—emerged as another important city-state, with a population of as many as 10,000. (Kilwa, for all its influence, had only about 4,000 inhabitants.) Also significant were Pate, ruled by the Nabahani tribe from Oman, and Malindi. In all these cities, Muslim rulers intermarried with the native Bantu population, creating a distinctive Swahili culture, and the appearance of visitors from far-flung ports of Asia served to enhance the international character of this area.

One notable visit was from the Chinese fleet under Admiral Cheng Ho or Zheng He (c. 1371-c. 1433). He went on a series of expeditions in the service of the Ming emperor Yung-lo (1360-1424), whose purpose was not so much trade as to display the wealth and power of China to the other nations of the world. On his fourth voyage (1413-15), Cheng Ho landed at Mogadishu and Malindi, and as a result the rulers of these city-states sent ambassadors to Yung-lo's court.

They also sent gifts, including ostriches, zebras, lions, and tigers. Perhaps the most notable of these was a giraffe, which the Chinese received as "a happy portent ... of Heaven's favor and proof of the virtue of the emperor." Cheng Ho also visited the East African port of Brawa on his fifth expedition (1417-19). Soon, however, the region would receive visitors from a completely different country, one destined to have an enormous impact on East Africa: Portugal.

Impact

After rounding the Cape of Good Hope on November 22, 1497, Vasco da Gama (c. 1460-1524) sailed up the eastern coast of Africa, arriving at Malindi on March 29, 1498. The crew were in dire straits when they arrived, suffering from scurvy and badly in need of supplies. Malindi proved to be a friendly port for the Europeans, and not only did they take on the stores they needed—including fresh fruit to deal with the scurvy—but there they also contracted the services of a skilled pilot, Ahmed ibn Madgid. With his help, they were able to sail successfully on to India, where they arrived on May 18.

Pedro Alvares Cabral (1467-c. 1519) landed at Sofala, in Mozambique, in July 1500 before sailing up the coast. His crew received a hostile reception at Kilwa on July 26—the city's rulers took them for pirates—but like the expedition under Gama before them, they were welcomed at Malindi on August 2.

The years that followed the Portuguese arrival saw a protracted struggle between the Europeans and the Arabs for control of East Africa. The Arabs, however, were much more firmly established than the Europeans, and benefited from the help of their allies in Oman. By 1740 they had driven out the Portuguese, and under Sayyid Sa'id (r. 1806-56), the East African city-states achieved a degree of unity with one another and with Oman.

Until the time of Sayyid Sa'id, slavery had not been a significant factor in East Africa. Certainly the Arabs had enslaved Zanj during the Middle Ages, and the population of slaves in the Middle East became so great that during the ninth century they staged a protracted revolt against their Arab masters in Iraq, but slavery in East Africa never reached the proportions it did in West Africa during the period from the fifteenth to the eighteenth centuries. In the late eighteenth century, however, French demand for slaves in their New World colonies led to the expansion of the slave trade in East Africa.

During the nineteenth century, the cultural and political climate of East Africa became increasingly complicated. Migrant Indian workers constituted a new ethnic element, while Britain and even Germany emerged as colonial powers in the region. Then, in the period after World War II, the Europeans began to let go of their colonies in East Africa, until in 1975 the Portuguese in Mozambique became the last to go. Arab power had long since been eclipsed, as had the influence of the trading city-states of the Indian Ocean coast.

JUDSON KNIGHT

Further Reading

Books

Clark, Leon E., editor. *Through African Eyes: Cultures in Change.* New York: Praeger, 1969.

Davidson, Basil. *African Kingdoms.* Alexandria, Virginia: Time-Life Books, 1978.

Davidson, Basil, with J. E. F. Mhina and Bethwell A. Ogot. *East and Central Africa to the Late Nineteenth Century.* London: Longmans, 1967.

Internet Sites

Celebi, Joan E. "The Indian Ocean Trade:" A Classroom Simulation." http://www. bu.edu/afr/Outreach/resources/handouts/indian.html.

Mansa Musa Makes His Hajj, Displaying Mali's Wealth in Gold and Becoming the First Sub-Saharan African Widely Known among Europeans

Overview

Though the modern nation of Mali is a landlocked country that, like much of Africa, suffers under extreme poverty, the medieval empire of Mali was quite a different place. Not only was its location along the Atlantic coast to the southwest of present-day Mali, it enjoyed considerable wealth, power, and prestige. The greatest of Mali's emperors was Mansa Musa, a devout Muslim who in 1324 made a pilgrimage to the Islamic holy city of Mecca. Along the way, he stopped in the Egyptian capital of Cairo and spent so much gold that he nearly wrecked the Egyptian economy. As tales of his wealth spread, he became the first sub-Saharan African leader to gain notoriety among western Europeans—some of whom later came southward, spurred by visions of gold in West Africa.

Background

Mali is not the only African geographical term that needs some clarification. Likewise, the Sudan is not to be confused with the modern nation of Sudan: *the* Sudan is an arid region of some 2 million square miles (3.2 million square km), about the size of the United States west of the Mississippi River. Located just south of the Sahara desert, it stretches from the Atlantic coast in the west almost to the Red Sea coast in the east. Farming is difficult there, but during the Middle Ages the region became home to a number of empires spanning trade routes across Africa.

The first of these was Ghana, yet another land not to be confused with the country that today bears its name: the modern nation of Ghana is located to the south of the medieval empire of Ghana. The latter came into existence during the fifth century in what is now southern Mauritania. Despite the climate, its people were originally farmers, but over time Ghana began to acquire wealth through conquest: by the eleventh century the empire had an army of some 200,000 men. Ghana became rich in gold, a metal so plentiful that the king's advisors carried swords made of it. The horses bore blankets of spun gold, and even the dogs had gold collars. The king, whose people considered him divine, held absolute control over the gold supply, and further increased his wealth by taxing trade caravans that passed through the area.

Ghana's capital was Kumbi-Saleh, formed from two neighboring towns. One of these municipalities became a center for Islam, a faith brought into the region by merchants from across the desert, while the other remained faithful to the native religion. Islamic practices were not permitted in public because they might challenge the spiritual authority of the king, and as it turned out, Muslims did destroy Ghana, though not from within: in 1080, the Almoravids of Morocco moved southward, bringing the kingdom to an end.

The next great power in the region was Mali, whose name means "where the king lives." The realm took shape under the leadership of Sundiata Keita (d. 1255), a figure whose biography is so filled with mythological elements—for instance, that he was crippled from birth but miraculously cured in his twenties—that it is hard to discern the exact details of his story. What is known is that beginning in about 1235, Sundiata led his people on a series of conquests, and established a capital in the town of Niana. By the fourteenth century, his dynasty ruled perhaps as many as 40 million people—a population two-fifths that of Europe at the time—in a region from the upper Niger River to the Atlantic.

Mansa Musa—or rather, Musa, since "Mansa" was a title equivalent to *highness*—was either the grandson or the grandnephew of Sundiata, and became Mali's ninth ruler in about 1307. As for his early life, little is known, though it appears likely that he was educated in the Muslim faith.

In his early years as a leader, Musa's devotion to Islam put him at odds with groups in Mali who maintained the traditional African religions. The latter were pagan, involving many gods, most of whom had some connection with nature (such as a sun god.) For the most part, however, Musa was able to avoid the sort of conflicts over religion that had affected the political climate in Ghana, primarily because he was a strong ruler and an effective administrator. His armies were constantly active, extending the power of Mali throughout the region.

Undergirding that power was the wealth of the nation's gold, wealth that in turn owed something to events far away. For many centuries following the fall of the Western Roman Empire in 476, Europe's economy had been weak; but beginning in about 1100—ironically, in part as a result of the Crusades against Muslims in the Middle East—the European economies had begun growing again. This growth created a need for gold coins, which drove up gold prices and, thanks to trade with Arab caravans on the Sahara, increased Mali's wealth. Like the rulers of Ghana before them, the dynasty of Sundiata Keita established a monopoly over the gold supply.

Gold wealth in turn spurred cultural advances under Musa's reign. Upon his return from Mecca, Musa brought with him an Arab architect who designed numerous mosques, as well as other public buildings. Some of those mosques still stand. Musa also encouraged the arts and education, and under his leadership, the fabled city of Timbuktu became a renowned center of learning. Professors came from as far away as Egypt to teach in the schools of Timbuktu, but were often so impressed by the learning of the scholars there that they remained as students. It was said that of the many items sold in the vast market at Timbuktu, none were more valuable than books.

In 1324 Musa embarked on his famous *hajj,* the pilgrimage to Mecca that all Muslims are expected to make at least once in their lives if they can afford to do so. He, for one, could certainly afford the trip: attended by thousands of advisors and servants dressed in beautiful garments, riding animals adorned with gold ornaments, Musa must have made a splendid figure when he arrived in Cairo. The Egyptian historian al-Omari later quoted a friend as saying, "This man spread upon Cairo the flood of his generosity: there was no person ... who did not receive a sum of gold from him. The people of Cairo earned incalculable sums from him.... So much gold was current in Cairo that it ruined the value of the money." Indeed, by spending so much gold, Musa caused an oversupply of the precious metal, and as a result, the value of gold plummeted throughout much of the Middle East for several years.

It is a hallmark of Musa's power as a leader that he could afford to be gone on the hajj, which took several years. In fact, while he was gone, his armies conducted a successful campaign against the powerful Songhai nation to the east. However, after his death in 1337 (some sources say 1332), none of his successors proved to be his equal, and later kings found the vast empire increasingly difficult to govern. Furthermore, they were plagued by religious and political conflicts, and by the mid-fifteenth century the Songhai, who rejected Islam in favor of their tribal religions, broke away from Mali and established their own state.

Impact

Yet even more powerful forces had been awakened far away—another unintended, and far more sinister, result of Musa's profligate spending. Europeans already had some idea of the vast gold supplies in Mali, but when rumors from Egypt began spreading westward, this sealed the fate of the African kingdom. Previously, European cartographers had filled their maps of West Africa with pictures of animals, largely creations of their own imaginations intended to conceal the fact that they really had no idea what was there. But beginning in 1375, these maps showed Musa seated on a throne of solid gold. It was the beginning of the end of West Africa's brief flowering.

When Portuguese sailors came to West Africa in the early to mid-fifteenth century, gold was among the commodities they found: hence the name for the region that became the modern nation of Ghana, the "Gold Coast." They also found ivory, and that too became part of a national title, Ivory Coast or Côte d'Ivoire. The name for the region to the west of the Gold Coast, however, would never become official for any country, though it serves to illustrate the worst outgrowth of the Europeans' arrival: the Slave Coast.

Elsewhere the fate of West Africa was symbolized by that of the fabled Timbuktu, which reached its peak under Songhai rule. Beginning in the fifteenth century, Europeans became increasingly fascinated by tales of a great city on the edge of the desert, which housed both wealthy merchants and scholars wealthy in knowledge. In 1470 an Italian journeyer became one of the first Europeans to visit, and more information surfaced with the publication of *Description of Africa* by Leo Africanus (c. 1485-c. 1554) in 1550. A series of wars and invasions by neighboring peoples during the early modern era, however, robbed Timbuktu of its glory. In 1828 a French explorer went to find the legendary Timbuktu, and in its place he found a "mass of ill-looking houses built of earth." Today it is known as Tombouctou, a town of some 30,000 inhabitants in Mali, and the name *Timbuktu* has long since become a synonym for a remote place.

JUDSON KNIGHT

Further Reading

Bovill, E. W. *The Golden Trade of the Moors.* London: Oxford University Press, 1968.

Davidson, Basil. *African Kingdoms.* Alexandria, Virginia: Time-Life Books, 1978.

Polatnick, Florence T. and Alberta L. Saletan. *Shapers of Africa.* New York: J. Messner, 1969.

Trimingham, J. Spencer. *A History of Islam in West Africa.* London: Oxford University Press, 1962.

Chinese Exploration: The Voyages of Cheng Ho, 1405-1433

Overview

Between 1405 and 1433 admiral Cheng Ho (1371-1433) commanded seven grand voyages from China to southeast Asia, India, Arabia, the Persian Gulf, and the eastern coast of Africa. To some western scholars, versed in the European voyages of exploration that profoundly affected much of the world's history, the voyages of Cheng Ho appear enigmatic. The voyages of Cheng Ho were significant undertakings that demonstrated China's impressive maritime technology and expertise, yet their impact was only short-lived. Undertaken with the aim of spreading China's imperial majesty to distant lands, these endeavors were to have very different consequences than the early European voyages of discovery, which took place soon after. Having achieved the aim of opening up trade and the flow of tribute from distant lands, the voyages suddenly ceased and private overseas trade banned by royal edict as China withdrew into itself.

Background

Cheng Ho was a eunuch and a military commander who had assisted the Yongle emperor, Zhu Di (1360-1424) to overthrow his nephew and become emperor. The fleets he commanded on the seven voyages were comprised of up to 317 ships, the largest of which were treasure ships, estimated to have been between 390 and 408 feet (119 and 124 m) long and more than 160 feet (49 m) wide. Some of the voyages included a crew of as many as 28,000 men. Although Cheng Ho was nominally in charge of all seven expeditions, he did not personally participate in all of them.

Historians suggest a number of reasons for the voyages. Part of the immediate impetus for the expeditions ordered by the Yongle emperor is said to have been the search for his nephew and predecessor, the Jianwen emperor, Zhu Yunwen (1377-?), whose throne Zhu Di had seized in 1402. There were rumors that Zhu Yunwen was still alive and living abroad, so, according to an unofficial history of the time, the emperor ordered Cheng Ho to search for him across the seas.

The purpose of the expeditions is best described as diplomatic. The size and grandeur of the expeditions, designed to inspire awe, expressed the majesty and power of Zhu Di and the dragon throne to distant lands. Although their mission was primarily peaceful, most members of the crew were troops who were well equipped to defend the fleet and its interests. The most dramatic example of this was the Chinese military victory in Sri Lanka on the third voyage (1409-1411) after a refusal to pay tribute. However, the presence of military weapons and soldiers was no doubt intended to display the might of the emperor and gain the allegiance and tribute of peoples without the use of actual force, as was indeed the case in the majority of places visited.

After the Yongle emperor died, the voyages of the treasure fleet ceased for six years. Then the Xuande emperor, Zhu Zhanji (1399-1435), ordered one final voyage in 1430 that also served a diplomatic purpose. As well as encouraging peace between Siam and Malacca, it intended to reverse a decline in the tribute trade and again display the majesty of the Chinese Empire, reinforcing the authority of the new emperor.

The voyages of Cheng Ho need to be understood in the wider context of Chinese seafaring and relationships with outsiders. Although his voyages were impressive for their scale and grandeur, they were not unique as diplomatic expeditions. Twelve centuries before his voyages, China carried out a diplomatic mission which spanned two decades and included visits to southeast Asia and the Arabian Sea, reaching as far as the eastern Roman Empire. Part of Marco Polo's (1254-1324) famous voyages can also be regarded as a precursor to the voyages of Cheng Ho, as Polo undertook a diplomatic mission as far as Persia in 1292 for Khublai Khan (1215-1294). This great Mongol ruler sent emissaries to Sumatra, Sri Lanka, and southern India and the Yongle emperor possibly attempted to emulate him.

There was no sudden technological breakthrough in Cheng Ho's time that made his voyages possible. Although his journeys demonstrated technology on an impressive scale, they used ship design and navigational techniques that had been developed in China many years earlier. The enormous treasure ships were based on earlier ship designs and were built in drydocks, which were used in Chinese shipbuilding some five centuries before their appearance in Europe at the end of the fifteenth century. The ships' hulls were divided into watertight compartments to give them strength, an invention that the Chinese had perfected by the end of the twelfth century. They also featured balanced rudders which gave them gave additional stability and facilitated steering. European shipbuilders did not use these innovations until the late eighteenth or early nineteenth century. Similarly, the compass had been commonly used as a navigational aid by Chinese seafarers since the thirteenth century.

Impact

The immediate impacts of Cheng Ho's voyages were primarily diplomatic and economic. He established the flow of overseas tribute from as many as fifty new places, underscoring the radiance of the emperor and the dragon throne, as well as stimulating China's overseas trade—indeed the voyages have even been credited with signaling an age of commerce in southeast Asia. Cheng Ho took with him cargoes including silk, porcelain, silver, and gold to offer as gifts to foreign rulers and exchange for luxuries, including spices and rare woods. He even built a transfer station in Malacca for trading purposes, an event unique in China's history. The spectacular porcelain pagoda built by Zhu Di at Nanjing from 1412, considered to be one of the seven wonders of the world by later European observers, is said to have been built using revenue from the voyages.

Cheng Ho had two tablets erected in 1431 documenting the achievements of his voyages. According to one of these, the Changle tablet, Cheng Ho believed that the achievements of "[t]he Imperial Ming Dynasty, in unifying seas and continents" surpassed those of previous dynasties. He added that "[t]he countries beyond the horizon and at the ends of the earth have all become subjects.... Thus the barbarians from beyond the seas, though their countries are truly distant ... have come to audience bearing precious objects and presents" for the emperor. The tablet also suggests that the voyages had made a significant contribution to Chinese geographic knowledge, allowing "the distances and the routes" of foreign lands to be calculated, "however far they may be."

However, the long-term consequences of the voyages were less impressive. Just at the point at which the Chinese had demonstrated their superior seafaring capabilities, the voyages ceased and the empire withdrew into itself. The strength of the Ming navy was greatly reduced over the following century and overseas trade outside the tribute system was banned. The tribute system itself declined. In 1477 another powerful eunuch named Wang Zhi wished to mount an expedition. When he asked for the official records from the voyages of Cheng Ho, the records were declared "lost" and his efforts were frustrated.

Such behavior may seem inexplicable to western scholars but it accorded with contemporary Chinese cultural beliefs and political climate. Internal conflict at court between the eunuchs and Confucian officials played a major role in creating this climate. Seafaring was traditionally the domain of the eunuchs while the Confucians adhered to an ethical code that regarded foreign travel and commerce as distasteful. By successfully stopping the voyages, the Confucians were striking a blow at their rivals. Moreover, they regarded the voyages to be a waste of the empire's resources and believed that China had no need of foreign curiosities. Indeed, there were economic and political factors that made the voyages seem less practical. There was severe inflation in the mid-fifteenth century and the empire's tax base shrank by almost half from what it had been at the turn of the century. In addition, the increased Mongol threat along the northern frontier diverted the empire's military resources away from coastal areas.

Unlike the European nations whose voyages of discovery gained rapid momentum in the sixteenth and seventeenth centuries, the Chinese were not interested in colonization. The difference between the experiences of Europe and China were economic and cultural rather than technological. As the voyages of Cheng Ho demonstrate, the Chinese certainly possessed the maritime technology and expertise to undertake long voyages of discovery. However, the Chinese were not interested in the wholesale exploitation of the resources of foreign lands, unlike subsequent European voyages of discovery. In Europe, such behavior was driven in part by the fierce competition between nation-states, which had fostered an attitude that encouraged the appropriation and adaptation of ideas and material resources from outside lands. China, however, believed itself to be self-sufficient and culturally superior to foreign lands, which meant it had no real need of outside resources, a belief that the voyages of Cheng Ho appeared to confirm.

PHILIPPA TUCKER

Further Reading

Duyvendak, J. J. L. *China's Discovery of Africa.* Arthur Probsthain: London, 1949.

Gang Deng. "An Evaluation of the Role of Admiral Cheng Ho's Voyages in Chinese Maritime History." *International Journal of Maritime History,* vol. VII, no. 2 (December 1995), pp.1-19.

Goodrich , L. Carrington, and Chaoying Fang (eds).*Dictionary of Ming Biography 1368-1644,* vol.1, Cambridge University Press: New York & London, 1976.

Levathes, Louise. *When China Ruled the Seas: The Treasure Fleet of the Dragon Throne, 1405-33.* Simon & Schuster: New York, 1994.

Mirsky, Jeannette (ed.). *The Great Chinese Travellers.* George Allen & Unwin Ltd: London, 1964.

Snow, Philip. *The Star Raft: China's Encounters with Africa.* Weidenfeld & Nicolson: New York, 1988.

Willets, William. "The Maritime Adventures of Grand Eunuch Ho." *Journal of Southeast Asian History*, Vol. 5, No. 2 (Sept. 1964), pp.25-42.

Foreign Exploration and Descriptions of India

Overview

During the latter part of the Middle Ages, India, China, and the European countries embracing Christianity were the prominent centers of the known world. For a brief period, exploration brought new geographic understanding and an exchange of ideas between the people and cultures of the East and the West. Arabic travelers and Christian pilgrimages provided important geographic and cultural information about China and India. Jordanus of Séverac, a French Catholic monk, traveled to India and wrote *Mirabilia* (Book of marvels, c. 1330), describing the geography and people. Biruni, a Persian scholar and scientist, wrote a book that discussed and described the history, geography and religion of India. Biruni's book was called *Kitāb fī tahqīq ma li'l-Hind*, translated and published as *India* (c. 1030), and later, *Al-Beruni's India*. Jordanus's account is still considered among the most valuable Western descriptions of India during the Middle Ages, and Biruni's work is considered today one of the greatest medieval works of travel and social analysis, and is still of great interest to scholars.

Background

The peoples of Asia and Europe had contact and knowledge of one another since the time of Alexander the Great (reigned 336-323 B.C.). Indirectly, goods from Asia arrived in Europe; spices were traded from Ceylon and Java, and carpets from Persia. Islam effectively lay like a wall between Europe and all of the trade routes to the East. However, by the beginning of the thirteenth century, Italian priests were established in the ports and cities of India and China. Because of the conquests of the nomadic Mongol people (known as the Tartars) and their roadbuilding, traders, merchants and monks moved unhindered on the great silk route across Central Asia or traveled through Persia to the coastal seaports to embark on the sea route. The Europeans could now take their Christian faith and trade directly to India.

One of the most interesting descriptions of India was given by Jordanus of Séverac, a French monk who traveled to India in 1321. His book, *Mirabilia*, or, *The Book of Marvels* discusses Jordanus's attempt to reestablish contact with the Christians living in India at the time, known as Nestorian Christian Missionaries. The book also contains a series of geographical as well as biographical sketches of his travels in India, Armenia, and Persia. Of India, he writes of the fabulous riches, the teeming bazaars, the weaving of cotton and muslin, the pearl fisheries of Ceylon, the abundant pepper and spice gardens, and the harbors filled with oceangoing ships. He describes India as being very hot, with exotic fruits like mangoes and coconuts, with multicolored parrots, elephants and crocodiles. He also speaks about the Hindu religion, with its sacred cows, enormous bronze idols and the Brahman caste system.

Jordanus also gives readers an interesting description of the Chinese boats, known as junks, that traveled from India to China. Constructed of fir and single decked, with fifty or sixty cabins and four to six masts, and powered by sail and oar crews of up to 400 men, these boats took travelers to Java, Sumatra, and Indochina on voyages lasting two years or more. The book contains accounts of the East Indies and Indochina, but it appears that Jordanus never personally visited these places. Instead, he repeats both accurate and inaccurate information that he had heard from other travelers.

A scholar from the East, Biruni (973-after 1050) from Persia, wrote an account of his travels in India nearly 300 years before Jordanus. He is among the most revered and renowned of the Persian scholars and one of the most outstanding scientists of all time. His scientific contributions to the body of scientific knowledge span the fields of history, geography, physics, chemistry, and astronomy. However, his most famous book, translated and published as *India*, was a study of India written early in the eleventh century.

Biruni served as the court astrologer and astronomer to the Mongol king, Mahmud of Ghazna. Biruni accompanied Mahmud on his conquests in India and for 20 years, he traveled throughout the area. It was during this time that Biruni learned Sanskrit and enjoyed a deep and scholarly exchange of ideas with Indian philosophers and scientists about the geography, chronology, medicine, literature, weights and measures, marriage customs, and astrology of the Indian people.

Impact

Early in the European medieval times, the arts, sciences, and philosophies dwindled and interest in exploration and discovery ceased. There were no expeditions sent out to acquire new geographic knowledge. Commerce and trade shrank, while people pursued theology and became involved in holy wars and crusades. What was left of knowledge and science found refuge in the monasteries. For most of Europe, the idea of a round earth, which the Greek astronomers had contemplated, became fiction as the world became a flat disc once again. Yet, this growing Christian Church and these conflicts and crusades introduced the exchange of ideas and cultures that would shatter the medieval world.

The opening up of Asia during the latter part of medieval times brought the rich cultures of India and China back into the European mind. This time period, although relatively brief, became the driving force for later exploration and offered unprecedented opportunities for geographic, cultural, and scientific contact between East and West. Travelers, monks, and scholars attempted to convey the richness and complexity of the geographic landscape, people, and religions of India, despite the fact that the worldviews of the Indians and the Europeans were very different.

Though the information about this period is scanty, it appears that Biruni served as a major catalyst in the exchange of East-West knowledge and philosophies early in the eleventh century. Biruni went into greater depth in his study of Indian religion, philosophy and literature than anyone before or after him. Biruni traveled and studied the Hindu culture for nearly 20 years. In return, he taught the Indian pundits Greek, Muslim sciences, and philosophy. Whether he was writing about astronomy, medicine, geography, or religion he attempted to see the world from the Indian perspective—an unusual attitude for his time.

Biruni was also a scientist, and he made constant use of observation measurement and, when possible, experimentation. Traveling through the plains of north India, he was struck by the nature of the soil and concluded that the entire area had once been a sea that had been filled in with alluvium. An interesting combination of observation, wide reading, and imagination led him to express an idea similar to Darwin's theory of natural selection. He also worked out the circumference of the earth to be 42,778 miles (about 68,844 km)—remarkably close to its true circumference. Another interesting characteristic of his method of studying Indian civilization was to compare Greek and Indian ideas. He felt that the Greeks were able to distinguish scientific truth and the Hindus were not.

As far as we know, Biruni was the first scholar to study the *Puranas*, vast collections of Indian stories about myths and gods. He was familiar with the *Mahabharata*, with its account of a great war, and the *Bahagavad Gita* finding it, as many others past and present, a guide to understanding the complex Indian religion.

Despite Biruni's unparalleled knowledge of Indian science, religion, and geography, his work in India had very little influence until the Persian historian Rashid-al-Din used it as a major reference in his book on the history of the world written in 1305. Not for over eight hundred years would another writer examine India with such thoroughness and understanding. His work, translated by the German scholar, E. C. Sachau in the late 1800s, from Arabic to German, and then to English has allowed many Westerners to appreciate the rigor of Biruni's writing.

Jordanus also displays a surprising ability for careful observation and description including a capacity for investigating and sifting facts, which was not very common among fourteenth-century writers. From him we do not get the sound and practiced scholarship of Biruni, but we do get a glimpse of the European medieval mind—one that blends mythical descriptions with realistic ones. He blends descriptions of dragons with elephants and unicorns while giving some of the first accurate descriptions of native plants and animals. Jordanus offered a gateway through which Europe was able to imagine the vast landscape of India and parts of Asia.

Through the writings and travels of individuals like Biruni and Jordanus, medieval Europeans had the opportunity to experience another part of the world completely different from their own. This intermingling of Eastern and Western ideas would provide fertile ground and a driving force for the coming centuries of exploration and discovery of the world.

LESLIE HUTCHINSON

Further Reading

Beazley, Raymond C. *The Dawn of Modern Geography,* Vol. III. New York: 1949.

Jordanus Catalani (de Séverac). *Mirabilia Descripta: The Wonders of the East.* Trans. Sir Henry Yule. London: Hakluyt Society, 1863.

Kimble, George. *Geography in the Middle Ages.* New York: Russell and Russell. 1968.

Newton, Arthur Perceival. *Travel and Travellers of the Middle Ages*. New York: Books for Libraries Press. 1926.

Sachau, Edward, translator. *Al-beruni's India*. New York: Norton. 1971.

Dinís Dias and Cape Verde

Overview

Dinís Dias was a Portuguese explorer who in 1445 conducted the first of two trade missions for Prince Henry the Navigator. In the course of his first voyage he explored the west coast of Africa, which was then unknown territory, sailing past Cape Bojador and the mouth of the Sénégal River to discover Cape Verde, the westernmost point of Africa. Dias called the area the "green cape" in honor of its tall trees and abundant plants. Dias significantly increased geographic knowledge of the West Coast of Africa.

Background

Prince Henry, son of the Portuguese King John I, was a key figure in Portugal's Age of Discovery. In 1415, he assisted in leading the Portuguese army in the conquest of Ceuta, a Muslim stronghold in North Africa. There he became interested in the geography, history and precious metals trade of western Africa. Prince Henry's impressions of the continent were based on rudimentary accounts and maps made by Arab geographers, which were sketchy and often wrong. For example, early maps showed the Sénégal River originating in a lake in central Africa, and claimed that this river was the source of the Egyptian Nile. The land beyond Cape Bojador, a tiny cape south of the Canary Islands, was marked *terra incognito*—unknown territory.

The Arabs called it *Abu Khatar,* "the father of danger," and Portuguese sailors feared it greatly. They knew its coast was too shallow for navigation, and believed that the water surrounding it "boiled." Conventional wisdom said that there were no water, vegetation, or people beyond that point, and that anyone foolish enough to venture beyond the cape would be lost forever. We know now that the constant north-south flow of the Canary current along the coast made it possible for ships to sail southward to West Africa but prevented a return voyage along the same route.

Undaunted, Prince Henry began sending expeditions southward along the west coast of Africa. Between 1424 and 1434 fifteen expeditions tried unsuccessfully to round the Cape. Finally, in 1434, Gil Eannes found that by sailing westward into the open sea, then turning east, he was able to round the Cape. This successful passage effectively ended the superstitions and fueled Prince Henry's confidence in his scientific approach to navigation. His crews continued to venture farther south towards the Sénégal River and the yet-to-be-discovered Cape Verde. New navigational aids such as the compass, astrolabe, and more accurate maps contributed to the success of later voyages.

After the breakthrough voyage around Cape Bojador Portuguese expeditions were immediately sent to explore rivers and coast to the south. Rivers such as the Sénégal and Gambia were important because they were an interconnected system of protected waterways that linked the rich interior trade routes of Africa to the coast. Some of these expeditions were successful in bringing back commodities such as oils and skins, but most were used for slave raids in the early 1440s.

In 1445 Dinís Dias was commissioned by Prince Henry to lead an expedition of caravels to the west coast of Africa. He sailed southward past the delta of the Sénégal River and reached a point of land he named Cape Verde, "the green cape," after seeing the first lush vegetation in nearly 800 miles (about 1,300 km). Although Dias did not realize it at the time, Cape Verde was the westernmost point on the African continent.

Sailing further south, Dias noted that the coast began to curve eastward, inspiring false hopes that he had rounded the African continent. When the Portuguese tried to land to go ashore, they were beaten back decisively by the natives. They returned a year later in another expedition sponsored by Prince Henry to explore the Sénégal River, then considered a western branch of the Nile.

Impact

Prince Henry the Navigator was a driving force behind Portugal's Age of Discovery, the period of

African coastal exploration that was to culminate with Vasco da Gama's voyage around the Cape of Good Hope and on to India. The number of works translated into Portuguese during the fifteenth century attest to Prince Henry's interest in all aspects of exploration. In 1419 he established a center for sailors, mapmakers, and other interested in navigation and discovery at Sagres, a town on the Portuguese coast.

At the school seamen were given instruction in new navigational techniques. Henry's captains were probably taught to use compasses, astrolabes, and maps. After each voyage they related all pertinent navigational and geographic information they'd acquired on the journey in order to update the charts for subsequent voyages. This system established a reservoir of useful and accurate information.

The Caravel

This kind of knowledge was necessary because sailing the open Atlantic was a different and far more difficult proposition than had ever been attempted. The Mediterranean was a long, narrow, enclosed sea where sooner or later a familiar landmark would come into view. The prevailing westerly and northwesterly winds made sailing along the coasts of Europe and Northwest Africa fairly easy. As exploration moved southward, however, reaching a specific destination meant learning about new wind patterns and ocean currents, as well as acquiring sophisticated sailing techniques and incorporating new ship designs.

Early Portuguese explorers sailed *barcas* and *barinels,* square-sailed ships that were clumsy and slow to respond. Prince Henry realized that a new type of ship and sail were needed to travel southward. Dias captained a newly designed ship known as a *caravelas* or caravel. This innovative design incorporated an axled rudder and three to four masts with triangular lateen sails, which gave the ship both speed and agility. The deck planks were with coated with shredded hemp and a layer of tar or pitch to improve water resistance.

This change in ship design was accompanied by improved navigation techniques. Prince Henry and other explorers had rediscovered Claudius Ptolemy's (c. 85-c. 165) *Geography,* a compilation of latitudes and longitudes of the known world known first published in 151 that had been translated from Greek into Latin in the early 1400s. Fifteenth-century astronomers used the *Geography* and other Arab works to produce almanacs showing the positions of stars and planets for each day of the year. This information became very useful in navigation during Dinís Dias's time.

The Astrolabe and Navigational Aids

Another important navigational tool was the *stella maris,* or astrolabe, that allowed navigators to read the height of the North Star. The quadrant, a quarter circle measuring 0° to 90° marked around its curved edge, was used to determine latitude. Its straight edges had tiny holes or sights located on each end and a plumb line hung from the top. The navigator lined up the holes and the plumb line would hang straight down over the curved area at a particular point. This would indicate the height of the stars in degrees and give the latitude. If, for example, a star measured 50° above the horizon, it meant they were at a latitude of approximately 50° north. Once mariners had gone beyond where the North Star was visible, they began to use the Southern Cross for reference.

Sailors would not be able to determine longitude until the chronometer was developed in the eighteenth century. In Dias's time sailors used an ancient technique known as "dead reckoning," to make a crude estimate of longitude by measuring the ship's speed, time, distance traveled, and direction. Dead reckoning also took observations of known landmarks, cloud formations, and wind and wave patterns into account.

In addition to better navigation, sea captains kept detailed notes of their expeditions for future use in *portolanos* or charts that showed bodies of water, landmasses, and ports. Compiled by masterful navigation and personal experience, these maps made the successful exploration of the coast of west Africa possible.

LESLIE HUTCHINSON

Further Reading

Bell, Christopher. *Portugal and the Quest for the Indies.* Constable and Compan, 1974.

Thorton, John. *Africa and Africans in the Making of the Atlantic World, 1400-1800.* Cambridge University Press, 1998.

University of Calgary. European Voyages of Exploration. "Technical Advances in Shipbuilding and Navigation." www.ucalgary.ca/hist/tutor/eurvoya/ship.html.

Diffie, Bailey W., and George D. Winius. *Foundations of the Portuguese Empire, 1415-1580.* Europe and the World in the Age of Expansion, volume 1. Minneapolis : University of Minnesota Press, 1977.

Biographical Sketches

Ingólfur Arnarson

fl. 874

Norwegian Explorer

Ingólfur Arnarson and his entourage were the first permanent settlers of Iceland, the only European country whose history has a definite beginning.

Two Norwegian Viking brothers, Bjornolf and Hroald, settled in the late eighth or early ninth century in Dalsfjord, Fjalar Province, Norway. Bjornolf's son, Orn, had a son, Ingólfur, and a daughter, Helga. Hroald's son, Hrodmar, was the father of Leif. The second cousins, Ingólfur and Leif, were the best of friends. They swore blood brotherhood, a solemn Viking bond by which each promised always to protect the other. Leif was also in love with Helga.

The blood brothers went on several raids with the three sons of Earl Atli the Slender. They got along well together until one of the earl's sons, Holmstein, began making advances to Helga. After Holmstein swore to marry either Helga or no one, Leif and Ingólfur killed him. Later they killed his brother, Herstein. They offered legal compensation, blood money, to the earl and his third son, Hastein, but Atli and Hastein demanded all their possessions instead. Rather than be reduced to poverty, Leif and Ingólfur chose to become Earl Atli's outlaws. That meant that either the earl or anyone in his service could kill Leif or Ingólfur anytime with impunity.

The outlaws sailed for Iceland, which was then uninhabited. Through stories of the voyages of two Swedes, Naddoddur and Gardar, and a Norwegian, Raven Floki, Norwegian Vikings had known of Iceland's existence for about 20 years. After preliminary reconnaissance, Leif and Ingólfur spent a winter at Ingolfshofdi in southeastern Iceland and determined that it was fit land for settling. Ingólfur returned to Norway to gather money, relatives, and friends for the emigration, while Leif raided in Ireland, plundering money, slaves, and a magnificent sword. There after he was called Hjorleif (Sword-Leif). About this time he married Helga. Hjorleif and Ingólfur joined forces in Norway, then sailed again for Iceland around 874.

A custom of Viking chieftains about to settle a new land was to throw overboard their highseat pillars, an emblem of rank, and then to settle wherever the pillars washed ashore. Sometimes this procedure would take quite a while. Meantime, the settler would either stay in his ship or erect a temporary dwelling. He would send slaves or servants to scout for the pillars. It was considered a very bad omen if the pillars drifted out to sea.

Ingólfur released his pillars as soon as he sighted land. He went ashore at Ingolfshofdi and awaited the report of his scouts. Hjorleif continued sailing west until he landed at Hjorleifshofdi near Myrdalsjökull in southern Iceland. The following spring his Irish slaves revolted, killed him, and fled to the Vestmanna Islands, where Ingólfur tracked them down and killed them.

Ingólfur and Hjorleif disagreed about religion. Ingólfur frequently sacrificed to Thor and the other Norse gods, but Hjorleif never sacrificed. Ingólfur rationalized that Hjorleif's murder was the just deserts of someone who refused to sacrifice.

Two years after the death of Hjorleif and three years after landing in Iceland, Ingólfur's scouts found his highseat pillars on the south shore of a beautiful fjord in southwestern Iceland. He named the place Reykjavík (Smoky Bay) and built his homestead there. Within the next 60 years about 30,000 more Norwegian refugees settled in Iceland.

Ingólfur freed the two slaves, Vifil and Karli, who located his pillars. He married Hallveig, daughter of Frodi, and became a beneficent chieftain in the new country. His son, Thorstein, founded the Kjalarness Assembly, one of the regional assemblies that preceded the national assembly, the Althing.

ERIC V.D. LUFT

Rabban Bar Sauma

c.1220-1294

Chinese Nestorian Monk and Explorer

In the late 1280s, a Nestorian Christian monk named Rabban Bar Sauma took the opposite route of many of his contemporary explorers by venturing from his homeland in China to western Europe. He and a student also made a trip to Persia and Iraq.

Statue of Ingolfur Arnarson. *(Nik Wheeler/Corbis. Reproduced with permission.)*

Bar Sauma was born around 1220, but accounts differ about his place of birth, noting it as either Chung-tu, modern-day Beijing, or Khanbaligh, which lay to the northeast. He followed his Christian upbringing and when he was in his twenties, became a monk in the Church of the East, or the Nestorian Christian Church. The church traces its origins to around the time of Christ's crucifixion, when the disciple Thaddeus (Addai) went to Edessa, an ancient city in Mesopotamia (now in Turkey), to spread the Christian faith. The religion radiated east from there. Although it had a Christian foundation, Nestorianism was not revered in the West, where the religion and its creed were historically viewed as heresy by the Western Christian church. Nonetheless, the Nestorian Church was influential in the East and had many followers.

After taking his vows as a monk, Rabban Bar Sauma gained a reputation as a religious teacher and as an ascetic. Eventually, he set out with a student, a young monk named Markos (Rabban Markos), to make a pilgrimage to the religious center of Jerusalem. Although they never arrived at their destination, due to violent local skirmishes, they did venture to eastern Persia (now Iran) to meet the Catholicus, leader of the Nestorian Church. After a short visit to Iraq, Bar Sauma and Markos followed the instruction of the Catholicus and embarked on a trip to see the Persian Mongol ruler Il-Khan. During these travels, the Catholicus made Markos a Nestorian bishop. As Markos and Bar Sauma were preparing for a return trip to China as emissaries of the Nestorian Church, the Catholicus died, and the Nestorian bishops elected Markos as his successor.

Several years later, a Mongol leader named Arghun met with Markos to seek ways to expand Christian support while suppressing any Muslim stronghold, particularly in Jerusalem. Markos recommended that Bar Sauma represent the church on a diplomatic expedition to Europe. Now in his sixties, Bar Sauma began the trip from Iraq in 1287. He traveled by land through Armenia, then by water to Constantinople and later to Naples. Once he finally set foot in Rome, he wasted no time in reaching the Holy See where he engaged in many religious debates with church leaders. During this expedition, he also made stops in Tuscany, Genoa, Paris, and Gascony, and met world leaders, including King Philip IV of France, King Edward I of England, and the new Pope Nicholas IV. As reported in "By Foot to China" by John M.L. Young, "The pope sent back a reply with Rabban Sauma to Markos, who ruled under the patriarchal name of Mar Yaballaha III, in which he confirmed the latter's 'patriarchal authority over all the Orientals.' For one brief period of history, sovereigns either espousing Christianity or friendly to it reigned from the Atlantic to the Pacific across Europe and Asia."

While on this mission, Bar Sauma confidently reported to Markos of firm Christian support from the West. While various leaders, including King Edward I, rallied behind the cause proposed by Markos, they were unable to generate a united front among the European nations and took no formal action against the Muslims.

Bar Sauma's diplomatic, political, or religious maneuvers, however, are not as important to history as his writings. During his 1287-88 trek, he wrote a detailed account of his adventure. This account is one of the only known, nonnative descriptions of Europe during the period, and provides a unique look at the medieval West.

LESLIE A. MERTZ

Jean de Béthencourt

c. 1360-c. 1422

French Explorer

Known as "the Conqueror of the Canaries," Jean de Béthencourt claimed the Canary Islands for Spain. His 1402 expedition, which he undertook with Gadifer de La Salle, was the opening chapter in a series of voyages southward and westward that would ultimately take the mariners of Spain and Portugal to sub-Saharan Africa, India, and the New World.

At that time people believed that the Canaries represented the edge of the known world—and indeed the islands, situated in the Atlantic Ocean off the northwestern coast of Africa some 823 miles (1,324 km) from the southwestern coast of Spain—were just that. The ancient Greeks and Romans had known them variously as the Garden of Hesperides, the Elysian Fields, and the Fortunate Isles, and during the Middle Ages sailors from the Arab world, as well as France and the Iberian peninsula, had visited the islands. There they had found a Stone Age people known as the *guanches,* who practiced very basic agriculture and herding but used sophisticated Egyptian-style embalming techniques.

Béthencourt himself was neither Spanish nor—though he hailed from France—purely French: he was a descendant of the same Norman stock that had produced William the Conqueror. On a crusade in Tunis in 1390, he met French soldier Gadifer de La Salle (fl. c. 1340-1415). Together the two men developed the idea of an expedition to the Canaries, and Béthencourt pledged his finances to pay for the voyage.

The two men and their crew set sail from La Rochelle, France, on May 1, 1402. They arrived in the Canaries in June, and though they had little trouble subduing the natives, they soon found themselves in need of supplies. Therefore both agreed that Béthencourt should go to Spain and seek help. France, to its detriment, took little interest in exploration during those early years.

Béthencourt returned 18 months later, having received the financial support he needed from King Henry III of Castile. During the preceding centuries, Castile had been emerging as one of the dominant powers among the various Christian principalities of the Iberian peninsula, and would eventually join forces with Aragon to unite all of Spain. No doubt Henry saw the Canary expedition as a means of flexing Castilian muscle, and he even obtained a bull from the antipope Benedict XIII recognizing the Canaries as the property of Castile.

Henry also bestowed on Béthencourt the title "King of the Canary Islands," which did not sit well with Gadifer. The latter had spent the preceding year and a half exploring, and consolidating European control over, the islands. Now he was so incensed he demanded that Béthencourt go with him back to the king, who would arbitrate the matter. Henry decided in Béthencourt's favor, and Gadifer went home to France.

The recognition Béthencourt has received as "Conqueror of the Canaries" is thus not entirely deserved. He later used Norman peasants from France to colonize the islands, but even this idea had been Gadifer's. In 1406 he placed his nephew, Maciot de Béthencourt, in charge of the islands, and returned to France.

JUDSON KNIGHT

Giovanni da Pian del Carpini

1182-1252

Italian Explorer

Long before Marco Polo (1254-1324), there was Giovanni da Pian del Carpini, an Italian priest sent by Pope Innocent IV on the first European mission to the court of the Mongol's Great Khan. Two years before Polo was born, Carpini returned from a journey that had taken him to many of the same far eastern territories later visited by the Venetian adventurer.

Born in the village of Pian del Carpini near Perugia in Tuscany, Carpini was the same age as St. Francis of Assisi, and when the latter formed the Franciscan order of monks in 1209, Carpini became one of its first members. He later served on Franciscan missions that took him all over Europe: to Germany, Spain, Bohemia (now the Czech Republic), Hungary, Poland, and Scandinavia. By the age of 63, he had become exceedingly obese, which made traveling difficult for him; yet it was then, in 1245, that the pope commissioned him to the task for which he became famous.

In 1241, Mongol invaders had very nearly taken Vienna, and though they had relented—their Great Khan, or leader, had died, and the choosing of a successor took the Mongols' attention from the attack on Europe—the pope understandably feared that the invaders would return. Therefore he decided to send a mission eastward, protesting the invasion of Europe and offering to send Christian missionaries to the Mongols. Though Carpini was on in years, his wisdom and experience made him an ideal candidate to lead the expedition.

Despite his infirmities, there is no record that Carpini complained about the hardships of the journey, which began in Lyons, France, on Easter Sunday, April 16, 1245. The immediate destination was Russia, controlled by a Mongol force led by Batu Khan and known as the Golden Horde, and to get there the delegation passed through Bohemia, Poland, and Ukraine. Finally on April 6, 1246, the group reached Batu's camp on the banks of the Volga.

Carpini was surprised to find Batu little-impressed by the fact he had come on a mission from the pope, and the Mongols' treatment of the delegation was far from gracious. This was unusual, given the Mongols' habit of respecting the religions and religious leaders of other cultures. In any case, Batu did provide the mission with an introduction to the court of the Great Khan, and they set out again later in April.

The travelers' route took them through inhospitable lands: the plains of southern Russia, the deserts north of the Caspian Sea, harsh central Asian regions controlled by Muslims who submitted to Mongol rule, and finally across the Altai Mountains. After passing the latter, located on what is now the border between Russia and Kazakhstan, the expedition entered territories directly controlled by the Great Khan, whose court they reached on July 22, 1246.

Not long afterward, the Mongols elected a new Great Khan, Kuyuk, and Carpini reported on the lavish celebration held in his honor, attended by some 4,000 representatives of nations from throughout the known world. After a few months' stay, on November 13, 1246, the mission set out on its return journey with a message from Kuyuk to the pope. The response was not the one Innocent had hoped for: in his letter,

Kuyuk demanded that the pope submit to *him* as the representative of God on earth.

Though they endured considerable hardships in their wintertime mountain crossing, the delegation found themselves received much more warmly than before when they reached Batu's camp on May 9, 1247. A month later, they also found a joyous reception among the Russian Christians in Kiev. Carpini reported to the pope in Lyons on November 18.

In later years, Carpini went on an unsuccessful papal mission intended to persuade France's Louis IX (later St. Louis) to postpone a Middle Eastern crusade. He was also appointed bishop of Antivari on the Adriatic coast, but, after his appointment was disputed by the local archbishop, he returned to Italy and died there on August 1, 1252. As for the feared Mongol invasion of Europe, it never came: when the Mongols under Kuyuk resumed their attacks, their target was the Muslim world, where in Nazareth in 1260 they suffered their first serious defeat at the hands of the Mamluks.

JUDSON KNIGHT

Cheng Ho

1371-c. 1433

Chinese Admiral and Explorer

Cheng Ho was one of the greatest early explorers, expanding the predominance of his homeland China into many foreign lands, particularly those in or bordering on the Indian Ocean. After his death, however, powerful Chinese officials became increasingly isolationist, eventually destroying many of the records documenting his travels.

Cheng Ho (also known as Zheng He) was born in 1371 in China's Kunyang, Yunnan province, under the name of Ma Sanpao. Ming troops captured him when he was ten years old and sent him as a household servant to a prince named Chu Ti (Zhu Di). Like the other local children captured by the army, Cheng was castrated.

Over the next two decades, Cheng's duties grew, and he soon began following the prince into battle against the Mongols. Through these engagements, Cheng made a name for himself as a military leader. In 1402, the prince led a successful revolt against the throne and became Ming emperor, with the new name of Yung-lo. Yung-lo repaid Cheng for his accomplishments by giving him the title of Imperial Palace Eunuch and making him commander of an overseas expedition that, at least in part, was to search for the fleeing previous emperor Hui-ti.

Cheng prepared for the voyage and set sail in 1405 with his Grand Fleet of 28,000 men and more than 300 vessels, some nearly 450 feet (137 m) long. Carrying various valuable commodities, including fine silk, the fleet left the Yangtze River and headed into the South China Sea for what would become a two-year trading expedition. During the voyage, Cheng entered into a conflict with Chen Tsu-i, a pirate and ruler of Sumatra. Cheng was victorious in battle, and sent the pirate as a prisoner to meet his fate back in Nanjing.

Over the next four years, Cheng made two more voyages. In these trips, he revisited the spice capital Calicut in India, made side trips to Thailand, Java, Malacca, Sumatra and Sri Lanka, and offered gifts from the Ming royalty to the foreign leaders. After a two-year break from his naval expeditions, Cheng made his fourth voyage from 1413-1415 to Sri Lanka, Bengal, Maldive Islands, Persia and Arabia. On his next two voyages in 1417-1419 and 1421-1422, his fleet sailed to Ryukyu Islands (near Japan), Borneo, Kenya, Zanzibar, Tanzania, Mozambique and Somalia.

During these voyages, Cheng traveled an amazing 30,000 miles (48,280 km) across the ocean's waters. It was not until the voyages of Portuguese and European adventurers nearly 100 years later that these seas would be so thoroughly explored. In all, Cheng visited nearly three dozen countries and opened diplomatic relations with the leaders of many of them. His travels greatly extended the Chinese influence throughout the world and helped to make the country a world power. Despite these achievements, the expeditions had their detractors.

When Emperor Yung-lo died in 1424, Cheng lost support for his travels. The Chinese officials who took control felt that the benefits of the voyages were outweighed by their high costs, and halted further expeditions. They could not see any advantage to forging relationships with the governments or the trading industries of foreign lands. The officials made Cheng commander of a garrison in Nanking, where he remained for several years.

Cheng's days of adventure weren't over, however. In 1431 new Emperor Chu Chan-chi (Hsüan-te) approved one last voyage for the Grand Eunuch. Cheng prepared to set sail the following year for travel to the east coast of

Africa. While on this voyage, Cheng died. His crew brought his body back to China for burial in Nanjing. Conservative Chinese government leaders, who disfavored eunuchs, opposed trade and championed isolationism, later destroyed all official records of Cheng's voyages. With China no longer a naval power, European countries now had access to the Indian Ocean and began their explorations.

LESLIE A. MERTZ

Niccolò de' Conti

c. 1396-1469

Italian Explorer

For a quarter-century beginning in 1419, Venetian merchant and adventurer Niccolò de' Conti engaged in an adventure recalling that of his countryman Marco Polo (1254-1324). Conti's journeys took him primarily along sea routes, from the Middle East to India to the East Indies, and when he returned to Europe he found himself an object both of fascination and (because he had renounced Christianity) disapproval.

Conti came from a wealthy trading family in Venice, then Europe's center for overseas trade. In the past his family had done a brisk trade with Egypt, but in 1419 Niccolò resolved to extend the Contis' reach by traveling to Syria. He remained in the latter country long enough to learn Arabic, one of the essential languages for Eastern trade in those days; then he traveled on to Baghdad, in what is now Iraq, and Persia (modern-day Iran). In the latter he established a trading company, and added a second key trading language, Farsi, to his repertoire.

At some point it must have become clear to Conti that he was not returning to Europe any time soon, because over the course of many voyages from Persia along the Indian Ocean, he married an Indian wife. They raised a family, and he renounced Christianity, though whether he did this out of necessity or conviction is not known.

In time he wound up much further east, where he traveled throughout the East Indies (modern-day Indonesia) and Southeast Asia. There he encountered numerous cultures and religions: parts of the East Indies were Hindu, others Buddhist—as was Burma, which he also visited. The Malay Peninsula, home to the modern-day nation of Malaysia, was fiercely Muslim. He ended up on the island of Java, from whence he finally resolved to return to Venice.

After a return trip that took him up the Red Sea and the Gulf of Suez, Conti arrived in his hometown in 1444. There his Indian family attracted widespread curiosity, but his renunciation of Christianity earned him a rebuke from Pope Eugene IV. The punishment Eugene meted out as penance was not a harsh one, however: the pontiff ordered Conti to write an account of his journeys. These records, dictated to Vatican secretary Poggio Bracciolini, were not publicly released until their 1723 publication as *Historiae de veritate fortunae.*

JUDSON KNIGHT

Gil Eannes

fl. 1433-1445

Portuguese Explorer

Viewed purely in geographical terms, Gil Eannes's 1434 rounding of west Africa's Cape Bojador was a small achievement. From a psychological standpoint, however, his voyage had an enormous impact, opening the way for other European expeditions further southward.

A native of Lagos, a town in southern Portugal, Eannes had grown up in the service of Prince Henry the Navigator (1394-1460), in whose household he served as a squire. Henry in 1420 established a navigational "school," sponsoring voyages of exploration southward and westward to islands such as the Madeiras, the Canaries, and the Azores, as well as the western coast of Africa.

The latter presented a challenge, however, in the form of Cape Bojador, a 25-mile (40-km) westward projection from the mainland just below latitude 27° north. Hazardous weather conditions and shallows associated with the cape made it difficult to pass, and Portuguese sailors were inclined to believe Arab geographers who called the region beyond Cape Bojador the "Green Sea of Darkness"—that is, the point of no return.

Frustrated in his attempts to find a mariner brave enough to attempt the voyage, Henry in 1433 ordered Eannes to undertake the expedition. No doubt spurred on more by a sense of duty than of curiosity or adventure, Eannes came as close as the Canary Islands before returning to Portugal. In the following year, 1434, Henry sent him out again with exhortations to ignore all the legends and establish a name for himself: "Make the voyage from which, by the grace of God, you cannot fail to derive honor and profit."

This time Eannes rounded the cape, where he found not the edge of the world, but a calm sea along a desert coast. He landed and collected what little plant life he could find—a species thenceforth known to the Portuguese as "St. Mary's roses"—and brought them back to Henry's court at Sagres in Portugal.

A year later, Eannes sailed a bit further down the coast past Cape Bojador, where he and fellow explorer Alfonso Gonçalves Baldaya saw signs of life: human footprints and camel tracks. Baldaya continued further south, where he traded for sealskins, the first commercial cargo brought back to Europe from West Africa. By then, Baldaya was not the only man daring to venture ever southward along the African coast. Thanks to Eannes's pioneering voyage, the way was opened for more and more Portuguese mariners to make their way past Cape Bojador without fear of annihilation.

As for Eannes, he last appears on the pages of history on August 10, 1445, when he left Portugal with an armada of caravels bound for the island of Tidra off the coast of what is now Mauritania. There the Portuguese forces did battle with the Muslim inhabitants, taking 57 captives. Eannes himself may have died in the fighting; or he may have been part of the group that left Tidra after the battle to sail south to Cape Verde, at that time the furthest point of Portuguese exploration along Africa's west coast.

JUDSON KNIGHT

Erik the Red

c. 950-1004

Icelandic Explorer

First European to land on Greenland, Erik the Red gave that forbidding island its somewhat deceptive name, and founded a European colony there that would remain operative for more than three centuries. His son, Leif Eriksson, launched an even more ambitious expedition from Greenland, to what became known as the New World. The two men rank as the greatest Viking explorers.

Not long after Erik was born, near the Norwegian town of Stavanger, his family had to leave their home and move hundreds of miles west. His father, Thorvald Asvaldsson (Erik's surname was Thorvaldson) had become involved in a blood feud, and had killed a man. As punishment, he and his family were banished to Iceland, which had been settled by Vikings in the 870s.

Growing up in a remote area of western Iceland no doubt helped Erik learn the survival skills that would later stand him in good stead. He did not, however, learn from his father's mistakes: after he grew up, married, and moved to another part of the island, he too became involved in a blood feud, and killed two men. He was therefore banished from Iceland—itself originally a place of banishment—for three years.

Having heard tales of islands to the west of Iceland, and of a larger island beyond those, Erik set sail with a group of other men in 982. After passing the islands, Gunnbjörn's Skerries, he landed on eastern Greenland. The effects of the currents' flow made the island's eastern portion less hospitable than the western half, and Erik soon sailed around the southern tip of Greenland at Cape Farewell before landing on the southwestern coast. There he founded what came to be known as the Eastern Settlement, modern-day Julianehåb or Qaqortoq.

Erik wintered at a place he named Erik's Island; then in the spring of 983 he sailed up what came to be known as Erik's Fjord. After another winter on Greenland's southern tip, he sailed up the eastern coast before spending the winter of 984-985 on Erik's Island. His period of banishment ended, he returned to Iceland. His reappearance seemed to spark the blood feuds again, but this time he appears to have learned from past mistakes, and began promoting colonization of the newly discovered land as a way of getting away from the violence in Iceland. In order to make his discovery sound more hospitable, he gave the large island the somewhat deceptive name of "Greenland."

In 986, Erik set sail for Greenland with 14 ships and nearly 500 people, along with their animals and belongings. He and his family took up residence in the Eastern Settlement, while other colonists established the Western Settlement (near modern-day Godthåb or Nuuk) and another one between the two larger colonies.

By then Erik had three sons, Leif, Thorvald, and Thorsteinn, as well as an illegitimate daughter, Freydis, who married a man named Thorvard. Though Leif was destined to become the most famous of Erik's children, all would play a role in later Norse expeditions to North America.

Set in his ways, the father maintained his belief in the old Norse gods, and took issue with Leif's conversion to Christianity, which occurred

Erik the Red. *(Corbis Corporation. Reproduced with permission.)*

around 999. But in 1001 or 1002, when Leif set off for Vinland, as he called North America, it stirred the spirit of adventure in Erik, who hoped to go with him. On his way to the ship, however, Erik fell off his horse, and the resulting leg injury prevented him from making the trip.

Erik died in the winter of 1003-1004, but the settlements he had founded outlasted him by many years. Changing climatic conditions, however, eventually threatened Europeans' livelihood from fishing, and increased ice made ocean travel more difficult. Finally in 1350 the last settlers gave up to the Inuit, Greenland's native inhabitants. It would be more than 200 years before Sir Martin Frobisher (c. 1535-1594) in 1576 became the next European to set foot on Greenland.

JUDSON KNIGHT

Genghis Khan

1167-1227

Mongol Conqueror

In 1206, the nomadic Mongol tribes of northern Central Asia united for the first time, and chose as their leader a chieftain, nearly 40 years old, by the name of Temujin. Long before, he had been given the title by which history knows him: Genghis Khan, or "rightful ruler"; now his compatriots declared him "ruler of all men." During the two decades that followed, Genghis would live up to that title, laying the foundations for the largest empire ever known.

According to legend, Temujin came into the world grasping a lump of clotted blood, a sign of the forcefulness and violence that would dominate his life. His father, a chieftain named Yesugei, was poisoned by Tatars, a rival nomadic group in the region, and thereafter Temujin's mother Ho'elun managed to keep her family alive only through sheer will and resourcefulness. As designated heir to Yesugei, young Temujin was in a particularly vulnerable position, and for many years the family hid from Targutai, a leader of another clan who had seized all their possessions.

At the age of 14, Temujin survived an assassination attempt at Targutai's hands, and in the process recruited his first subordinates, men who would later hold places of honor in his army. He soon married a girl named Borte, a marriage that had been arranged long before by his father. When Borte was kidnapped by enemy tribesmen, Temujin called on the aid of two men: Jamukha, a childhood friend; and Toghrul, a powerful chieftain who had once been an ally of his father's. The three succeeded in rescuing Borte, but when a group of clans in 1187 proclaimed Temujin "Genghis Khan" (sometimes translated as "rightful ruler"), this created tension with Jamukha.

Over the years that followed, this enmity grew, and in time Jamukha and Toghrul formed an alliance against Genghis. Genghis had difficulty organizing an army to deal with his two enemies, but with a small force he eliminated first Toghrul, then Jamukha. It was at that point, in 1206, that he united all Mongols under his rule.

It was as though the years of conflict with his countrymen had created an unstoppable momentum in Genghis, who then embarked on a campaign of conquest that seems to have had no immediate cause. His first target consisted of enemy tribes around him; then in 1211, his forces assaulted China. By 1213, they had crossed the Great Wall, spreading out through northern China; and in 1215, Genghis's hordes sacked Peking.

In those early years of conquest, Genghis had simply killed everyone who stood in his way. But the gradual absorption of China added an aspect of sophistication previously lacking in the Mongol tactics. In particular, a former official of the Chinese emperor pointed out to Genghis that if he allowed some people to continue living in the lands he conquered, they could pay him valuable tax money to finance further warfare. Genghis accepted this sound advice.

Certainly Genghis was much more of an empire-builder than he was an administrator, and he placed Mongol-controlled portions of China under the control of a general while he moved into Central Asia. In their pursuit of Küchlüg, a rival chieftain who had fled to Afghanistan, the Mongols had found themselves faced with Sultan Muhammad, who controlled much of Persia, Afghanistan, and neighboring Central Asian territories. After Muhammad's forces executed a group of Genghis's ambassadors, war was virtually inevitable.

Genghis began moving westward in 1219, and by the spring of 1220 had taken Bukhara and Samarkand. After they destroyed Sultan Muhammad, Genghis's troops kept going, swarming into the Caucasus and thus beginning the centuries-long occupation of Russia under the Golden Horde. Genghis himself occupied what is now Tajikistanin the winter of 1220-21, and in the latter year laid waste to the ancient city of Balkh in Khurasan. By summer he was on the shores of the Indus River, dealing a decisive blow against Muhammad's son Jalal al-Din.

Genghis returned to Mongolia in the spring of 1225, and was soon embroiled in another conflict with yet another tribe. It was during this campaign that he died at age 65 on August 18, 1227, of complications resulting from falling off a horse. By this point, the Mongols controlled a region that stretched from the borders of Turkey to Russia to northern India to China—an empire already as large as that of Rome at its peak. In the years that followed Genghis's death, the Mongol realms would stretch from the Korean peninsula to the outskirts of Vienna, and from Siberia to the Indian subcontinent.

Thus Genghis, known to history as a bloodthirsty and merciless conqueror, actually provided the world with an orderly, unified governmental system. For the century that followed, Mongol rule made travel between East and West relatively safe, which it had not been since the declining days of the Roman Empire. This in turn made possible journeys of discovery—most notably that of Marco Polo (1254-1324)—that added greatly to Europeans' emerging knowledge of the world around them.

JUDSON KNIGHT

Prince Henry The Navigator

1394-1460

Portuguese Prince, Explorer, and Navigator

Prince Henry of Portugal did not earn the title "The Navigator" because he himself sailed the seas. In fact, Henry did not venture out on any of his country's many expeditions. But his quest to establish Portugal as one of the wealthiest trading nations in the world drove his country's ships further down the African coast than any previous European missions. During his lifetime, Prince Henry not only made significant advances in navigation and shipbuilding, he helped establish Christian Europe's authority over Africa and Asia, while breaking down Muslim control over trade and sea routes.

Prince Henry was born in 1394, the third son of the Portuguese monarch King John I and his wife Queen Philippa, sister of England's King Henry IV. He was taught to be a great statesman and soldier, and proved himself as both in 1415, when he led the Portuguese army in the conquest of Ceuta, a Muslim stronghold in Morocco. At the time, the Moors of North Africa ruled most of the Iberian peninsula (Portugal and Spain). Europe's Christians, including the Portuguese, were determined to drive these Muslims out.

Portugal's conquest of Ceuta was not only a defeat over the Muslims—it established Portuguese control of shipping lanes across the Mediterranean Sea. All trading vessels from Africa and Asia would now have to pass through Portuguese-controlled waters. To celebrate their victory, the king knighted Prince Henry, along with his brothers, Princes Duarte and Pedro.

But once King John gained control of Ceuta, traders who had no desire to do business with the Christians began to take their business elsewhere. And there was more trouble for Portugal: Ceuta's Muslim neighbors were threatening the Portuguese stronghold. The king sent his son Henry to aid the embattled city.

In 1418, while serving as governor of Ceuta, Prince Henry began learning the trade routes to Africa and Asia, believing that control of these sea routes could offer Portugal unprecedented wealth. That year he returned to Portugal, setting up his new home on the tip of the Iberian peninsula, the Cape of Sagres. There, he began to plot a Portuguese expedition along Africa's west coast, but knew that no European ships had ever sailed as far down the coast as he was planning.

Henry the Navigator, Prince of Portugal. *(Library of Congress. Reproduced with permission.)*

As a child, the prince had heard the legend of a man named Prester John, a Christian king who was rumored to rule a huge empire in either Africa or Asia. If he could find this man he could, he thought, create an alliance that would conquer Muslim influence in Africa, and make Portugal one of the richest nations on earth.

Besides securing wealth, Henry had an additional motive in his quest to conquer Africa. As a devout Christian, he strove to convert the region's pagans to Christianity. He was promoted in this endeavor by the Order of Christ, a supreme ministry under the Pope himself. The order made him a grand master, which required him to lead a chaste and ascetic life, but in exchange, they promised to bankroll his voyages.

Also aiding Henry were the best mapmakers and navigators in Europe, who helped him make significant improvements to several navigational tools. Henry and his scholars invented a portable version of the circular astrolabe, which measured the angle of stars above the horizon, and improved upon the triangular quadrant, which measured the height of the sun and stars above the horizon, as well as the compass. The latter two devices helped sailors pinpoint their correct latitude, or position, in relation to the equator.

Prince Henry was also frustrated with the slow, clunky ships that were available at the time. He had his shipbuilders create a faster, more maneuverable ship that would travel the ocean with ease, called a *caravel*.

After 1418, Henry began his explorations in earnest, sending his ships to the south, where his captains discovered the island of Santo Porto. On the following voyage, his vessels traveled just beyond Santo Porto and found the island of Madeira.

Henry sponsored voyage after voyage, each of which traveled a bit further down Africa's coast. In 1427, Henry's ships discovered and took control of the Azores, a group of islands approximately one thousand miles (1,600 km) due west of Sagres. He yearned to conquer even more of Africa, but his captains were always afraid to venture further south than Cape Bojador for fear that their ships would be caught up in the region's dangerous currents.

Finally, in 1434, Henry convinced his squire Gil Eannes to complete the voyage. With Cape Bojador in sight, Eannes steered his ships westward, then made an arc back towards land, finally making it to the south side of the Cape. Now that the boundary had been crossed, Prince Henry's expeditions began to travel faster and further down the African coast.

In 1441, a caravel returned to Portugal with gold dust and slaves. Henry's captain Dinís Dias in 1445 reached the mouth of the Sénégal River, and a year later Nuño Tristão discovered the Gambia River. By 1448, Henry had established an armed fort in Arguim Bay to handle Portugal's growing gold and slave trade along the African coast.

By the time of his death in 1460, Henry the Navigator had sent his ships further down the coast of Africa than any previous European sponsored expedition. He had established for Portugal a thriving sea trade with Africa, and the nation boasted colonies all across the Atlantic.

STEPHANIE WATSON

Bjarni Herjólfsson

fl. 985

Icelandic Explorer

Bjarni Herjólfsson was probably the first European to see North America. Blown off-course during a voyage from Iceland to Greenland, he saw the lands that Leif Eríksson would later explore and that Thorfinn Karlsefni would later try to colonize.

Bjarni was the son of Herjolf Bardarson the Younger, the son of Bard Herjolfsson, the son of Herjolf Bardarson the Elder. The family lived at the far western end of Reykjaness peninsula in southwestern Iceland on land granted to the Herjolf the Elder in the late ninth century by Ingólfur Arnarson, the original settler of Iceland. Like most Icelanders of the time, they made their living by farming, fishing, and raiding. Bjarni became a successful merchant.

About 900, Gunnbjorn Ulfsson, a Norwegian-born Icelander, discovered what was later called Greenland, but nothing resulted from this discovery. The Gunnbjarnar Skerries, small islands just east of Ammassaalik Fjord, were named for him. About 978, Snaebjorn Galti Holmsteinsson, fleeing Icelandic law with about two dozen companions, spent a terrible winter in the Gunnbjarnar Skerries. After murdering Snaebjorn and his foster father, Thorodd, for having brought them to that dreadful land, the rest returned to Iceland to face vengeance.

Snaebjorn's tragedy was still fresh in the minds of Icelanders when Eiríkr Raudi Thorvaldsson (Erik the Red), likewise in trouble with the law, decided to flee Iceland about 982. Erik sailed due west, as both Gunnbjorn and Snaebjorn had done, but when he saw the Gunnbjarnar Skerries, he turned south and followed the coast around the southern tip of Greenland, now called Nunap Isua or Cape Farewell, and landed near modern Tunulliarfik, which he called Erik's Fjord. Having promised his friends in Iceland that he would return for them if he found a suitable site for colonization, he named the land Greenland to make it sound more attractive, gathered 25 ships full of colonists in Iceland, and set sail about 985. A natural disaster, probably a submarine earthquake, sank 11 of the ships, but the rest arrived safely.

Herjolf accompanied Erik on this colonizing expedition. As one of the first settlers, Herjolf named Herjólfsfjord (modern Narsap Sarqaa) and Herjólfsness (modern Ikigaat) after himself.

Bjarni was in Norway on a trading voyage when his father and Erik moved to Greenland. When he arrived home in Eyrar, Iceland, he was surprised to learn that Herjolf was gone. He decided that he would not break his old habit of staying with his father each winter, and immediately sailed for Greenland.

Viking ships, whether the long, slender warships or the broad, stout merchant vessels, were all powered by only oars and a single rectangular sail, and steered by only a starboard tiller. They could not sail close to the wind, had difficulty holding course in adverse winds, and were especially at the mercy of crosswinds. Their main advantage was that they were the fastest ships of their time.

Bjarni fell victim to the unmanageability of his heavily laden merchant ship. Winds forced him southwest, where he sighted a strange land. Since it was heavily wooded, he knew it was not Greenland and set a new course north. The second land he saw was also wooded, so he continued north. The third land was rocky, mountainous, and covered with glaciers. Sailing east from there, he landed at Herjólfsness four days later.

Bjarni gave up trading as soon as he was reunited with Herjolf, and farmed at Herjólfsness for the rest of his life, except for a few voyages to Norway. Bjarni was sharply criticized and lost some honor for failing to drop anchor and explore the new lands he had seen from offshore, but he continued to talk about his discovery, even as far as the Norwegian courts of King Olaf Tryggvason (995-1000) and Earl Erik Haakonarson (1000-1015). Earl Erik made Bjarni one of his retainers.

Fourteen years after Bjarni sighted North America, Leifur Eiríksson (Leif the Lucky) decided to explore these lands. He bought Bjarni's ship and hired 35 sailors. Thus, about 1000, it was Leif who first landed in North America and coined the names Helluland (Slab-Land, probably Baffin Island or northern Labrador), Markland (Forest-Land, probably southern Labrador, Newfoundland, or Nova Scotia), and Vínland (Wine-Land, probably New England or Long Island).

ERIC V.D. LUFT

Ibn Battuta

1304-1368

Arab Traveler

In spite of the fact that Marco Polo (1254-1324) is much better known outside the Arab world, in fact Ibn Battuta traveled much more widely. Over the space of 29 years from 1325 to 1354, he covered some 75,000 miles, or about 120,000 kilometers—three times the distance around the Earth at the Equator.

His full name was Abu 'abd Allah Muhammad ibn 'abd Allah al-Lawati at-Tanji ibn Battuta; fortunately for non-Arabic speakers, however, he is known to history simply as Ibn Battuta. A member of a wealthy family in the Moroccan city of Tangier, Ibn Battuta planned to study law and become a judge in one of the city's Islamic courts. First, however, he planned to undertake a *hajj*, a pilgrimage to the Muslim holy city of Mecca in what is now Saudi Arabia. All Muslims are encouraged to make the hajj at least once if they can afford to do so, but it would be Ibn Battuta's remarkable achievement to complete the hajj a total of four times.

On his first hajj (1325-27), Ibn Battuta made a side trip into Persia, then returned to Mecca, thus completing a second hajj. He then sailed along the east African coast to the trading city of Kilwa in the far south before returning to Mecca yet again in 1330. But he was just getting started: over the next three years, he journeyed through Turkey, the Byzantine Empire, and southern Russia, at that time part of the Mongols' Golden Horde. He then passed through Afghanistan and other parts of central Asia before entering India from the north.

Eventually Ibn Battuta wound up in the court of the ruthless sultan Muhammad ibn Tughluq (r. 1325-51) in the great Indian city of Delhi, center of the Delhi Sultanate. Despite Tughluq's bloodthirsty reputation, Ibn Battuta managed to remain in his service as a judge for eight years. Tughluq sent him on an official visit to the Mongol emperor of China, but Ibn Battuta was shipwrecked, and never returned to Tughluq's court.

During the next few years, Ibn Battuta visited Ceylon, Southeast Asia, and China—possibly even as far north as the capital at Beijing. He then made the long journey home, stopping in Mecca a fourth time; but he quickly headed out again, this time to Muslim Spain and then south, across the Sahara into the splendid African empire of Mali.

Ibn Battuta stopped traveling in 1354, after which he sat down to write the record of his journeys in a volume entitled the *Rihlah* (Travels), later published in English as *The Travels of Ibn Battuta.* Along with all the other activity that filled his life, Ibn Battuta had multiple wives and children, and died when he was more than 60 years of age.

JUDSON KNIGHT

Ahmad ibn Fadlan

fl. 920s

Arab Traveler

A theologian in the court of the Abbasid Caliph al-Muqtadir (fl. 908-932), Ahmad ibn Fadlan in the early 920s participated in a diplomatic mission from Baghdad to what is now Russia. Over the course of his journey, he encountered a number of Turkic peoples, as well another group that left a strong impression on him: the Vikings. He recorded these events in a volume that has yet to be fully translated into English; yet thanks to best-selling novelist Michael Crichton (1942-), Ibn Fadlan—at least, a fictionalized version of him—has become known to a number of Western readers.

The circumstances of Ibn Fadlan's life prior to 921 are almost entirely unknown. By judging from certain specifics in his writing style, it has been surmised that he was not an Arab, and it appears certain that prior to his departure on his historic mission, he had already been serving for some time in the court of al-Muqtadir. The rest, however, is a mystery.

On June 21, 921, a diplomatic party led by Susan al-Rassi, a eunuch in the caliph's court, left Baghdad. Ibn Fadlan served as the group's religious advisor, a crucial role: among the purposes of their mission was to explain Islamic law to the recently converted Bulgar peoples, a Turkish tribe living on the eastern bank of the Volga River. (These were the Volga Bulgars; another group of Bulgars had moved westward in the sixth century, invading the country that today bears their name, and became Christians.)

The travelers made their way along established caravan routes toward Bukhara, now part of Uzbekistan, but instead of following that route all the way to the east, they turned northward in what is now northeastern Iran. Leaving the city of Gurgan near the Caspian Sea, they crossed lands belonging to a variety of Turkic peoples. Among these were the Khazars, who in the previous century had adopted Judaism. Ibn Fadlan provided a rare portrait of the Khazar Khanate, one of the few places other than ancient and modern Israel where Judaism was the majority religion. He also chronicled his encounters with the Oghuz on the east coast of the Caspian, the Pechenegs on the Ural River, and the Bashkirs in what is now central Russia.

On May 12, 922, the group arrived at the Volga Bulgars' capital. There Ibn Fadlan read aloud a letter from the caliph to the Bulgar khan, and presented the latter with gifts from the caliph. The Bulgars in turn introduced the Arab visitors to the Varangians, local Vikings who had come to be known by a term that would eventually become the name of the country itself: Rus.

Ibn Fadlan provided a memorable account of these Vikings, for instance describing a ship burial for a dead chief. The most shocking part of the funeral ceremony involved ritual sexual intercourse between various Viking males and a female slave, who was then stabbed to death and placed in the boat. After launching the vessel bearing the dead chief and his slave girl, the Vikings set the craft alight to send it and its contents into the next world.

His description of the Vikings, who he called the "filthiest of God's creatures" yet the most physically beautiful people he had ever seen—"tall as date palms, blond and ruddy"—was but one of many notable passages in the writings of Ibn Fadlan. He also discussed the existence of Gog and Magog, beastly creatures mentioned in the biblical Book of Revelation and associated with the end of the world. Throughout the Middle Ages, travelers and pseudo-authorities claimed to have located Gog and Magog somewhere in Central Asia; Ibn Fadlan, at least, reported this tale merely as a legend he had heard from others.

Upon his return to Baghdad, Ibn Fadlan wrote an account of his journey. The final portion—the part that presumably would have told about his journey back and his later life—has been lost, but the fragments that survive make for highly informative and sometimes powerful reading. The version known in the West today comes from the work of a Russian scholar, C. M. Fraehn, who in 1823 translated the text from Arabic to German. More than 150 years later, in 1976, Crichton published *Eaters of the Dead,* a novel that features Ibn Fadlan as its main character and uses fragments of his account as a point of departure for a fictional narrative.

JUDSON KNIGHT

Ibn Hawkal
920-990
Arab Traveler

Over a series of journeys that lasted some 30 years, Ibn Hawkal saw almost the entire Muslim world, from Spain to Central Asia, from the cool mountains of Afghanistan to the hot sands of West Africa. He set down an account of his travels in *On the Shape of the World,* a text noted for its accuracy, its scientific approach, and the author's attention to detail.

Abu al-Kasim ibn Ali al-Nasibi ibn Hawkal was born in the city of Nisibis in upper Mesopotamia (now Iraq). As with many other premodern figures, virtually nothing is known about his life other than the circumstances surrounding his greatest achievement. In Ibn Hawkal's case, he was able to cite not only the date—7 Ramadan, A.H. 331. (May 15, 943, by the Christian calendar)—but even the day, Thursday, when he set out from Baghdad.

Baghdad was then the leading city of the Islamic world, capital of the Abbasid caliphate that ruled most Arab countries. But other forces were at work in the once-united Muslim realm. In the farthest western corner of North Africa, and in Spain, the Umayyad dynasty, which had controlled the caliphate until the Abbasids seized power in 750, held sway. In Egypt, the Fatimid sect of Shi'ite Muslims were growing in power and influence, and in fact certain aspects of Ibn Hawkal's narrative have led some scholars to speculate that he was traveling as a spy for the Fatimids.

On the first leg of his travels, Ibn Hawkal made his way westward across Tunisia to Morocco, then in 948 sailed across the Straits of Gibraltar to Spain. Later he would write with obvious admiration about the great city of Cordoba. He noted the intellectual achievements and wealth of the Spanish Arabs—but also, in a move that helped fuel speculation regarding his role as a secret agent, observed how vulnerable the Umayyads were to attack by a foreign power.

Returning to North Africa, Ibn Hawkal moved southward through Morocco, and by 951 had arrived in the African kingdom of Ghana. He was one of the first outsiders to visit this extraordinarily wealthy realm, a land where gold was plentiful and the king ruled as a god, and certainly he was the first to write about it. He was also the first to describe the country's capital at Kumbi (later Kumbi-Saleh), and probably the first to write about the Niger River. Because the latter flowed to the east, he incorrectly surmised that it served as the headwaters of the Nile.

In time Ibn Hawkal would reach the real Nile, having crossed the Sahara and made his way to Egypt. It so happened that at the time he arrived in Egypt, the Fatimids were in the process of invading, another fact that has encouraged the belief that Ibn Hawkal was more than a merely curious traveler. By 969, at which point Ibn Hawkal had finished his journeys, the Fatimids completed their decades-long conquest of Egypt, and would maintain control there for two centuries.

By 955, Ibn Hawkal was in Armenia and Azerbaijan, thousands of miles to the northeast of Egypt. In the years that followed, he covered much of western Asia, first moving back to

Syria, then making his way to Iraq and Iran before crossing the Oxus River into Central Asia. Among the most memorable parts of his account is his recollection of Samarkand, the legendary city located in what is now Uzbekistan. There he watched as gardeners trimmed bushes and trees to look like animals, observed local methods of irrigation, and provided an inside look at the city's bureaucracy.

The final leg of Ibn Hawkal's journeys took him in 973 to Sicily. (The latter had been an Arab possession since 825, and would remain under Muslim control until 1061.) Already in 967 he had written an account of his travels called *Of Ways and Provinces,* and in 977 when he returned to Baghdad, he produced a second edition. With time to reflect, he created an entirely new work, *On the Shape of the World*, which has remained the definitive version of his travelogue.

Scholars have long noted the care with which Ibn Hawkal discussed the specifics of each place he visited, and he is remembered as the first Arab geographer to systematically provide such details. Whether he was a spy or simply a traveler with an eye for pivotal facts, he contributed greatly to knowledge of the medieval Muslim world.

JUDSON KNIGHT

Al-Idrisi

1099-1166

Arab Geographer and Cartographer

Al-Idrisi (also known by his short name Al-Sharif Al-Idrisi al-Qurtubi) was one of the most renowned Arab geographers of his day. Under the service of King Roger II of Sicily, he dispatched draftsmen around the world and used the information they returned to compile the most up-to-date world maps of his time.

Al-Idrisi was born in Sabtah, which is now the Spanish port city of Ceuta in Morocco. He is believed to be a descendant of the Prophet Muhammed's eldest grandson. Al-Idrisi graduated from the University of Cordova and spent many years traveling extensively throughout Europe and North Africa. Over the course of his many journeys, he gathered detailed information on each region and soon gained international prominence as a geographer.

King Roger II of Sicily (1097-1154) heard of al-Idrisi's cartographic expertise, and summoned him to the court in Palermo in 1145. The king wanted to compile an updated world map, and hired the Arab geographer to assist him with the task. In his research, al-Idrisi drew upon contemporary world maps drawn up by other Muslim geographers, as well as works by ancient Greek geographers such as Ptolemy. He brought the ancient material up-to-date by dispatching travelers and draftsmen to various points around the world, instructing them to make careful records of what they saw. The result was a silver planisphere on which was etched a map of the world, complete with major world cities, lakes and rivers, and mountains and trade routes.

As well as the planisphere, Al-Idrisi also made a world map which divided Earth north of the equator into 70 sections and wrote a book titled *Kita Rujar* ("The Book of Roger" or "Roger's Book"). The book was completed shortly before King Roger's death.

Al-Idrisi later wrote two comprehensive geographical texts, apparently for Roger's son William II. The first was titled *Kitab nuzhat al-mushtaq fi ikhtiraq al-afaq* (The delight of him who desires to journey through the climates), which contained detailed information on Europe, Africa, and Asia. Later he compiled another volume, titled *Rawd-Unnas wa-Nuzhat al-Nafs* (Pleasure of men and delight of souls). Al-Idrisi's works contained remarkably precise depictions of Africa and the Nile River; for example, this description of the lost city of Ghana:

> *From the town of Ghana, the borders of Wangara are eight day's journey. This country is renowned for the quantity and abundance of gold it produces. It forms an island 300 miles long by 150 miles wide: this is surrounded by the Nile on all sides and at all seasons.*

Many of his books were translated into Latin and circulated throughout Europe, where they gained great popularity. His works continued to be studied in the East and the West for several centuries.

Al-Idrisi also made numerous contributions to the science of botany. He wrote many books on medicinal plants, the most well known titled *Kitab al-Jami-li-Sifat Ashtat al-Nabatat*. In his works, he combined the written knowledge of the day with research he gathered on his journeys. The names of drugs he included were translated into several languages, including Persian, Hindi, Greek, Latin, and Berber.

Very few details are known about the last years of al-Idrisi's life, possibly because Arab scholars regarded him as a traitor for having served in the court of a Christian king. The exact date of his death is not certain, but he is believed to have died around 1165 or 1166.

STEPHANIE WATSON

Jordanus of Séverac

1290-1354

French Missionary

A Dominican missionary to India, Jordanus of Séverac was the first European to visit that region since Giovanni da Montecorvino (1246-1328) landed on the subcontinent some 30 years before. He produced an account of his experiences in the East,*Mirabilia descripta,* (Description of marvels), which discusses Armenia, Persia, and India, as well as lands Jordanus himself had not actually visited. He was the first writer to identify Prester John as the emperor of Ethiopia, thus adding a key aspect to a legend that fascinated Europe—and drove much of Europeans' exploration efforts—from the twelfth to the fifteenth centuries.

It appears that Jordanus spent the early part of his career serving as a Dominican missionary in Persia; then in 1320 or 1321, he and a group of four others set out by boat for China. They intended to join the Catholic mission established years earlier by John of Montecorvino, and India was to be only a stopover along the way. Events took a different turn when they reached the subcontinent, however.

The first change of plans came when they were forced to land at Thana near present-day Mumbai or Bombay. They had hoped to land further south on the Malabar or western coast of India, to visit the home of Christians whose ancestors had supposedly been converted by the disciple Thomas. Storms, however, drove them to Thana, where they were met by a group of Nestorians, eastern Christians whose sect had broken with the mainstream church nine centuries before. Because Jordanus spoke Farsi or Persian, then an important trade language in the region, the Nestorians asked him to come with them to the city of Broach as an interpreter. So he did, and while he was away, a group of Muslim zealots attacked and killed the other four missionaries.

Alone now, Jordanus went to Surat, where the putative first Christian missionary to India, St. Thomas, had allegedly landed many centuries earlier. In the months and years that followed, Jordanus wrote two of the three important documents (the third being the *Mirabilia descripta* itself) to emerge from his time in India. These were two letters to the Dominican missionaries in Persia, the first dated October 12, 1321, and the second January 24, 1323. They are painful accounts of a lonely man, stranded in an alien country and forced to make the best of a situation in which he could never be entirely certain who his friends were.

Jordanus spent more than a decade in India, and in 1330 the pope declared him bishop of Columbum, or Kulam, on the southern tip of the subcontinent. During the early 1330s, he returned to Europe, where he wrote the *Mirabilia descripta.* The book, though it is not lengthy, offers detailed recollections of the lands he had visited, accounts that include the first written descriptions of numerous plant and animal species.

Other aspects of Jordanus's narrative are less reliable. This is particularly the case in those sections where he discussed areas he never actually visited, such as the East Indies and Indochina. He did, however, go to great pains to remind readers that the information he provided was the product of hearsay and not direct experience.

From European merchants he had met in India, Jordanus heard accounts of what geographers of the time commonly referred to as a distant part of India itself: Ethiopia. Though the two regions are in fact thousands of miles apart, the warm climates and the dark skin of the inhabitants convinced Europeans that they were somehow connected, and thus Jordanus referred to Ethiopia, where he said he hoped to one day go on a missionary journey, as "the Third India."

In his discussion of "the Third India," Jordanus identified the emperor of the Ethiopians with Prester John. The latter was a mythical Christian king whose alleged existence had first been rumored in the mid-twelfth century. For some time thereafter, Europeans believed that Prester John's wealthy kingdom, a land full of bizarre plants, animals, and minerals, was in central Asia. Jordanus was the first to direct the quest for Prester John to Africa, opening a new chapter in the saga. Many years later, when Prince Henry the Navigator (1394-1460) inaugurated the Renaissance era of exploration by sending Portuguese mariners southward to Africa, he would be driven in part by a desire to find the lands of Prester John.

JUDSON KNIGHT

Thorfinn Karlsefni Thordarson

fl. 1000-1015

Icelandic Explorer

Thorfinn Karlsefni attempted the first permanent European settlement in America.

Thorfinn was the great-grandson of Thord Bjarnarson, one of the original settlers of Iceland. Thord's third son, Snorri Thordarson, married Thorhild Ptarmigan Thordardóttir, daughter of the powerful chieftain Thord Gellir Olafsson. Snorri and Thorhild's son, Thord Horse-Head Snorrason, and Thorunn, his wife, were the parents of Thorfinn Thordarson. The sobriquet "Karlsefni," by which he is commonly known, must have been given him when he was quite young, because it means something like "auspicious boy." The family's home was at Höfdi on Skagafjord in the North Quarter of Iceland.

Karlsefni became a wealthy merchant. Sometime in the first decade of the eleventh century, he and his partner, Snorri Thorbrandsson, and Bjarni Grimolfsson and his partner, Thorhall Gamlason, all Icelanders, sailed two merchant ships with 80 men to the settlement of Erik the Red at Brattahlid in southwestern Greenland. Because they arrived at Erik's Fjord in the autumn, they stayed with Erik over the winter. Karlsefni married Gudrid Thorbjarnardóttir, the widow of Erik's son, Thorstein.

Karlsefni heard directly from Bjarni Herjolfsson and another of Erik's sons, Leif the Lucky, about their respective voyages to the three coasts of North America that Leif had named Helluland (Slab-Land), Markland (Forest-Land), and Vínland (Wine-Land). Leif had erected temporary shelters in Vínland. Karlsefni asked Leif if he could use those shelters as a base camp from which to found a permanent colony there. Leif agreed.

The next spring, probably between 1004 and 1010, Karlsefni, Gudrid, and either 65 (according to *Graenlendinga Saga*) or 160 (according to *Eirik's Saga*) men and women sailed for Vínland. They located and expanded Leif's camp. Gudrid gave birth that autumn to Snorri Karlsefnisson, the first European child born in the New World.

The first year of the new settlement passed peacefully and even the winter was comfortable, but trouble began the following spring when the Europeans encountered Native Americans for the first time. The Vikings cheated them at trading, killed them without provocation, and called them *skraelings,* which means something like "wretches," "savages," or "those shriveled up by the sun." Eventually the native (perhaps Algonquin) counterattacks drove the settlers out. Snorri was three years old when Karlsefni's expedition abandoned the colony.

Karlsefni returned to his farm in Iceland. Gudrid survived him, made a pilgrimage to Rome, and became a nun. Snorri and their other son, Thorbjorn, both became prominent men in Iceland, and both the grandfathers of bishops.

In 1960 the Norwegian archaeological team of Helge Ingstad, his wife Anne, and their daughter Benedicte discovered the ruins of a Norse settlement at L'Anse aux Meadows, Newfoundland. Their excavations over the next eight years proved the Icelandic sagas correct about Vikings reaching America five centuries before Christopher Columbus. But there is no warrant for the hasty judgment that L'Anse aux Meadows is Leif and Karlsefni's Vínland. There could well have been many Norse settlements in America. L'Anse aux Meadows could have been any of them. For example, the account in *Eyrbyggja Saga* of another such settlement, that of Bjorn Breidavik-Champion Asbrandsson, tallies much better with L'Anse aux Meadows than do the stories of Vínland.

No one knows where Vínland was, but because it had wild grapes, it must have been south of Passamaquoddy Bay, and because it had salmon, it must have been north of the Hudson River. Some who believe it was farther north, perhaps in Newfoundland or Nova Scotia, have tried to argue away the Icelandic word "vín" (wine), substituting "vin" (fertile); but such attempts fail because there is no confusion in the Icelandic language between these two words and because the sagas specifically mention grapes and wine in this land. Moreover, in 1075 a German priest, Adam of Bremen, wrote that many voyages had been made to Vínland and that it yielded excellent wine.

ERIC V.D. LUFT

Leif Eriksson

c. 970-c. 1020

Icelandic Explorer

Half a millennium before Christopher Columbus's ships landed in the New World, Leif Eriksson and his crew of Vikings became the first Europeans to reach North America. They landed at several places on the coast of what is now Canada, and even established colonies; however, because they possessed no great technological advantage over the Native Americans, they were not able to hold on to the lands they had acquired.

Leif was the eldest of four children, all destined to travel to North America, born to Erik the Red. The father later moved his family from Iceland to Greenland, where in 986 he founded a permanent colony. Around this time, a Viking named Bjarni Herjolfsson was sailing from Iceland to Greenland when his ship was blown off course. Historians now believe that he was the first European to catch sight of North America; but he did not land.

Unlike his father, who clung to the Vikings' old pagan traditions, Leif accepted Christianity. Though he was probably one of the first Vikings in Greenland to espouse the new faith, Leif's role as missionary has been overstated by legend. In any case, it was as an explorer that he would attain his greatest notoriety.

In 1001, when he was about 30 years old, Leif sailed westward with a crew of 35 men. It is believed that they landed first on the southern part of Baffin Island, then sailed along the coast of Labrador. They reached what may have been Belle Isle, an island between Labrador and Newfoundland which they dubbed Markland, or "forest land." From Markland they sailed to a place they called Vínland, or "land of the vine" (i.e., grapes), probably a spot on Newfoundland's northeastern tip. There they established a settlement they called Leifrsbudir, or "Leif's booths."

Leif's party returned to Greenland in 1002, but his brother Thorvald made a second journey to Vínland in 1003. Thorvald and his men fought with Native Americans, whom they called *skraelings,* and Thorvald was killed by an arrow in 1005. His crew sailed home the following year, but another brother, Thorstein, returned to the area to recover Thorvald's body. He ran into storms and died upon his return to Greenland.

In 1010, Leif's brother-in-law Thorfinn, who had married Thorstein's widow Gudrid, founded a settlement on Vinland. Thus Gudrid and the other females on this voyage were the first European women in North America, and her son Snorri, born in the summer of 1011, was the first European child born on the continent. The Norsemen traded furs with the skraelings, but later they fell into conflict, and warfare drove them back to Vinland.

Leif's half-sister Freydis (illegitimate daughter of Erik) also traveled to Vinland, where she set up a trading partnership with two Norse brothers, Helgi and Finnbogi. She double-crossed her partners, however, and had them murdered along with their families. When Freydis returned, Leif did not have the heart to punish her, so he allowed her to go free.

By that time, he had settled into his role as leader of the colony in Greenland, and he never sailed westward again. Nor did any of the other Vikings, but their legends were recorded in *Erik the Red's Saga* and other epic poems describing their voyages.

For many centuries, historians regarded these tales as merely fanciful stories, but in the twentieth century, evidence began to mount that indeed Norsemen had landed in the New World five centuries before Columbus. In the 1960s, nearly a thousand years after the founding of Leifrsbudir, archaeologists found remains of a Norse settlement in Newfoundland.

JUDSON KNIGHT

Mansa Musa

c. 1280-1337

Malian Emperor

Mansa Musa, emperor of Mali in West Africa, was the first African ruler to become widely known throughout Europe and the Middle East. In particular, he was celebrated for his pilgrimage to the Muslim holy city of Mecca, during which he lavished so much gold on his hosts in Cairo that he nearly wrecked the Egyptian economy.

The modern nation called Mali is poor and landlocked, but the medieval empire by that name, located to the southwest of present-day Mali, enjoyed considerable wealth, as well as access to the Atlantic. The source of Mali's wealth, like that of Ghana, an earlier kingdom in the region, was gold. The kings of Ghana had exerted tight control over the gold supply, and Mali's ruling dynasty, established c. 1235 by Sundiata Keita, was equally strong.

Mansa Musa—or rather, Musa, since "Mansa" was a title equivalent to *highness*—was either the grandson or the grandnephew of Sundiata, and became Mali's ninth ruler in about 1307. As for his early life, little is known, though it appears likely that he was educated in the Muslim religion. Though Islam had taken hold in Mali around 1000, a great number of Malian leaders maintained traditional African religions even in Musa's time. For the most part, however, Musa avoided serious conflicts over religion, primarily because he was a strong ruler and an effective administrator.

Leif Eriksson. *(Library of Congress. Reproduced with permission.)*

Musa's armies were constantly active, extending the power of Mali throughout the region. Even while he was away on his pilgrimage to Mecca, they captured a stronghold of the powerful Songhai nation to the east. Eventually his empire would control some 40 million people—a population two-fifths the size of Europe at the time—over a vast region nearly the size of the United States.

Undergirding Musa's power was his nation's wealth in gold, which owed something to events far away. For many centuries following the fall of the Western Roman Empire in 476, Europe's economy had been weak; but beginning in about 1100—ironically, in part as a result of Europe's crusades against Islam—the European economy had begun growing again. This growth created a need for gold coins, which drove up gold prices and in turn increased Mali's wealth.

Gold wealth in turn spurred cultural advances under Musa's reign. Upon his return from Mecca, Musa brought with him an Arab architect who designed numerous mosques, as well as other public buildings. Some of those still stand in present-day Mali. Musa also encouraged the arts and education, and under his leadership, the fabled city of Timbuktu became a renowned center of learning. Professors came from as far away as Egypt to teach in the schools there, but were often so impressed by the learning of Timbuktu's scholars that they remained as students. It was said that of the many items sold in the vast market at Timbuktu, none was more valuable than books.

In 1324, Musa embarked on his famous pilgrimage to Mecca, on which he was attended by thousands of advisors and servants dressed in splendid garments, riding animals adorned with gold ornaments. He stopped in Cairo, and spent so much gold that he caused an oversupply of the precious metal. As a result, the value of gold plummeted throughout much of the Middle East for several years.

Musa died in 1337 (some sources say 1332), and none of his successors proved his equal. Later kings found the vast empire difficult to govern, and they were plagued by religious and political conflicts. By the mid-fifteenth century the Songhai, who rejected Islam in favor of their tribal religions, broke away from Mali and established their own highly powerful state.

But even more powerful forces had been awakened far away. Europeans had some idea of the vast gold supplies in Mali, but when rumors from Egypt began spreading westward, Europeans' interest in sub-Saharan Africa increased dramatically. Previously, European mapmakers had filled their maps of West Africa with fanciful illustrations, largely creations of their own imaginations; but beginning in 1375, maps showed

Musa seated on a throne of solid gold. Eager to help themselves to the wealth of the distant land, Portuguese sailors began making their way southward. It was the beginning of the end of West Africa's brief flowering.

JUDSON KNIGHT

Muhammed ibn Ahmad al-Maqdisi

945-1000

Arab Traveler

Muhammed ibn Ahmad al-Maqdisi, an Arab traveler and writer, is best known for his geographical work *Ahsan al-taqasim fi macrifat al-aqalim*. This extremely influential text was designed for readers at all levels of society, and appealed to merchants as well as to commoners. Al-Maqdisi, who was born in Jerusalem, traveled extensively in order to gain material for this work, which provided a clear and detailed account of the medieval Muslim world. In the process, he developed an almost scientific method for collecting information and examining the human geography of the region.

Al-Maqdisi constructed *Ahsan al-taqasim fi macrifat al-aqalim* as an articulation of a human geography. His prose depicts the lives and customs of the people he met. Likewise, al-Maqdisi wrote in an elegant style. Much of the book is in rhymed prose and skillfully intertwines personal observation with popular fables. Above all, the work was intended to instruct and entertain. To this extent, al-Maqdisi's volume is a good example of an *adab*, the polite literature of the medieval Muslim world.

But al-Maqdisi's work is significant for more than its broad appeal. His writing guided and entertained because of his personal experience in fact-finding. He compiled the information that became the *Ahsan al-taqasim fi macrifat al-aqalim* on travels that spanned nearly 20 years. He felt that he would be unable to provide a clear illustration of life in the Islamic kingdoms and territories until he had traveled them all.

During the tenth century, the Islamic kingdoms and territories covered a considerable expanse and ranged from the Middle East through northern Africa and into Spain. And, while these areas may have been united under a single ruler and religion, the local cultures of these places maintained their distinctive features. Indeed, in the territories, different aspects of Muslim culture merged with native customs and produced distinct cultural variants based on the Muslim model.

Al-Maqdisi was interested in depicting all of these social variations and reveled in such cultural diversity. He visited public preachers and assemblies where stories and legends were narrated. He was careful to associate with people of all classes, asking them questions concerning their habits and beliefs. He viewed his information collection process as a science: he did not compile or arrange his material in a random fashion. For al-Maqdisi, the lives and experiences of people were qualities that could be measured in the same way that a writer could examine the climate or landforms of a given region.

While this information gathering process signals the development of a new method for geographical inquiry, al-Maqdisi's writing is also distinctly medieval. Indeed, al-Maqdisi was the last of the geographers who were part of the classical school. His focus on metaphysics and geometry echoes the focus of medieval science and mathematics, disciplines which, in the medieval era, closely bordered geography.

Al-Maqdisi was indebted to his predecessor, al-Balkhi, who more closely fit the mold of the classical school. Al-Balkhi created a set of 20 maps that outlined the world. Al-Maqdisi evidently consulted this work for his own geographical constructions. However, al-Maqdisi's socioeconomic descriptions of the regions he explored helped to expand the geographical possibilities of the classical school.

And, while he helped to create a human geography, al-Maqdisi's topographical information was also more precise than that presented by his predecessors. The methodical and scientific approach that he applied to the collection of personal information also enabled him to create a work that was much more reliable than any previously available on the subject. While his work was not devoid of error, al-Maqdisi's emphasis on accurately recording his observations helped to ensure the popularity of his book.

DEAN SWINFORD

Al-Mas'udi

895-956

Arab Historian and Geographer

A pioneer of scientific geography, al-Mas'udi traveled throughout the Muslim world in the course of writing several books, the most notable of which is known as *The Meadows of Gold and Mines of Gems*. Admired as the "Herodotus of the Arabs," al-Mas'udi is generally considered the greatest Arab historian of the medieval period.

Born in Baghdad, now the capital of Iraq, al-Mas'udi descended from a close friend of the prophet Muhammad (c. 570-632). He set out on his travels at the age of 30, heading eastward through Persia and Khurasan (part of modern Afghanistan), and deep into the Deccan plateau of central India. Along the way, he took extensive notes regarding plant and animal life, and reported information from travelers he met who had been to China and Ceylon (modern Sri Lanka).

Al-Mas'udi moved westward in 916 or 917, sailing to Oman and then East Africa. His writing provides some of the first written accounts of sub-Saharan Africa, in particular the coastal trading cities in the area stretching from what is now Somalia to modern Mozambique.

Beginning in 918, al-Mas'udi spent a decade traveling around Iraq, Syria, the Arabian peninsula, and Palestine. During this time, he studied a number of different sects and ethnic groups. Between 932 and 941 he roamed throughout the Caucasus and the southwestern edge of Central Asia, traveling among the Khazars and other peoples. He also provided the first written description of the Aral Sea, and became the first geographer to correctly note that the fresh-water Caspian Sea is not connected to the Black Sea. In 941 al-Mas'udi made the *hajj* or pilgrimage to Mecca required of all Muslims, then moved southward to what is now Yemen. By 942 he was in Egypt, where he spent time among the Coptic Christians.

Most of al-Mas'udi's books, of which there were 20 or more, have been lost. Contemporary scholars generally ignored his *Akhbar az-zaman* (History of time), a 30-volume work, and when another book met the same fate, he decided to compress his universal history into a single volume. Thus was born *Meadows of Gold,* or *Muruj adh-dhahab wa ma'adin al-jawahir.*

Meadows of Gold is a work that goes deep into the earliest recesses of recorded time, relying on religion and folklore for its account of Earth's origins. It is equally vast in its breadth, containing detailed observations on the plants and animals, as well as the cultures, customs, and nations, of the world. In this it resembles the *History,* by the man to whom al-Mas'udi has been compared, Herodotus (c.484-c.420 B.C.)

Among al-Mas'udi's achievements as a geographer was a map that depicted numerous significant features, many for the first time. These included the meeting of the Indian and Atlantic oceans at the southern tip of Africa; the correct position of the Nile valley; the locations of the Indus and Ganges rivers of India, with Sri Lanka at the subcontinent's southern tip; and the outlines of the Caspian and Aral seas.

Al-Mas'udi spent his last decade dividing his time between Egypt and Syria. He died in Egypt in about 956.

JUDSON KNIGHT

Montezuma I

1398-1468

Aztec Emperor

Though he is not as well-known today as Montezuma II (r. 1502-20), last of the Aztec emperors, Montezuma I was among the greatest of the empire's rulers. It was under the reign of Montezuma I, who became noted for his conquests, that Aztec lands first extended to the Gulf of Mexico, or the "Sea of the Sky".

Like most Aztec rulers, Montezuma (sometimes rendered as Moctezuma) was born into the nobility, and rose to the empire's highest position not by means of a hereditary claim, but after winning an election among the nobles. After taking the throne in 1440, when he was 42 years old, he set about consolidating the gains of his predecessor, Itzcoatl (r. 1427-40), first to establish full Aztec control over the Valley of Mexico. In 1445, Montezuma led the Aztecs in their conquest of the neighboring state of Oaxaca.

Montezuma's reign was fraught with natural disasters. In 1446, an attack of locusts destroyed most of the crops in the Valley of Mexico. Three years later, the Aztec capital at Tenochtitlán was flooded. Then in 1450, the region experienced the first of four years of bad harvests caused by drought and early frost. The famine became so bad, it was said, that people sold themselves into slavery for a few ears of corn. When the cycle of bad harvests ended in 1454, the Aztecs took what they believed was the obvious lesson: the gods were unhappy because they had not sacrificed enough victims. Therefore the pace of human sacrifices increased dramatically.

From the Zapotec, founders of Oaxaca, the Aztecs had adopted a dual calendar system, with one calendar based on the 365-day year, and another on a 260-day religious cycle. Once every 52 years, the first days of both matched up, and that was a day of celebration for the renewal of the earth. The year 1455 marked the beginning of a new cycle, and thus despite the misfortunes they had suffered, the Aztecs took heart at what they considered a good sign from heaven.

In 1458, Montezuma led a new series of conquests, expanding the boundaries of the empire so that his people could call themselves "Neighbors of the Sea of the Sky." This title referred to their control over lands between the Valley of Mexico and what is today known as the Gulf of Mexico. It is believed that under Montezuma's reign, Aztec territories first extended to the Pacific Ocean as well.

Montezuma's court was plagued with the same sorts of intrigues that affected his counterparts in the Old World. His half-brother Tlacaelel may have opposed him for leadership in his early years, though some historians believe he was given an opportunity to take the throne and simply declined. In any case, he was happy to hold power from the sidelines, and after Montezuma's death in 1460, he took control over the empire.

Despite his quest for authority over many lands, Montezuma himself seems to have had doubts about the value of power. Aztec records quote him as saying, "Rulers of many peoples eat the bread of sorrow," and he encouraged his children to seek careers in the trades and crafts, far from the headaches of rulership. Yet after the death of Tlacaelel in 1469, Montezuma's son Axayacatl took the throne. Just fifty years later, under the reign of Montezuma's namesake Montezuma II, the splendid Aztec Empire came to an end.

JUDSON KNIGHT

Odoric of Pordenone

1286?-1331

Italian Missionary

Odoric of Pordenone was a Franciscan missionary who traveled extensively throughout Asia. He was the first European traveler to describe distinctions between Oriental and Occidental culture accurately and in detail. While Marco Polo (1254-1324) is perhaps better known to the modern reader as the first European explorer to provide accounts of the Far East, Polo did not provide as many details regarding what Europeans viewed as the peculiarities of the Chinese people. Furthermore, Odoric was the first European to enter Lhasa, the capital of the Dalai Lama, the spiritual leader of the Tibetan Buddhists.

Before Odoric's death, a manuscript was prepared that detailed his travels. John Mandeville (c. 14th century) used this account and added extensive elaborations and fabrications in his own work. *The Travels of Sir John Mandeville,* loosely based on Odoric's journeys, was particularly popular in the later Middle Ages and was used as a manual by other travelers and geographers of the period.

Odoric was born at Villanova, a village close to Pordenone, Italy, around 1286. He came from a Czech family by the name of Mattiussi. In his early teens, he entered the Franciscan Order at Udine. At that time the Franciscans were primarily responsible for conducting missionary work in central Asia, as directed by the Holy See during the middle of the thirteenth century. The missionaries followed trade routes that had been recently established between Europe and Asia. In 1318 Odoric was sent to follow in the footsteps of missionaries such as Willem van Ruysbroeck (1215?-1295?), Giovanni da Montecorvino (1247-1328), and Giovanni da Pian del Carpini (1180?-1252).

Odoric began in Padua and reached western India in 1321. From there he traveled to China. After visiting China, Odoric sailed in a junk to Sumatra. He also visited Java and the coast of Borneo before returning to China.

Odoric then remained for three years in Peking, where Montecorvino, archbishop of the city, provided hospitality for the weary traveler. Odoric returned to Italy in 1330, traveling by land through Tibet, Badachschan, and Armenia. Upon Odoric's return, Giudotto, his superior, requested that Odoric dictate his travels to a fellow Franciscan, Brother William of Solagna.

Odoric's account is viewed as mostly factual, though its credibility is damaged by the inclusion of many fantastic tales. However, his detailed recollections of the specifics of Chinese culture help to verify its truthfulness. Some of the details of Chinese life that Odoric documented, but were omitted by Marco Polo, include cormorant fishing, the extremely long fingernails of some of the natives, and the custom of binding the feet of women.

Odoric told his tale in a clear narrative style. He emphasized equally the variety of goods to be found in particular cities and his encounters with bizarre personages—for example, his recollection of the "Old Man of the Mountain." According to Odoric, the Old Man had "built a wall to enclose [two mountains]. Within this wall there were the fairest and most crystal fountains in the whole world: and about the fountains there were most beautiful virgins in great number, and goodly horses also, and in a word, everything that could be devised for bodily pleasure and delight, and therefore the inhabitants of the country call the same place by the name of Paradise." Odoric's narrative is filled with similarly fantastic characters and places.

Odoric's recollections have remained popular because of their clear evocation of both the everyday and the unusual. His account, along with those of his fellow Franciscans, Ruysbroek and Carpini, were plagiarized and combined in *The Travels of Sir John Mandeville*. This book, which was extremely popular in medieval Europe, focused on the most extraordinary events and places that the Franciscans recounted. Despite its almost purely fantastic content, the book was used as a travel guide by medieval explorers and merchants.

DEAN SWINFORD

Pachacutec Inca Yupanqui

d. 1471

Inca Emperor

Pachacutec Inca Yupanqui was not only the first Inca emperor whose dates and existence are firmly established in history; he is also widely considered the greatest Inca ruler—if not one of the greatest leaders in world history. An empire builder who began with a kingdom of perhaps 25 square miles (65 sq. km) and shaped it into a vast realm, Pachacutec created a system of roads and a highly organized, efficient government. These were particularly impressive achievements in light of the fact that the Inca had not discovered the wheel; had no pack animals other than the diminutive llama; and possessed no system of writing.

Pachacutec, whose name meant "he who transforms the earth," was born the son of Viracocha, a semi-legendary ruler whose name was taken from that of the principal Inca deity. But he was neither the first nor the favorite son. At some point in the 1430s, the Incas were attacked by a neighboring tribe, and both Viracocha and his designated heir fled the capital at Cuzco for the safety of the mountains. Pachacutec, however, held his ground, and marshalled his army to drive back the invaders. With victory secured, he took the throne in 1438. The latter year is the beginning of Inca history, inasmuch as events after that point can be dated with relative certainty.

Pachacutec set about first strengthening his hold on the region around Cuzco. Then he extended his reach to parts of the Amazon valley, and to the Andean highlands as far as Lake Titicaca. Rather than simply fighting battles of conquest, Pachacutec was pursuing a clearly defined strategy aimed at building a strong and unified empire. Wherever possible, he and his advisors won over neighboring tribes through diplomacy; if this failed, the Inca army—by far the most powerful in the region—won victory by force. Most tribes wisely agreed to bloodless conquest by the Inca.

As a means of ensuring that his ever-widening empire developed a common culture, Pachacutec saw to it that the Inca language, Quechua, became the regional lingua franca. To reduce threats from potentially hostile groups, Pachacutec sometimes ordered tribes to relocate, thus separating them from homelands where they might develop a base of support for future resistance. In line with his policy of not making Inca rule too harsh on the conquered peoples, however, Pachacutec's government pursued its relocation policy with care. For instance, he avoided moving people from the lowlands to the high mountains where the thin air and cold climate might cause deaths.

Roads were another key element of Pachacutec's program to solidify his empire. Under his reign, the Inca constructed some 2,500 miles (4,000 km) of stone roads, many of them across high mountain passes and others through steaming swamps. Though these were extremely well-built, with tightly fitted stones, they were not roads as Europeans would understand them: most were only about three feet (1 m) wide, which was sufficient to accommodate pedestrians or load-bearing llamas.

Along with the roads, the Inca built way stations placed at intervals equal to a day's travel, so that travelers could rest and obtain supplies. Trained runners traversed the road system, keeping the emperor abreast of events throughout his empire. Compared to the slow postal system of Europe (which, like that of the Inca Empire, was only for the use of the government, not ordinary citizens), the Inca messenger service was extraordinarily fast and efficient. Thanks to the relay runners, who could transport a message at the rate of 140 miles (224 km) a day, Pachacutec's army was never caught unaware by rebellions on their borders. In addition, the emperor kept troops stationed throughout the empire, ready to go into action whenever the alert was sounded.

Particularly impressive was the means by which the Inca overcame the limitations imposed by illiteracy. In place of written records, Pachacutec's bureaucrats used the *quipu*, an ingenious system of strings in varying lengths and colors, for recording numerical information. For mathematical calculations, they made use of the abacus.

After years of administering his empire, Pachacutec turned over the reins of leadership to his son Topa. He continued to be actively involved in governmental affairs, however, particularly a program to rebuild Cuzco from the devastation or earlier attacks. He created a plan for the city, and initiated vast building projects, including a huge central plaza surrounded by temples. By the time of Pachacutec's grandson Huayna Capac (r. 1493-1525), the Inca controlled an area equal to that of the U.S. eastern seaboard, with some 16 million subjects, a number equal to the population of France at the time.

JUDSON KNIGHT

Marco Polo
1254-1324
Italian Merchant

Marco Polo's extensive travels led him to China, where he spent 17 years and established himself as an official in the Mongolian court. He recorded his experiences in *Il milione,* or "The Million," a classic of travel literature known in English as "The Travels of Marco Polo."

Polo was the son of an itinerant merchant. His father, Niccolò, and uncle, Maffeo, spent many years traveling and trading before Marco was born. In fact Marco was around the age of 15 or 16 when he first met his own father. By then Niccolò had established great wealth for his family, though whether or not he was an actual member of the aristocracy is unclear.

Little is known of Polo's early years in Venice, however, details are abundant once his travels began. Polo's father had years earlier established a good relationship with the Mongol court and the Emperor Kublai Khan. Niccolò and Maffeo had originally been sent back to Europe as papal ambassadors for the emperor. In 1271 the Polos, with Marco in tow, set out to return. Accompanied by two friars, they departed from Acre to deliver papal letters to Kublai Khan. However, the friars soon abandoned the party, leaving the three Polos to continue on their own to China.

In 1272 the Polos most likely made their way through the territory that is now eastern Turkey and northern Iran. To avoid taking the sea passage to India, the party chose to travel over land to the Mongol capital. It took them a couple of years, but in around 1274 they reached their destination and presented their patron, Kublai Khan, with sacred oil from Jerusalem as well as papal letters.

The Polo family stayed in the Mongolian empire for the next 16 or 17 years. It was there that Marco seemed to endear himself to the emperor. This personal association afforded him certain responsibilities, including being sent on fact-finding missions to remote parts of the territory. Research suggests that aside from his travel assignments, Marco was also a sort of government official.

Around 1292 the Polos, hoping to once again see their home and their families, offered to accompany a Mongol princess to Persia. Although Kublai Khan was reluctant, he finally granted his permission, and their fleet of 14 ships set sail, eventually arriving in Venice in 1295.

In 1298, while on another voyage, Marco was apprehended by the Genoese. He was sent to Genoa and locked in a prison. It was there that he met Rustichello, a prisoner from Pisa who was a writer. As he was not at ease with the Venetian or Franco-Italian language, Marco dictated the story of his travels, the basis of *Il milione,* to Rustichello.

Because Polo's account was written before the time of the printing press, copies of the book were made by hand. This left the door open for subsequent scribes and translators to take liberties with the content. Today there is no known authentic copy of *Il milione.*

Once completed, the book, later published as *Divisament dou Monde* (Description of the world), was an instant success. It included details of the politics, agriculture, military power, economy, sexual practices, religions, and other customs of the Far East. Polo lived a reclusive existence sustained by a modest fortune until he died at the age of 70.

AMY LEWIS MARQUIS

Johann Schiltberger
1381-1440
German Nobleman

Johann Schiltberger's travel narrative, *The Bondage and Travels of Johann Schiltberger,* indicates that he was an unwitting and unwilling explorer of the Middle East. Indeed, he was a prisoner for the large majority of the time period covered in his account. However, his narrative is significant in that it provides numerous details pertaining to medieval Muslim culture. In order to survive for so long as a prisoner, Schiltberger had to, more than likely, either reject or carefully conceal his Christian faith. His depiction of aspects of the Muslim faith, such as "Of the Infidels' Easter Day," "How a Christian Becomes an Infidel," and "Of a Fellowship the Infidels Have among Themselves" suggests a familiarity with the Muslim faith denied to Christians of that time.

Schiltberger was born near Munich in 1381. Little is known of his family except that they were probably a well-placed family of Bavarian burghers, marshals, and dukes. Schiltberger left his home at an early age; in 1394, at the age of 14, he set off with his master, Leonard Richartinger, and immediately entered into military service. Richartinger served in the auxiliary forces under Sigismund, the Hungarian king. At that time Hungary was threatened by invasion from Turkey, prompting King Sigismund to summon as many Christian warriors as possible to help defend him from the Muslim Turks.

Two years after his departure from Bavaria, Schiltberger entered into a battle that would exert a lifelong impact. In 1394 Sigismund instigated a military engagement with the Turks over Nicopolis, a city on the Danube River. While Sigismund had been successful in earlier battles, he was quickly overwhelmed at Nicopolis and forced to retreat. The Turkish Duke of Iriseh, also known as the despot, overwhelmed the Christian soldiers, who fled in disorder. Many of the soldiers fighting under Sigismund were either slaughtered or drowned.

Schiltberger was taken prisoner and narrowly escaped execution. The Turkish king, overwhelmed by grief at the numbers of his army that had been slain, was determined to avenge their deaths. Many of the prisoners were led to the battlefield and beheaded. Schiltberger, who was 16 at the time, escaped death; the king's son ordered that he be left alive because he was still a youth.

Schiltberger's narrative is filled with humble depictions of his numerous close brushes with death. Several years after his initial capture, for instance, a group of 60 Christians decided to escape from King Weyasit. They made a pact between themselves that they would succeed, or die in the attempt. When Weyasit learned of their escape, he sent 500 horse-mounted soldiers to capture the group, who defended themselves in order to avoid capture and death. The commander of the king's troops vowed to the prisoners that he would die at the king's hand before allowing the king to execute the prisoners. The group consented and returned with the commander. Upon their return, the king ordered that the prisoners be killed immediately, but the commander interjected. Instead, they were imprisoned for nine months.

Schiltberger remained a slave until 1427. His position in Muslim society was not to be envied. However, this position enabled him to participate in Muslim culture to a much greater extent than may be expected. He witnessed many military conquests during his travels, but was also present for events such as the marriage festivities of the nobility. His narrative provides a fairly accurate account of Muslim beliefs and details the daily lives of his captors.

However, after his escape Schiltberger expressed little sympathy for the Muslims. While in Mingrelia, a Christian country situated along the Black Sea, he secured passage on a European ship, fled to Constantinople, and, in 1427, finally returned to Bavaria. Upon his return he thanked God for his escape "from the Infidel people and their wicked religion," and was honored as the chamberlain and commander of the bodyguard to Albrecht III.

DEAN SWINFORD

Nuño Tristão

d. c. 1447

Portuguese Sailing Captain and Explorer

Nuño Tristão is known for his travels to Cape Blanco, Sénégal, and Gambia in Africa from 1441-1446 under the command of Prince Henry of Portugal (1394-1460). These and other voyages of Prince Henry opened the West African coast to exploration, provided new geographical information for mapmakers, and helped confirm Portugal's place as a navigational power. They also, however, marked the beginning of the slave trade.

History first records Tristão for his involvement with Prince Henry. Although known as Prince Henry the Navigator, the prince never actually joined any of the expeditions associated with his name. Instead, this son of Portuguese King John I opted to plan and fund the trips while remaining onshore. In particular, Prince Henry focused on the exploration of Africa, developing an overall strategy and establishing a navigational school to ensure its success. He invited cartographers, astronomers and other experts to share their knowledge at this navigational school. As a result, navigational skills improved greatly and the boundaries of naval exploration widened.

Over the next four decades, the prince oversaw and funded some two dozen expeditions, three of which were captained by Tristão. On his first voyage under the prince, Tristão set sail in 1441 for Cape Blanco, which now falls within Mauritania below Western Sahara on Africa's west coast. Concurrently, another voyage led by captain Antão Gonçalves, left for the same location. Gonçalves landed just south of Cape Blanco. While there, he captured a dozen African men and returned with them to Portugal to show Prince Henry. One of the 12 men was the chieftain Adahu, who exchanged freedom for himself and a young male relative for the slavery of other men in Africa. Adahu also told Henry about the central and southern reaches of Africa that had yet been unexplored by Europeans.

Within two years, Gonçalves had set up a fort on Arguim, an African island. The fort became a slave-trading center. Over time, the trade grew and within a decade, thousands of African people had been captured and sent to Portugal as slaves.

In the meantime, Tristão was making additional trips by caravel, a commonly used fifteenth-century sailing ship with triangular sails, broad bows and a high raised deck (poop) at the stern. He ventured about a third of the way down Africa's west coast to Senegal in 1444, and to Gambia, a narrow slice of land within present-day Senegal, in 1446. While in Gambia, he died at the hands of the native people in 1446 or 1447. Luiz de Cadamosto and Antoniotto Usidimare followed in the footsteps of Tristão, and led the second voyage to Gambia in 1455. During this excursion, the two men conducted a more thorough exploration of this recently discovered land.

Prince Henry continued to fund and to plan expeditions to Africa until the time of his death in 1460. By then, Portuguese sailors had made landfall along Africa's coast all the way to Sierra Leone, which lies more than 400 miles (644 km) south of Gambia. In addition to introducing Europe to western Africa, Tristão and Prince Henry's other captains applied the knowledge gained at Henry's navigational school and allowed the Portuguese to maintain and to build upon their dominion over the seas.

About two decades after Prince Henry's death, Portugal established an outpost, named São Jorge da Mina on the Guinea coast in Africa. This armed outpost became the first permanent slave-trading center constructed along the continent's western edge. Although the center changed owners, it remained in business for three centuries. As the Americas developed, they became some of the center's best customers, trading for about 30,000 slaves by the eighteenth century.

LESLIE A. MERTZ

Willem van Ruysbroeck

c. 1200-1260?

French Missionary and Writer

Willem van Ruysbroeck made an extended missionary and diplomatic journey to see the Great Khan Möngkhe and attempt to deter the Mongols from invading Western Europe. During his travels, which encompassed Constantinople, Crimea and southern Russia to Mongolia, Willem wrote one of the most detailed travel accounts of the Middle Ages.

Born around 1200 in the Flemish town of Ruysbroeck in northern France, Willem was a Franciscan monk. He is mainly known for his travels as envoy of King Louis IX of France in response to a feared, major invasion from the Mongols. The king, also known as St. Louis, decided to send Willem to the great khan with the goal of learning about this leader and his people, and defusing the threat, perhaps by converting them to Christianity.

Willem set out in 1252 and headed for Constantinople, where he made a year-long stop and enlisted a traveling companion, Friar Bartholomew of Cremona, and a guide, who also served as a translator. These three and others in the crew set sail to Crimea, where they then traveled across southern Russia toward the Mongol domain.

Among the often-nomadic Mongols, Willem carefully recorded his observations. "As for their food and victuals," he wrote, "I must tell you they eat all dead animals indiscriminately, and with so many flocks and herds, you can be sure a great many animals do die.... They feed 50 or 100 men with the flesh of a single sheep, for they cut it up in little bits in a dish with salt and water, making no other sauce; then with the point of a knife or a fork especially made for this purpose—like those with which we are accustomed to eat pears and apples cooked in wine—they offer to each of those standing round one or two mouthfuls, according to the number of guests." He made similarly explicit observations about the Mongol's

Willem van Ruysbroeck.

lifestyle, including their shelters, furnishings and clothing, as well as their customs.

Willem and his party eventually met up with the great general Batu, who granted Willem the opportunity to see the Great Khan. Along the way, Willem made unsuccessful attempts to convert the Mongols to Christianity.

After more than three months of travel, Willem arrived at his destination, only to wait another week before meeting the Great Khan. The initial meeting was unproductive for Willem, who was forced to remain rather tight-lipped as the Khan prodded him for information about France's military strengths and weaknesses. Later, Khan brought Willem and his men to Karakorum, the capital city, where Willem again initiated efforts to convert Mongols to Western Christianity.

About six months after meeting the Great Khan, Willem saw him for the last time on May 31, 1254. During this encounter, Willem tried to persuade the Khan to allow him to remain and promote his religion. Instead, the Great Khan insisted Willem return to France and deliver to the King Louis IX a correspondence threatening attack unless French ambassadors met with Mongol leaders. Willem left Karakorum without his traveling companion Friar Bartholomew, who had taken ill. The party returned via a northern route through Asia Minor and arrived in Tripoli in the spring of 1255.

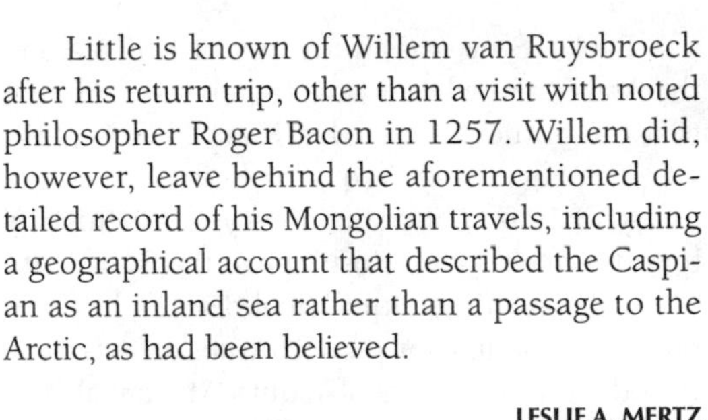

Little is known of Willem van Ruysbroeck after his return trip, other than a visit with noted philosopher Roger Bacon in 1257. Willem did, however, leave behind the aforementioned detailed record of his Mongolian travels, including a geographical account that described the Caspian as an inland sea rather than a passage to the Arctic, as had been believed.

LESLIE A. MERTZ

Yaqut

1179-1229

Syrian Geographer, Historian, and Ethnographer

Yaqut is known primarily for two works, *Kitab mu'jam al-budan* and *Mu'jam al-udaba'*. The former is a summation of historical, geographical, and ethnographic information in the Arab world of his time, the latter a collection of biographical sketches concerning important men of the era. These, the first Muslim works organized encyclopedically, have provided scholars with invaluable insights concerning the Middle East and central Asia in the twilight years of the Abbasid Caliphate.

His full name was Yaqut ibn 'Abdallah, but because his parents were Greek he was often known as al-Rumi, "the Roman" or "the Byzantine." He was also called al-Hamawi, a reference to the fact that he came from the town of Hama in Syria. As for the name Yaqut, it means "ruby." Slaves, of whom Yaqut was one, were often given the names of gemstones.

The life of a slave in medieval Islam was not unremittingly harsh; indeed, Yaqut's master, recognizing his talents, arranged for him to be educated. In 1199, the master moved his household to Baghdad, then capital of the Muslim under Abbasid rule. In Baghdad, Yaqut reportedly married and fathered several children before his master released him.

Many slaves, upon their release, were granted property and a place of security, but Yaqut had only his freedom, and he spent many years wandering, making a living by copying and selling manuscripts. He traveled to far corners of the Muslim world, including Oman, northwestern Iran, Egypt, Palestine, and Syria. His support for the radical Kharijite sect got him into trouble in Damascus in 1215, and he fled to the peaceful scholarly town of Marw in northeastern Iran. There Yaqut spent two years in libraries, absorbing much of the knowledge he would later put into his books.

By 1218 he was in the city of Khiva along the Aral Sea in what is now Uzbekistan. Hearing that the armies of Genghis Khan (1167-1227) were on their way, Yaqut rushed back to the safety of Iraq. The remaining years of his life found him either in Mosul, or in the Syrian city of Aleppo. During Yaqut's final decade, he composed the *Kitab*, later partially translated as "The Introductory Chapters of Yaqut's 'Mu'jam al-buldan,'" and the *Mu'jam al-udaba'*, translated as "Yaqut's Dictionary of Learned Men."

The *Kitab*, the final draft of which he completed in 1228, is significant not only for its encyclopedic quality, but also for its synthesis of Greek and Arabic views on science and cosmology. In addition, Yaqut was one of the last scholars to gain access to the libraries east of the Caspian Sea, in what are today the former Soviet republics of Central Asia—libraries that would be, along with the lands and cities around them, devastated in the Mongol onslaught.

Indeed, Yaqut's contribution to learning is primarily that of a preserver and a synthesizer, rather than that of an original thinker. In his *Dictionary of Learned Men,* he provided priceless information on figures who might have been completely lost to history without his efforts at maintaining their memory.

It is said that at the end of his life, Yaqut had acquired enough wealth to provide assistance to the widow and orphans of his former master, who had fallen on hard times. This is particularly remarkable in light of the fact that unlike most scholars in medieval times, Yaqut had no patron.

JUDSON KNIGHT

João Gonçalves Zarco
fl. 1420s
Portuguese Navigator

In 1418 Portuguese navigator João Gonçalves Zarco sighted one of the Madeira Islands off the coast of Morocco. Within two years, he had claimed the Madeiras for Portugal, whose possession they remain. Zarco's "discovery"—the Madeiras had actually been known to mariners before, then forgotten—was the first notable achievement credited to the school of explorers founded by Prince Henry the Navigator (1394-1460).

Little is known about Zarco's life aside from his adventure in the Madeiras, though it appears he was Jewish. This in itself is interesting, because while anti-Semitism was strong in all of Western Europe during the Middle Ages, it was rampant in the Christianized lands of the Iberian peninsula. Yet the Zarco family, which apparently came from the Portuguese city of Tomar, managed to flourish under Catholic rule. João Gonçalves himself became a noble in the house of Prince Henry, known to the Portuguese as Dom Henrique, the Infante.

In 1415 the Infante (so named because he never ruled) took part in a successful crusade for Ceuta, a city in northern Morocco. Most likely Zarco himself participated in this crusade, a turning point in Henry's life: not only did his victory win him considerable honors, but his experience in North Africa influenced his desire to learn more about the African continent and outlying islands.

On the heels of his success in Ceuta, Henry began to gather explorers, mariners, and cartographers around him in an informal "school" he established in Sagres, a city in southwestern Portugal. Zarco went out on one of the very first voyages Henry sponsored, during which he caught sight of the Madeiras. The islands, some 560 miles (about 900 km) west of Morocco, had probably been known to the Phoenicians in ancient times. Certainly Genoese sailors of the fourteenth century had known about the island group, but the knowledge seems to have been lost; thus when Zarco landed on the island of Porto Santo in 1419, it was as though he were finding the islands for the first time.

Bereft of human habitation, the Madeiras were filled with extremely dense vegetation. On orders from Henry, Zarco and others began to colonize the islands, practicing a form of slash-and-burn agriculture. It was said that the fires raged for seven years. They planted grapes brought from the eastern Mediterranean, and today the Madeiras abound with crops that include not only grapes (Madeira wine is famous), but sugarcane and bananas. At least this is true for the inhabited islands of Madeira and Porto Santo: the Desertas, southeast of Madeira, are home only to rabbits and wild goats, while the Selvagens are a trio of uninhabited rocks.

The Madeiras were the first major fruit of Henry's efforts, and their rediscovery would soon be followed by the discovery of the Azores by Diego de Sevilha beginning in 1427. As for Zarco, he dropped from the pages of history, and it is unknown whether he died on the Madeiras or lived out his days in Portugal.

During the 1980s and 1990s, a U.S. medical doctor named Manuel L. da Silva put forward a controversial theory, which he discussed in books such as *The Religious and Mythological Powers in the Name of Cristovao Colon*, that linked Zarco with Christopher Columbus (1451-1506). According to da Silva, who had studied a number of documents from the era, Columbus was actually Zarco's grandson. It was his conjecture that Dom Fernando, first Duke of Beja, had an affair with Isabel Gonçalves Zarco, the explorer's daughter, and out of this relationship came a son, Salvador Fernandes Zarco. The latter supposedly later changed his name to Colon (the Portuguese version of Columbus) to hide his Jewish ancestry. This theory, while interesting, has not received widespread support among historical scholars.

JUDSON KNIGHT

Biographical Mentions

Adahu
fl. 1430s-1440s

African chieftain, leader of the first group of slaves taken from Africa in 1441. Captured along with 11 others by the Portuguese in Cape Blanco in 1441, Adahu was chief of the Azanaghi people, a Muslim tribe. He spoke Arabic and thus was able to communicate with his captors through a Bedouin interpreter. The group was brought to Portugal and presented to Prince Henry the Navigator (1394-1460), who questioned Adahu concerning land routes in Africa. Later, Adahu was allowed to return to Africa with one other male, on the promise that the Portuguese would receive 10 others in their place.

Anawrahta
r. 1044-1077

Burmese king who was first to unite his nation. From Anya or Upper Burma, Anawrahta conquered the Mon states in Lower Burma, and eventually subdued much of the surrounding region. He converted to Theravada Buddhism and built numerous pagodas in his capital at Pagan.

Alfonso Gonçalves Baldaya
fl. 1430s

Portuguese sailor who in 1436 brought Europe its first cargo from West Africa. In 1435, Baldaya sailed along the west coast of Africa with Gil Eannes (fl. 1433-1445). A year before, the latter had been the first to sail past Cape Bojador, which Europeans had previously regarded as a point of no return. Baldaya and Eannes, exploring the shoreline, saw human and camel tracks—proof that people lived south of the cape. Baldaya continued southward to Cape Blanco on what is now the border between Western Sahara and Morocco. There he traded for sealskins, which he brought back to Europe.

Baldwin of Boulogne
1058?-1118

French nobleman and leader in the First Crusade (1095-99). In 1098 he conquered the Syrian city of Edessa and surrounding regions, which became the first crusader state. After the death of his brother Godfrey of Bouillon (c. 1060-1100), Baldwin was elected "protector of the Holy Sepulchre," or king of the crusader state of Jerusalem. He then set out to conquer other lands in the area, an effort that put him into conflict with erstwhile fellow crusader Tancred (1078?-1112). Baldwin died on a raiding expedition into Egypt.

Barlaam the Calabrian
c. 1290-c. 1350

Byzantine scholar who greatly influenced Western Europe's growing interest in Greek culture, which would blossom during the Renaissance. A humanist, Barlaam was distinguished by his strong opposition to Hesychasm, a radical mystic movement among the monks of the Byzantine Empire. From 1339 he served as Byzantine envoy to the court of Pope Benedict XII (r. 1334-42) at Avignon, and during this time he taught Greek to Petrarch (1304-1374). It was the latter who first used the term *Renaissance* to describe the changes taking place in thought and society at that time.

Batu Khan
c. 1203-c. 1255

Mongol military commander and ruler who extended the Mongol Empire in Europe. The grandson of Genghis Khan (founder of the Mongol Empire), Batu was chosen in 1235 by his uncle Ogodei Khan, Genghis Khan's successor, to command the western part of the Mongol Empire. During 1237-1241, he conquered Russia, Poland, and Hungary. Batu succeed Ogodei Khan in 1242 and established Kipchak Khanate, known as the Golden Horde because of its beautiful tents.

Al-Bekri
fl. 1060s

Spanish Arab traveler who visited West Africa in about 1067, and wrote about his travels in *Al-Masalik wa'l-mamalik.* The latter is notable for its description of the kingdom of Ghana, particularly its great wealth, the power of its king, and the potential for trouble posed by the society's division between centers of Islam and traditional African religions. Al-Bekri's account is particularly valuable because in 1080, just a few years after his visit, Moroccan invaders conquered Ghana and destroyed its civilization.

Bohemond I
c. 1050-1111

Norman knight and one of the leaders of the First Crusade (1095-1099). Bohemond was the son of Robert Guiscard (c. 1015-1085), who with his brother Roger (1031-1101) controlled much of Italy in the eleventh century. Bohemond first distinguished himself by helping his father take Rome from Emperor Henry IV (1050-1106) in 1094. On his way to the Holy Land, Bohemond stopped in Constantinople, a visit later reported in detail by the Byzantine princess Anna Comnena (1083?-1148), the first female historian in European history. In 1098 Bohemond led the crusaders in the capture of Syrian Antioch and went on to become its ruler, but in the following year he was captured by the Turks while trying to take Sebastea. Released in 1103, Bohemond spent his latter years in an unsuccessful campaign to defeat the Byzantines.

Brian Boru
c. 941-1014

Irish king who briefly united his land against the Danes. In 999 he won the capital city of Dublin, and in 1002 became king of all Ireland. By then more than 60 years old, he turned his attention from battle to administration, seeking to solidify his position through a strong alliance with the Catholic Church. Brian instituted a system of church schools to strengthen education throughout the country, and modeled his reforms on those undertaken by Charlemagne (742-814). He was killed in a battle with the Danes, and his vision of Irish unity died with him.

Geoffrey Chaucer
1340?-1400

English poet and civil servant who wrote the literary classic *The Canterbury Tales,* among other works, and held numerous government posts during the reigns of Edward III, Richard II, and Henry IV. Fluent in French (the language of the wine trade), Chaucer held royal appointments as customs controller, forester, and clerk of the king's works. He also went on (sometimes secret) diplomatic missions to France and Italy, where he discovered and assimilated the works of Boccaccio, Dante, and Petrarch. This experience, combined with his acute powers of observation, humor, and discernment, allowed Chaucer to capture the complexity, vitality, variety, and contradictions of the medieval world in his writings, which are still vivid after six centuries.

Dinís Dias
fl. 1440s

Portuguese navigator and explorer who discovered the westernmost point of Africa. In his quest to establish Portugal as a world leader in sea trade, Prince Henry the Navigator sent Dinís Dias, his captain, on an expedition to Africa in 1445. Dias reached the mouth of the Sénégal River, which the Portuguese later believed to be the western branch of the Nile, then went on to discover Cape Verde in western Africa. Dias named the area "Green Cape" because of its lush vegetation. On a later voyage, he sailed a caravel along the Sénégal.

Eleanor of Aquitaine
1122-1204

French noblewoman and later queen who traveled to the Holy Land during the Second Crusade (1147-1149). Not only was she the best-known European woman to participate in a crusade—though not, of course, as a combatant—Eleanor also had the distinction of serving first as queen of France, and later of England. She married Louis VII of France and later divorced him to marry Henry of Anjou, who in 1154 became Henry II of England. Among their sons were Richard I (the Lion-Heart; 1157-1199) and King John of England (1167-1216).

Abu al-Fida'
1273-1331

Arab historian and geographer. Al-Fida', who served as a local sultan under the Mamluk rulers of Egypt, wrote two important works, *Mukhtasar ta'rikh al-bashar* (Brief history of man), which covered pre-Islamic and Islamic history in the Middle East up to 1329; and a geography called *Taqwim al-buldan* (Locating the lands, 1321.) Neither work represented original scholarship, but rather consisted of writings by other authors compiled, edited, and expanded by al-Fida'; both, however, became highly influential among

European Orientalists of the eighteenth and nineteenth centuries.

Freydis
fl. c. 1013

Viking explorer and half-sister of Leif Eriksson (c. 970-c. 1020). In 1013 Freydis, the illegitimate daughter of Erik the Red (c. 950-1004), took part in an expedition to Vínland, a region in eastern Canada discovered by her half-brother some years earlier. She established a trading partnership with two Norse brothers, Helgi and Finnbogi, but later turned against them and had them murdered along with their families before returning to Greenland.

Jean Froissart
1333?-1405?

French historian and poet whose renowned *Chronicles* detail the history of Western Europe from the early 1300s to 1400, supplying valuable contemporary information about medieval life and society, noble courts, and warfare. A scholar and traveler, Froissart ventured to seats of power throughout England, Scotland, France, and Italy, living among Europe's kings, queens, and princes, and studying their history and the chivalry of their courtiers. The ideals of knighthood and the heroic deeds of the time were a part of the romantic accounts in his *Chronicles*.

Giovanni da Montecorvino
1246-1328

Founder of the Catholic mission in China. After the Nestorian priest Rabban Bar Sauma (c. 1220-1294) visited from China, Pope Nicholas IV decided to send Giovanni, then serving as a missionary in Persia, further east. Giovanni departed in 1289 with letters to a number of rulers along the way, from the king of Armenia to Kublai Khan (1215-1294) himself. He stopped in India for more than a year and converted some 100 people, but by the time he arrived in China in 1294, Kublai was dead. In 1299 and 1305 he established two churches in Peking, and during this time bought from slavery around 150 boys. These he educated, using the Chinese language as well as Latin and Greek, in Christianity and the Catholic liturgy. He eventually translated the New Testament and Psalms into Chinese, and by the time he died, the See of Zaiton was well established.

Godfrey of Bouillon
1060-1100

French nobleman and a leader of the First Crusade (1095-1099). In 1099, Godfrey gained the title "Protector of the Holy Sepulchre," and defended the crusaders' gains against an invading force from Egypt. Godfrey, who was handsome, dashing, and died young, later became the focus of legends that portrayed him as a perfect Christian knight.

Diogo Gomes
1440-1482

Portuguese navigator who sailed the west coast of Africa exploring and establishing trading relationships with the Africans. He sailed up the Gambia River in 1458 and traded with the leader of the region. In 1462 Gomes and Antonio de Noli discovered the Cape Verde Islands; Gomes explored all 10 of the islands with Noli and other captains.

Antão Gonçalves
c. 1400s

Portuguese captain and trader who, in 1441 sailed along the western coast of Africa to just south of the Río de Oro where he explored the area, collected sealskin and seal oil, and took two African prisoners. On a second and third voyage, both to the Río de Oro and the last in 1445, Gonçalves did more to further the evergrowing trade of slaves than exploration.

Gudrid
fl. 1000s

Viking explorer and sister-in-law of Leif Eriksson (c. 970-c. 1020). Gudrid first married Leif's brother Thorstein, who in 1006 led an unsuccessful expedition to Vínland, a colony in what is now eastern Canada discovered earlier by Leif. Thorstein died on the return voyage, and Gudrid married another brother, Thorfinn (980-c. 1007). The two took part in another expedition to Vinland, making Gudrid the first European woman to set foot on the North American continent, and in 1005 she gave birth to the first European child in North America, a son named Snorri. Later the family returned to Greenland.

Hayam Wuruk
1334-1389

Javanese ruler of Majapahit who extended the influence of his Hindu kingdom throughout much of what is now Indonesia. Using sea power, Hayam maintained a monopoly over trade throughout the region, but his division of lands between his sons led to the rapid decline of Majapahit and thus of Hindu influence in Java.

Hugh of Jabala
fl. 1140s

French prelate who first related the story of Prester John to the historian Otto of Freising (c. 1111-1158). Hugh served as bishop of Jabala, now the city of Jubayl in Lebanon, a coastal town that in ancient times had been the Phoenician city of Byblos. Sent by the crusader king Raymond of Antioch to gain support from Pope Eugenius IV for a new crusade, he told the pope about a Christian king in the East who had supposedly attempted to assist the crusaders in their efforts against the Muslims of Palestine. Otto later included the story in his *History of the Two Cities* (1143-46). In 1165 a forged letter, allegedly from Prester John, appeared in Europe. The quest for the legendary king would inspire numerous European efforts at exploration through the time of Prince Henry the Navigator (1394-1460) and beyond.

Ibn Jubayr
1145-1217

Spanish Muslim journeyer who wrote a popular book, the *Rihlah,* about his 1183-85 pilgrimage to Mecca. The *Rihlah* provides valuable information about the Christian and Muslim lands along the Mediterranean, as well as the practices of Genoese mariners. In modern times, it was translated into both French and English.

al-Istakhri
c. 900s

Arab geographer who is known for his contribution to the study of geography with his book *Routes of the Provinces,* a publication of regional maps marked by detailed illustrations. This book was later revised, expanded, and renamed *Book of Roads and Provinces.*

Jean de Joinville
1224?-1317

French historian who accompanied King Louis IX on the Seventh Crusade (1248-1254) to Egypt. His *Histoire de saint-Louis,* for which he is best remembered, chronicles Louis IX's life and provides an important account of the crusade. Joinville was originally a member of Champagne's lower aristocracy and worked his way up to expert status in the matters of court procedures.

Kublai Khan
1215-1249

Mongolian general who conquered China and went on to rule its Yüan, or Mongol, dynasty. Kublai Khan was the grandson of Genghis Khan, leader of the nomadic Mongols. Compared to his grandfather, who conquered with an iron fist, Kublai became known for his great humanity. Kublai succeeded his brother Mangu as leader of their grandfather's empire in 1260, and in 1279 conquered the Sung dynasty, thus gaining control of both North and South China. While continuing his rule of Mongol dominions in southern Russia, Persia and Mongolia, Kublai became the first emperor of China's Yüan dynasty. He established a magnificent capital city at Cambuluc, which is now Beijing.

Ragnar Lothbrok
fl. 800s

Scandinavian monarch whose military exploits, including a battle with Charlemagne, became legendary in medieval European literature. Lothbrok's story was recounted in several Icelandic sagas and the twelfth-century *Gesta Danorum* by Danish historian Saxo Grammaticus, in which Lothbrok is reported to have been captured by the Anglo-Saxons and thrown into a snake pit to die. Anglo-Saxon legend maintains that it was Ragnar's three sons who led the Viking invasion of East Anglia in 865 to avenge their father's murder.

Louis IX
1214-1270

French king and crusader, Louis was the son of King Louis VIII and his wife, Queen Blanche of Castille. Although fourth in line, the deaths of three of his brothers made Louis heir to the throne. When his father died in 1226, Louis IX became king at 12. Under the guidance of his mother, Louis set out to end the long struggle between France and the Plantagenets of England for control of French soil. At 15, he commanded French troops in a battle against Henry III, forcing the British ruler to withdraw. Louis led two crusades during his reign, the first, to the Holy Land, was an attempt to capture Jerusalem from the Muslims. After a valiant effort, his troops, wearied from battle and decimated by the plague, were forced to retreat. Louis's second crusade took him to Tunisia. After a series of victories, his troops again fell to disease. The ailing king did not survive the crusade, and passed on his kingdom to his son Philip before dying in August, 1270. Despite his defeats, Louis IX was renowned for his courage and wisdom, and was often asked by other monarchs to arbitrate disputes. He was canonized by Pope Boniface VIII in 1297, becoming the only French king to achieve sainthood.

Otto of Freising
c. 1111-1158

German bishop and historian who furnished the first written mention of the Prester John legend. Otto was present at a meeting in 1145 between Hugh of Jabala and Pope Eugenius IV when Hugh related the story of "a certain John, a king and priest who lives in the extreme Orient, beyond Persia and Armenia, and who like his people is a Christian...." The tale, which appeared in Otto's *History of the Two Cities,* (1143-46), inspired numerous European efforts to locate the mythical king's realms, first in Asia and later in Ethiopia.

Maffeo Polo
fl. 1200s

Venetian merchant and trader who made several extended trips into central Asia to the court of Kublai Khan. He made the first trip, 1253-1269, with his brother Niccoló; the second, 1271-1295, with Niccoló and Niccoló's son Marco. They were among the first Europeans to visit this part of the world. The account of their travels published by Marco Polo encouraged Europeans to extend their explorations and trade during the subsequent Age of Exploration.

Niccoló Polo
fl. 1200s

Venetian merchant and trader who made several extended trips into central Asia to the court of Kublai Khan. He made the first trip, 1253-1269, with his brother Maffeo; the second, 1271-1295, with his son Marco and Maffeo. They were among the first Europeans to visit this part of the world. The account of their travels published by Marco encouraged Europeans to extend their explorations and trade during the subsequent Age of Exploration.

Nicholas IV
1227-1292

Italian pope (born Girolamo Masci) who sent missionaries around the world, and received Rabban Bar Sauma (c. 1220-1294) from China. The first Franciscan pope, Nicholas was elected in 1288. Three years later, when the last crusader stronghold at Acre fell, he attempted unsuccessfully to organize a new crusade. Around this time, the Nestorian priest Bar Sauma arrived, and the two men joined in a second unsuccessful effort to launch a holy war against the Muslims. Nicholas sent missionaries to the Bulgars, Ethiopians, Tatars, and Chinese, the most celebrated of these being Giovanni da Montecorvino (1246-1328), who established the Catholic mission in Peking.

Rajaraja I
r. 985-1014

Tamil ruler of the Chola Empire in southern India who conquered Ceylon (modern Sri Lanka) and the Maldive Islands. Rajaraja enlarged and unified his realm, defeating a number of other kingdoms along the way. He was also the first significant ruler of India to employ naval forces, which he used in his conquests of the islands.

Rajendra
r. 1014-1044

Tamil ruler of the Chola Empire in southern India who conquered large areas and conducted extensive foreign trade. Son of Rajaraja I (r. 985-1014), Rajendra extended Chola rule as far north as the Ganges River, and traded with Hindu lands in Southeast Asia.

Raymond of Toulouse
1042-1105

French nobleman and leader in the First Crusade (1095-1099), who established the Latin County of Tripoli in the Holy Land. Known variously as Raymond IV and Raymond de Sainte-Gilles, Raymond took part in the successful attacks against Antioch and Jerusalem in 1098 and 1099 respectively. During the period 1100-1105, he fought to prevent fellow crusader Bohemond I (c. 1050-1111) from expanding southward. Almost alone among all crusaders, Raymond maintained favorable relations with the Byzantine Empire.

Richard I, the Lion-Heart
1157-1199

English king who led the Third Crusade to the Holy Land. Richard was the third son of Henry II and Eleanor of Aquitaine. At the age of 11, he was given control of his mother's Duchy of Aquitaine, and was named Duke of Poitiers in 1172. Richard held no allegiance to his father, and joined his brothers Henry and Geoffrey in a rebellion against the king in 1173. In 1189, he joined forces with King Philip II of France to force his father out of power and placed himself on the throne. In 1190, Richard embarked on the Third Crusade, hoping to regain Christian authority of the Holy Land from the Muslim sultan Saladin. His forces nearly took Jerusalem twice, but in the end Saladin prevailed. In 1192, Richard received word that his brother John was conspiring with Philip II against him and set off for home. On the way, a storm sent his ship aground near Venice, and he fell into the hands of the Austrian Duke Leopold, who held a grudge against him. The duke held Richard prisoner for a time, then turned him over to the

German king, Henry VI, who kept him locked up at various castles until England paid a hefty ransom in 1194. Richard spent his last few years in battle against Philip II. In 1199, his desire for wealth led to his early death. Seeking treasure, he attacked the castle of the Vicomte of Limoges and was mortally wounded. Although not a particularly successful ruler, Richard's reputation as a courageous soldier earned him the nickname Coeur De Lion (the lion-heart).

Rurik
d. 879

Varangian prince who in 862 established the Rus dynasty in Novgorod, regarded as the historic foundation of the Russian state. According to the Russian *Primary Chronicle,* compiled at the beginning of the twelfth century, the people of Novgorod, tired of political unrest, invited the Varangians (Scandinavian interlopers also known as the Rus) to establish a stable government. Rurik, along with his two brothers, responded by seizing control of Novgorod and the surrounding region. Whether Rurik was an invited monarch or simply a conqueror is debatable. The Rurik dynasty ruled Russian until the death of Fyodor I in 1598.

Tancred
1078?-1112

Norman knight and one of the leaders, along with his uncle Bohemond I (c. 1050-1111), of the First Crusade (1095-1099). After fighting alongside Bohemond at Antioch in 1098, Tancred went on in the following year to lead the savage assault on Jerusalem that ensured victory for the crusade. In the years that followed, he held positions as Prince of Galilee and regent of Antioch and Edessa, and became the most powerful European in the Middle East. He spent much of last two decades in war against the Turks and Byzantines.

Tristão Vaz Teixeira

Portuguese military adventurer who sailed with João Gonçalves Zarco in the Atlantic Ocean, where off the northwest coast of Africa they discovered an unclaimed island with fresh water and fertile soil, which they named Porto Santo, meaning "holy haven." Around 1420 Teixeira returned with Zarco to the island to form a colony, whose first governor, Bartolomeu Perestrello, later became the father-in-law of Christopher Columbus. Though unsuccessful, the colony was a starting point for further Portuguese explorations commissioned by Prince Henry the Navigator and would lead to the discovery of Madeira, sighted nearby.

Gunnbjorn Ulfsson
fl. c. 920

Icelandic voyager who was the first European to report sighting a land to the west of Iceland, which turned out to be Greenland. In about 920 Gunnbjorn's ship was blown off course during a voyage from Norway to Iceland, and he saw what appeared to be an island. More than 60 years later, in 982, Erik the Red (c. 950-1004) sailed westward in search of the land that had been described by Gunnbjorn, and soon found it.

Ugolino and Vadino Vivaldi
fl. 1200s

Italian merchants who made the first European attempt to reach Asia by sailing westward. In 1281 or 1291, more than two centuries before the historic first voyage of Christopher Columbus (1451-1506), the Genoese Vivaldi brothers set sail for the Indies in two ships. The crafts left Genoa, passed through the Straits of Gibraltar, and landed briefly on the Moroccan coast before sailing westward, never to be heard from again. In later years the Vivaldi brothers became the subjects of legends that depicted them as having circumnavigated Africa before being captured by the mythical Christian king Prester John.

William of Malmesbury
c. 1090-c. 1143

English historian whose writings are a major source of information regarding the Anglo-Saxon period and the subsequent Norman invasion. His first notable work was *Gesta regum Anglorum* (c. 1125), an account of England's kings modeled on the writings of the noted English historian Bede (c. 672-735). He followed this with *Gesta pontificum Anglorum* (c. 1126), and *Historia novella,* which covered events in England up to 1142. Among the other items of historical or geographical significance mentioned in William's work were tales of a British king who served as the model for the legendary Arthur, as well as a reference to a shrine to St. Thomas in India.

Al-Ya'qubi
d. 897

Arab historian and geographer whose *Kitab al-buldan* was the first scientific treatment of historical geography produced by the Arab culture of the Middle Ages. At various times al-Ya'qubi lived in Armenia and Khorasan (modern Afghanistan), and he traveled both in India and the Maghrib region of North Africa. His *Ta'rikh ibn wadih* was a history of the world to 872.

Yeh-lü Ta-shih
1098-1135

Khitan (northern Asian) king whose defeat of a Seljuk Turk leader in Persia probably served as the basis for the Prester John legend. A "barbarian" nation to the north of China, the Khitan had established a dynasty called the Liao, but in 1125 they were driven out by the rival Jurchen people. Yeh-lü Ta- shih, a member of the ruling house, escaped with some 200 subjects to central Asia, where with the support of numerous Turkish tribes, he founded the Kara-Khitai empire. His forces defeated those of Sanjar, Turkish ruler of Persia, in battle in 1141, and as tales of the conflict moved westward, the story changed dramatically. Eventually Yeh-lü Ta-shih was transformed into Prester John, a Christian ruler eager to aid the crusaders in Palestine. In 1145 Bishop Hugh of Jabala—who presumably knew nothing of Yeh-lü Ta- shih—reported this story to the historian Otto of Freising (c. 1111-1158), and in the years that followed, Europeans engaged in a quest to find Prester John and his kingdom. As for the Kara-Khitai, they were overthrown in 1211 by Mongol invaders.

Yung-lo
1360-1424

Chinese Ming Dynasty emperor who sent Admiral Cheng Ho (c. 1371- c. 1433) on a series of naval expeditions to Southeast Asia, India, Persia, Arabia, and even Africa. Also known as Chu Ti (Yung-lo was his reign title), the emperor in 1405 commissioned what would turn out to be six voyages during his lifetime. Though the purpose of these expeditions was to display Chinese power, they also yielded some exchanges, including the delivery of an African giraffe to Yung-lo's court. In 1421, Yung-lo moved the imperial court from Nanjing to Beijing, where China's government has remained ever since. At Beijing, he built the "Forbidden City," a palace some 5 mi (8 km) in circumference, containing approximately 2,000 rooms in which more than 10,000 servants attended the imperial family.

Antonio Zeno
fl. 1300s

Venetian navigator who, with his brother and perhaps Scottish explorer Sir Henry Sinclair, is reputed to have followed the route of European fisherman to North America in 1398, nearly a century before Christopher Columbus, Amerigo Vespucci, and other "professional" explorers discovered the New World. Sinclair's voyage in the northeast Atlantic purportedly resulted in a visit to Nova Scotia, and was documented in letters written home. Zeno's letters remained unpublished until several centuries after his death, surfacing in 1558 when published by his great-great-great-grandson, Nicolo Zeno. The plausibility of Zeno's story was damaged by the imagination of Nicolo, who embellished the manuscript with maps and other editorial materials not originally included in Antonio's journals.

Bibliography of Primary Sources

Bar Sauma, Rabban. *The Monks of Kublai Khan* (c. 1290). A record of the author's journey from present-day Beijing, China, to France, on a religious mission on behalf of his Nestorian sect of Christianity. The work did not appear in English until 1928, in a translation by Sir E. A. Wallis Budge.

Bekri, al-. *Al-Masalik wa'l-mamalik* (c. 1070). Al-Bekri, a Spanish-Arab traveler who visited West Africa in about 1067, wrote about his travels in this work. It is notable for its description of the kingdom of Ghana, particularly its great wealth, the power of its king, and the potential for trouble posed by the society's division between centers of Islam and traditional African religions.

Biruni, al-. *Kitāb fī tahqīq ma li'l-Hind* (Al-Biruni's India). This book, written nearly 300 years before the travels of Jordanus, discusses and describes the history, geography, and religion of India. Biruni's work is considered one of the greatest medieval works of travel and social analysis, and is still of great interest to scholars today.

Conti, Niccolò de', and Poggio Bracciolini. *Historiae de veritate fortunae* (c. 1450). A record of Conti's travels throughout India, Indonesia, Africa, and the Middle East. During his travels Conti renounced Christianity; to atone for this act, Pope Eugenius IV demanded that Conti write an account of his journeys upon his return to Venice. Vatican secretary Poggio Bracciolini acted as scribe. The work wasn't published until 1723.

Froissart, Jean. *Chronicles*. Renowned work that detailed the history of Western Europe from the early 1300s to 1400, supplying valuable contemporary information about medieval life and society, noble courts, and warfare. Froissart was a French historian, poet, scholar and traveler. The ideals of knighthood and the heroic deeds of the time were a part of the romantic accounts in his *Chronicles*.

Giovanni da Pian del Carpini. *History of the Mongols Whom We Call the Tartars*. An excellent firsthand account of the Mongols, this work has often been cited as the best reference on the subject from this time period. It described the Mongols' character, history, foreign policy, and military tactics, and included a section on the best way to defeat or resist the Mongols in case of attack.

Ibn Battuta. *Rihlah* (Travels) (1350s). Details Battuta's travels to India, Southeast Asia, China, Muslim Spain, across the Sahara, and into the splendid African empire of Mali. The book was later published in English as *The Travels of Ibn Battuta.*

Ibn Fadlan, Ahmad. *Risala* (900s). Account of his diplomatic mission to Russia to explain Islamic law. The text includes details of his visits and experiences among various Turkic peoples as well as the Vikings.

Ibn Hawkal. *On the Shape of the World.* An account of the author's 30 years of travel throughout the Muslim world, from Spain to Central Asia, and Afghanistan to West Africa. The text is noted for its accuracy, scientific approach, and attention to detail.

Ibn Jubayr. *Rihlah* (c. 1200). A journey of the author's 1183-85 pilgrimage to Mecca. The *Rihlah* provides valuable information about the Christian and Muslim lands along the Mediterranean, as well as the practices of Genoese mariners. In modern times, it was translated into both French and English.

Idrisi, al-. *Kitab al-jami-li-sifat ashtat al-nabatat.* The best-known of Idrisi's many contributions to the science of botany and medicinal plants. He compiled the written knowledge of the day with research gathered on his journeys. The names of the drugs included in his books were translated into several languages, including Persian, Hindi, Greek, Latin and Berber.

Idrisi, al-. *Kitab nuzhat al-mushtaq fi ikhtiraq al-afaq* (The delight of him who desires to journey through the climates), also known as *Kita Rujar* (The book of Roger) (1154). One of the greatest geographical works of the Middle Ages, this book was written as a companion to a silver planisphere on which was etched a map of the world, complete with major cities, lakes and rivers, mountains and trade routes. The book compiled information from existing Greek and Arabic works, and also included drawings and reports from contributors that had been sent to foreign lands for the project.

Joinville, Jean de. *Histoire de saint-Louis.* Chronicles Louis IX's life and provides an important account of the Seventh Crusade (1248-1254) to Egypt.

Jordanus of Séverac. *Mirabilia descripta* (Description of marvels) (early 1330s). A Dominican missionary to India, Séverac was the first European to visit that region since Giovanni da Montecorvino. His account discusses Armenia, Persia, and India, as well as lands Jordanus himself had not actually visited. He was the first writer to identify Prester John as the emperor of Ethiopia, thus adding a key aspect to a legend that fascinated Europe—and drove much of Europeans' exploration efforts—from the twelfth to the fifteenth centuries.

Mandeville, John. *The Travels of Sir John Mandeville* (c. 1300s). Loosely based on Odoric of Pordenone's journeys, this book was particularly popular in the later Middle Ages and was used as a manual by other travelers and geographers of the period.

Maqdisi, al-. *Ahsan al-Taqasim fi Macrifat al-Aqalim* (The best division for knowing the provinces) (c. 985). This work recounts the journeys of al-Maqdisi, who traveled throughout much of the Arab and Muslim world during his lifetime. Al-Maqdisi's painstaking attention to matters such as climate, local economies, ethnicities, cultures, and units of weight and measurement make this one of the great works of Islamic geography.

Mas'udi, al-. *Muruj adh-dhahab wa ma'adin al-jawahir* (The meadows of gold and mines of gems)(c. 942). In this landmark work, al-Mas'udi examined the grand sweep of history from the creation of the world to the present time. The book includes detailed observations on culture, customs, flora and fauna, and ethnicity. Pearl-diving in the Persian Gulf, the great temples of the world, hazards posed to mariners by waterspouts, and the burial customs of the Hindus are but a few of the topics covered in the work.

Odoric of Pordenone. Travel narrative as dictated to Brother William of Solagna. Odoric of Pordenone was a Franciscan missionary who traveled extensively through Asia. He is the first European traveler to describe distinctions between Oriental and Occidental culture both accurately and in detail. Odoric's account is viewed as mostly factual, though it includes many fantastic tales. His recollections have remained popular because of his clear evocation of both the everyday and the unusual, and they, along with those of his fellow Franciscans, Giovanni da Pian del Carpini and Willem van Ruysbroeck, were plagiarized and combined in *The Travels of Sir John Mandeville.* This book was extremely popular in medieval Europe, and focused on the most extraordinary events and places that the Franciscans recounted. Despite its almost purely fantastic content, the book was used as a travel guide by medieval explorers and merchants of the period.

Otto of Freising. *History of the Two Cities* (1143-46). The first written mention of the Prester John legend. Otto was present at a meeting in 1145 between Hugh of Jabala and Pope Eugenius II when Hugh related the story of "a certain John, a king and priest who lives in the extreme Orient, beyond Persia and Armenia, and who like his people is a Christian...." The tale, which appeared in Otto's book, inspired numerous European efforts to locate the mythical king's realms, first in Asia and later in Ethiopia.

Polo, Marco. *Divisament dou Monde* (Description of the world) (1298). Originally called *Il milione*, and now known generally as "The Travels of Marco Polo," this book was thought by contemporary readers to be a work of fiction. Only later was it realized that most of its contents are quite accurate. In any case, it served to excite Europeans about the riches in trade and culture that might be found in unfamiliar areas of the world and to encourage them to venture out in search of them.

Schiltberger, Johann. *The Bondage and Travels of Johann Schiltberger.* A significant description of medieval Muslim culture based on the author's experiences in the Middle East, which began when he was taken prisoner by the Turks during a 1394 battle.

Ya'qubi, al-. *Kitab al-buldan* (c. 890). The first scientific treatment of historical geography produced by the Arab culture of the Middle Ages.

JOSH LAUER

Life Sciences and Medicine

Chronology

c. 700 Chen Chuan, a Chinese physician, notes the sweetness of urine in patients with diabetes mellitus.

c. 700-c. 900 Arab doctors systematically translate and adapt ancient Greek medical texts, ensuring that the advanced learning of antiquity is preserved.

c. 825 Ophthalmology flourishes in Islamic medicine, with works such as *Disorder of the Eye* by Ibn Massawaih (Mesue the Elder).

896 Ar-Razi (also known as Rhazes), a Persian physician, differentiates between the measles and smallpox, and later becomes the first scientist to classify all substances as either animal, vegetable, or mineral.

975 Abu Mansur al-Muwaffaq writes a treatise in which he classifies some 600 cures.

1010 The School of Salerno, founded two centuries earlier, flourishes; the first institution in Europe to grant medical diplomas, it brings in students from around the Mediterranean, and in 1170, an affiliated physician named Roger of Salerno publishes the first book on surgery in the West.

1037 Ibn Sina (also known as Avicenna), the great Persian physician, dies; his medical encyclopedia is destined to become an accepted authority for many centuries.

c. 1175 Ibn Rushd (also known as Averroës), an Arab physician in Spain, compiles Galen's medical works, which have an enormous influence on medieval science.

1224 Holy Roman Emperor Frederick II issues laws regulating the study of medicine, thus elevating the status of real physicians and diminishing the number of quacks; later, in 1241, he becomes the first major European ruler to permit dissection of cadavers, formerly prohibited by religious law.

c. 1250 Albertus Magnus, a German scholar, writes *De Vegetabilibus,* destined to become the most significant work on natural history written in western Europe during the Middle Ages.

1270 Ibn an-Nafis, an Arab physician, correctly describes the pulmonary circulation.

1285 Probable date for the invention of eyeglasses, most likely in Venice.

1345 The first drug store, or apothecary shop, opens in London.

1348 Military medicine first begins with the treatment of wounds from newly invented firearms.

Overview: Life Sciences and Medicine 700-1449

The Ancient World

The Greeks had made important contributions to the life sciences, including writings on medicine by Hippocrates (c.460-c.377 B.C.), work on plant and animal classification by Aristotle (384-322 B.C.), and books on anatomy by Galen (c.130-c.200). In the Roman world, however, interest in science and medicine declined. With the final collapse of the Roman Empire in A.D. 476, Europe entered a long period when scientific knowledge and inquiry were largely absent. The Church assumed a rigid mantle of authority, refusing to recognize any scientific explanations for natural phenomena that did not match its own particular interpretation of Scripture, and frowned on inquiries that might prove otherwise.

Medicine

From the eighth to the eleventh centuries, the quality of medicine in Europe was so poor that many sick people sought relief from miracle healers and by making pilgrimages to holy shrines. They also consulted astrologers who studied the supposed influence of stars and planets on human conditions.

In the Arab world, however, medical science had made great strides. The founding of Islam in the seventh century had established a sophisticated culture that valued learning and inquiry. Arab scholars translated the works of ancient Greek writers into Arabic, keeping the work of earlier physicians alive. Several of these translators also made important contributions of their own, particularly in medicine.

Abu Bakr Muhammad ibn Zakariya' ar-Razi, known in the West as Rhazes (c.865-c.923) was a physician who carefully observed the characteristics of measles and smallpox and noted how they differed from each other, introducing the idea that there were different infectious diseases with identifiably different symptoms.

The Arab physician, Ibn an-Nafis (c.1210-1280), studied anatomy and discovered that blood circulates between the heart and the lungs. He criticized some of Galen's anatomical observations at a time when Galen was still considered infallible in the West. Yet another Arab physician, ibn Sina, known in Europe as Avicenna (980-1037), wrote the *Canon of Medicine,* which became an important medical text for hundreds of years, not only in the Middle East, but in Europe after it was translated into Latin.

Many ancient Greek and Roman medical texts were not reintroduced into Europe until the Crusades brought contact with the Arab world; these texts were eventually translated from Arabic into Latin. The works of Galen, Hippocrates, and Aristotle once again became important influences on the study of medicine and the life sciences in Europe, especially from the eleventh to the thirteenth centuries.

Medical Schools and Hospitals

Two institutions were crucial to the limited scientific and medical progress made in during the Middle Ages: the university medical school and the hospital. The first medical school was founded at the University of Salerno in Italy in the ninth century. Around 1170, one of its professors, a physician named Roger of Salerno, wrote a practical guide to surgery. Several other medieval surgeons wrote important texts, including Guido Lanfranchi in Italy and Henri de Mondeville (1260-1320) and Guy de Chauliac (c.1300-1368) in France. But despite these studies, physicians performed less and less surgery, often leaving it to barbers. Not until the Renaissance was surgery firmly reunited with other medical practices.

Salerno remained the most distinguished medical school for over 200 years. Its success led to the establishment of medical schools in other cities, including Paris and Montpellier in France and Padua in Italy, each of which would play an important role in the development of medical knowledge in the Renaissance. Toward the end of the Middle Ages, it was at such institutions that anatomical dissections of human cadavers became a part of medical education. This led to more empirical observation and less reliance on the writings of ancient physicians such as Galen—discoveries that ultimately corrected numerous errors. This trend was slowed by the fact that dissection was illegal in many places and cadavers often had to be obtained clandestinely. Even where it was legal, the main source of cadavers was limited to the bodies of executed criminals.

Exposure to the Arab world during the Crusades was a major factor in the development of hospitals. Islamic ideals of charity and public welfare encouraged the establishment of institutions where the sick could be given aid and where those with infectious diseases could be isolated to prevent contagion. Using this model, hospitals were established throughout Europe by the Church, where religious orders cared for the sick and the poor. Conditions in these hospitals, however, were extremely crude. It was not uncommon, for example, for more than one patient to share a bed. In thirteenth-century London the first institution for the mentally ill, Bethlem Royal Hospital was created. Its name, commonly shortened to "Bedlam," has come to mean a chaotic, disordered place.

Infectious Diseases

Contagious diseases created severe problems throughout this period. Leprosy was common, though sometimes confused with other skin ailments. Lepers were shunned and, since the disease was considered by many to be the result of sinful behavior, were often considered not only physically but spiritually diseased. Smallpox, measles, and other diseases were also common.

Beginning in 1348, recurring epidemics of bubonic plague killed one third of the entire European population, often wiping out whole villages and causing enormous social and economic upheaval. Though the role of rats and lice in transmission of the plague was not understood (and would not be known until the end of the nineteenth century), improved sanitation did seem to slow its spread. This observation led to the establishment of sanitary laws in 1388 by Richard II of England (1367-1400). It was also at this time that the practice of quarantining arose, preventing those who had been infected or who had contact with the infected, from mixing with the rest of the population until they were no longer considered contagious.

Botany and Zoology

During the Middle Ages, botany and zoology were at a very rudimentary stage of development. Plants were of interest primarily as sources of medicine. Most of the texts upon which physicians and others relied on for information about plants were copies of a book written by a first-century Greek writer named Dioscorides (c. 40-c. 90), who described the medicinal uses of about 500 plants. These books, called herbals, were often illustrated, but with images so crude that it was frequently impossible to identify a plant on the basis of its picture. There were also books called bestiaries that described and pictured exotic animals, some of which, like the unicorn, were imaginary or based on erroneous descriptions brought back to Europe by travelers.

By the twelfth century, however, interest in the living world had reawakened. Plants began to be studied not only for their medicinal uses but also for their structures and relationships to other organisms; there was also increased interest in animals. One of the most important figures in the life sciences during the Middle Ages was Albertus Magnus (c.1200-1280), a Dominican bishop and teacher of Thomas Aquinas (1225-1274). Like others interested in the living world, he was greatly influenced by Aristotle's biological writings, but Albertus Magnus also made many original observations and was a source of useful information for later scholars. His research legitimized study of the natural world as a science within the Church; so revered was Albertus that he came to be known as "the Great One" (Magnus) during his own lifetime. Another keen observer of animals was the Holy Roman Emperor Frederick II (1194-1250), who wrote a book on falconry, the art of hunting of with birds of prey called falcons.

The Future

By the middle of the fifteenth century, interest in the natural world was increasing. Universities became mature institutions that served as centers of new studies in anatomy, botany, and zoology. With the dawn of the Renaissance, one of the driving forces for studies of the natural world would come from artists such as Leonardo da Vinci (1452-1519) and Albrecht Dürer (1471-1528), who had interests in all these areas of the life sciences.

ROBERT HENDRICK

The Medical Influence of Rhazes

Overview

The Persian physician known as Rhazes (c. 865-c. 923), or ar-Rhazi (Abu Bakr Muhammad ibn Zakariya' ar-Razi) is primarily remembered for his encyclopedia of medicine and for his pioneering work on differentiating between smallpox and measles. His great synthesis of Greek and Arabic medical learning was first published under the title *Kitab al-hawi*, but it is better known in the form of a Latin translation published in 1279 as the *Liber continens*. The work was considered quite controversial at the time because of the author's willingness to criticize the Greek physician Galen (c. 130-c. 200), generally considered an infallible source of medical knowledge. For almost three centuries, the *Liber continens* served as the main source of Western therapeutic knowledge. Rhazes' book *A Treatise on Smallpox and Measles* has become a landmark in the development of the concept of specific disease entities and the value of diagnostic precision.

Background

The Middle Ages of European history roughly correspond to the golden age of Islam, the religion founded by the prophet Muhammad (570-632). Just as Latin served as the common language of learning for students throughout Europe, Arabic was the language of learning throughout the Islamic world and Persians, Jews, and Christians took part in the development of Arabic medical and scientific literature. Pharmacology, optics, chemistry, and alchemy were of particular interest to Arab scientists.

For many European scholars, so-called Arabian medicine was significant only in terms of the role it played in preserving Greek philosophy during the European Dark Ages. Until rather recent times, European scholarship generally dismissed evidence of originality in the works of medieval Arabic medical and scientific writers and assumed that the primary accomplishment of Arabic science, medicine, and philosophy was the preservation and transmission of ancient Greek learning. In general, however, medieval scholars, physicians, and philosophers accepted the writings of the ancients as true and authoritative. Since the 1970s scholars have redefined "Arabian medicine" as "Islamic medicine," in reference to the translation, assimilation, and transformation of the texts, theories, and concepts of the ancient Greek philosophers which were introduced into Arab countries in the ninth century. For contemporary scholars, Islamic medicine, therefore, reflects the assimilation, integration, and development of the many elements that formed Islamic culture.

The writings of Islamic physicians and philosophers, which were often presented as commentaries on the work of Galen, were eventually translated from Arabic into Latin and served as fundamental texts in European universities. Medieval physicians and scholars, whether Muslims, Jews, or Christians, generally shared the assumption that Galenism was a complete and perfect system. Thus, many new insights into the history of science and medicine have been revealed through attempts to study the work of Islamic writers, such as Rhazes, on their own terms.

Impact

Although medieval physicians, Muslim and Christian alike, generally assumed that Galenism was a complete and perfect system, the great sages of Islamic medicine are worth studying in their own right, not just in terms of their role in preserving classical medicine. Latin translations of the medical writings of Rhazes, Avicenna (Ibn Sina, 980-1037), Haly Abbas, Averroës (Ibn Rushd, 1126-1198), and Albucasis (al-Zahrawi, 936-1013), were most influential in Europe, but many of these writers were also well known as philosophers and alchemists.

Rhazes has long been honored as one of the greatest physicians of the Islamic world, as well as one of the most scientifically minded physicians of the Middle Ages. A man of remarkable energy and productivity, Rhazes was the author of at least 200 medical and philosophical treatises, including the famous *Continens*, or "Comprehensive book of medicine." Although Rhazes claimed that the work was still incomplete, it was so massive that a two-volume Latin edition printed in 1486 weighed over 20 pounds.

In answer to charges that he had overindulged in life's pleasures, Rhazes published a book called *The conduct of a philosopher*. Here Rhazes described himself as a man who had always been moderate in everything except acquiring knowledge and in writing books. He claimed that he worked day and night and damaged his eyes and hands by writing over 20,000 pages in just one year. Nevertheless, Rhazes taught that a middle road between extreme asceticism and overindulgence was the most healthy way of life. Many biographers state that Rhazes became blind near the end of his life, probably as a result of his alchemical experiments. Although his colleagues urged him to undergo surgery to correct his loss of vision, the great physician told them that he was weary of seeing the world and he refused to undergo any medical or surgical treatments. Later biographical accounts, however, generally claimed that Rhazes became blind after his patron al-Mansur had the physician beaten on the head with one of his books for failing to provide proof of his alchemical theories.

After traveling widely and mastering a broad range of subjects, including philosophy, music, poetry, and logic, Rhazes became interested in medicine after a chance encounter with an apothecary in Baghdad. Thus, Rhazes was already over 40 by the time he began to practice medicine, but he soon mastered the art and established a great reputation as a healer. In competition with hundreds of candidates, Rhazes was chosen as physician-in-chief of one of the first major hospitals in Baghdad. Rhazes selected the healthiest location for the hospital by hanging pieces of meat at various sites in order to find the site where there was the least putrefaction. Through his private practice and his supervision of the hospital Rhazes compiled many intriguing case histories. These offer insight into the range of complaints for which his contemporaries consulted physicians, which signs and symptoms the physician thought significant, the kinds of treatment used, the occupations and family background of his patients, and the relationship between patient and physician. Among the discoveries credited to Rhazes are the identification of the guinea worm (*Dracunculus medinensis*), the recurrent laryngeal nerve, and spina ventosa. According to Rhazes, both physicians and patients were bound by ethical duties. In order to prevent and cure disease, patients were obligated to trust and cooperate with the physician. According to Rhazes, a learned physician and an obedient patient could vanquish illness. Unfortunately, not all patients were obedient and many quacks and impostors claimed to cure diseases.

Rhazes' book *A Treatise on Smallpox and Measles* has become a landmark in the history of medicine. Smallpox (*Variola*) is an acute viral disease usually transmitted by airborne droplets. Generally, the virus entered the body through the upper respiratory tract. The characteristics and virulence of the virus have apparently varied over time, but virologists have recognized two forms of smallpox: *Variola major*, which has a mortality rate of about 30%, and the mild form known as *Variola minor*, which has a fatality rate of about 1%. *A Treatise on Smallpox and Measles* provides valuable information about diagnosis, therapy, and concepts of diseases during the Middle Ages. At the time, in keeping with classic tradition, diseases were generally defined in terms of major symptoms, such as fevers, eruptive fevers, diarrheas, and skin lesions. Therefore, Rhazes' treatise on smallpox and measles is a major landmark in establishing the concept of specific disease entities. According to Rhazes, smallpox was essentially a stage in the transition from infancy to adulthood during which the blood fermented like wine. By suggesting that this change was a natural aspect of aging, Rhazes attempted to explain why almost all children contracted the disease. These observations indicate that smallpox was a common, perhaps ubiquitous, childhood disease at the time. Measles, which Rhazes recognized as a separate disease, was caused by very bilious blood. However, Rhazes admitted that even an experienced physician might have trouble distinguishing smallpox from measles. To protect his reputation, the physician should wait until the nature of the illness was obvious before giving his diagnosis. Both smallpox and measles could be described as eruptive fevers, but smallpox was more dangerous and almost invariably left survivors with pockmarks and scarring.

Translated into Latin, Rhazes' book had a profound influence on the treatment of smallpox in Europe into the seventeenth century. According to Rhazes, the physician might lessen the virulence of the disease by proper management at the onset, but once the disease was established the physician should encourage eruption of the pox by rubbing, steaming, purging, and bleeding. For many centuries, physicians accepted the "heat treatment" prescribed by Rhazes and wrapped patients in blankets in order to increase perspiration and promote the eruption of the pox. Various recipes were supposed to remove pockmarks, but the nearly universal presence of smallpox scars suggests that these remedies were useless. By differentiating between smallpox and measles Rhazes provided a paradigmatic case for thinking in terms of specific disease entities.

LOIS N. MAGNER

Further Reading

Books

Hopkins, D. R. *Princes and Peasants: Smallpox in History*. Chicago: University of Chicago Press, 1983.

Khan, M. S. *Islamic Medicine*. London: Routledge & Kegan Paul, 1986.

Meyerhof, M. *Studies in Medieval Arabic Medicine: Theory and Practice*. London: Variorum Reprints, 1984.

Rhazes. *A Treatise on the Smallpox and Measles*. Translated by W. A. Greenhill. London: Sydenham Society, 1848. Reprint, *Medical Classics* 4 (1939): 22-84.

Siraisi, N. G. *Medieval and Early Renaissance Medicine*. Chicago: University of Chicago Press, 1991.

Articles

Magner, L. N. "Smallpox: Most Terrible of All the Ministers of Death." *International Journal of Dermatology* 24 (1985): 466-470.

Early Medieval Medicine in Europe

Overview

Early medieval medicine in Europe saw little change since antiquity. The collapse of the western Roman Empire brought barbarian invasions and the rise of warrior fiefdoms to Europe, both of which hampered civilization and its amenities—including the practice of scientific medicine. Medical care was provided in a practical fashion based upon ancient ideas, with little regard for scientific methods. Religious influences crept into medicine, as those confined to monastic cloisters struggled to keep medical studies alive by copying and preserving the few original medical manuscripts of the Dark Ages. Not until the second millennium, around 1100, would the scholarly pursuit of medicine in western Europe experience a rebirth, when Greek and Arabic medical texts were brought to southern Italy and translated for the Latin-speaking cloistered West.

Background

Medieval European physicians (physics) based their medical care mainly upon the teachings of Galen of Pergamum (born c. A.D. 130 in what is now Bergama, Turkey). Galen's writings were prolific, and were based upon the relationship among the body's four humors, or bodily fluids (blood, phlegm, yellow bile, and black bile), and the four elements of external nature (earth, air, fire, and water). Galen taught that illness stems from an imbalance of the humors and the elements, and that restoring this balance would effect a cure. Galen described over 300 pharmaceutical remedies (mostly herbal concoctions), utilized bloodletting (phlebotomy), and depicted disease as an individual susceptibility rather than a general affectation. Galen took a reasoned approach to medicine based upon direct observation of the patient. Although Galen incorporated philosophy into medicine, his practical approach saw the physician as an attendant of the patient, rather than an intermediary seeking healing from spiritual sources. Galen's writings remained the dominant influence in medicine for almost a thousand years.

By the end of the fourth century, the Roman Empire was formally split, with the two halves ruled by separate emperors. By the end of the sixth century, the West was further broken into fragmented kingdoms, repelling attacks from each other, as well as continued invasions from the East. The economy of the East thrived, while the great cities of the West declined. Amid the constant atmosphere of war, the development of medical science was given little encouragement to flourish, and eastern and western medical thought parted ways.

In western Europe, as the Latin language declined, cloistered monks became responsible for the care-taking and dissemination of medical literature. The *Lorch Book of Medicine*, written about 795 in a German Benedictine abbey, discussed the humors, and contains a brief text on simple anatomy and prognostics. The book also contained therapeutic recipes and dietary treatments, and other practical advice for tending the sick. Years earlier in England, monks in then-remote North Umberland produced a medical book in the non-Latin language of Anglo-Saxon. The monks (among them, Bede, the Venerable [c. 672-735]) referred to the English healer as laece, or leech, and cited extensive plant remedies. Bede's writings also attributed certain diseases to bad luck, darts shot by elves, snakes, insects, or dragons.

In approximately 950, England's King Alfred was convinced by a British nobleman to commission a manual of the established medical

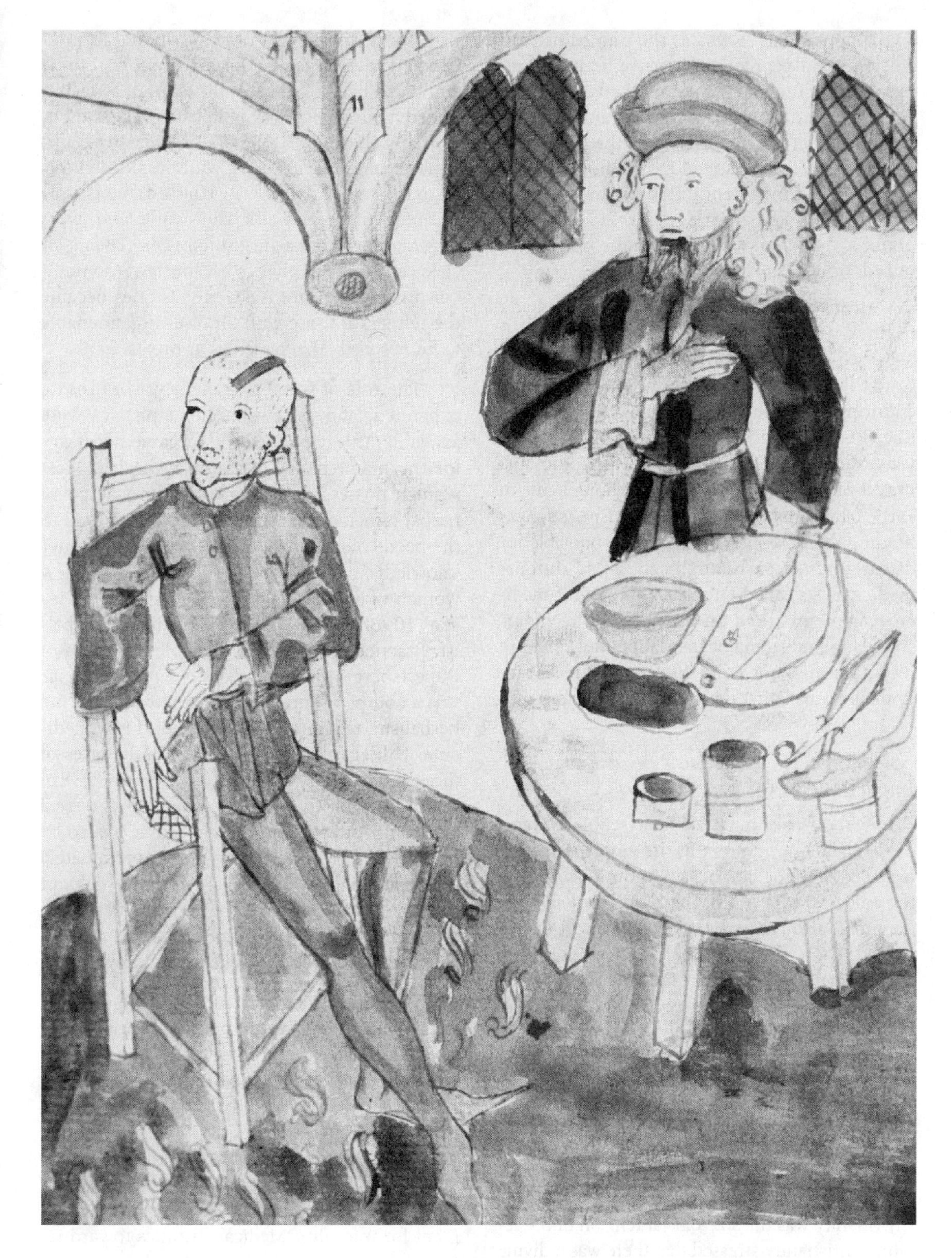

Medieval doctor with patient. *(Archivo Iconografico, S.A./Corbis. Reproduced with permission.)*

treatments of the day. The *Leech Book of Bald* combined the herbal practices of the Celts and Anglo-Saxons with that of the Greco-Romans and Arabs. Compiled by Bald, and scribed by a monk named Clid, the *Leech Book of Bald* is the oldest surviving Anglo-Saxon medical text. The book contains simplified Latin recipes with local ingredients replacing the exotic ones found in Mediterranean or Arab lands. Some scholars assert that the simplified language implies the book was intended to be shared with any of the literate population, and was meant to serve as a layman's manual, as well as the trained healer's aid. The book contains extensive herbal remedies containing mugwort, periwinkle, violets, vervain, wood betony, and yarrow, among other botanical treatments. Included in the *Leech Book* are remedies for constipation containing psyllium, a form of treatment still in use today. Other remedies invoked charms, prayers, drinking of potions made of water in which frogs were boiled, or walks on a moon-lit night.

Meanwhile, by the time of Muhammad's death in 632, almost all of Arabia had been won over to Islam, as well as Egypt and Spain, where Córdoba became Islam's European capital. After enduring prolonged warfare between the Byzantine (Roman) and Persian Empires, social chaos was exacerbated by 200 years of periodic outbreaks of bubonic plague. Although medicine in early Islam also included mysticism, such as holding the "evil-eye" and "jinn" responsible for disease, by the early ninth century, traditional medicine was called into question. Islamic medicine then embarked on the pursuit of scholarship, launching a major translation movement to recapture the knowledge of classical Greece and Rome by translating the works of Galen and others into Arabic.

Impact

While medicine maintained a scholarly (and mostly Galenic) course in the eastern empire, learned medicine languished in the West during the early Middle Ages. The number of schools declined, and trained physicians almost disappeared. Latin, the universal language of scholars of the time, took sanctuary in the church. Mysticism and ritual gained favor. The *Leech Book of Bald* is representative of the spirit of early medieval Europe—a classical base overwhelmed by a mysticism that impeded scientific discovery.

Along with the break-up of the Mediterranean civilizations, the rise of Christianity coincided with the stagnation of medical science in Western Europe. From the early fifth century, Christianity was the sole official religion of the region. Christianity stressed that there was a divine plan for every occurrence, and that Christian rituals and sacraments covered every significant event in the life of a believer, even illness and suffering. Christianity also maintained that the body and soul were separate, and demanded obedience to its doctrines of subordination. The physic only treated the body, while the Church cured the eternal soul. As the soul was prized above the flesh, Christian doctrines became ingrained in medicine.

While disease was often attributed to punishment by God, and suffering was a trial to be embraced before paradise, the Church also continued a mission of healing in early Medieval Europe. By the year 700, many monasteries contained makeshift hospitals, with monks dedicated to the care of the sick. By contrast, the Islamic hospitals that followed a short time later, were larger, more elaborate institutions. As Galenic, or classically trained, physics became fewer in number during the Dark Ages, monasteries became the refuge for those with illnesses not amenable to homestyle herbal remedies or rituals.

The role of women in medieval healthcare expanded, and as fewer trained physics were available, this expanded role became necessary for the maintenance of everyday health. Most women possessed a rudimentary knowledge of herbal remedies and first aid in order to tend to the needs of her household. Those with greater knowledge or skill became village healers. A few women wrote medical texts. Hildegard von Bingen (1098-1179), a German Benedictine herbalist, practiced medicine in her role as abbess of Rupertsberg. Hildegard's *Book of Simple Medicine* was a compendium of traditional lore featuring herbalism, religion, superstition, and folk medicine. Hildegard wrote on the natural causes of disease, advising treatment on the principle of opposites. Hildegard claimed that herbs were a gift from God, and those unable to be healed by herbs were willed by God to die. Approximately 300 years later, as men regained the prominent role in healthcare delivery, women healers were ostracized from the professional practice of medicine and, on occasion, were executed as witches.

Medieval texts revealed childbirth to be an all-woman affair. The mother was supported by a female healer/midwife, relatives, and neighbors. The social status of midwives began to rise, and some villages paid the local midwife to act in an official capacity for cases involving women's illnesses, childbirth, and care of infants. Midwives were also called upon to certify virginity and infant deaths, and to provide treatment for infertility. Medical attitudes toward reproduction (known as "generation" during medieval times) were not puritanical. The village midwife or healer had many remedies at her disposal (herbs and potions believed to contain the qualities of aphrodisiacs) for restoring humoral balance through sexual release.

Medical mathematics and astrology were reintroduced in early medieval European medicine, and combined the mathematical tradition of the ancients with the mysticism of the Church. The physic used tangled calculations and handy charts to illustrate the significance of the motions of the heavens on health and illness. Zodiac charts included the 12 signs, along with areas of the body each sign was said to influence. Often, herbs and minerals were included on the charts to denote the optimum time for their effectiveness. Because it was felt that God endorsed the powers of celestial bodies (God sent a star to announce Christ's birth), images of Christ sometimes appeared on zodiac charts, or Christ was portrayed with a halo of 12 sun rays, symbolizing the zodiac's 12 heavenly bodies. Physics used the zodiac charts to help determine the imbalance of the four elements in an ill patient.

The Islamic translations of ancient medical works by Galen and others set the stage for the reawakening of scholarly medical thought in the West. The Archbishop of Salerno, Alphanus (d. 1085) traveled to Constantinople in 1063 where he became familiar with Greek medical texts. Alphanus wrote *Premnon Physicon*, a philosophical approach to medicine that was inspired by the texts he studied in Constantinople. Alphanus's writings introduced the West to a Christianized Galenism, and advocated a scholarly approach to medicine that put the physic above the everyday village healer. Salerno soon became a known center for academics, funneling vast numbers of Arabic and Greek translations to the West. For the first time in 400 years, Latin scholars could share in contemporary medical thought. The academic rigor at Salerno provided impetus for the founding of the great western universities (Paris, Bologna, Oxford) that would, in turn, play a fundamental role in the great awakening that was the Renaissance.

BRENDA WILMOTH LERNER

Further Reading

Braziller, George. *The Medieval Health Handbook*. New York: Tacuinum Santatis, 1976.

Lindberg, David, ed. *Science in the Middle Ages*. Chicago: University of Chicago Press, 1978.

Porter, Roy. *The Greatest Benefit to Mankind: A Medical History of Humanity*. New York: W. W. Norton & Co., 1997.

Siraisi, Nancy. *Medieval and Early Renaissance Medicine*. Chicago: University of Chicago Press, 1990.

The Emergence of University Education

Overview

Medical education did not always take place in universities. Until the tenth century, medicine was practiced by individuals and taught to apprentices, or passed down from father to son. Those who practiced medicine were more like tradespeople but they had a higher status than "lowly" midwives, herbalists, stone-cutters, bone-setters, cataract-couchers, or tooth-pullers. The latter traveled from town to town to ply their specialty because there were no cities large enough to support them. Itinerant healers had varying degrees of knowledge, experience, and education but some were charlatans and magicians who preyed on a patient's gullibility.

Barber-surgeons, apothecaries, and surgeons were organized into guilds that had rules for membership, training, practice, and fees. These individuals had more status than herbalists but less than physicians. In fact surgeons were suspect individuals, granted little respect.

Background

The school of Salerno, Italy was created by a group of students who studied with a group of physicians composing a faculty. The word *universitas*, meant an aggregate of persons at first but later came to designate learned people or scholars, not a place to study. The city of Salerno had been an ancient Greek colony which came to be known as *civitas hippocratica* because a mix of clerics, lay people, natives, and foreigners joined forces to practice medicine. It was near a monastery, Monte Cassino, which had a reputation for healing by miracles performed by saints and the use of magic herbs.

Geographically, it was an ideal place for healing. The Bay of Naples was an attractive respite for noblemen injured in the crusades. All trade routes pointed to Salerno. With Normans from the west, Sicilians from the south, Greeks, Benedictine monks, Jewish physicians, and Arabs, it was an ideal learning environment. This city is the oldest center of lay medical instruction and the first school to confer diplomas and the title of doctor.

There were female physicians for female patients and female physicians to train them. The most famous female physician in the twelfth century was Trotula but her name survives more than factual information about her. Other women physicians at Salerno were Constanza, Calendula, Abella, Mercuriade, Rebecca Guarna, and Louise Trencapilli.

However, few facts are known about Salerno until 1071 when the African slave, Constantine (Constantinus Africanus; c. 1020-1087), who spoke Greek and Babylonian came to study at Monte Cassino. Through his travels through Syria, India, Egypt, and Ethiopia, he accumulated many medical manuscripts. After converting to Christianity from Islam, he translated Arabic, Greek, and Indian medical documents for the university, making new knowledge available.

The *Pantegni* is the treatise for which he is best known. It contains mostly Galenic medical information and if one were to "blame" anyone for the perpetuation of Galen's misinformation, it would be Constantine. Soon that information became available to the medical people in Salerno.

Prior to the development of the medical school at Salerno, the Arabs excelled in medical knowledge. They had translated Greek texts. Their rich tradition boasted of Rhazes (c. 865-c. 923), who wrote more than 200 medical and philosophical treatises and Abu Ali al-Husayn ibn Abd-Allah ibn Sina (Avicenna; 980-1037), a genius whose acute observations led him to discover how mental phenomena affected physiology as when a quickened pulse betrays one's feelings. Al-Zahrawi (Albucasis; 936-1013) wrote *On Surgery and Instruments*, an early comprehensive book on cautery. The Arabs had expertise in pharmacology, ophthalmology, and cardiology. Interestingly, the mystery of pulmonary circulation was known to Ibn an-Nafis long before Miguel Serveto (Michael Servetus; 1511-1553) or William Harvey (1578-1657) expounded on their discoveries. One of the most famous people who lived among the Arabs was Abu Imran Musa ibn Maymun ibn Ubayd Allah (Maimonides, or Moses ben Maimon, 1135-1204), a Jewish-Spanish scholar and physician. Born in Córdoba, he had fled that city to avoid persecution and lived in Morrocco and Cairo. He wrote in Arabic, his best known works being *Poisons & Antidotes*, *Regimen Sanitatis*, and *Medical Aphorisms*. Interestingly, Jewish scholars had the skills to translate Arab manuscripts into Latin which aided the spread of Arabic science and medicine to Salerno and other European centers of learning.

The theoretical basis of medical training was Socratic and depended on the student being able to answer questions such as "What is medicine? Is it an art or not? If so, what other disciplines are important?" just as the method used in religious institutions in the past. Through these questions, the student would come to understand comprehensively rather than by rote, although memorization played a large part in knowing the answers to these questions. Galenic knowledge was extremely important as were inductive reasoning and logic. The logical syllogism was the method used in teaching physicians. A syllogism is a structure of language where a fact or truth can be arrived at in a simple three-part sequence using deductive reasoning. This is not to say that logic was more important than authority. Galen was the authority, and although the modern student may wonder why there were so many errors if medical education was logical and inductive, it is because no one dared to question discrepancies. This was a time when the senses and observation were not trusted.

Philosophically, to depend on human perception was a fearful practice, irreligious and definitely discouraged. Thus, a five-lobed liver, a two-horned uterus with seven chambers, and the "rete mirable" were repeated for many years. Also, even though a syllogism was supposed to be a way to organize information, teaching in certain cases consisted of taking a sentence, breaking it down into words and then analyzing each word that made the subject difficult to identify. For example, in talking about a symptom such as ringing in the ears, the professor would talk about sound, the theoretical aspect related to hearing. One can see that this was a less focused and more circumferential way of presenting knowledge.

As medical knowledge became institutionalized, it could be transmitted more effectively. It was now group knowledge rather than individual thus its distribution was greater and a tradition had begun. Standards of competency were set for physicians. Students had to finish three years in the humanities and then four years of medical school. After that, they were required to work with a physician for a year. When the "internship" was completed, only then could the graduate practice medicine alone. By 1140 examinations were required for those who wished to practice medicine.

Manuscript illustration of a lecturer at the University of Bologna, Italy. *(Gianni Dagli Orti/Corbis. Reproduced with permission.)*

Students attended four lectures a day. In the morning they were taught theory, in the afternoon, practice. Year one depended on the preliminary book of Avicenna and anatomy, although there was more anatomy taught at Bologna than Salerno. The first two lectures included diseases, children's ailments, dietetics, and regimen. The next about fevers and Galen. But as the reader now knows, Galen's information about anatomy was based on his dissection and vivisection of the pig, the dog, and the Barbary ape. And Galen had the endorsement of the church which meant that his authority was not to be questioned.

In the second, third, and fourth years, there was a great deal of review. The tradition spread from Salerno to Bologna, Montpellier, Paris, and Padua. Hippocrates' *Aphorisms*, Galen's *De diebus criticus*, Averroës' *Colliget*, Avicenna's *Canon*, and Gilbertus Angelicus' (1180-1250; he worked in Montpellier) *Compendium medicinae*, a seven volume work, were texts known to have been taught during the time of developing university medical education. Later, on the practical side, Arnau de Villanova (c. 1235-1311), also called Arnaldus Villanovanus, wrote *De regimine sanitatis*, a poem that recommended common sense preventive treatment. It can be considered a folk epic guidebook. For example the translation of one line reads, "If thou to health and vigor woudst attain, Shun weight cares, all anger deem profane, From heavy suppers and much wine abstain." He later taught at Montpellier.

To the Greek theory of humors, the Salernitan doctors attributed personality types to go with each imbalance of an element. Bloodletting, a practice which rid the body of sanguinous humor, was said to "soothe rage, bring joy to the sad, and save all lovesick swains from going mad." Although physicians developed the theory of phlebotomy (bloodletting), barbers performed it. The red stripes on white poles outside barber shops are reminders that long ago barber-surgeons bled people who suffered from an excess of sanguinous humor.

At Bologna, *Anathomia*, the first manual of dissection was written by Mondino dei Liucci (c. 1270-c. 1326). For over 500 years this book was the only guide to human anatomy. From Salerno came Taddeo Alderotti (c. 1223-1295), who developed a new tradition of collecting and reporting on clinical cases.

Also at Bologna was Hugh de Lucca, the surgeon who taught that "laudable pus" was not necessary for healing. In the thirteenth century, wound infection was so common that it was mistakenly believed that healing had to go through a pus-producing stage. We know now that pus is a sign of infection rather than health. It is laudable that de Lucca recognized that wounds could heal without becoming infected.

The center for university medical education in France was Montpellier. There, students required three and a half years of arts education, and six months of practical experience in order to obtain a bachelor of medicine. The most famous person associated with Montpellier was Guy de Chauliac (1300-1368), a surgeon whose *Chirurgia magna* and writings on dentistry were translated many times. He employed anesthesia and noted that loss of brain tissue was not necessarily fatal.

The medical school at Paris was influenced by both Salerno and Montpellier. Its requirements were complex and ongoing. First the study, then examinations, votes by faculty, then back to lectures, practical experience, and more exams. Paris is said to have been less innovative than Bologna but more important as far as developing the role of medicine as a profession.

The school at Padua, Italy (famous for both Galileo and Vesalius) was philosophically influenced more by Aristotle than theology. It is reasonable that natural science and scientific methodology (albeit early science) was a better foundation for the study of medicine than religion. Jacopo da Forli and Hugo of Siena both taught at this university, later to be the epitome of intellectual achievement during the Renaissance.

Impact

Thus medical education started as a small group of lay teachers and students who eventually standardized a curriculum and created examinations of competency to insure that those who practiced medicine would first "do no harm." The qualifications and examinations became more stringent as medicine became part of a university setting starting in Italy and spreading throughout Europe by the end of the fourteenth century.

LANA THOMPSON

Further Reading

Bettmann, Otto. *A Pictorial History of Medicine*. Springfield, IL: Charles C. Thomas Pub., Ltd., 1979.

Bullough, Vern. *The Development of Medicine as a Profession*. Canton, MA: Watson Pub. Intl., 1966.

Garrison, Fielding H. *History of Medicine*. Philadelphia: W.B. Saunders Company, 1929.

Lyons, Albert S., and R. Joseph Petruzelli. *Medicine: An Illustrated History*. New York: Harry Abrams, 1987.

Magner, Lois. *A History of Medicine*. New York: Marcel Dekker, Inc., 1992.

Marti-Ibanez, Felix. *The Epic of Medicine*. New York: Clarkson N. Potter, Inc, 1962.

McGrew, Roderick. *Encyclopedia of Medical History*. New York: McGraw Hill, 1985.

Talbott, John H. *A Biographical History of Medicine*. New York: Grune and Stratton, 1978.

Walsh, James J. *Old-time Makers of Medicine*. New York: Fordham University Press, 1911.

Ibn an-Nafis and Pulmonary Circulation

Overview

During the thirteenth century, the Egyptian physician Ibn an-Nafis (c. 1210-1280; also known as Ali ibn Abi al-Harma al-Qurayshi Ibn an-Nafis) described the minor, or pulmonary circulation of the blood. However, this discovery was apparently ignored by his contemporaries and was forgotten until the twentieth century. Historians of medicine generally believed that pulmonary circulation was first described by Miguel Serveto (Michael Servetus; c.1511-1553) and Realdo Colombo (c. 1516-c. 1559) in the sixteenth century.

Background

The Middle Ages of European history roughly correspond to the golden age of Islam, the religion founded by the prophet Muhammad (570-632). Just as Latin served as the common language of learning for students throughout Europe, Arabic was the language of learning throughout the Islamic world. Thus, Arabic texts need not have Arab authors; Persians, Jews, and Christians took part in the development of Arabic scientific literature. From the ninth to the thirteenth century Arab scholars could justly claim that the world of scholarship had been captured by Islam. Scholars at ancient centers of learning were encouraged to assist in the immense task of translating ancient manuscripts into Arabic. Pharmacology, optics, chemistry, alchemy, and medicine were of particular interest to Islamic scholars.

Until about the second half of the twentieth century, European historians generally looked at so-called Arabian science, medicine, and philosophy in terms of a single question: were the Arabs merely the transmitters of Greek achievements or did they make any original contributions? Eventually, historians rejected this approach and concluded that this question was inappropriate when applied to a period in which the very idea of a quest for new secular knowledge was virtually unknown. Medieval physicians, philosophers, and scientists accepted the writings of the ancients as essentially true and authoritative. Their goal was to preserve, analyze, and clarify the writings of the ancient. Nevertheless, Arabic translators, physicians, and philosophers eventually absorbed, assimilated, and transformed the writings, ideas, and theories of the ancient Greeks. The work of the Greek physician Galen (c. 130-c. 200) was especially revered and medieval physicians and scholars, whether Muslims, Jews, or Christians, generally shared the assumption that Galenism was a complete and perfect system.

The writings of Islamic physicians and philosophers, which were often presented as commentaries on the work of Galen, were eventually translated from Arabic into Latin and served as fundamental texts in European universities for hundreds of years. But for many European scholars medieval Arabic writers were significant only in terms of the role they played in preserving Greek philosophy during the European Dark Ages.

Much about the history of science and medicine, however, has been revealed through attempts to study the work of Islamic writers on their own terms. Western scholars assumed that the major historical contribution of Islamic medicine was the preservation of ancient Greek wisdom and that medieval Arabic writers produced nothing original. In the West, during the late Middle Ages and the Renaissance, the Arabic manuscripts that were sought out for translation into Latin were those that most closely followed the Greek originals. Arabic manuscripts that criticized Galen, or introduced novel ideas were dismissed as corruptions of ancient Greek wisdom. Therefore, the original premise of European scholars, that medieval Arabic scholarship lacked originality, was confirmed.

Impact

The story of the thirteenth century Egyptian physician Ibn an-Nafis and his discovery of pulmonary circulation demonstrates the unsoundness of previous assumptions about medieval Arabic literature. The writings of Ibn an-Nafis were essentially ignored until they came to the attention of twentieth century science historians. Ibn an-Nafis is now regarded as the first physician to recognize the minor, or pulmonary, circulation of the blood, that is, the movement of blood between the heart and the lungs.

Contemporaries honored Ibn an-Nafis as a learned physician, ingenious investigator, and prolific writer. According to one biographical sketch, he completed a treatise on the pulse between ablutions at the public bath. In addition to a commentary on Avicenna's (Ibn Sina; 980-1037) *Canon*, Ibn an-Nafis had plans for a comprehensive 100-volume compendium of medicine. About 30 volumes were apparently completed when he died, but only a few survived. While serving as chief of physicians in Egypt, Ibn an-Nafis became seriously ill. Other physicians urged him to take wine as a medicine, but he refused for reasons of piety. He told his colleagues that he did not wish to meet his creator with alcohol in his body.

It is not clear how Ibn an-Nafis actually discovered pulmonary circulation. Human dissection was generally prohibited by Islamic law and custom and a man as pious as Ibn an-Nafis was unlikely to challenge such laws. In his writings, Ibn an-Nafis frequently invoked religious law to explain why he did not perform dissections. Indeed, in his *Commentary on the Anatomy of Avicenna*, Ibn an-Nafis explained that he could not practice human dissection because of the prohibitions of religious law and his own natural charity. Systematic human dissection was not acceptable in the Muslim world, because orthodox Muslims believed that mutilation of a cadaver was an insult to human dignity. Islamic law forbids the practice of the kinds of ritual mutilation that had apparently followed warfare in ancient times. Muslim legal experts argued that scientific dissection was essentially the same kind of violation of the human body. However, some anatomical research was conducted on various animals, such as monkeys and sheep. Despite his reluctance to perform dissections, Ibn an-Nafis was eventually led to question Galenic physiology.

As was customary, an-Nafis began his account of the structure and function of the heart with a fairly conventional reflection on traditional theories. However, after finding inconsistencies in Galen's account of the functions of the lungs, heart, and associated vessels, an-Nafis went on to propose a novel theory of the movement of the blood between the heart and the lungs. Rejecting the usual belief that Galen was virtually infallible, an-Nafis expressed his criticism of traditional Galenic physiology and insisted that there were no visible or invisible passages between the two ventricles of the heart. Moreover, he argued that the septum between the two ventricles was thicker than other parts of the heart. The thick-walled septum would, therefore, prevent the harmful and inappropriate passage of blood or spirit between the right and left parts of the heart. This position was actually stronger than that taken by the sixteenth century European anatomists, because Ibn an-Nafis rejected the possibility of invisible, as well as visible, pores in the septum of the heart. To account for the presence of blood in both sides of the heart, Ibn an-Nafis argued that after the blood had been refined in the right ventricle, it must be transmitted to the lungs where it was rarefied and mixed with air. The finest part of this blood was then clarified and transmitted from the lung to the left ventricle. In other words, the blood must move in a circle between the heart and the lungs because the blood could only get into the left ventricle by way of the lungs.

The writings of Ibn an-Nafis were essentially ignored until 1924 when Dr. M. Tatawi, an Egyptian physician, presented his doctoral thesis to the medical faculty of Freiburg. Fortunately a copy of Tatawi's thesis came to the attention of professor Max Meyerhof, an eminent historian of medicine who recognized the significance of the long-neglected writings of Ibn an-Nafis. Without the attention of Tatawi and Meyerhof, the first account of pulmonary circulation might have been forgotten again.

There is no direct evidence that Ibn an-Nafis could have influenced later writers, but the fact that he was able to describe his theory so clearly in the thirteenth century provides an interesting challenge to previous assumptions about progress and originality in the history of science. Since only a small percentage of Arabic manuscripts from this period have been studied, edited, translated, and printed, the possibility remains that some other medieval manuscript could contain a commentary on the unorthodox theory of Ibn an-Nafis.

Since World War II, scholars have examined many Arabic manuscripts that were previously unknown, ignored, or unpublished. This has changed the way historians of science and medicine and other students of Islamic studies view the development of Islamic science. Although scholars generally agree that a decline in Islamic science and medicine did occur after its so-called golden age, it occurred later than the eleventh century, which was previously considered part of the period of decline, and might be placed as late as the fifteenth or sixteenth century. New studies of Arabic manuscripts reveal that between the twelfth and thirteenth centuries, for example, important work in astronomy, mathematics, optics, pharmacology, and medicine was conducted. The research on Ibn an-Nafis is a prominent example of this scholarship. Some historians believe that hundreds, or even thousands, of manuscripts written between 500 and 1,000 years ago are still gathering dust in various archives throughout the world. Further studies of such manuscripts might clarify ambiguous aspects of Islamic scholarship and might lead to new discoveries.

LOIS N. MAGNER

Further Reading

Books

Bittar, E. Edward. *The Influence of Ibn Nafis: A Linkage in Medical History*. Ann Arbor: University of Michigan, 1956.

Doby, T. *Discoverers of Blood Circulation: From Aristotle to the Times of da Vinci and Harvey*. New York: Abelard-Schuman, 1963.

Meyerhof, M. *Studies in Medieval Arabic Medicine: Theory and Practice*. London: Variorum Reprints, 1984.

Articles

O'Malley, Charles Donald. "A Latin translation of Ibn Nafis (1547) related to the problem of the circulation of the blood." *Journal of the History of Medicine and Allied Sciences* 12 (1957): 248-253.

Persaud, T. V. "Historical development of the concept of a pulmonary circulation." *Canadian Journal of Cardiology* 5 (1989): 12-16.

Temkin, Owsei. "Was Servetus influenced by Ibn an-Nafis?" *Bulletin of the History of Medicine* 8 (1940): 731-734.

Astrology and Medicine

Overview

Astrology is a form of divination based on the assumption that the motions of the heavenly bodies influence human affairs. Therefore, astrologers claim that they can predict the future by observing and interpreting the positions of the stars, sun, moon, and planets. During the Middle Ages, the relationship between medicine and astrology was very close. Medieval astrologers blamed disease epidemics on dangerous combinations of the planets and studied the motions of the heavenly bodies as a guide to the treatment of individual patients. The practice of astrological medicine required knowledge of the astrological correspondences among the seven planets, the seven metals, and parts of the body. Even in the twenty-first century medical astrology practitioners continue to claim that they can predict potential illnesses and select the best time for surgery.

Background

Astrologers believe that by interpreting the influence of the heavenly bodies they can predict the future of individuals, groups, or even nations. Astrology originated in ancient Mesopotamia as a system of observing and characterizing celestial omens. Various forms of astrological doctrine eventually found their way into Greek, Roman, Indian, Egyptian, and Islamic worlds. Plato's (428-348 B.C.) *Timaeus* became the foundation of nature-mysticism by developing ideas about the divinity of the heavenly bodies and their direct relationship to life on Earth. The *Timaeus* expressed the concept of man as a microcosm. Despite his general opposition to mysticism, Aristotle (384-322 B.C.) aided the development of Western astrological theory by proposing that the world of movements in the sublunary sphere (beneath the orbit of the moon) was determined by the movement of the celestial bodies, which were guided by divine influences. The group of philosophers known as the Stoics provided further grounding for astrological ideas.

During the Middle Ages, Western Europe was strongly influenced by classical Greek astrological doctrines that had been incorporated into Islamic scholarship. European interest in astrology was stimulated by the translation of Arabic texts and astronomical tables into Latin during the twelfth and thirteenth centuries. By the fifteenth century, European scholars were able to study new translations made directly from the classical Greek texts.

Throughout the Middle Ages, the terms astronomy and astrology were used synonymously in reference to knowledge of the stars. The modern distinction between astronomy (the science of the stars) and astrology (the art of divination by the stars) evolved very gradually. Some astrological theories assumed the existence of a totally mechanistic universe, which left no room for free will or intervention by a deity. Such forms of astrology were rejected by Christian and Muslim theologians because it went against their religious beliefs. Other forms of astrology postulated that the movement and position of the heavenly bodies merely provided information about possibilities that could be altered either by divine intervention or resisted by human will. Christian theologians argued that the determinism inherent in astrology was incompatible with the Christian belief in freedom of will, but they accepted the general concept that astral influences affected the sublunary sphere. Fatalistic or deterministic forms of astrology were rejected and banned. Human beings might be influenced by the stars, but by wisdom and will they could escape the fate predicted. Broad areas of astrology, including medical astrology, were accepted along with the system of rules that determined the proper moment for undertaking various actions and for distinguishing between propitious and unpropitious days.

The special relationships that existed among the heavenly bodies and the earth were regarded as extremely complex. Thus, errors in astrological prediction were to be expected. Astrology was used to inform individuals about the course of their lives, based on the positions of the planets and the zodiac signs (the 12 astrological constellations) at the moment of birth or conception. This form of astrology was called genethlialogy or judicial astrology, the casting of horoscopes. Other astrological techniques were used to determine whether or not a particular moment was propitious for the success of a given course of action. This form assumes that an individual could choose to act at astrologically favorable times and avoid misfortunes predictable from the casting of a horoscope.

The revival of ancient learning and the growth of universities during the European Middle Ages stimulated interest in astrological doctrines and the practice of astrological medicine. In the late European Middle Ages many distinguished universities, including Paris, Padua, Bologna, and Florence, had chairs of astrology. Classical astrologers assumed the existence of a geocentric universe in which the sun, moon, planets, and even the stars revolved in circular orbits around the earth. Throughout the European Renaissance and the Reformation astrology remained a part of the university curriculum, despite the challenge that Copernican theory posed for the geocentric worldview.

Impact

During the Middle Ages, the ties between astrology and medicine were quite strong. The belief that the heavenly bodies influenced human fortunes was widespread. Physicians were expected to take astral influences into account when dealing with each patient. In addition, the physician had to understand the general influence of the heavenly bodies on medications and parts of the human body.

Medical astrology, which was also known as iatromathematics, was based on the assumption that the motions of the heavenly bodies influenced human affairs and health. An important premise of medical astrology was the correspondence between the 12 signs of the zodiac and the organs or parts of the human body. Each section of the zodiac acted on the corresponding part of the body by means of its association with a particular stellar force. Particular positions of the planets therefore affected the functioning of the body. Such astrological considerations could be applied to individuals or to whole populations. For example, the outbreak of bubonic plague in 1348 was attributed to a malign conjunction of Saturn, Jupiter, and Mars.

Astrological medicine required knowing the exact time at which the patient became ill. With this information and a study of the heavens, the physician could prognosticate the course of illness with mathematical precision and avoid dangerous tendencies. Doctors were guided by their knowledge of the astrological correspondences among the seven planets, the seven metals, and parts of the body. Physicians could cast retrospective horoscopes to identify the configurations of the planets that had caused diseases and epidemics. Physicians also had to incorporate the astrological concept of the "medicinal month" into their plans for treatment of the patient. The concept of critical days could be traced back to Hippocrates (460-377 B.C.), who frequently marked the number of days after the onset of illness on which certain zodiac signs appeared. Eventually, computations involving numerical relationships among calendar dates, critical days, the motions of the moon, and favorable and unfavorable days became part of astrological medicine. The complexity of the system and uncertainties in determining the exact moment of conception, birth, or the onset of illness could be used to explain errors in prognosis.

In therapeutics, astrological considerations determined the nature and timing of treatments, the selection of drugs, and the use of charms. For example, the sun ruled the chronic diseases, Saturn was blamed for melancholy, and the moon (a very changeable planet) influenced the outcome

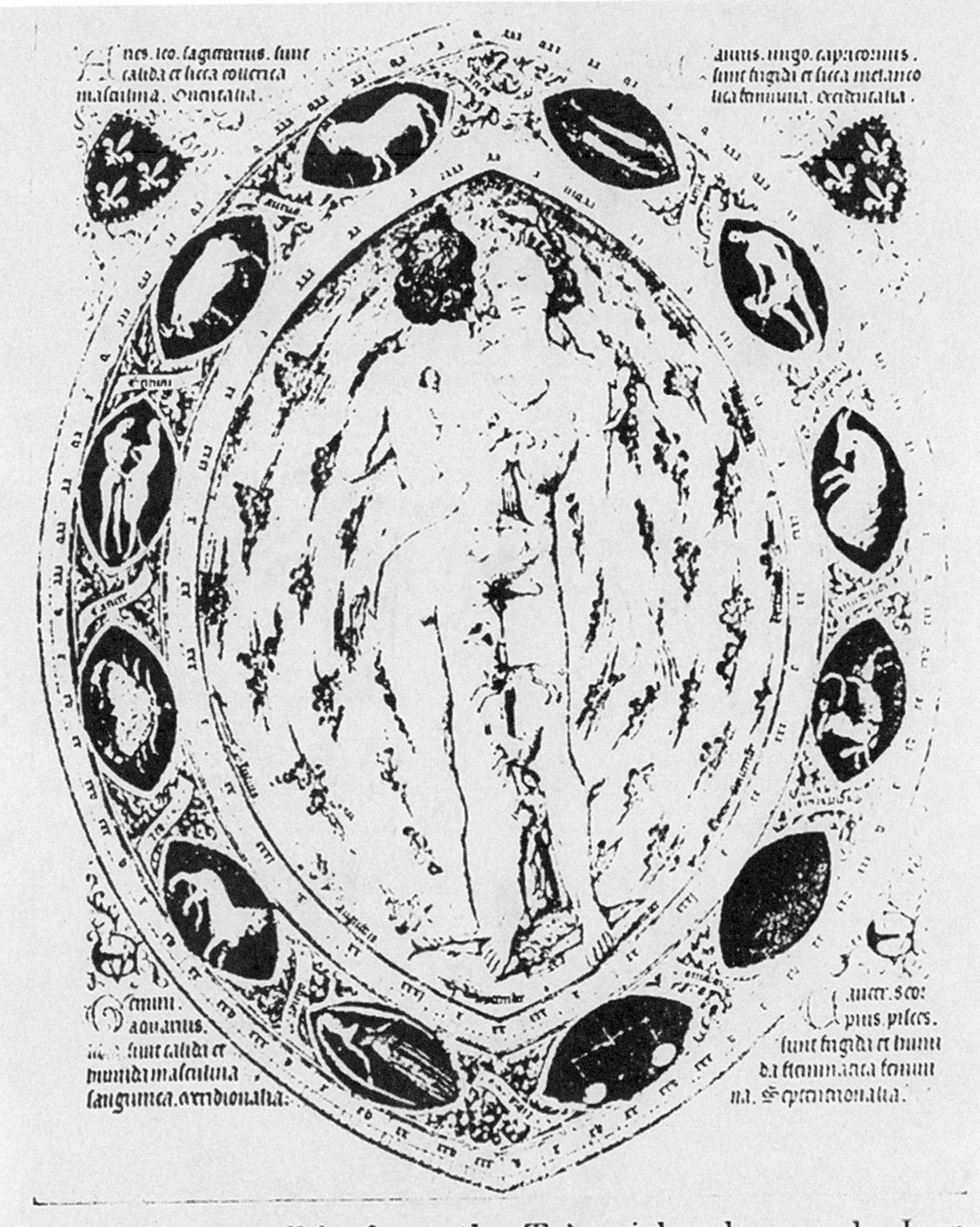

Zodiac-manikin from the Très niches heures de Jean de France, Duc de Berry (1340–1416), Chantilly. From Sudhoff's Beiträge zur Geschichte der Chirurgie im Mittelalter, Leipzig, J. A. Barth, 1914.

During the Middle Ages the zodiac was thought to influence human health. *(Corbis Corporation. Reproduced with permission.)*

of surgery, bleeding, purging, and acute illness. Because the moon governed the tides and the flow of blood in the veins, the surgeon was well advised to consult an almanac in order to study the configuration of the heavenly bodies before performing therapeutic bleedings. Astrological practitioners might also recommend that a patient should wear a specific talisman containing an appropriate astral image engraved on a precious stone. However, the presumed relationships between the heavenly bodies and the human body were so complex, numerous, and contradictory

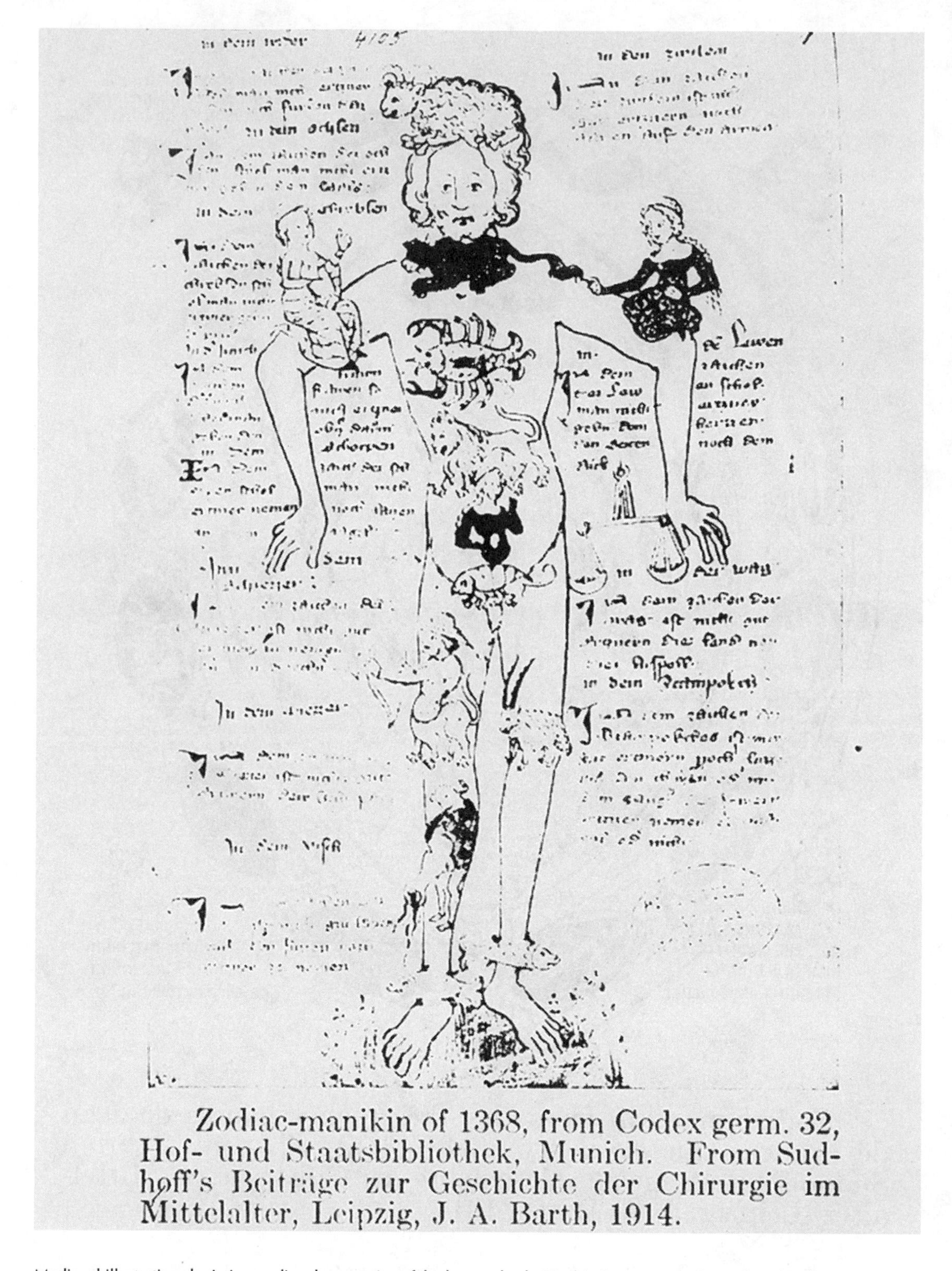

Medieval illustration depicting zodiacal properties of the human body. *(Corbis Corporation. Reproduced with permission.)*

that in practice it was impossible to carry out any operation without breaking some rule.

European acceptance of Greek and Islamic astronomy and astrology during the twelfth century encouraged the development of medical astrology. Until sophisticated planetary tables were available, the application of complex astrological calculations to medical practice was inhibited. The peak of medical astrological practice in Europe probably occurred between the fourteenth and sixteenth centuries. Far from being seen as superstition, astrology seemed to link the practice of medicine to the world of science and mathematics. Astrology was seen as part of the intellectual basis of medical education. Learned medicine included medicine, logic, natural philosophy, and astrology. Physicians studied Latin texts at the university, but even surgeons, who were generally not associated with universities, studied medical and astrological texts.

Medicine and astrology were often combined in the careers of the most prestigious and prosperous physicians, in particular those who served wealthy and powerful patrons. Demonstrating competence in astrological computations was considered one of the ways that a learned physician could separate himself from an empiric or quack. The most ambitious medical students were particularly likely to study astrology and mathematical calculations at the most advanced level possible. Graduates of the university went on to write treatises on astronomy and astrology, improve the design of astronomical instruments, and prepare new astronomical tables.

By the seventeenth century, as the world of science was transformed by the works of Copernicus (1473-1543), Galileo (1564-1642), Johannes Kepler (1571-1630), René Descartes (1596-1650), and Isaac Newton (1643-1727), astronomers rejected and repudiated astrology. Since the Scientific Revolution, astrology and alchemy have been defined as occult systems whose basic doctrines are incompatible with modern science. Popular belief in astrology, however, has continued undiminished into modern times, as is evident in the articles and horoscope found in the newspapers and the flood of almanacs and astrology manuals found in bookstores.

LOIS N. MAGNER

Further Reading

Books

Barton, Tamsyn S. *Power and Knowledge: Astrology, Physiognomics, and Medicine under the Roman Empire*. Ann Arbor, MI: University of Michigan Press, 1994.

Cornell, Howard Leslie. *Encyclopaedia of Medical Astrology*. York Beach, ME: S. Weiser, 1992.

French, Roger. *Practical Medicine from Salerno to the Black Death*. Cambridge: Cambridge University Press, 1994.

Kibre, Pearl. *Studies in Medieval Science: Alchemy, Astrology, Mathematics, and Medicine*. London: Hambledon Press, 184.

Klein-Franke, Felix. *Iatromathematics in Islam*. Hildesheim, NY: G. Olms, 1984.

McBride, M. F. *Chaucer's Physician and Fourteenth Century Medicine: A Compendium for Students*. Bristol, IN: Wyndham Hall Press, 1985.

O'Boyle, Cornelius. *Medieval Prognosis and Astrology*. Cambridge: Welcome Unit for the History of Medicine, 1991.

Shumaker, W. *The Occult Sciences in the Renaissance; A Study in Intellectual Pattern*. Berkeley, CA: University of California Press, 1972.

Articles

French, Roger. "Foretelling the Future: Arabic Astrology and English Medicine in the Late Twelfth Century." In *International Review Devoted to the History of Science and Its Cultural Influences* 87 (1996): 453-480.

Hare, E. H. "Medical Astrology and Its Relation to Modern Psychiatry." In *Proceedings of the Royal Society of Medicine* 70 (1977): 105-110.

The Development of Surgery during the Twelfth through Fourteenth Centuries

Overview

Early in twelfth century pre-Renaissance European surgery and medical practice began to mature, in large part through the heavy influence of ancient Greek texts and the work of Arabic physicians and surgeons. Medical scholarship and practice became centered in southern Italy at Salerno, where a medical school was established in 1140. In the thirteenth century the medical school at Bologna, Italy, was the most influential center of medical expertise, anatomical teaching, and competent surgical practice. Surgeons who trained at Bologna took their practice and scholarship into northern Europe. By the close of the fourteenth century France became the preeminent seat of European surgical practice.

Surgery, as practiced by Theodoric (1205-1296), Guido Lanfranchi (1250-1306), Henri de Mondeville (1260-1320), and Guy de Chauliac (1300-1368), became a respected and important part of medicine rather than the craft of barbers and butchers. The surgical masters of the fourteenth century paved the way for the great European surgeons of the Renaissance and beyond.

Background

In pre-Renaissance Europe surgery was not considered part of legitimate medical practice. Performed by barbers and butchers in Europe, surgery was not even a noble profession. Surgery had, however, been an important aspect in the practice of medicine as conducted by the great Islamic physicians, such as al-Zahrawi (930-1013). Europe's pre-Renaissance surgeons were to be heavily influenced by medicine and surgery as practiced in the Islamic world.

Al-Zahrawi, born near Cordova, Spain, attended the University of Cordova to study medicine. He became a great surgeon and served as court physician to King Abdel-Rahman III. Al-Zahrawi wrote a surgical encyclopedia that influenced surgical practice and training well into the 1600s. In his works he stressed the importance of knowledge of anatomy, particularly the function of organs, along with functional knowledge of bones, nerves, and muscles. He demonstrated and practiced complicated surgical procedures, such as stripping vericose veins (by a procedure not unlike modern practice), closing skull fractures, and performing dental extractions as well as dental implants. In his writing, al-Zahrawi described 200 surgical and dental tools and wrote chapters advising midwives.

Impact

European surgery during the time of al-Zahrawi was still in the Dark Ages. However, in Salerno, Italy, a seaside town long known as a health resort, a school of medicine was established in 1140. The sick (poor and wealthy, young and old), wounded crusaders, and both teachers and students of healing had visited Salerno since the ninth century. The formal school of medicine, Civitas Hippocratica, was established at Salerno by Roger II of Sicily, who introduced the first system of examination for physicians at Salerno. Roger's grandson, Frederic II, later instituted regulations for the practice of medicine and approved a curriculum for medical education.

One legend says that the school at Salerno was created by four masters—a Greek, a Jew, an Arab, and a Latin. Some sources say that the school was heavily influenced by Arab medicine, others contend that its downfall in the thirteenth century was due to the rise of Islamic medicine. The Salerno school reached its zenith during the twelfth century.

At the Salerno school—not far geographically or culturally from Islamic influence—medical texts, such as those written by Al-Zahrawi, and ancient Greek medical texts by Galen (130-200), were translated and widely read. Salerno instructors also produced medical literature, such as the "Regimin Sanitatis Salernitanum" (Salernitan Guide to Health). Roger of Salerno and his star pupil, Roland de Parma, wrote *Surgery of the Four Masters,* a work that influenced physicians all over Europe.

As the traditions of Arab science and medicine and Greek healing arts were carried on at Salerno during the twelfth century, seats of medical learning, including both anatomical and surgical study and instruction, soon spread northward. One of the first great surgeons of northern Italy was Hugh of Lucca (1160-1257). Influenced by Arab and Greek medicine, Hugh founded what was to become the famous medical school at Bologna, Italy. Most of Hugh's accomplishments and thoughts on surgical practice are best known from accounts written by his student, Theodoric, a cleric-surgeon. Theodoric wrote a book entitled *Chirurgia* (Surgery) that built upon the teachings of Hugh. It featured a treatise on wound management that debunked the previous century's surgical opinion that the production of pus in a wound was desirable. Hugh and Theodoric felt that wounds should be immediately joined by cautery (the application of a hot blade to seal open wounds), as recommended by Islamic surgeons. Theodoric was also known for experimenting with antiseptics and anesthesia.

Surgery and wound management was also the central concern of other Bologna-trained Italian surgeons, such as Mondino dei Liucci (1270-1326) and William of Saliceto (1210-1277). Anatomy and surgery were separate chairs at Bologna. Mondino, famous for teaching anatomy through public dissection, is credited for having revived the art of dissection as practiced by Islamic anatomists. William of Saliceto, who hoped to unify the practice of medicine and surgery, taught surgery at Bologna and later practiced medicine in Verona. William described the difference between venous and arterial hemorrhage and discouraged cautery. He wrote on both internal medicine and surgery, with a book on surgery published in 1275. William also taught two notable students—Lanfranchi and Henri de Mondeville.

Lanfranchi was part of a movement that saw the "capital" of surgical study and practice shift from Italy to France. Born in Milan, Italy, Lanfranchi studied under William of Saliceto at Bologna but fled to France after civil war in Italy in 1290, settling in Lyon to practice surgery before finally taking residence in Paris in 1295, where he died in 1306. In his last 20 years, Lanfranchi was credited for writing his *Chirurgia Parva* (Little Surgery) and teaching the next generation of surgeons.

Lanfranchi tried to unify surgery and medicine, asserting that "no man can be a good physician who knows no surgery and no one can be a good surgeon without a knowledge of medicine." Yet, Lanfranchi's surgery has been considered "conservative." For example, he did not believe in surgery for a hernia, preferring trusses and bandages. He preferred to surgically remove bladder stones only as a last resort, yet was considered the father of French surgery.

No French surgeon (save his successor, Guy de Chauliac) is more important to the history and development of surgery than Henri de Mondeville. Remembered as the "first great French master of surgery," Henri, after learning surgery from Theodoric in Bologna, served Louis X, taught anatomy at the school at Montpellier, and was the first French surgeon to write a book on surgery. In *Chirgurie,* written between 1306 and 1320, Henri cites Theodoric on the treatment of wounds and Lanfranchi on the treatment of ulcers and other diseases. Henri's teaching was largely about the treatment of wounds.

Like Theodoric, Henri taught that the production of pus in a wound was not desirable. He preferred to remove foreign bodies from wounds, control bleeding by any means possible (usually cautery), and to close a wound as quickly as possible by bandages or suture. Henri's later books focused on surgical problems other than wound management and influenced the next generation of French surgeons, most importantly Guy de Chauliac.

Guy de Chauliac is often considered the second great French surgeon of the pre-Renaissance. Born near Avignon, France, Guy became the physician and surgeon to the popes who established residency at Avignon during this period. Guy studied medicine at Montpellier and is best known for his huge text *Chirurgie Magna* (Grand Surgery). In its many chapters, Guy discussed anatomy as indispensable for the surgeon's art, quoting Mondeville, but going into greater anatomical detail. To his credit, Guy stayed in Avignon rather than flee when bubonic plague, or Black Death, ravaged Europe in 1348 and 1360. Though infected by the plague, he survived and wrote objective accounts of its symptoms and kept track or mortality figures.

A proponent of Mondevillian methods of wound treatment, Guy went further—and many of his contemporaries and successors thought too far—in his treatment of wounds. His practice was often characterized as "meddlesome." Guy advocated removing foreign bodies from wounds, cleaning them with preparations of wine, turpentine, or brandy, packing or draining wounds, cauterization to close wounds, and bandaging or suturing when advisable. He also differentiated types of wounds, such as those "altered by air," contusion wounds, abscessed wounds, and even bites. Guy shocked the fourteenth-century medical world by stating that it took more than nature to heal wounds.

Jean Yperman (1260-1310), a Belgian student of Lanfranchi in Paris, practiced both medicine and surgery and, after returning to his native town of Ypres, was often called to serve as a military surgeon. Because of his military service, he became an expert in the ligature of arteries. Although he left no disciples, he is regarded as the father of Flemish surgery, having invented and developed techniques for ligature that two centuries later were attributed to French master surgeon Ambroise Paré (1510-1590).

The first great English surgeon was John of Arderne (1306-1390). Although the source of his training is not known, John served as a military surgeon during the Hundred Years War, during which time gunpowder was first used. After his military service, John began practicing in Nottinghamshire, where records show he specialized in diseases of the rectum, using surgery to repair anal fistulas, hemorrhoids, and to treat cancer. Records show that John did not use harsh suppositories after his surgery, preferring simple cleansing. Later in his career he moved to London where he was admitted to the Guild of Surgeons. In London, toward the end of his career, he wrote on medical practice.

RANDOLPH FILLMORE

Further Reading

Baas, John Herman. *The History of Medicine.* Robert E. Krieger, 1971.

Ingus, Brian. *A History of Medicine.* New York: World Publishing Company, 1965.

Meade, Richard Hardaway. *An Introduction to the History of General Surgery.* Philadelphia: W.B. Saunders Company, 1968.

Wangenstein, O.H., and Sarah D. Wangenstein. *The Rise of Surgery.* Minneapolis: University of Minnesota Press, 1978.

Ziegler, Philip. *The Black Death.* Phoenix Mill, Gloucestershire: Sutton, 1997.

Zimmerman, Leo M., and Ilza Veith. *Great Ideas in the History of Surgery.* New York: Dover, 1967.

The Emerging Practice of Human Dissection

Overview

According to many sources, human dissection was not allowed by the church, so any dissection performed was primarily on animals. Despite the informative works from authors such as Aristotle (384-322 B.C.), *Historia Animalium,* and Galen (A.D.130-200), *On Anatomical Procedure,* their experience was primarily with pigs, dogs, and Barbary apes. Two early practitioners that did use humans in their works are Herophilus of Chalcedon (335-280 B.C.) and Erasistratus (fl. c. 250 B.C.). Herophilus wrote *On Anatomy* where he described parts of the brain, the uterus, arteries, and veins. The only extant proof that he dissected a human is in his description of the liver. Erasistratus was another early dissector who was accused by two religious writers as having performed dissection procedures on the living, not the dead. Altogether, few accurate images of the interior of the body remained.

Background

One reason there was so much controversy surrounding dissections prior to the fourteenth century had to do with bans and prohibitions. Many sources state that dissection was prohibited by the church but more detailed research reveals information that helps to explain this statement.

At least three reasons for this misinterpretation existed. The first was the fear that mutilation of the body to great extents would result in an inability to be resurrected. The second was the inability of monks to perform surgery, since the church allegedly abhorred bloodshed. Since cutting is done in both surgery and dissection, confusion between the two may explain this. A third reason for the misinterpretation is that the word anatomization was a pejorative word. It meant that the body of a criminal (or other disenfranchised individual) would be mutilated to create a punishment beyond execution, the final indignity: the denial of a Christian burial.

By the twelfth century, the persisting religious rejection of dissection had started to ease. Medical schools in Salerno and Naples were performing and teaching others the skill of dissection. The Salerno school, Civitas Hippocratica, was established in 1140 by Roger II of Sicily. It was here where the first enforced exams were given to surgeons and other dissectors.

A problem to the dissectors was the lack of refrigeration or storage space. Bodies decayed rapidly and during normal time periods there were few deaths. Since towns were small, when someone did die they were neighbors, relatives, or friends. It is reasonable to assume that a certain amount of reticence was involved when one might be required to dissect the body of a friend or relative.

Of course there were instances later in the seventeenth century when it was helpful to know the person and his or her pathology before death, so that an investigation could correlate changes in the postmortem body with symptoms prior to death. But during medieval times, dissection was done to learn about structure and to a lesser degree function. When there was an epidemic such as the plague, dissection was discouraged because of the potential for contamination.

Impact

Against this background, came the professionalization of medicine and university medical education. Many medical schools started in institutions that already had a law school. A renewed interest in anatomy was encouraged by its forensic application.

The medical school in Bologna was a place that fostered the growth of understanding the structure of the human body and investigation of unexplained death. It can boast of students such as Henri de Mondeville (1260-1320) and the teacher Taddeo Alderotti (Thaddeus). Mondino dei Liucci, (1270-1326) writer of *Anothomia* in 1316, studied with de Mondeville under Alderotti and received his degree at Bologna in 1290

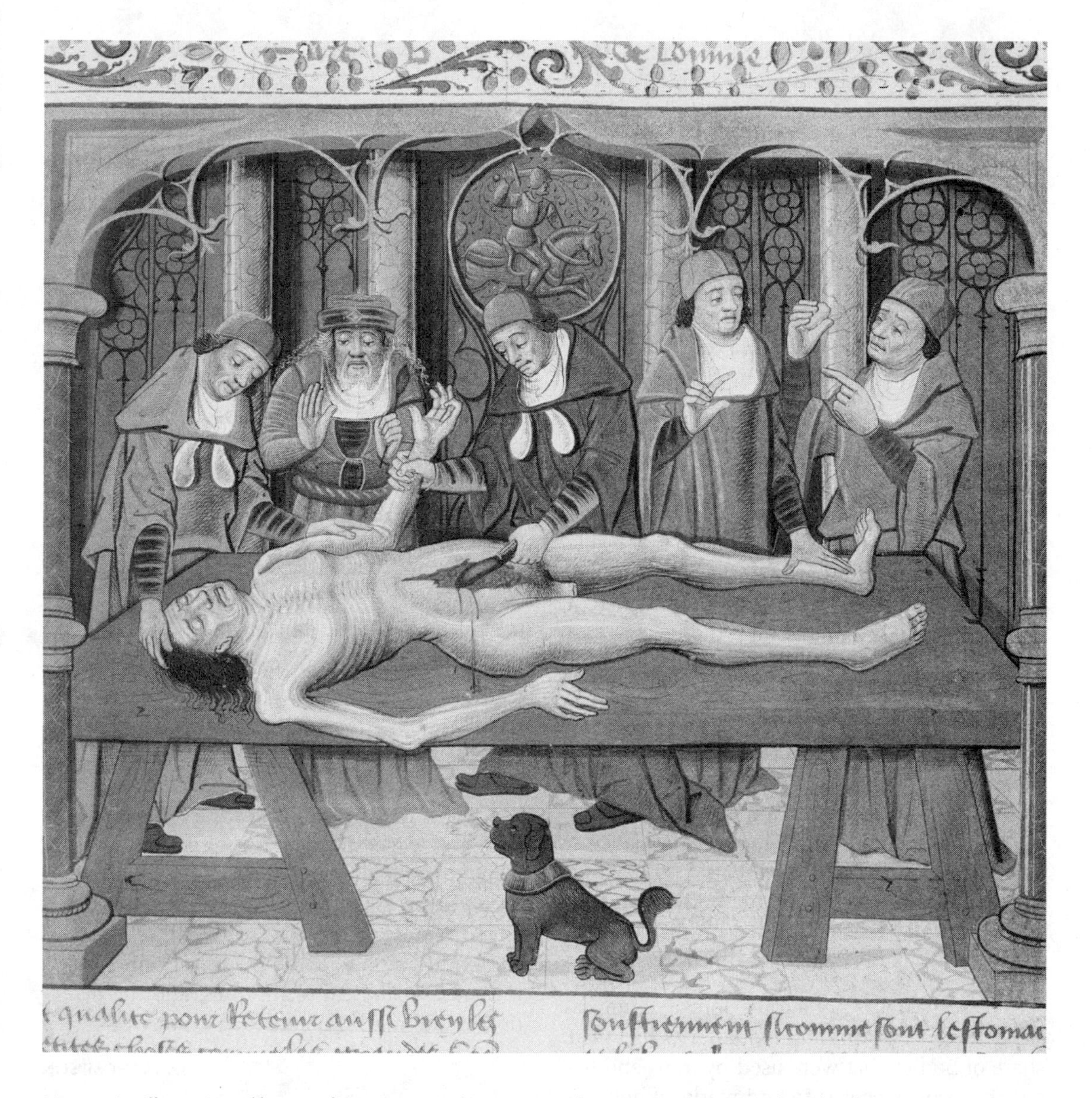

Manuscript illustration of human dissection. *(Archivo Iconografico, S.A./Corbis. Reproduced with permission.)*

(1300 according to some sources). Although Mondino named many organs with peculiar and sometimes redundant names, it was his method of dissection that was unique.

Mondino lectured and demonstrated dissections, but unlike other anatomists he did not read from Galen while dissecting. This was an advantage because none of Galen's errors were repeated. Mondino also performed the cutting himself. The custom had been for the professor to sit far above the dissection bench, allow another person to point to a part, and a dissector (barber or surgeon) perform the cutting. He did make use of a prosector, a woman who prepared parts of the body for demonstration. Alessandra Giliani was a talented individual who cleaned the body and then injected vessels with colored liquids in order to make them easier to study. Then the exterior of these vessels was delicately painted with appropriate colors, red for arteries and blue for veins.

Mondino had to work fast, first demonstrating the organs of the abdominal and thoracic viscera, then the head, and lastly the extremities. Again, without refrigeration or preservation techniques, one can imagine that during the summer, a body after four days would be quite unusable for future demonstrations. The problem of obtaining adequate study material and how best to use it was tantamount to continuing this groundbreaking practice.

Mondino's language was convoluted and difficult to understand. During a time when medical knowledge was borrowed from both Greek and Arabic sources, Mondino's nomenclature was confusing. For example, the sacrum was referred to by several names which appear to be of Arabic origin (e.g. alchatim alhavius, allannis), but other bones of the pelvis, hips, acetabulum, and corpora quadrigemina (four rounded eminences, a part of the midbrain) were all referred to by the same word, anchoe. It is believed that he introduced the term matrix for uterus. Some of his etymological information was untrue also. He stated that the word, colon was derived from the Latin cola (cell) but it was really from a Greek word which meant organ.

A student might only get experience with one cadaver during his entire medical education. This posed the question of how a student might see more examples of the mysteries of the human body. Mondino's book was the only one available and although texts are useful, nothing can take the place of direct observation. The practice of grave robbing began because there were not enough criminals to donate their bodies to each student. Students and professors exhumed bodies from nearby cities, believing that the decedents would be unknown to the dissectors. For skeletons, which were less identifiable, graves still had to be disturbed but there was less anxiety and care involved. Sometimes a skeleton could be obtained from a gibbet, a post with a projecting arm from which the body of an executed criminal was suspended on display. Still there was a lack of suitable specimens to study.

Dissection was not always done in a university setting. Convents and monasteries had their share of bodies that were used by nuns and friars; their objective was to find marks of holiness on the deceased. Although the holy were said to have rubies instead of blood clots and precious gems rather than gallstones, the world view of the prior was quite different than that of the anatomist. Religious individuals were interested in finding the place of the soul and how long it took after death for the soul to leave the body. There was also the belief that a luz bone existed, a structure that could clone or resurrect a new individual from itself.

What seems to be quite unusual to the modern student is the fact that regular public dissections were also performed. A stage was erected with a table and instruments. An announcement was made, sometimes posted in the center of town, and all citizens were invited. The tenor of the audience, rather than serious or scholarly, was ready for entertainment. As the tradition grew, the stages became more elaborate with curtains, elegant lamps, and decor. People dressed for the occasion and youngsters sold refreshments. Dissection theaters by the late Renaissance were akin to operatic productions.

Towards the end of the seventeenth century, rather than prohibit dissection, the church endorsed and encouraged the appropriate study of the human body. It did, however, watch closely to assure that vivisection (dissection of a living person) was not performed. According to Jewish law, nowhere in the Bible—Talmud, post Talmudic, or later rabbinical writings—was there evidence that postmortem examinations to benefit the living were prohibited.

Changes that occurred in both the study of medicine and the public's acceptance of dissection as a valid scientific inquiry were also philosophical. The centuries of crude, early dissection paved the way for surgeons to learn about the way the body works and to help patients who otherwise may not have had a chance at life. A shift from the theological to the humanist now was in its earliest beginnings. Observation was being documented and although few images existed, the emerging practice of human dissection paved the way for the study of physiology, anatomic pathology, the preservation of specimens, new surgical techniques, and the detection of foul play in unattended deaths.

LANA THOMPSON

Further Reading

Gonzalez-Cruzzi, F. *Suspended Animation.* New York: Harcourt Brace and Company, 1995.

Rabinowitch, I. M. *Postmortem Examinations and Jewish Law.* Montreal.

Sawday, Jonathan. *The Body Emblazoned.* London: Routledge, 1996.

Singer, Charles. *The Evolution of Anatomy.* New York: Alfred A. Knopf, 1925.

Thompson, Lana. *Death, Dissection and Disease in the Sixteenth Century.* Talk at IAFS, Los Angeles, 1999.

Walsh, James J. *Old-Time Makers of Medicine.* New York: Fordham University Press, 1911.

White, A.D. *A History of the Warfare of Science with Theology in Christendom.* New York: Dover, 1960.

The Black Death

Overview

The pandemic of bubonic plague that swept across Europe between 1347 and 1353 is known today as the Black Death, though contemporaries called it the "Great Pestilence," and the disease itself was generally known as *peste.* During these years, plague affected the lives of all Europeans, and killed nearly half of them. Its impact was enormous, not only because of the tremendous loss of life, but because of the pessimism, fear, suspicion, and even persecution of Jews (who were blamed for the disease) that followed.

In the long term, the Black Death may have increased economic opportunities and promoted a higher standard of living for those who survived. Its rapid spread gave rise to the medical theory of contagion. This scientific observation, in fact, is one reason that the epidemic is often cited as a turning point from the medieval era to the Renaissance.

Background

Plague, caused by the bacterium *Yersinia pestis,* is usually a disease of rats, not of man. Named for Alexander Yersin, the nineteenth-century scientist who first isolated it, the bacillus is found naturally in rodent populations, among which a small number of cases at any given time is common. Occasionally, however, the disease becomes endemic, killing off large numbers of rats. When this happens, the rat flea, *Xenopsylla cheopis,* which normally feeds on rodent hosts, turns to people instead. Their bite transmits the plague from infected rat to man.

In medieval times, plague was most often carried by the common black rat, *Rattus rattus,* which lived among the populace, feeding on grain stores and other foodstuffs. Some historians argue that the human flea, *Pulex irritans,* may also have played a significant role in transmitting the disease, as it will feed on any available blood source, moving indiscriminately between rats and humans.

Symptoms of plague develop quickly after infection. In man, the disease takes one of three forms: *bubonic* (involving the lymphatic system), *pneumonic* (centered in the respiratory system), and *septicemic* (involving the bloodstream).

The best-known symptom of bubonic plague were *buboes*—hard, extremely painful, swollen lymph nodes—which filled with blood and pus, turned black, and often burst, giving the disease its common name. The buboes were accompanied by a high fever, headache, chills, body aches, and sensitivity to light. At least half the people who contracted this form of the plague died.

Those who suffered from pneumonic plague usually had no buboes, but their lungs filled with fluid and blood, and they too endured raging fevers, sweats, and pain. Almost no one survived infection with this form, and unlike bubonic plague, pneumonic plague could be transmitted directly from one person to another. The septicemic form of the disease, which occured when the bacillus invaded the bloodstream, often killed its victims so quickly that symptoms rarely even had time to develop.

When plague exploded in 1347, it was an unknown disease. According to contemporary accounts, it was brought to Europe aboard Genoese grain ships, which had been trading for grain in the Black Sea port of Kaffa when Mongols attacked the city. When the disease broke out among the attacking troops, they are reported to have catapulted the dead bodies of plague victims over the city walls in an effort to spread disease among the besieged. The tactic worked, and Genoese ships fleeing the city carried the disease back to Italy.

While these Genoese ships certainly helped disseminate the disease into the Mediterranean basin, they were not the sole cause of plague's spread. Modern scholars have established that plague was actually spreading beyond its natural reservoirs in central Asia before the Genoese arrived in Kaffa. Mongol conquests and new trade routes likely disrupted these reservoirs, allowing plague to spread. The disease seems to have moved through the Middle East into the Upper Nile River, then traveled to the islands of Cyprus and Rhodes in 1348. From there, it spread eastward, infecting Eastern Mediterranean coastal cities. At the same time, the pestilence reached across the Mediterranean to infect Sicily, spread outward to Mediterranean port cities and then moved rapidly northwards. Over the next few years, it raced across all of Europe, reaching Scandinavia by 1351 and traveling eastwards to Moscow by 1353.

Woodcut of doctor attending to a plague victim.

Impact

The plague wreaked enormous and long-lasting consequences. After the initial pandemic, known as the Black Death, it remained an active health threat for over 500 years. (The last pandemic started in Asia in 1894; by the time it ended in 1908, over 6 million people had died.) In the centuries that followed, port cities were most often affected, but all areas faced at least some risk. Subsequent epidemics prompted many negative but predictable reactions, including fear, blame, suspicion, and isolation.

Firsthand accounts of the Black Death refer repeatedly to the social breakdown that occurred as people tried to protect themselves, neglecting traditional ties and obligations to friends, neighbors, and even children and family. Plague victims and their families were isolated, sometimes even walled up inside their houses and left to die. It is clear that contemporaries were profoundly fearful, not only of the disease itself, but of the changes it produced in morality, beliefs, and social relations.

The people of the time believed that one or more factors had caused the plague, particularly divine punishment for mankind's sin. Many communities prayed, made pilgrimages, and held ritual processions in attempts to appeal for God's mercy. Patron saints of plague victims emerged, the first being the ancient martyr Saint Sebastian; later Saint Roch, himself a victim of the disease, was canonized. An extreme religious group, the flagellants, roamed the cities and towns of Central Europe holding public confessions and performing displays of piety in which they used whips, known as *flagella* to scourge themselves.

The most extreme response to the terror of the plague was the scapegoating of Jews, who were rumored to have poisoned communal wells to spread disease. This produced a hysterical campaign of ferocious violence against Jewish communities, many of which were entirely destroyed in mass executions.

Not everyone succumbed to ignorance, however. Physicians wanted a scientific explanation for the plague, and although no one would know the truth for another 600 years, they at least sought physical causes. The Paris Consilium, a committee of 49 medical experts from the University of Paris appointed by Philip VI in 1348, issued a treatise saying that absolute knowledge of the plague's origin was impossible. It suggested, however, that the alignment of stars and planets had produced earthquakes and storms. These, in turn, spread putrefaction and noxious fumes ("miasmas") arising from swamps, rotting garbage, and unburied corpses. When these poisonous vapors were inhaled, they went to the heart, corrupting the entire body from within.

While little could be done to change the heavens, attempts were made to purify air by publicly burning herbs such as rosemary, and by cleaning streets and removing garbage. Individuals tried to protect themselves by carrying handkerchiefs or posies (bouquets of flowers) that were thought to block out these poisonous fumes, and doctors devised elaborate protective costumes similarly meant to prevent breathing any corrupt air.

An important result of the Black Death was the development of a crude theory of contagion. Until the advent of germ theory in the nineteenth century, disease was believed to result from an imbalance of the four basic humors within the body (blood, phlegm, black bile, and yellow bile or choler). The humoral theory held that just as the world was composed of four basic elements (earth, air, fire, water), the human body was composed of four constituents, called humors, which were maintained in individual proportions in each body. Since disease was thought to result from humoral imbalance, there was little thought that one person could "give" disease to another.

When plague began to spread in the mid-fourteenth century, observation and experience seemed to point to a form of contagion. The dis-

BUILDING A BETTER RAT TRAP

Rats have bedeviled humans throughout recorded history. They eat food needed by the hungry, they can inflict serious bites and injury, and they can spread disease, including the bubonic plague that decimated Europe several times in the past millennium. Finding a way to catch and kill rats is one of the biggest challenges in public health, and has been for millennia. In general, rats are adaptable, intelligent, and breed quickly.

For centuries, there just weren't many ways to catch rats. The story of the Pied Piper aside, most rats were killed with shovels and hoes. Although cats can be efficient rat killers, in the Middle Ages cats were often thought to be associated with witches. Thus, rather than using cats to their advantage in the battle against rats, people tended to kill cats or drive them away. In the twelfth century rat traps finally began to find common use. These traps had many advantages over previous methods of rat-killing. One person could set many traps, and the traps were automatic, catching rats even when humans were asleep. Rat traps did not, of course, eliminate rats, and bubonic plague continued to pose a threat. But for the first time, people had a weapon to use. Rats continue to be a nuisance today, especially in urban areas.

P. ANDREW KARAM

ease spread quickly within households, often taking entire families. Those in closest contact to the sick, such as caretakers, clergy, and medical professionals were frequently the next to fall ill, seemingly because of their simple proximity to the disease. Thus, a belief in the *transmissibility* of plague developed long before a formal medical theory was proposed. It was not until 1546 that the Italian physician Girolamo Fracastoro (c. 1478-1553) argued that illness could be spread directly from person to person via "seeds" that could travel short distances or become embedded in textiles for longer trips.

This belief in contagion reinforced people's natural tendency to flee in the face of in impending epidemic, but it also gave weight to municipal responses that emphasized exclusion and isolation. By the late fourteenth century, the Italian town of Ragusa required arriving ships to wait at sea for a period of 40 days in order to confirm the health of the crew. Thus the quarantine (from the Italian *quaranti giorni,* or 40 days) was born. In subsequent decades, cities and towns began to restrict entry in times when plague threatened, often requiring health "passports" for admittance. Once plague broke out within cities, they employed a practice of isolation, building plague hospitals, called *lazarettos,* outside the city walls and placing those diagnosed with the disease in them, using force if necessary. While certainly a rational response to contagion, it did little to prevent the movement of rats and their fleas, which continued to roam the city freely.

If any good can be said to have come from the Black Death, it's that those who survived were able to improve their place in society afterward. The tremendous loss of population created much economic opportunity, and many scholars believe that it hastened the end of serfdom by making labor both scarce and valuable. The plague's most surprising result, however, was the intellectual and artistic flowering of the Renaissance, which followed quickly on its heels. The intellectuals who emerged as the first generation of Renaissance humanists, such as Frances Petrarch (1304-1377), were survivors of the Black Death; their successors continued to strive and achieve despite the constant threat of plague.

KRISTY WILSON BOWERS

Further Reading

Biraben, Jean-Noël. *Les Hommes et la peste en France et dans les pays européens et méditerranéens,* 2 volumes. Paris: Mouton, 1975.

Campbell, Anna Montgomery. *The Black Death and Men of Learning.* New York: AMS Press, 1969.

Carmichael, Ann G. "Contagion Theory and Contagion Practice in Fifteenth-Century Milan." *Renaissance Quarterly* (1991): 213-256.

Carmichael, Ann G. *Plague and the Poor in Renaissance Florence.* New York: Cambridge University Press, 1986.

Cipolla, Carlo. *Fighting the Plague in Seventeenth-Century Italy.* Madison, WS: University of Wisconsin Press, 1981.

Dols, Michael. *The Black Death in the Middle East.* Princeton: Princeton University Press, 1977.

Herlihy, David. *The Black Death and the Transformation of the West.* Cambridge, MA: Harvard University Press, 1997.

Shrewsbury, J.F.D. *A History of Bubonic Plague in the British Isles.* Cambridge: Cambridge University Press, 1970.

Slack, Paul. *The Impact of Plague in Tudor and Stuart England.* London: 1985.

Twigg, Graham. *The Black Death: a Biological Reappraisal.* London: Batsford Academic and Educational Press, 1984.

Williman, Daniel, ed. *The Black Death: The Impact of the Fourteenth-century Plague.* Binghamton, NY: Center for Medieval and Early Renaissance Studies, 1982.

Zeigler, Philip. *The Black Death.* New York: 1969.

The Western Revival and Influence of Greco-Roman Medical Texts

Overview

Ancient Greeks and Romans had extensive knowledge of and innovative ideas about medicine. Scientists such as Hippocrates and Galen wrote sophisticated books about medicine between 400 B.C. and A.D. 200. For the next several hundred years, advances in medicine were fairly minimal. Beginning in about A.D. 850, key texts of ancient Greece and Rome were rediscovered by Islamic scientists. The subsequent translation and dissemination of these works throughout western Europe and the Middle East led to a revival of Greco-Roman ideas that influenced medicine well into the Renaissance several hundred years later.

Background

The need to alleviate ills and cure man's diseases has always existed, and ancient scientists made important discoveries about anatomy and physiology. Building upon the knowledge of ancient Egyptians and others, the philosophers of ancient Greece produced sophisticated writings about medicine. Among these scientists was Hippocrates (460-377 B.C.), often called the "Father of Medicine." He lived on the Greek island of Cos and was known for using thoughtful treatments and opposing the use of magic or witchcraft. He originated the ethical standards to which physicians are still held accountable today. His "Physician's Oath"—while not actually written by him—is still administered to new medical practitioners in modified form. Other Greek medical practitioners and teachers, like Herophilus (335-280 B.C.) and Erasistratus (fl. 320 B.C.), are less well known but come from the same intellectual tradition.

Besides Hippocrates, Galen (A.D. 130-200) was another well-known ancient physician whose work survived into later centuries. He was born in what is now Turkey but moved to Rome to practice medicine. He did experiments to further his understanding of the human body and wrote many treatises on medicine, anatomy, and physiology. His major work, *Anatomical Procedures*, was the standard medical text in western Europe and the Middle East for 1,500 years. Many of his theories, as well as those of Hippocrates, were only overturned when advances in dissection, experimentation, and observation were made.

Both Hippocrates and Galen exerted tremendous influence on medical practice for many centuries. One reason that Hippocrates' ideas were never superseded was that the scientific activity of the Greeks in ancient times decreased when the power and dominance of Athens and other Greek cities faded. Defeated and taken over by the Romans, the Greeks not only lost their independence but also their incentive to engage in innovative thought. Following the work of key Roman thinkers, notably Galen, there was little further innovation in medicine for several hundred years.

About A.D. 850, an Arab physician named Hunayn ibn Ishaq (808-873) became interested in ancient medical literature. The Arab world, revitalized by the religion of Islam after about 650, had conquered the Near East, North Africa, and Spain. This energy fostered the leap to pursue and expand on the knowledge of the ancients in history, geography, science, and medicine. Thinkers such as Hunayn ibn Ishaq translated Galen, Hippocrates, Plato (428-348 B.C.), Aristotle (384-322 B.C.), and others into Arabic or Syriac, a language of the Bible. Thus the Arabic world became heir to ancient Greek and Roman medical ideas. Hunayn studied medicine in Baghdad and became chief physician to the court of the Caliph. In particular, he was fortunate to work in the "House of Wisdom" established by the Caliph al-Ma'mun, a center of learning that emphasized the translation of ancient texts. Hunayn's translations of Greco-Roman texts were widely disseminated throughout the Islamic empire.

Impact

Since many original Greco-Roman texts have since been lost, the importance of the translation work of Hunayn and others of his era is hard to overestimate. In the short run, the revival of interest in the Greeks that he influenced helped to spark the prodigious scientific advancements of the medieval Arab world. Building on the knowledge of ancient scientists, Arab scientists made fundamental advances in all areas of science, from astronomy to physics to the life sciences.

In the eleventh and twelfth centuries, European scholars such as Gerard of Cremona (1114-1187) discovered the Arabic translations of ancient Greco-Roman texts and began translating them back into Latin. Thus, these works became the foundation of medical education in early medieval universities like Oxford in England, Bologna in Italy, and the University of Paris. Subsequent scholars who helped preserve and pass on the ideas of these early physicians included Robert Grossteste (1168-1253), the first chancellor of Oxford, who published a number of commentaries on Aristotle and helped draw attention to the newly restored texts of the ancients.

Similarly, influential philosopher Roger Bacon (1220-1292) relied on Latin translations from the Arabic. This drew further attention by Western intellectuals to the large body of Arabic literature and sparked a desire to correct extant versions of ancient literature. Greek texts, when available, were then used to correct Latin versions. Bacon also made observations of his own and was known as the founder of experimental science in the Middle Ages. His *Opus Maius* of 1267 is a summary of all knowledge to that time.

Another medieval physician who benefited from the preservation of Greco-Roman texts was German thinker Albertus Magnus (1200-1280). He made many contributions to medieval philosophy and was known for discoveries and observations in natural science. His work in botany and zoology was important to the development of those disciplines. Like most other medieval philosophers, he considered Aristotle the greatest philosopher and writer on natural history who ever lived. Magnus wanted to create works that would fill in what was lacking or had been lost of Aristotle's work. The medical writings of Galen and Hippocrates helped him do so.

At the dawn of the Renaissance, Nicholas of Cusa (1400-1464) provided another boost to the recovery of ancient Greco-Roman texts. He was interested in mathematics, science, and philosophy as well as medicine. He believed that knowledge was gained from the study of ancient works and through experiment. He was a collector of manuscripts and recovered many that survived to his day. Not only did he discover the lost comedy plays of Roman playwright Plautus, but in 1426 he also discovered an important work called *De Medicina* by an ancient Roman medical writer named Celsus. Written in the first century A.D. in Rome, it shows the status of medicine and the level of medical knowledge achieved by the Romans at that time. Romans washed wounds and treated them with antiseptic substances like vinegar and thyme oil. They also performed surgical procedures and did plastic surgery using skin from one part of the body to restore another. Celsus's work also includes the first accounts of heart disease, dietetic treatments, insanity, and the use of ligatures or stitches in the treatment of arterial bleeding.

Along with the greater socio-cultural currents sweeping through Europe, the recovery of key texts by ancient scholars helped fuel the tremendous intellectual and artistic advances of the Renaissance of the fifteenth and sixteenth centuries. In no field was this more true than in medicine, where the writings of Hippocrates, Galen, and others held enormous sway throughout the Renaissance. The survival of these texts is due in large part to the Arabic translations of Islamic scientists and to the later Latin translations of medieval European thinkers.

LYNDALL BAKER LANDAUER

Further Reading

Jouanna, Jacques. *Hippocrates*. Baltimore: Johns Hopkins University Press, 1998.

Magner, Lois N. *A History of Medicine*. New York: M. Dekker, 1992.

Sigerist, Henry. *The Great Doctors: A Biographical History of Medicine*. New York: W.W. Norton, 1933.

Singer, Charles. *The History of Biology*. New York: Abelard-Schuman, 1959.

Westacott, Evalyn. *Roger Bacon in Life and Legend*. New York: Norwood Editions, 1953.

The Significance of Ibn Sina's *Canon of Medicine* in the Arab and Western worlds

Overview

Ibn Sina, (980-1037), whose name was Abu al-Hussayn ibn Abdullah ibn Sina, was an outstanding medical writer and physician. His *Al-Quanun fi al-Tibb,* was a masterpiece of Arabic systemization, in which he sought to collate and organize all known medical knowledge. When the work was translated into Latin, it became known as the *Canon of Medicine* and was the dominant text for the teaching of medicine in Europe. It went through many versions and was later translated into the vernacular of several nations. The *Canon* was used as a medical text for over 800 years, continuing in some areas until well into the nineteenth century.

Background

The prophet Muhammad, born at Mecca in 570, created a religion—Islam—that quickly spread and unified the tribal people of the Arabian peninsula. Mohammed recorded his beliefs in the Koran, the holy book of Islam. He was concerned with health and addressed many issues in his writings. During the seventh and eighth centuries, Islam spread eastward and north into regions as far away as parts of the future Soviet Union. Dynasties of rulers held various kingdoms of Islamic influence. In the area that is presently Uzbekistan, the Samanid princes developed a strong empire with a distinguished court and extensive library of Greek writings and other texts. Ibn Sina was born in this environment in the tenth century.

Impact

Ibn Sina was to the Arab world what Aristotle (384-322 B.C.) was to ancient Greece. He was a scholar not only in medicine but in law, mathematics, physics, and philosophy. The name Avicenna, as he was known in Europe, is a Latinized form of his Arabic name.

Born in Bukhura, in what is now Uzbekistan, Ibn Sina had memorized the complete Koran by the time he was 10. He was 16 when he decided to turn to medicine, which he found very appealing and quite easy. While in his teens he knew enough about medicine to treat the sick Samanid ruler, Nuh ibn Mansur. This gave him what he wanted—access to the royal library, which had a large collection of Greek philosophy and sciences. At age 20 he was appointed court physician and traveled widely, but somehow found the time to write 20 books in other fields and 20 on medicine. After traveling to several places and meeting famous contemporaries, he moved to Hamadan in present-day Iran and completed many monumental writings. His *Kitab al-Shifa* (The book of healing) is a medical and philosophical encyclopedia.

His great work, the *Canon of Medicine* was written before his twenty-first birthday. At more than one million words long, it represents a comprehensive collection of all existing medical knowledge during his time. He summarized Hippocrates (460-377 B.C.), Galen (130-200), Dioscorides (40-90), and late-Alexandrian physicians, adding Syro-Arab and Indo-Persian knowledge along with his own notes and observations. He tried to put anatomy, physiology, diagnosis, and treatment into proper categories. The work is not only a systematic digest of all medical information, but is clearly arranged, organized, and written. As a result, Ibn Sina's work became preferred to those of Rhazes (865-923), Ali ibn Abbas Maliki, and even Galen.

The five books of the *Canon of Medicine* are organized with summaries and comments. Book One begins with the general principles of humors. Ibn Sina elaborates on the versatility of the humoral theory and how it fits into the four elements, four ages of man, and the four temperaments. He then moves on to anatomy, physiology, hygiene, etiology (the origin of diseases), and symptoms and treatments of disease.

Book Two deals with *materia medica,* or pharmaceuticals. He details all known information on the physical properties of simple drugs and discusses how to collect and preserve herbals. In a separate section he lists 740 different types of medicinals.

Book Three zeroes in on specific diseases that he catalogs from head to toe. He outlines the etiology, symptoms, diagnosis, prognosis, and treatment of each one of these diseases.

Book Four tackles conditions that affect the entire body, including fevers, infections, ulcers, abscesses, pustules, fractures, and injuries. Avicenna also discusses poisons and includes a section on anorexia and obesity.

Book Five ends with a discussion of compound drugs, using terms like theriacs, electuaries, emetics, pessaries, and liniments, along with their medical uses.

In its basic conception the *Canon* follows Galen in the ancient tradition of elements and humors. The elements are air, water, fire, and earth. The theory of the four humors includes blood, phlegm, choler or yellow bile, and melancholy or black bile. However, Ibn Sina incorporated many observations not found in Galen's.

Ibn Sina's original contributions recognized the contagious nature of phthsis and tuberculosis, the distribution of diseases through soil and water, and the interaction of mind and body. He suggested a treatment for lacrymal fistula and introduced a medical probe for the channel.

He also stressed the importance of prevention. Discussing diet and the influence of climate and environmental factors on health, the book also discusses rabies, hydrocele, breast cancer, tumor, labor, and treatments for poisons. He showed the difference between meningitis and the meningismus of other acute diseases. Other conditions he described include chronic nephritis, facial paralyses, ulcer of the stomach, and types of hepatitis and their causes. He discussed in great length the dilation and contraction of the pupils and how they were used in diagnosis, as well as the six motor muscles of the eye and the function of the tear ducts. Noting the contagious nature of some diseases, he attributed this to "traces" left in the air by the sick person.

In addition to a description of 740 medicinal plants and drugs made from them, he laid out a series of basic rules for clinical drug trials, many of which continue to serve as the basis of modern clinical trials. Ibn Sina stipulated that: 1) The drug must be pure and must not have anything that would lessen the quality; 2) The drug must be used on a "simple" disease, not one with several complications (Today, this is called targeted treatment); 3) The drug must be tested on at least two different types of diseases, as sometimes a drug cures one disease by the essential quality and cures another by accident (Ibn Sina was proposing the parameters for a controlled experiment); 4) The quality of the drug must correspond with the strength of the disease (This is based on the Greek view that the "heat" of some drugs is less than the "coldness" of certain diseases, so they would not have an effect on them); 5) The observation of the time must be recorded so as not to confuse with natural healing; 6) Several trials must be made and the effect of the drug deemed to be consistent, otherwise the effect may be considered an accident; 7) While animal testing on a lion or horse may begin the experiment, it must be tested on the human body, as reactions on an animal may not be the same as the effect on man.

Ibn Sina encouraged observation and a close study of the human body; he condemned conjecture and presumption in anatomy. Some of his observations were ahead of his time. For example, he observed that the aorta, at its origin, contains three valves that open as blood rushes into it from the heart during contraction and closes during relaxation so the blood may go back into the heart. He determined that muscles could move because of nerves and that pain in the muscles is also due to nerves. He found that the liver, spleen, and kidney do not contain nerves, but that nerves are present in the coverings of these organs.

The *Canon* was translated into Latin as early as the twelfth century by Gerard of Cremona (1114-1187) and published in Venice in 1493-1495. The *Canon* became a standard textbook for medical education in the emerging medical schools of Europe. The fact that it was reissued 16 times in the last 30 years of the fifteenth century underscores its influence; 15 editions were in Latin and one in Hebrew. During the sixteenth century, it was revised more than 20 times. In 1390 Cameron Gruner translated part of the book into English as "A Treatise on the Canons of Medicine of Avicenna." From the twelfth to the seventeenth centuries the *Canon* was the chief guide to medical science in the West. As the great medical scholar William Osler (1849-1919) noted, the *Canon* has remained a medical bible for longer period than any other work.

Although some of the science of the elements and humors are no longer accepted, the contributions of Ibn Sina have helped set the standard for current medical practice. His encouragement of observation is one of the basic tenets of the scientific process, his description of the anatomy and some of the muscles of the eye are still used, and his advocacy of trial and control in drug testing forms the basis of modern pharmacological discovery.

Ibn Sina and his *Canon of Medicine* has become so important that no discussion on the history of medicine can be complete without referring to him. He has earned many honorary titles, notably the "Galen of Islam." His preeminence in the Latin West was even enshrined by Dante (1265-1321), the fourteenth-century Italian poet, who ranked Ibn Sina between Hippocrates and Galen.

Nevertheless, some critics have alleged that Ibn Sina stifled independent thought through his writings. However, some of his treatises and comments are highly critical of the works of past writers, opening the door for subsequent physicians to question his own writings and those of other scientific authorities.

EVELYN B. KELLY

Further Reading

Clendening, Logan. *Source Book of Medical History.* Dover: Henry Schuman, 1942.

Khan, Muhammed Salim. *Islamic Medicine.* London: Routledge & Kegan Paul, 1986.

Porter, Roy. *The Greatest Benefit to Mankind: A Medical History of Humanity.* New York: W.W. Norton, 1997.

Pouchelle, Marie-Christine. *The Body and Surgery in the Middle Ages.* New Brunswick, NJ: Rutgers University Press, 1990.

The Development of Arab Medicine During the Eighth through Thirteenth Centuries

Overview

During the Middle Ages, Arabic medicine developed and filled a major gap left by the fifth-century collapse of the Roman empire in the West. At first Islamic physicians sought to preserve knowledge by collecting, then translating, the classical Greco-Roman medicine that Europe had lost. Then they began adding information from other cultures, giving their own comments and interpretations. Arab physicians laid the foundations of modern medicine as well as that of important medical institutions.

Background

After the collapse of the Roman empire, very little knowledge of Greek medical science was available in the West. The Church became the center of society and greatly influenced the development, or stagnation, of medicine in the West. The Church made no pretense of its mission, which was to minister to the soul and not to the body. Monastic orders ran hospitals, but they were places for the seriously ill, and people were expected to recover or die as God willed. No medicines were given, nor did physicians attend to them. Without studying disease, and with the physical health of people considered relatively unimportant, medicine as a craft vanished. In the middle of the seventh century the Catholic church banned surgery by monks.

The Eastern empire at Byzantium experienced turmoil in the form of ethnic tensions and bitter splits among Christian sects. The Greek heritage weakened and declined, though medical learning was perpetuated in centers like Alexandria in North Africa.

At about this time a different development was occurring on the Arabian peninsula. In this region a tribal society rife with injustice, promiscuity, and disease was transformed with the rise if Islamic religion. The teachings of the prophet Muhammad, born at Mecca in 570, united the scattered tribes of Bedouins under one set of rules. Through Mohammed the concepts and practices of health became incorporated in the general body of Islamic religious teachings. Muhammad was concerned with the health of his followers and even wrote about treatment of many conditions, such as leprosy and smallpox. In the seventh and eighth centuries the Islamic empire spread eastward to the Indus River, west along the Mediterranean coasts of Africa, and north to the Pyrenees in western Europe.

During the Umayyad dynasty (660-750), which was based in Damascus, translations of ancient medical works began. Prince Khalid bin Yazid had a passion for medicine and alchemy and instructed Greek scholars in Egypt to translate Greco-Egyptian medical literature into Arabic. In addition, after Plato's Academy was closed in 529, some of the scholars moved to the University of Jundishahpur in the old capital of Persia. This city also became home to excommunicated Nestorian Christian physicians. When Persia became part of the Islamic empire in 636, the Arab rulers supported the medical school and for the next 200 years it was one of the greatest centers of medical learning in the Islamic world. Under the Umayyads, the Hispanic Muslim area of Cordoba, Seville, and Granada also became great centers of learning, which later peaked in the twelfth century.

In around 750, political rule of the Muslim world passed into an era known as the Abbasid period. In 754 Abu Jafar al Manser founded Baghdad on the banks of the Tigris River in the fertile area of Iraq. The climate was ideal, and there was an absence of mosquitoes. In 750 the empire divided into the Eastern Caliphate, with Baghdad as the capital, and the Western Caliphate, with Cordoba, Spain, as capital. The Muslims had conquered the eastern Byzantium empire. At first the Arabs were indifferent to the learning of their infidel subjects, but gradually grew to appreciate it. In around the middle of the eighth century they began to undertake Arabic translations of Persian, Hindu, and especially Greek scientific, medical, and philosophical works. Most translations were made by Syrian-Christian and Sabian scholars.

The Muslim world developed an attitude of relative tolerance for Jews and Christians; as Islam was considered a superior religion by its adherents, Muslims did not make proselytizing a priority and these groups got along very well in that region. It was not until the papacy launched the Crusades to drive the Muslims out of the Holy Land that hostilities developed.

Impact

Medicine was the first of the ancient Greek sciences to be revived and studied. The Abbasid caliph Harun al-Rashid (786-809) and his son al-Mamun (813-833) recognized the importance of translating Greek works into Arabic and established a translation bureau, called the Bayt-al-Hikmah or House of Wisdom, in Baghdad. They sent people throughout the old Greek empire to collect scientific works. Islamic physicians studied the works of Hippocrates (460-377 B.C.), Galen (130-200), and other Greek physicians. At the same time, they gathered the knowledge of Byzantium, Persia, India, and China. This ushered in the first era of Islamic medicine—the period of translation and compilation.

One of the important translators was Hunayin ibn Ishaq (808-873), known in the Latin West as Johannitius. A profound scholar of Greek, he went to Baghdad to study medicine and soon became recognized for his great ability as a translator after translating a work by Galen. He was sent by al-Mamum to Byzantium to obtain manuscripts and to work on translations of the great Greek philosophers and physicians, such as Hippocrates, Galen, and Dioscorides (40-90). He made long journeys into Mesopotamia, Syria, Palestine, and Egypt to find the manuscripts. He strove to render the Greek text as clearly as possible and was impatient with poor translations. His methods are still used in modern philology.

Hunayin's medical works were divided into three areas: translations of ancient texts, summaries and paraphrases of these texts, and original treatises. He and his pupils translated 129 works of Galen into Arabic. Since Hunayin preferred Galen, he provided the Arabic world with more translations of Galenic texts than are available today in Greek.

Islamic medicine largely accepted Galen's humoral theory, by which he proposed that the human body was made up of the four elements that comprise the world—earth, air, fire, and water. When these elements were mixed in various proportions, the differing mixtures gave rise to different temperaments or "humors." When the humors were in perfect balance, the body was healthy; when the humors were out of balance the person would be ill, and this must be corrected by the doctor's healing arts. Thus, physicians viewed it as their task to preserve health if possible and then to heal if necessary.

During this period of translation, advances were made in other fields. Hospitals, or bimaristans, were established where the sick could be treated and health promoted. Medical schools and libraries were attached to hospitals, and the hospitals gave examinations and diplomas. There were traveling clinics that would go to places inaccessible to the hospitals. The Islamic hospital was the prototype of the modern teaching hospital today. Also, pharmacies were developed. The Arab alchemist Jabir ibn Hayyan (721-815) is considered the father of pharmacy. A large number of new drugs were introduced into practice. The first private apothecary shop opened in Baghdad at the start of the ninth century. Spain had a favorable climate and diverse array of flora, which aided in the development of botanical medicine. Ibn al-Baytar, born in Malaga in 1197, worked in southern Spain on different plants and wrote a commentary on Dioscorides. In general, Hispanic Islam had a high level of medical practice and surgery.

As the ninth century drew to a close, the first major Arabic medical work was produced by Abu Bakr Muhammad ibn Zakariya' ar-Razi (865-923), known in the West as Rhazes. Rhazes is regarded as Islamic medicine's greatest practitioner and original thinker. Some have dubbed him the father of pediatrics because of his treatise on the diseases of children. He wrote a medical encyclopedia consisting of 25 books.

Rhazes also knew the importance of the doctor-patient relationship. He developed a philosophy of medicine that asserted the primacy of reason and condemned slavery to authority in medicine. He later became one of the most respected medical scholars in the West and his notebook of medical writings, the *Kitab al-Hawi* (Comprehensive book), was translated into Latin in 1279 under the title *Continens.*

Not long after Rhazes' death, another great physician, Ibn Sina (980-1037), known in the West as Avicenna, was born. Some have said that Avicenna was to the Arab world what Aristotle (384-322 B.C.) was to ancient Greece, Leonardo da Vinci (1452-1519) to the Renaissance, and Johann Wolfgang von Goethe (1749-1832) to Germany. Avicenna's major work, the *Canon of Medicine,* was used as a standard reference until into the nineteenth century.

During the tenth century, Arab texts began to be translated into Latin. Europeans began to see the intellectual riches of the Arabs and were inspired to seek out their own heritage. Galen and Hippocrates returned to the West by way of the Arab medical classics. Salerno, at the crossroads of East and West, became one of the first great centers of medical scholarship in Europe.

The development of Arab science reached its high point in the tenth and eleventh centuries, and, after some sporadic and short-lived but important revivals in the thirteenth and fourteenth centuries, came to a virtual halt by the end of the fifteenth century.

Despite the fact that humors and miasmas have been displaced by other theories, the medicine of Islam made great contributions to techniques of diagnosis, knowledge of preventive health, and the use of quarantine to limit the spread of disease. The Islamic world proved an essential source for the European rediscovery of "lost" Greek medical knowledge and the promotion of scientific observation and experimentation in the West, where such fundamental precepts of medical practice had been abandoned.

EVELYN B. KELLY

Further Reading

Carmichael, Joel. *The Arabs Today.* New York: Doubleday, 1977.

Hoyt, Edwin P. *Arab Science: Discoveries and Contributions.* Nashville, TN: T. Nelson, 1975.

Khan, Muhammed Salim. *Islamic Medicine.* London: Routledge & Kegan Paul, 1986.

Porter, Roy. *Medicine: A History of Healing.*New York: Barnes and Noble, 1997.

The Emergence of Hospitals in the Middle East, Constantinople, and Europe During the Tenth through Twelfth Centuries

Overview

During the tenth century, hospitals began to emerge along the routes taken by European crusaders to the Middle East, as their travels were accompanied by microorganisms, discord, and holy wars. The Crusades, a series of military campaigns from 1096 to 1291, were undertaken by Christian groups to take back the Holy Land from the Muslims. Trauma, either due to injuries, war, or poor living conditions, was a constant companion to those who traveled.

Then, as now, a stable, adequate diet, availability of clean water, and sanitary lodging were necessary for the maintenance of health. In the past pestilence and disease were frequently more lethal than actual warfare. The first hospitals were military institutions built along the well-worn routes of traders and the crusaders. Later, as populations increased, both Christians and Muslims found the need to build hospitals, and their respective motivations were both charitable and noble.

Background

The Latin word for the place where a guest was received is "hospitium," and the word "hospice" was used to mean places for permanent occupation by the poor, infirm, incurable, or insane. "Hospitalis," the adjective, came into use to describe a temporary place for occupation by the sick. "Hospital" then started to be used both in the sense of a permanent retreat for the disenfranchised and a temporary place for sick people; it later came to take the second meaning exclusively.

Prior to the Middle Ages, many hospitals had been built at the time of Emperor Constantine's conversion to Christianity in 325. Roman military *valetudinaria* served a similar purpose. Early Christian hostels, built to shelter pilgrims, were more like inns or hotels than institutions for study or teaching. However, they also cared for the sick and infirm. At the time, medicine was considered a work of charity, and helping the sick a duty.

In Persia, the city Jundishapur became the cross-cultural center for exchange of information among Greeks, Indians, Jews, and Nestorians. The hospital there is said to have been more like a medical school than a hostel. It was associated with a university where scholars translated Hippocrates (460-377 B.C.) and Galen (130-200), the mainstays of medical knowledge, from Greek into Arabic.

The Muslims, followers of the religious prophet Muhammad (570-632), became prime movers in the spread of culture and medical information throughout Asia, Persia, Egypt, North

Africa, and Spain. Wherever Muslims went, so did Arab medicine. Although there were no hospitals during Muhammad's lifetime, a multiplicity of educational institutions was established as part of Islamic culture. One of the most significant achievements of the Golden Age of Islamic medicine was the development of hospitals and hospital-based clinical training of medical practitioners. By the end of the ninth century, more than 30 hospitals had been built.

Impact

Hospitals in the Islamic World

The first true Islamic hospital was built, in the style of Jundishapur, during the reign of Caliph Harun-ul Rashid (786-809) in Baghdad. The next hospital was the Audidi, built under the direction of ar-Razi (865-923), or Rhazes as he was known in the West. One story recounts that, to select the site, Rhazes hung pieces of meat around the city to determine which took the longest to rot. The site where the meat became least putrefied was chosen for the Audidi. Audidi had a staff of 24 physicians, including bone setters, surgeons, oculists, and physiologists.

As Islam grew, the Muslims conquered west Asia, Egypt, North Africa, and Spain. The resulting medical system drew the best medical knowledge from these cultures. When hospitals were built, they were large complexes, employing 25 physicians and divided into specialized wards. They served as teaching hospitals as well, giving instruction to physicians.

Thirty-four of these early hospitals were in Muslim cities. It is believed that the elaborate hospitals were reconstructed palaces, the best known being the Mansuri in Cairo, built in 1283. This model hospital had male and female nurses in addition to physicians, and special wards for women, fever cases, eye diseases, and mental patients. Its water supply came from the Tigris River and it had the capacity to take care of 8,000 patients. It was endowed and had a pharmacy, library, and lecture hall.

However, the splendor of the environment was not matched by the professionalism of the staff. Standards had not been developed for the type of knowledge and skills necessary to work in a hospital and there were not formal examinations for membership. Many lay healers, magicians, and mountebanks plied their trades within those walls. Unfortunately, women, many of whom learned medicine from physicians in their family, were put in the same category as quacks and prohibited to practice; they could be nurses and midwives only.

Hospitals in Constantinople

In the early sixth century B.C. the Greeks founded a city called Byzantium, which was the beginning of Istanbul (now in Turkey). It was renamed Constantinople in A.D. 300. Captured by the armies of the crusaders in 1203, it remained Constantinople until 1454.

One of the most elaborate and complex hospitals in Constantinople was the Pantocrator, (which means "Christ the ruler") started by John II Comnenus, and completed in 1136. It was part of a complex of buildings that included a church, tombs, and a monastery. It had 50 rooms that were divided into five departments. Interestingly, these were divided into wards much like modern institutions, with a number of rooms designated for: surgery cases (5), acute illnesses (8), men (10), women (10), gynecological cases (12), and emergency rooms or miscellaneous (5). The hospital also trained students and had support services such as outpatient, a pharmacy, mill, and bakery.

Despite the named departments and similarity in words to our modern hospitals, the philosophy underlying the building complex was quite different. A hospital was not a primarily medical institution until the late Middle Ages. Until then, it was more like a small city where medicine was integrated with religion, healing, faith, and philosophy, rather than a distinct place of scientific care like it is now.

Hospitals in Europe

In Europe, hospitals developed near churches or as parts of them. There were also orphanages, almshouses, and residences called leprosoria and pesthouses for people with leprosy. A pesthouse was an abandoned building or structure, usually far from the living quarters, where sick people could go during epidemics with the hope of separating them from healthy people. While there was often no treatment for them, it did protect the other inhabitants of the city.

In England, the first hospital was built in York in 937, followed by others in Cherbourg, Bayeux, Caen, and Rouen in France. The oldest European hospital still standing is the Hôtel Dieu in Paris near Notre Dame. It is arranged like a large square with a courtyard in the center.

The Abbey of Saint Gall in Switzerland, built about 822, was a hospital within a monastery. Visitors were divided into groups based on their socioeconomic status. The Hospice of Pilgrims and Paupers consisted of a large dormitory without sanitary facilities. On the other hand, the House of Distinguished Guests had individual rooms and toilet facilities called "necessaria."

In the tenth century, under the patronage of Abd al-Rahman III (912-961), the Spanish town of Cordoba became a major cultural center of Europe, with no fewer than 50 hospitals. Some of the religious orders also built hospitals, including the Knights of Saint John of Jerusalem, the Knights Templar, and the Teutonic Knights.

The Influence of Religion

Religion and medicine are closely related in all cultures. However, the nature of this relationship is often very different. In certain cultures medicine, or healing, is inseparable from art or music. Sometimes it is the doctor who takes the drugs so that he or she can better perceive what is harming the patient. In Western society, influenced by the culture of Europe and the Judeo-Christian belief system, modern medicine is a distinct category of science, but not without its own spiritual legacy. Disease was originally seen to be a punishment from God. The Bible tells of plagues that were sent by the vengeful God of the Hebrews. Healing was the work of God, perhaps manifested through the acts of a physician, but most importantly through prayer.

As Christianity spread throughout Europe, the moral value of caring for the sick, the infirm, and the lonely and providing charity with regard to feeding the poor was the basis for building the earliest hospitals. According to Christian belief, if one cared for the sick, the good works would have their reward in the afterlife. However, salvation of the soul was more important than restoration of the body. It was not until the Renaissance that the philosophy of healing changed in its focus.

In Byzantium it was because of Constantine's conversion to Christianity that so many hospitals were built. In the Levant (Syria, Egypt, and the eastern Mediterranean), Jewish translators helped to bridge the gap between Arabic, Greek, Syriac, and Hebrew so that the knowledge base expanded exponentially.

Islamic ideas of charity and public welfare were probably based on the Koran, which states, "You shall not attend to virtue unless you spend for the welfare of the poor from the choicest part of your wealth." Stories about Muhammad told of his visiting the sick at home to give hope and comfort, and his mentorship encouraged these very necessary acts and deeds. They transferred easily to the value in having hospitals to care for those who needed charity and restoration of health. The Muslims, like the Christians, had conflicts concerning which values were more important—faith or health—but more with regard to the healer than the patient. The Muslims were able to rationalize their study of medicine by believing that as long as they acknowledged the primacy of faith, it was acceptable to practice medicine as a form of religious service to relieve suffering.

LANA THOMPSON

Further Reading

Bettmann, Otto. *A Pictorial History of Medicine.* Springfield, IL: Thomas, 1956.

Haeger, Knut. *The Illustrated History of Surgery.* New York: Bell Publishing Company, 1988.

Lyons, Albert S., and R. Joseph Petrucelli. *Medicine: An Illustrated History.* New York: Harry Abrams, 1978.

Magner, Lois. *A History of Medicine.* New York: Marcel Dekker, 1992.

Margotta, Roberto. *The History of Medicine.* New York: Smithmark, 1996.

McGrew, Roderick. *Encyclopedia of Medical History.* New York: McGraw-Hill, 1985.

Leprosy

Overview

Leprosy is a chronic infectious diseases that attacks the skin, peripheral nerves, and mucous membranes of the eyes and respiratory tract. The disfiguring disease has long been feared by the general public and, unfortunately, its victims have been misunderstood and despised throughout history.

Historically speaking, there are two types of conditions described as leprosy. One is the biblical leprosy described in the book of Leviticus, which was viewed as the result of terrible sin and Jewish law demanded that the afflicted be cast out. The other form is a bacterial leprosy caused by an organism related to tuberculosis called Mycobacterium leprae. During the Middle Ages, the two forms (moral and medical) were confused and the attributes of the bacterial variety acquired the stigma of the Biblical leprosy. In the Middle Ages those who were deemed lepers were cast out of their homes and left to wander alone. Later, those with leprosy were isolated in colonies and leprosariums to die slowly, and usually without treatment, from the illness. As leprosy declined during the twelfth and thirteenth centuries, the condition still occurred in some populations and was treated as a curse. In the nineteenth century the Norwegian scientist Armauer Hansen (1841-1912) discovered the bacteria that causes the disease.

Background

Until the rise of AIDS, leprosy was perhaps the most feared of all infectious disease. While effective treatment for leprosy is now available in the Western world, the disease is still a terrible scourge for millions in South America, Africa, and Asia. Even the Black Death would come and go in its attacks, but leprosy has continued to torment people throughout history into the present.

The disease is often considered not only a physical ailment, but also a spiritual condition. Therefore, the disease has deep-rooted mystical meanings in the human psyche. Still, many in Asia and Africa regard it as one of the most dreaded diseases because those who have it are thought to be cursed.

The origin of leprosy is unknown. An Egyptian papyrus produced between 1552-1350 B.C. describes a condition that may be leprosy. Writings from India dating from 600 B.C. also describe a similar disease. The disease appears in records following the return of Alexander the Great (356-323 B.C.) to Greece from India in 326 B.C. Likewise, Roman accounts show that the army of Pompey brought the disease back from Asia Minor in 62 B.C. It is assumed that the condition originated in Asia.

The Bible refers to a condition called leprosy, which appears 54 times in the text and links the disease with personal sin, as the disease was a punishment by God for transgression. Leviticus, a book of Hebrew law, was written around 1000 B.C. and focuses on the uncleanness of the leper. While the New Testament looks more kindly upon lepers, it sets them apart. When Jesus cleansed the leper, he also cleansed them of all moral stigma.

Some scholars contend that mistranslations of the Old and New Testament have caused much confusion about leprosy. The Hebrew word *tsara'ath* does not refer at all to leprosy as we know it today. In fact, the Biblical description of leprosy cannot be connected with any known contagious disease.

Thus, there appear to be two forms of leprosy, with bacterial leprosy being completely different from that described in the Bible. Bacterial leprosy can cause destruction of the upper lip and nose and damage to the nerves of the larynx so that the afflicted individual speaks in a hoarse, raspy voice. In addition, the parched skin of the afflicted may cause increased sensitivity to pain. This description does not fit the condition described in Leviticus.

Some think that the leprosy of Leviticus is more akin to elephantiasis, a disease caused by a parasitic worm. This condition was described by Hippocrates (460-377 B.C.) as "lepra" and translated as leprosy. During the early years of Christianity, those with leprosy were looked upon with revulsion. At the Council of Ancyra in 314, the church ordered restrictions upon lepers and defined them as unclean both in body and soul.

Impact

A great irony of history has connected bacterial leprosy to the leprosy of Leviticus. An outbreak of the bacterial version spread through western Europe at about the same time as the rise of Christianity. It was during the Middle Ages that the disease was deliberately linked to the biblical version of leprosy and the method for controlling the disease, which was to drive the afflicted out or to isolate them, was enforced. The clergy were among the few educated people with access to medical texts and translations, giving their authority on the matter even greater significance.

During the early Middle ages dreadful stories about lepers were used to set a moral example, which also justified their exclusion. Constantine, the first Christian Roman emperor, ordered the expulsion of lepers, but, in his zeal to care for the sick, assigned a minister to serve them.

As the Middle Ages continued, lepers were accused of a host of sins. They were said to anger easily, have terrible dreams, and to be deceivers and schemers. Their behavior was viewed as evil, and they were regarded as a threat to society—both through their the disease and by their evil minds. Leprosy was considered a venereal disease and priests sometimes gave the afflicted mock funerals before casting them out of their communities.

In 1179 the Third Lateran Council addressed the problem of leprosy. Laws were passed to guide the treatment of lepers. Priests were instructed to lead the afflicted into an open field and to tell them never to enter churches or houses. Lepers were required to wear a cloak of gray or black with a yellow cross emblazoned upon it. They were ordered to warn people of their unclean presence with the distinct sound of a horn, clapper, or bell. Begging was required to be done with a bag tied to the end of a long stick. They were forbidden to wash their hands or clothes in springs or streams, touch anything they wanted to buy, talk to others unless downwind from them, touch railings without gloves, or touch children or give them gifts.

In the eleventh century Britain experienced a terrible outbreak of leprosy. Lanfranc, the first Norman archbishop of Canterbury, founded a leprosy hospice at Harbledown. Queen Matilda, daughter-in-law of William the Conquerer, also founded a place for victims outside the gate of London in 1101, which she dedicated to Saint Giles, the patron saint of outcasts.

Many other hospices, called lazar houses, were founded before the end of the eleventh century. There were as many as 20,000 hospices in Europe, and over 200 in Britain alone. But even with such an abundance of hospices, many lepers did not find sanctuary there. More likely only the wealthy would go to these hospices, with the poorer individuals going into the forests.

The Church's attitude towards lepers was often contradictory, varying among priests and nobles from time and place between good and bad. Some lepers were provided for as a matter of charity, and some priests taught that the disease was a gift from God, who had chosen the lepers to bear one of the heaviest burdens of man. Others saw the leper as spiritually unclean, and thus unholy and sinful. In this view, God was thought to have given them special grace, but also had punished them for their sins.

There were some kings who hated lepers. Phillip V of France and Henry II of England had them burned at the stake without any religious rites. Henry's great grandson allowed them funeral rites, then took them to cemeteries and buried them alive. Apparently Henry II later changed his mind. He gave 40 marks to Harbledown hospital while on a pilgrimage to Canterbury to do penance for the murder of Thomas Becket. He also made a gift to the lepers and the poor while on a visit to Paris in 1158.

Some kings had relatives who were lepers. Henry II's daughter-in-law, Constance of Brittany, was a leper, and his cousin Baldwin IV became known as the leper king of Jerusalem.

At the close of the Middle Ages both the biblical and bacterial disease began to wane. Several reasons for this include better diagnosis, depopulation as a result of the Black Death and other diseases, and segregation of the lepers from the general community. By the eighteenth century leprosy still existed in the Shetland Islands, and the disease reached a late peak in Norway in the nineteenth century, where the last European leprosy hospital closed in the 1950s.

The association of leprosy with sin is not relegated to Western culture alone. Chinese and Hindu attitudes were similar. While the Muslims claimed more tolerance toward lepers, in 1253 the Saracens killed all of the lepers living in a hospice in Jerusalem.

The stigma associated with leprosy perished with the discovery that the disease occurs naturally in wild armadillos. Also, in 1950 a Danish pathologist related skeletons from a leprosaria of the twelfth and thirteenth centuries with those of bacterial leprosy. As of 1940 leprosy could be treated with the drug dapsone. Although the bacteria is seldom completely removed from the body, it can be halted with multi-drug treatment. A major center for the disease and its study in the United States is at Carville, Louisiana. In 1974 a major effort to develop a vaccine was supported by the World Health Organization.

There is evidence suggesting that when the influential texts of the medieval Arab scholar Avicenna (980-1037) were translated into Latin, biblical leprosy was confused with bacterial leprosy. While it had always been assumed that God visited leprosy upon humans for their sins, the disease has been definitively shown to be a bacterial infection devoid of any religious significance. Sadly, the suffering and stigma endured by many of the afflicted throughout history was unwarranted and mistaken.

EVELYN B. KELLY

Further Reading

Brody, Saul N. *The Disease of the Soul: Leprosy in Medieval Literature.* Ithaca, NY: Cornell University Press, 1974.

Gussow, Zachary. *Leprosy, Racism, and Public Health: Social Policy in Social Disease Control.* Boulder: Westview Press, 1989.

Kiple, Kenneth. *Plague, Pox, and Pestilence.* New York: Barnes and Noble, 1997.

Mental Illness During the Middle Ages

Overview

Mental illness remains a mystery wrapped inside a puzzle. Although much research has been done, mental disorders remain elusive, and their treatment is still disputed. No single paradigm for explaining mental illness exists.

It is no wonder that throughout history people of different cultures have explained deviant or abnormal behavior as the work of demons, external spirit forces, and poisons. As a result, magical approaches to therapy and rituals evolved. With the development of the Christian church during the Middle Ages, exorcism, shrines, and saints became of great importance for the treatment of mental illness.

During the early years of the Middle Ages the community took care of the mentally ill. Later, hospices, then asylums developed to house them. London's Bethlem asylum—better known as Bedlam—was founded in 1247, making it one of the oldest institutions of its kind. The term "bedlam" became associated with chaos, confusion, and poor treatment, which reflected the general attitude toward mental illness at the time.

Background

People with mental disorders have always been recognized as different and treated in various ways. The divergent responses have paved the way for lasting controversy that exists even today. Early medicine men, considering such individuals to be possessed by demons, introduced a technique called trephination. This procedure involved drilling a hole in the head of the individual to let evil spirits out of the body. Many other civilizations independently developed such a procedure. For example, among the remains of the Incas in Peru are skulls with holes and trephination devices.

Madness has been know and feared by society throughout history. Herodotus (484?-420? B.C.), the famous Greek historian, described the madness of King Cambyses of Persia who mocked the gods. Indeed, the idea of mental disorders as religious conflicts evolved early. In Greek mythology Ajax, in a deranged state, slaughtered sheep thinking they were enemy soldiers.

Hippocrates (460-377 B.C.) viewed the body as stable until illness perverted it. He developed the theory of the four humors: blood, yellow bile, black bile, and phlegm. These humors were connected to the four ages of man, the four elements, and the four temperaments. Hippocrates classified madness as either mania or melancholy. Mania was characterized by choler or yellow bile, and melancholy, or depression, by black bile. Hippocrates proposed that the treatment of mental disorders should involve restoring the balance of fluids or humors in the body. In addition to bleeding and purging, this could be done by rest, exercise, abstinence from alcohol, and sex.

Both the Greeks and Romans recognized that the mentally ill were capable of causing major social problems, as well as harm to themselves. They made provisions for guardians to take care of the insane. Realizing that these people would hurt themselves or others and could destroy life and property, laws were passed that set specific guidelines. Since there were no lunatic asylums, people with mental illness were a family responsibility. The seriously impaired were restrained at home, but others were permitted to wander in the hope that evil spirits might fly out of them.

In general, those considered crazed were feared and shunned. In the New Testament in is related how Jesus cured the Gadarene demoniac, who spent his years wandering among the tombs. The evil spirits were cast out into a herd of swine, the most despicable of animals for Jews.

Galen (130-200) later discussed hysteria marked by pain and breathing difficulties, which were caused by a wandering uterus in women. In fact, the Greek word for womb is "hyster." The term hysteria developed from the idea of the wandering womb. The cure was marriage.

Greek and Roman medicine served as the foundation of medical knowledge for the next thousand years. Practitioners of the Middle Ages would add their own unique twist to the subject of mental illness.

Impact

During the fifth century, Rome was overcome by waves of barbarian invaders from the north and the empire lapsed into decline. The Roman capital was moved to Constantinople, or Byzantium (now modern Istanbul in Turkey). Constantine, the Roman emperor for which the city was originally name, became a Christian in 313 and, by the early fifth century, Christianity became the official religion of the Eastern Roman empire. The Church then became the controlling force over medicine, with the spiritual taking precedence (at least where the Church was concerned) over the body. The Church blended folk superstitions and religious tradition, so that many old healing practices became incorporated into Christian beliefs. For example, Paul of Aegina (c. 625-c. 690), who studied and practiced medicine in Alexandria, recommended gentle treatment for mental illness, including music. He also alluded to satanic possession.

During the Middle Ages their were some many health problems that treatment and distinctions became overwhelming. Outbreaks of bubonic plague, smallpox, and leprosy would come in waves and decimate populations. However, mental illness was another major public concern. Madness, insanity, and lunacy were terms used to describe a variety of mental illness and mental handicaps. What caused these conditions and what to do about them was especially disputed.

Devotion to Galen's medical teachings led the people of the day to adopt four major categories of mental illness: frenzy, mania, melancholy, and fatuity. Each of these was purportedly caused by an imbalance in the humors. To restore balance was a goal of the physicians.

Folk beliefs and traditions, however, largely guided the perception of mental illness among the common people. The belief that the moon caused lunacy (the Latin word for moon is "luna") persisted well into the nineteenth century. The mentally ill person was thought to have slept where the moon beams hit his head, causing the erratic behavior.

The Church had a different interpretation of people with mental illness, viewing such disorders as evidence of sorcery or possession by a demon. Later, people who had degrees of insanity, especially women, were considered to be dabbling in witchcraft. However, some viewed the mentally ill as having a divine gift, perhaps like the gift of tongues. Many villages would take mentally handicapped people under their wings and treat them like children. Some of the troubadours or traveling musicians sang of tragic love madness.

Views concerning the treatment of mental illness were even more diverse. Bleeding was one of the primary treatments thought to balance the humors, but some physicians recommended drugs to sedate and calm the mentally unstable. A unique form of "shock" treatment was also tried—the mentally ill individual was hurled into the river to try to help him or her come to the senses.

Certain saints were thought to be more active in the domain of madness. In northern France the shines of Saint Mathurin at Larchant and Saint Acairus at Haspres were known for healing. In Flanders, now Belgium, citizens of Geel developed at shrine to Saint Dymphna that became a hospice to house the mentally ill. When there were too many people for the building, villagers took them in, forming a special family colony that still exists at Geel.

Attitudes toward the mentally ill varied from place to place. Some German communities cast out the mentally ill and mentally retarded by whipping them out of town or pointing them in the direction of other villages. Monasteries were often a welcome haven for such individuals.

Some towns had madmen towers, in which the mentally ill were incarcerated in chambers called "narrenturme." Some hospitals, like the Hôtel Dieu in Paris, had special rooms set aside for the mentally ill. The Teutonic Knights at Elbing had a madhouse called the Tollhaus, serving as a special place for the mentally ill. Under the influence of Islam in Spain, specialized hospitals developed at Granada (1365), Valencia (1407), Zaragoza (1425), Seville (1436), Barcelona (1481), and Toledo (1483).

In London, in 1247, Saint Mary of Bethlehem was established to house people "deprived of reason." By 1403, six people were housed there. The institution gained more and more patients and eventually developed into the infamous Bedlam, a perversion of the name Bethlehem. The asylum became notorious for its terrible conditions, under which people were chained and lived in squalor. Bedlam was like a living hell, and the name has come to signify conditions that are chaotic and hopelessly confused.

In some areas the "fool" was romanticized. A ritualized "feast of fools" developed during the Middle Ages, serving as a vehicle by which society came to grips with the idea of madness by becoming mad themselves for a short period of time. This festivity was accompanied by much drinking and debauchery.

As the medieval years progressed, insanity became linked to witchcraft and demon possession. Those considered to be possessed with demons were exorcised. This ritual, performed by a priest, would call upon the demon to come out of the individual and to transfer itself into an animal or inanimate object.

The Church, trying to find a scapegoat for the cause of plague and heresy, were convinced that these possessed people were the causes of their difficulties. The witchcraft delusion spread rapidly. In 1484 Pope Innocent VIII declared Germany full of witches that needed to be hunted out. The next 300 years were characterized by terrible witch-hunts designed to seize those thought to be possessed by the devil. Upward to 50,000 people, mostly women, were tortured and killed in these searches. People actually believed that witches existed and that they befriended the devil, brewed strange mixtures of toads, serpents, and poisons in cauldrons, rode broomsticks, and brought curses and plagues upon the earth. Witches were thought to be identifiable by the stigmata diaboli, or mark of the devil, on their body, providing the origin of the word "stigma." Worse still, a simple accusation of witchcraft was often enough for an individual to be found guilty and condemned.

The treatment of mental illness deteriorated in the late Middle Ages and remained poor through the eighteenth century. It was only in the nineteenth century that scientists and society began to reconsider deviant behavior from the perspective of mental illness rather than as a manifestation of evil spirits.

EVELYN B. KELLY

Further Reading

Alexander, Franz G., and Sheldon Selesnick. *The History of Psychiatry: An Evaluation of Psychiatric Thought and Practice from Prehistoric Times to the Present.* London: George Allen and Unwin, 1967.

Andrews, Jonathan, et al. *The History of Bethlem.* London: Routledge, 1997.

Porter, Roy, *The Greatest Benefit to Mankind: A Medical History of Humanity.* New York: W.W. Norton, 1997.

Public Health in the Middle Ages

Overview

Public health encompasses many aspects of disease prevention, but it has its roots in the prevention of communicable diseases. It is a science as well as an art, and has the goals of averting disease in order to prolong life and promote physical and mental well being. These goals are accomplished through various means such as medical and sanitary services, personal hygiene, control of infection, and organization of health services. Early in recorded history, governments became increasingly aware that it is in their best interest to promote health and prevention of disease. As understanding of the factors that affect disease grew, so did community programs designed to prevent them. The public health system draws heavily on medical science and philosophy, and concentrates its efforts on controlling the surrounding environment for the benefit of the public. Without the theoretical understanding of disease, any attempts to control it would likely be haphazard guesses based on intuition.

Personal hygiene practices have been utilized by most primitive societies, often in association with religious rituals. In many cases, it was a desire to be pure in the eyes of a god. As an example, the Old Testament has a tremendous amount of information regarding how a person should go about living a clean life. There were many sanitary regulations to follow, some of which dictated the preparation of food. For many people religion, law, and custom became woven into the fabric of their lives. However, cleanliness was usually the exception rather than the rule.

The concept of vitalism was rampant in early societies. This line of thought looked upon disease, epidemics, and natural disasters as a divine judgment from the gods penalizing early man for some transgression. In time, the idea that disease could be due to natural causes gradually took root and gained a foothold in Greece during the fifth and fourth centuries B.C. It was during this time that true science was born and the first attempts to explain the causes of disease were made. One of the first public health issues ever recorded comes from this period. While the mechanism was not known, the association between the disease malaria and swamps was first noted in fifth century B.C. The book *Airs, Waters, and Places,* most probably written by the Greek physician Hippocrates (460-377 B.C.) around the same time, noted a potential relationship between pestilence and environment. It provided the earliest theoretical constructs explaining the differences and causes of endemic (pertaining to a particular region) and epidemic (pertaining to large numbers of people) disease. It would only be supplanted by modern views of bacteriology and immunology.

The Romans furthered the application of public health with their engineering prowess. Aqueducts brought fresh water into cities, while sewers carried away waste. These waterways had a tremendous positive impact by helping to prevent disease. The availability of public baths and the encouragement to use them also helped to keep many diseases in check. The Romans also introduced the first health care system complete with physicians and hospitals by the second century. Unfortunately the fall of the Roman Empire in the fifth century halted many of these accomplishments.

Background

Diseases in epidemic proportions were often seen during the Middle Ages. The threat of disease was a constant problem confronting populations during this time. In fact, many authorities define this period as occurring between the sixth century Byzantine Empire plague and the bubonic plague of fourteenth century Europe that devastated the continent. There were many types of disease (such as leprosy and smallpox) seen in epidemic proportions, but the most severe disease of them all was the bubonic plague of 1348, known as the Black Death. While there were not many protocols in place to help prevent the spread of communicable diseases, the quarantine of people with leprosy was a common practice. People inflicted with leprosy were expelled from the community at large and made to live in separate housing. Although the mechanisms for the transfer of the disease were unknown, it was felt that the isolation of the person with the disease would prevent its transmission. This was one of the first recorded public health measures in England. This principle was later applied to people inflicted with the Black Death when Europe had to deal with the increased threat from that disease.

The Black Death reached the shores of southern Europe from the Middle East in 1348 via the trade routes. Within three years, it had taken an extreme toll on Europe and was responsible for an estimated 25 million deaths out of a population of 60 million. The chief method of combating the plague was to isolate those suspected or known to have contracted the disease. The period of isolation ranged from two weeks to 40 days. This practice of quarantine did not prevent the spread of bubonic plague because

Richard II. *(Library of Congress. Reproduced with permission.)*

fleas from rats, not direct human contact, spread the disease. However, quarantines did alleviate some transmission of the pneumonic plague that followed, since it was transferred by contact with an infected person. The Black Death exacted a huge toll on Europe, but had a tremendous impact on public health reform.

Impact

The immediate impact seen by the institution of public health practices was to help prevent the spread of certain diseases. In addition, certain practices directly improved the standard of living and made life more comfortable. As a readily seen example, it would obviously be much more comfortable to live in a household free of fleas and bed bugs, then one infested with them. However, many of the techniques used to control disease were ineffective because of a lack of understanding regarding transmission. Likewise, many practices, which could have been instituted had there been a better understanding of the disease process, were not put into place due to this lack of knowledge. It has been often written that Medieval Europe did little to advance sanitary conditions and hygiene practices, but given the state of knowledge at the time, it seems that reasonable attempts were made to rectify these conditions. Nowhere is this demonstrated as well as in the case of the Black Death outbreak.

Beginning in Italy just after the plague ended, new initiatives were aimed at raising the level of public sanitation and governmental regulation of public life. Perhaps more importantly, there was an increased emphasis on the scientific process used to investigate disease rather than the traditional reliance of ancient theories based on Greek thought. These new theories based on scientific investigation and hypothesis helped to explain the contagion theory of the plague rather than the historical theories of corruption. Finally, by the sixteenth century a debate over the causes of plague spread in the medical community as old corruption theories inherited from Greece and Rome were replaced by new ideas of infection and disease transmission. The technology at this time was insufficient to allow for significant advances in this area, but this new channel of thought helped led to scientific breakthroughs that revolutionized medicine in the nineteenth century.

Across the English Channel, a set of "sanitary laws" were put into use in England in 1388 by Richard II (1367-1400). These were a series of laws that were intended to combat some of the negative factors that were thought to have contributed to the spread of the plague. Soon, other public officials followed suit and created similar legislation in their areas. These laws created a system of sanitary control to combat contagious diseases. They made use of observation stations, isolation hospitals, and disinfecting procedures. Major changes were made to improve sanitation, including the development of pure water supplies, garbage and sewage disposal, and food inspection. These efforts were especially important in the cities, where people lived in crowded conditions in a rural manner with many animals around their homes. It was common to have multiple pests within a household that compromised the health of the inhabitants.

The Middle Ages are often cited as a time where there was little concern for personal hygiene and health education. As mentioned before, reforms were instituted but were not very successful due in part to the ignorance of the principles of public health. The conditions at the time were so horrendous, that any improvement was needed and most welcome. The lack of technology also made it difficult for many of these principles to be instituted by the society at large. As an example, compared to modern times, even taking a bath required an enormous amount of time and effort.

The Middle Ages witnessed a number of important steps in public health administration. These included attempts to cope with the unsanitary conditions of the cities and, by means of quarantine, to limit the spread of disease; the establishment of hospitals; and provision of medical care and social assistance. Attempts were made to increase the levels of health education and personal hygiene. While these practices bear little resemblance to modern day principles, they had the effect of improving the standard of living for people at that time and perhaps more importantly, these reforms provided the basis on which the modern public health system is built.

JAMES J. HOFFMANN

CLEANING UP IN EUROPE: THE INVENTION AND SPREAD OF HARD SOAP

Soap has been with us for over 2,000 years, but not in its current form. First invented by either the Phoenicians or the Celts (or, perhaps, simultaneously by both), soap was used as a medicine by some cultures and as a cleansing agent by others. However, early soap was soft, making it hard to transport and sometimes difficult to use. This changed in the Middle Ages with the invention of hard soap. Starting in France and Germany, hard soap manufacture moved across the English Channel by the twelfth century. In spite of this, the use of soap was not widespread for another several centuries and, in fact, the gift of a box of soap in the sixteenth century caused quite a stir in Germany. Although it took a few centuries for soap to become widely accepted (primarily because it took that long for the price of soap to come within the reach of a large part of the population), by the nineteenth century soap use was viewed as a measure of a nation's advancement by some European economists.

P. ANDREW KARAM

Further Reading

Porter, D. *Health, Civilization and the State: A History of Public Health from Ancient to Modern Times.* New York: Routledge Publishing, 1999.

Ranger, T., and P. Slack. *Epidemics and Ideas: Essays on the Historical Perception of Pestilence (Past and Present Publications).* Cambridge: Cambridge University Press, 1996.

Rosen, G. *A History of Public Health.* Baltimore: John Hopkins University Press, 1993.

Holy Shrines and Miracle Healers

Overview

During the Middle Ages, medicine and religion were interrelated in many ways. While some have criticized the medieval church for hindering medical progress through its opposition to dissection, the Church did provide comfort and assistance for many when medical knowledge was at its lowest ebb. However, it was clear that the Church always put divine interests over the human.

Holy shrines and miracle healers have been meaningful to people in many civilizations. Never were they more important than in the Middle Ages. In Europe, especially during the eighth through eleventh centuries, the Church helped transform historical locales related to the lives (and deaths) of martyrs into sacred healing places and shrines. Special saints became patrons of healing, and miracles were performed in their names. Pilgrimages to sacred places were increasingly viewed as a way of securing divine favor. The rich could go to the shrines of Rome and Jerusalem, while the poor had an abundance of local sites to visit.

Shrines and healers gained in popularity during this time in large part because secular medical treatment in Europe abysmally ineffective. The medical help that was available, primarily to the wealthy, was still based on the Greek system of humors. The peasants were left to care for themselves with folk remedies, superstition, and the healing shrines.

Even today shrines are very important to many worshippers. In a day of enlightened scientific medicine, healing shrines at Fatima and Lourdes draw many visitors and pilgrims.

Background

Throughout antiquity religious healing has been a part of many sacred religions. The treatment of physical or spiritual ailments by non-medical means has been a part of religion in many cultures. The relationship between healing and religion has developed because many ailments were thought to be caused by possession by demons or manifestations of evil.

Healing usually centers around a certain place or holy shrine, on certain operations, or on a particular person. There are many categories of sacred sites, including sacred mountains like Mount Olympus and sacred manmade mountains like the Great pyramids of Egypt or the Mayan pyramids. Ancient ceremonial sites such as Machu Picchu in Peru and Stonehenge in Britain were similarly places for worship.

From time immemorial, man has sought healing in the waters of natural springs and other bodies of water. Water, essential for life, indispensable for cleansing, became embodied in myth and legend and imbued with curative powers. From ancient times and continuing into the medieval era, people were drawn to water for healing. Rituals and legends evolved around these places, where the deities were believed to dwell. Cathedrals were often built at these ancient sites, with saints (or their relics) substituted for the pagan gods, and pilgrimages to these places became a source of comfort to the religious of many civilizations.

The mineral springs of Spa, Belgium, known to the Romans, were rediscovered in 1326, beginning centuries of popularity, especially among the rich and famous. The hot sulfur springs of Baden, Austria, were a Roman bath, or *aquae,* as were those at Bath, England; Aix-les-Bains, Grisy and Saint-Sauveur in France; Fiori in Italy; and Saint Moritz in Switzerland. Here those seeking to restore their health sought healing for both diseases and disabilities; others simply hoped to preserve the well being they already enjoyed.

During the early years of Christianity, Christians were persecuted and many martyrs were made at the hands of the Romans. In 313 Constantine, emperor of what remained of the Roman empire, converted to Christianity and worship changed dramatically. Already, in the second century, it was customary to commemorate the anniversary of a martyr's death by celebrating communion where they had been buried. Churches were built on the sites and eventually people began to think that worship was especially valid if it was celebrated in one of these holy places. If relics of the martyrs were also kept at such locations, it was even more sacred. The bodies of many martyrs were exhumed and placed under the altars of new churches that were being built. Other people began claiming visions of martyrs that were not known or forgotten.

Eventually, the places and relics of the saints and places of New Testament fame were said to have miraculous powers. Empress Helena, mother of Constantine, gave special impetus to the development of this idea when, in a pilgrimage to the Holy Land, she thought she had discovered the cross of Christ. Soon this cross was said to have miraculous powers and pieces of wood that were claimed to have come from it were found all over Europe.

While these ideas were developing, some leaders of the church viewed them with disfavor and tried to prevent superstition from becoming extreme. They contended that, to be a good Christian, visits to the Holy Land or to the place of martyrs were not essential and should not be exaggerated. However, this idea was overwhelmed by the demand of people flocking to the shrines and later such objections disappeared.

Impact

With a history of shrines, holy objects, and pilgrimages, it was not unusual for people of the Christian faith to incorporate these into their beliefs. In fact, one of the most widespread features of the medieval religious life was the pilgrimage. These pilgrimages were not only motivated by religious fervor, but by a desire to travel. Visits to such sacred sites also served as a beckon of hope in the dismal lives of many medieval worshippers. The wealthy found great solace in the sites of Palestine, and thousands flocked to the places of the life, death, and resurrection of Christ. Rome, with its churches and tributes to Peter and Paul, lured thousands. Poorer individuals settled for visiting tombs of local saints and their relics.

During this time, Christian monasteries evolved as key medical centers in Europe, more important than universities before the 1300s. Monks were involved in bleeding, purging, and surgery, the treatments of the day. However, the Lateran Council of 1215 forbade churchmen to shed blood, thus ending much of their function as healers. Thereafter, monasteries became places of shelter for pilgrims and for sick monks, and hospitals began to develop separately.

Almost every saint and holy person has been associated with the ability to heal. The same is true for certain unusual natural characteristics, scenes of epiphanies, and locations related to the life or burial places of holy men. The monastic orders throughout Europe also had as their primary function the care of the sick, as taken up by the Knights of Malta, the Augustinian Nuns, the Order of the Holy Ghost, and the Sororites Order. The ability to heal is often reserved for a select few, and some healers traced their knowledge back to the gods. The physicians of Myddval in Wales, active herbalists for more than five centuries, are one such group that claimed a divine origin.

The development of healing activities around certain healing saints resulted in specialized sacred organizations. Saint Cosmos and Saint Damian, for example, promoted organizations that required training, the need for special equipment, and libraries. Many religious leaders were also physicians and the origin of hospitals in both the East and West is linked to these special monastic orders.

One example of a place that became a famous shrine for healing is Canterbury in England. In 1170 Thomas Becket, the humble archbishop of Canterbury, was murdered in this cathedral for refusing to let the English king become involved in Church matters. The people of England were horrified, and Becket was immortalized as a saint. Even before his body was moved, people were dipping their fingers and rags into his blood to receive the benediction of the martyr. Miracles of healing were reported at once, and within 10 years there were claims for more than 700. The blood of Becket was supposed to cure blindness, insanity, leprosy, and deafness. The shrine became a famous place for pilgrimages in the Middle Ages. The idea of the pilgrimage was made famous in Geoffrey Chaucer's *Canterbury Tales,* in which a group of pilgrims relate their various stories during their travels.

Most of the healers possessed a sacred gift or commission. Scores of saints were called upon for healing. Each organ of the body and each complaint acquired a particular saint. Instead of the pagan Asclepius, Damian and Cosmos became the patron saints of medicine. These brothers lived in Cilicia in Asia Minor at the close of the second century. They performed a famous surgery involving the amputation of a white man's and its replacement with the transplanted leg of a black Moor. News of the transplant and the man with one white leg and one black leg spread near and far and they became celebrated for their healing powers. Emperor Diocletian stoned the men, burned them, crucified them, sawed them in half, but only after decapitation did they die. Many medieval paintings depict their miracle operation.

Saint Luke and Saint Michael may be called upon for all kinds of illnesses, but many saints had a specialty. Saint Anthony was called upon for erysipelas (or Saint Anthony's fire), Saint

Artemis for genital difficulties, Saint Sebastian for pestilence, and Saint Christopher for epilepsy. The story of Saint Roch emerged during the plague. This humble saint, who had ministered to many with the plague, contracted the disease himself and was healed by an angel. Throat complaints, such as goiter, were the realm of Saint Blaise, and backache was the domain of Saint Lawrence. Saint Apollonia became the patron saint for toothache because all of her teeth were knocked out during her martyrdom. Saint Margaret of Antioch was the helper of women in labor. While walking one day Margaret was swallowed whole by a dragon. She made the sign of the cross and the cross grew, bursting open the dragon and releasing the saint.

The healing shines of Europe acquired a wide range of relics and images, attracting visitors and providing hope to the people of the day. Many of the medieval shrines remain and are still visited by throngs of pilgrims seeking miracles.

EVELYN B. KELLY

Further Reading

Gonzalez, Justo L. *The Early Church to the Dawn of the Reformation.* San Francisco: Harper & Row, 1984.

Latourette, Kenneth Scott. *A History of Christianity.* 2 vols. San Francisco: Harper & Row, 1975.

Porter, Roy. *Medicine: A History of Healing.* New York: Barnes and Noble, 1997.

The Development of Medical Botany and Pharmacology During the Middle Ages

Overview

Once early humans developed the ability to reason, they began to experiment with various plants and herbs. Through the process of trial and error, early humans discovered which plants might be used as a food source, which could be used to flavor food or drink, which caused sickness or death, and which had medicinal value. Over time, nearly every type of herb and plant was classified into one of these categories. Pharmacy at that time was not a scientific discipline; rather, it was a mixture of medicine, superstition, and magic.

However, such primitive study of the medicinal value of plants and herbs established the roots of the modern branch of medicine called pharmacology. Traditionally, the field of pharmacology relies heavily on the subdiscipline of medical botany, the scientific study of the medicinal value of plant life. The vast majority of chemical compounds that have physiological actions in the human body derive from plants. Humans have used these types of remedies to alleviate common discomforts such as colds, allergies, and constipation. Interestingly enough, most of the present-day laxatives were used in similar fashion by societies during antiquity. While pain remedies such as aspirin would not be isolated for thousands of years, it was known that willow bark would help relieve pain. We now know that willow bark contains a substance that is similar to salicylic acid, the main active ingredient of aspirin.

It should not be assumed, however, that humans at that time had an understanding of the pharmacological actions of plants and their subsequent actions within humans. Drugs were administered with little concern for the patient, often with incantations, charms, and other paraphernalia that were believed to be necessary for action. As previously mentioned, drugs made from plants and other products were only a small portion of the cure. Many times, the administration of a specific compound was not based on reason, but misconceptions and superstitions. While these may have had some effect as placebos, they were often ineffective, or, even worse, the treatments were more dangerous than the condition that was being treated. Pharmacology as science really did not begin until the late nineteenth century. As a result, attempts were made to systematically describe the field of medical botany so that it could become more useful and reliable.

The ancient Greeks were the most successful in describing the medicinal value of plants. Theophrastus (c. 372-c. 287 B.C.) was the first scientific botanist who described and classified many different types of plants. Later, Dioscorides (40-90) applied this knowledge and described nearly a thousand different pharmacological treatments, with the vast majority of those using plant-based ingredients. He is often known as the first herbalist and is often referred to as the "father of pharmacology." This body of knowledge was expanded upon during the Middle Ages, with new remedies being described and formulated with increasing frequency.

Background

Despite the available body of knowledge in medical botany and pharmacology, it was quite common for a person to die during the Middle Ages from a contracted disease, especially if it was caught during a major epidemic. Over one million people died from the Black Death (bubonic plague) in England alone, and millions of others died throughout Europe. Certainly deficiencies in medical technology and practice played a major role in the low survival rates, but most people who were sick during those times would have fared better if they were able to visit a physician. While wealthy individuals could afford proper medical care if they were stricken ill, less well off individuals were often priced out of the market. A single visit to the doctor during that time would cost a laborer an entire month's pay. In order to bridge this service gap, apothecaries and "wise women" served the local populations in the capacity of the physician. In fact, in most cities the majority of the populous was served by apothecaries rather than trained physicians.

Apothecaries and "wise women" were individuals who made and dispensed medicines. Most of these people also made diagnoses and treated diseases as well. Nearly every town had at least one apothecary shop that sold medicines for many types of ailments. The drugs that were dispensed were either filled prescriptions according to a physician's direction, or a concoction of their own devising. These remedies were assembled using a wide variety of ingredients, which consisted mainly of herbs and flowers. Apothecaries also relied heavily on animal organs and tissues to make their potions, salves, and ointments. Most apothecaries learned their trade as an apprentice mixing medicines, so there was little formal education. They spent a large amount of their time mixing and dispensing plant and herbal derivatives as drugs. Some apothecaries sold their remedies while traveling from town to town, never staying too long in one place. This type of setup was a prime target for abuse and many charlatans practiced this way. Because of this type of practice and their lack of medical training, there was an attempt to make all apothecaries practice their trade while licensed. This was largely unsuccessful as many simply ignored the law. Over time, the influence and importance of apothecaries waned.

Christian monks were another important component of medieval pharmacy, as religion was tightly entwined with medicine at the time. It was common for a monk to know how to mix herbs and drugs, but put his faith in the healing power of God. However, it was a nun named Hildegard von Bingen (1098-1179) who made some of the most important contributions to pharmacology at this time. Hildegard was a practical medical practitioner who trained nurses and studied medical botany. She made use of over 100 medicinal plants grown in her own garden, gathered from the surrounding area, or imported for that specific purpose. She also published two important pharmacological works, which were compilations of current knowledge along with her own reflections and improvements. She even advocated many modern practices, such as the importance of a good diet and exercise.

The field of pharmacy took a significant step in the early fifteenth century when the Italian physician Saladin di Ascoli wrote the first European guide for pharmacists. In his work, he outlined the duties and obligations of a pharmacist and warned against the misuse of drugs. He also advocated following the orders of the physician literally, without changing any of the ingredients of the prescription. This was the first step in establishing pharmacy as a reliable and useful profession.

Impact

It is obvious that modern day medical practice and pharmacology have their roots, at least in part, in the Middle Ages. The application of plant and herbal medicines grew tremendously during that time. These drugs became an important component of treatment just as they are today. The wealth of medieval knowledge on this subject is much more than one would think, despite the general lack of systematic scientific inquiry. Perhaps no other person had as significant impact as Saladin di Ascoli. To a large degree, he made pharmacology a legitimate science and made the pharmacist a trusted and viable professional by his insistence upon reform. The use of drugs was a controversial subject at that time, and by encouraging legitimate use of these materials, pharmacy became a respected and useful science.

People like Hildegard von Bingen also had a significant impact on modern pharmacy and medical botany in two major ways. First, they kept alive the body of knowledge about medicinal plants in their everyday work, and some carefully recorded their uses. Second, their natural curiosity led them to greatly expand upon this knowledge through their own research and experimentation. Although many of these developments were somewhat hampered by superstition and myth, they nevertheless made significant strides in the field on medical botany. This quest for knowledge in the area of herbs and plants is continued in modern society. Each year there are about 4,000 newly described chemical compounds reported by scientists. Of those, about 75% are derived from plant life. All of these compounds have the potential for becoming a useful drug. One of these compounds, for example, might even hold the cure for cancer. This quest for knowledge is a direct result of the work in medical botany done hundreds of years ago. In fact, this has led to a new title for an old profession, bioprospecting. Bioprospecting involves searching for new sources of drugs and chemical compounds from living organisms (primarily plants).

Another important development is the increased use of physician assistants and nurse practitioners. These medical professionals can be likened to modern apothecaries, as they provide similar services. With increased medical costs, it seems ideal to provide inexpensive medical services through people who have medical training, but are not full-fledged doctors. These licensed medical professionals have significant schooling and training. They are well qualified to treat a variety of ailments and represent a cost effective alternative to a standard physician.

While much of our knowledge and practice of medicine has changed significantly since medieval times, the field of pharmacology has its roots in the Middle Ages. Medieval contributions to this body of knowledge proved to be important steps in the development of our modern knowledge and practice of pharmacology and medical botany.

JAMES J. HOFFMANN

Further Reading

Getz, Marie Faye, ed. *Healing and Society in Medieval England: A Middle English Translation of the Pharmaceutical Writings of Gilbertus Anglicus.* Madison: University of Wisconsin Press, 1991.

Grant, Edward, ed. *A Source Book in Medieval Science.* Cambridge: Harvard University Press, 1974.

Siraisi, Nancy. *Medieval and Early Renaissance Medicine: An Introduction to Knowledge and Practice.* Chicago: University of Chicago Press, 1990.

Urdang, George. *Pharmacy's Part in Society.* Madison: American Institute of the History of Pharmacy, 1953.

Botany in the Middle Ages, 700-1449

Overview

The ancient Greeks, especially Aristotle (384-322 B.C.) and his pupil Theophrastus (c.370-285 B.C.), made important contributions to botany, the study of plants, but there were few significant additions to that body of knowledge by the Romans. All this knowledge was lost to Europeans after the fall of Rome in 476, when Europe settled into a period called the Dark Ages during which there was little attention to science. This period lasted until about A.D. 1000 when a curiosity about the natural world began to increase slowly. Interest was spurred by the translation of Greek and Roman texts into Latin so that the learning of the ancients again became available in Europe. This meant that by the mid-fifteenth century when movable type was invented, the stage had been set for the reemergence of science in the Renaissance.

Background

During the Dark Ages, economic and social conditions were such that the energies of most people went totally into struggling through life from day to day. There was no time or energy left for scholarly pursuits. Also, the influence of the Church was predominant during this time. This religious focus meant that people paid more attention to preparing for the next life than investigating the present world around them. This was especially true until about A.D. 1000. After that time, economic and social conditions began to improve and the attitude of the Church towards the natural world changed to one of interest rather than neglect.

Through much of the Middle Ages, which stretched roughly from 500-1500, there was little attention to science, in the sense of curiosity about the natural world for its own sake. Any interest in living things, including plants, was solely practical. Most of what was written about plants concerned their medicinal uses, and most of these writings were based on the work of Greek and Roman writers. The Roman, Pliny the Elder (23-79), had written an extensive *Natural History* in the first century of Christian era, and it was the only satisfactory presentation of the botanical ideas of Theophrastus to survive into medieval times. Theophrastus's work itself wasn't rediscovered and translated into Latin until the end of the Middle Ages.

HOPS USED IN BEERMAKING

One of the defining characteristics of beers is the distinctive bitter flavor imparted by the hops added during brewing (hops are the dried female flowers of a vine from the hemp family). Adding hops to beer is a relatively recent innovation; the earliest beers, brewed in ancient Egypt, were a simple beverage made from fermented grain. Sometime in the tenth century, beer makers discovered that hops not only made beer taste better, but helped preserve it from spoiling, too. This became the famous India pale ale. Hops were used for brewing beer in Germany by the eleventh century and in Britain and Holland by the fifteenth century.

P. ANDREW KARAM

The botanical work that was most influential and most copied was that of Dioscorides (c. 20-90), a Greek physician who wrote a practical guide to medicine that included information on about 500 different plants. While no copies of his original text have survived, it is assumed that, to make identification easier, it included illustrations of the plants he described. Over the centuries many copies of this work were produced, and these too were often illustrated. Such books on plants came to be called herbals. As time went on they became less and less accurate and sometimes careless copying led to the introduction of more and more errors in the text and to simplification of the drawings to the point where they became of almost no help in identifying the plants described.

After Muhammad (c. 570-632) founded the Islamic religion in the seventh century, the Arab world became an important seat of learning. As Europe sank into the Dark Ages, the knowledge of the ancient Greeks and Romans was translated into Arabic by Islamic scholars; these works included the *Materia Medica* of Dioscorides and the *De Plantis* of Nicolaus of Damascus (c. first century B.C.). Building on the work of Dioscorides, the great Arab physician Avicenna (Ibn Sina, 980-1037) included 650 plants in his list of over 750 drugs. Thus the Arabs preserved and added to knowledge of medicinal plants, but they were not botanists in that they were not interested in plant structure and function outside of practical considerations.

Even before 1000, Europe showed some signs of an awakening interest in knowledge. An important medical school was founded at Salerno in central Italy in the ninth century, and it was strengthened by its connection to the nearby Benedictine monastery of Cassino where Greek texts on medicine and botany were available. It was at Cassino in the mid-eleventh century, that Arabic medical texts, including plant lists, were translated into Latin, and thus became accessible to European scholars.

Impact

In the twelfth century, two writers made contributions that indicated the reawakening of interest in intellectual pursuits, including those in botany. Hildegard von Bingen (1098-1179), who is considered the first woman to have written about plants, produced over 200 works, including a book called *Physica* that contained about 200 short essays on plants and their medicinal uses. She included folk remedies she had collected as well as information derived from other sources such as Dioscorides. The other writer is Peter Abelard (1079-1142). Though again his work on plants was not original, but primarily a rehash of the ancient Greeks, his writing style indicates an increased interest in logic and in the exploration of ideas.

Perhaps the most important figure in botany of the Middle Ages is Albert the Great or Albertus Magnus (1200-1280), who belonged to the Dominican Order of the Church. He studied at the University of Padua, which was the leading medical and scientific center in Italy at the time. He then taught at the University of Paris where Roger Bacon (1220-1292) gave a series of lectures that included a discussion of plants, and these seem to have greatly influenced Albert. Bacon had himself been influenced by the *De Plantis* of Nicolaus of

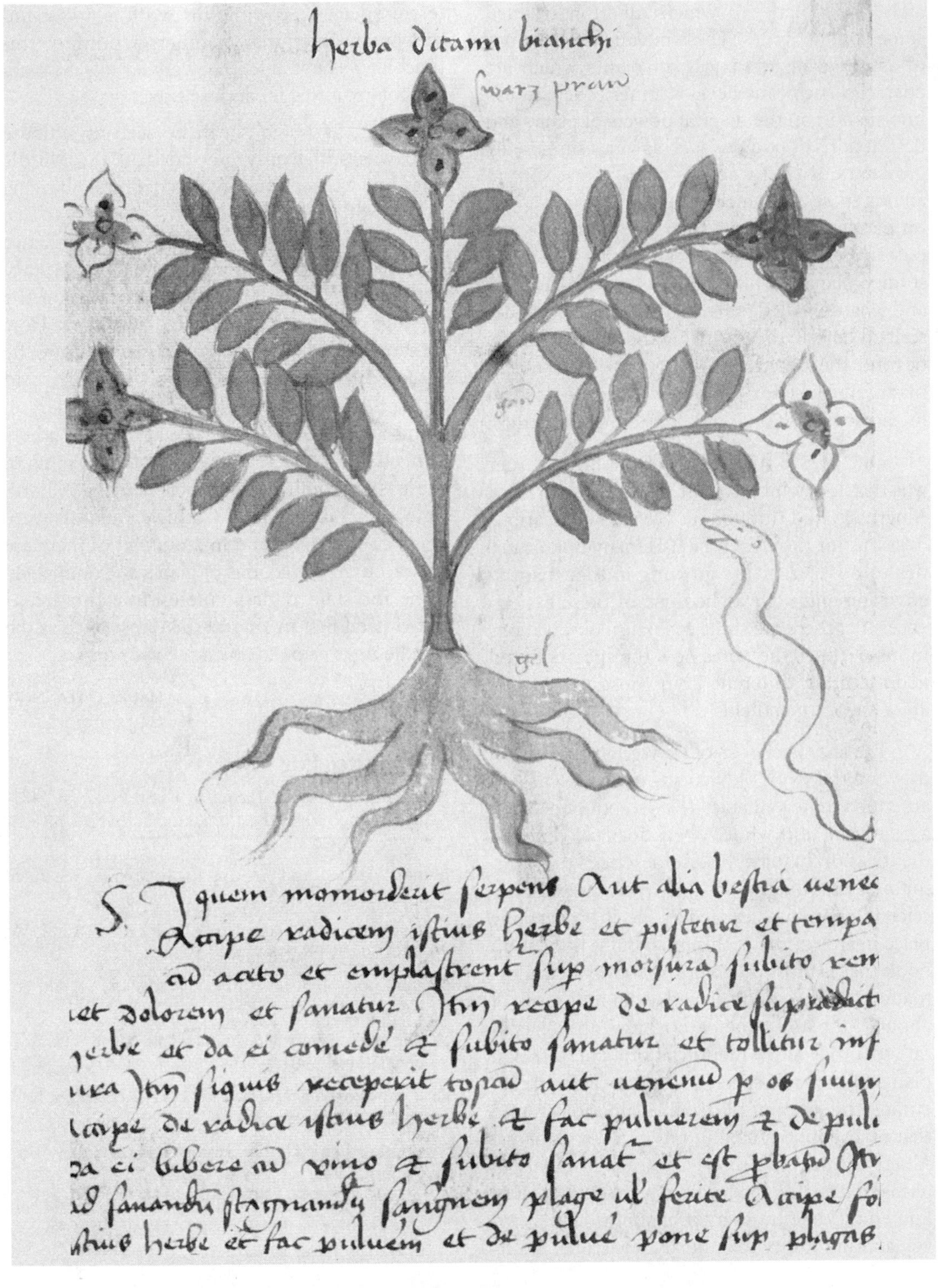

Fourteenth-century botanical illustration from Rizzardo's *Liber herbarius una cum rationibus conficiendi medicamenta.* *(Gianni Dagli Orti/Corbis. Reproduced with permission.)*

Damascus which had recently been translated into Latin. Some historians see it as unfortunate that the work of Nicolaus and not that of Theophrastus was translated at this time, because, though it contained a great deal of misinformation, it came to represent Greek learning on botany to European scholars. Bacon relied on it, and in turn, so did Albert who produced his writings on plants between 1250 and 1260.

The first five books of Albert's seven-book *De Vegetabilibus* are essentially a reworked version of Nicolaus, to which Albert has added some commentaries. These include explanations of astrological influences on plants which are characteristic of medieval writings. There is also information on the magical powers of plants and descriptions of oddities such as vines supposedly growing out of the acorns of oak trees. But in among these are some very accurate observations on plants that indicate that Albert was not just parroting from the writings of others, but had studied plants for himself. He wrote of structures and anatomical details that had not been described before, thus bringing the study of plants beyond the practical. Albert was interested in plants not just for how they could be used in medicine but for their own interesting properties.

One of the rules of the Dominican Order was that its members must travel on foot. Since Albert traveled throughout Germany on various missions for the Order, he had many opportunities to observe plants growing in their natural environments. It was because of these experiences that there are some ecological observations in his writings; he notes how the species found in forests are different from those in swampy areas and in open fields.

The last two books of *De Vegetabilibus* do not draw on the work of Nicolaus of Damascus and are much more valuable. The seventh book is on agriculture, and while Albert does use many of the ideas of the past, he also describes the farming practices of the day. This work shows how a scientific mind was excited by the changes taking place in agriculture at this time. But it is the sixth book that is most noteworthy; it provides information on the medical and economic uses of about 270 plants. The descriptions are so accurate that they allow identification of at least 250 plants to genus or species. In some cases, the accuracy and amount of detail is superior even to that of Theophrastus. But unlike Theophrastus, Albert did not develop a technical vocabulary of terms to describe the features he identified. This limited the amount of information he could convey and his ability to compare structures in different species. So while his work is a vast improvement over that of others writing at this time, he was not able to go very far toward making botany into a modern science.

Albertus Magnus can be seen as a figure who was still firmly grounded in the Middle Ages but whose thinking was definitely moving in new directions. In the introduction to the sixth book on plants, he emphasizes his reliance on direct observation and personal experience and notes that he is not just reporting on the findings of others but doing his own work. Here he shows that like the scientists of the future, he values direct experience of the natural world rather than accepting the authority of others. It is this attitude which separated him from others of his day who merely reworked the learning of the past without adding to it. During Albert's time, many of the great Gothic cathedrals were being constructed, and in a number of them are very accurate depictions of plants and animals in stone and stained glass. Interest in nature manifested itself first in art and then in science as the Middle Ages came to an end.

MAURA C. FLANNERY

Further Reading

Arber, Agnes.*Herbals: Their Origin and Evolution 1470-1670.* Cambridge, Great Britain: Cambridge University Press, 1986.

Blunt, Wilfred, and Sandra Raphael. *The Illustrated Herbal.* New York: Thames and Hudson, 1994.

Iseley, Duane.*One Hundred and One Botanists.* Ames, IA: Iowa State University Press, 1994.

Magner, Lois. *A History of the Life Sciences.* Second ed. New York: Marcel Dekker, 1994.

Morton, A.G. *History of Botanical Science.* New York: Academic Press, 1981.

Reed, Howard. *A Short History of the Plant Sciences.* New York: Ronald Press, 1942.

Serafini, Anthony. *The Epic History of Biology.* New York: Plenum, 1993.

Singer, Charles, and E. Ashworth Underwood. *A Short History of Medicine.* Second ed. New York: Oxford University Press, 1962.

The Contributions of Albertus Magnus and the Development of Zoology during the Thirteenth through the Fifteenth Centuries

Overview

The beginning of animal science, or zoology, is often traced to the influential naturalist Albertus Magnus (1200-1280). After many centuries of nearly complete reliance on superstition and on the writings of respected ancient scholars, this German theologian and philosopher proposed the notion that information about nature could and should be collected by actually observing the various aspects of nature. Other philosophers, particularly Thomas Aquinas (1225-1274) and Roger Bacon (c. 1220-1293) continued this call for a "natural" rather than "revealed" truth, and science began to develop independently from theology and superstition. This shift in thinking cleared the way for the development of zoology as a field of study.

Background

Compared to the numerous early scientific advances that were made in the philosophical centers of the world, scientific inquiry stagnated from about A.D. 200-1200. During this period, superstition often ruled. Bestiaries, which are collections of often-moralistic fables about animals, became very popular. Christian religious leaders adopted the bestiary and used this animal lore to teach their values. With the popularity of the bestiary, fact and fiction became inseparably entangled, and a strong link formed between religion and animal science.

During the same time, religious leaders and many philosophers felt the writings of ancient scholars held the truth, and that methods such as observation and experimentation were worthless because they could neither add nor subtract anything from this truth. The works of Aristotle (384-322 B.C.), in particular, became a nearly insurmountable hurdle to those who felt direct study of the natural world could yield important insights into scientific thought. As described in Ernst Mayr's (b. 1904) *The Growth of Biological Thought: Diversity, Evolution and Inheritance,* "When an argument arose as to how many teeth the horse has, one looked it up in Aristotle rather than in the mouth of a horse."

That began to change with Albertus Magnus (Albert the Great). A member of the Dominican order of Mendicant friars, he was assigned the job of editing and interpreting past written works as they related to the teachings of the Church. He not only made exhaustive studies of Aristotle's contributions, but delved into plant anatomy, animal diversity, and studies of chemistry and geology. He wrote of minerals, plants, and animals, respectively, in *De mineralibus, De vegetabilibus et plantis,* and *De animalibus.* The plant and animal books contained great detail about various plant and animal species, descriptions of the taxonomy of plants, and his thoughts on a variety of topics, such as the relationship between an animal's form and its environment. In addition, he suggested that the works of Aristotle and others should not be the end point in scientific study, but rather that observation and investigation were vital to scientific growth. He wrote, "Science does not consist simply in believing what we are told, but in inquiring into the nature of things."

At around this same time, Hildegard von Bingen (1098-1179) and Frederick II (1194-1250) also wrote important accounts of the natural world. Frederick's *Art of Falconry* focused on birds, while Hildegard's *The Book of Simple Medicine,* or *Physica*, described the medical uses for and natural history of animals, plants, and minerals. Thomas Aquinas, who taught at the University of Paris with Albertus Magnus, shared the opinion that science should have a rational basis rather than one dictated solely by theology. English philosopher Roger Bacon carried it further by suggesting that the traditional writings should not be accepted in the absence of scrutiny. At the urging of Pope Clement IV, Bacon wrote *Opus maius, Opus minus,* and *Opus tertium.* In these volumes, he proposed an educational reform that incorporated methods such as observation and measurement-taking in scientific enterprises. In one of his writings, he declared, "Cease to be ruled by dogmas and authorities; look at the world!"

Impact

Albertus Magnus, who eventually attained the title of saint, did more for the study of animals than provide additional detail about various species. His 26-volume *De animalibus* was an important, descriptive publication that included sections on reproduction and embryology. Like the tomes of the period, however, it was still mainly a summary, interpretation and explanation of the writings of such past philosophers as Aristotle. Albertus Magnus's most important contribution toward the advance of animal science was his belief that animal study should shed its human-placed, moralistic dressing, and instead should proceed objectively and through such techniques as direct observation.

This view marked a split from the belief that all questions could be answered by the ancient writings of great scholars. Without this new reasoning by Albertus Magnus and Thomas Aquinas, animal science would have continued to flounder under the firm hands of religion and tradition.

Building on this idea, Bacon advanced experimentation as an important scientific tool. His push for experimentation—which was unnecessary under the old conviction—combined with his rather brash personality brought accusations as a user of magic and witchcraft. As a result, Bacon lived most of his life in isolation. In the meantime, German scientists Hildegard von Bingen and Frederick II were doing the type of observational work suggested by Albertus Magnus and Thomas Aquinas. Frederick's *Art of Falconry* was one of the first books in hundreds of years to incorporate a great amount of new zoological information that was based on comprehensive personal observation.

Also during this period, Europeans were beginning to conduct grand explorations of what to them were largely unknown areas of the world. Marco Polo (1254-1324) told of China, Prince Henry the Navigator's (1394-1460) men returned with stories of Africa, and Christopher Columbus (1451-1506) had stories of the New World. Each described a diversity of newly discovered animals, the like of which were obviously missing from the works of the early naturalists. In addition, their tales refuted bestiary claims about fanciful creatures and fearsome monsters that lived in or on some of these foreign waters and lands.

These philosophers, naturalists, and explorers together helped catapult the study of animals from religion, bestiaries, ancient writings, and superstitions, and launch it as a new, objective field of scientific endeavor. After an interruption of some 1,000 years, from the time when the ancient scholars relied on observation and nature study, these methods again began to take their places as legitimate scientific activities.

Experimentation also began in earnest. One method of experimentation was particularly important to the fledgling field of zoology: animal dissection. Like observation, animal dissection basically ceased as a method of scientific inquiry for about 1,000 years, but following the shift in thinking toward objective zoological studies, it again gained prominence. Through dissection, naturalists and scientists were able to enhance their understanding of anatomy and physiology. Through comparative studies, they began to explore the similarities among various species and the differences between individual organisms of the same species—the understanding of which would become important to later work in species classification, evolution, and other areas.

In addition, the dissection of animals yielded basic knowledge about body function that could be applied to human medical care and treatment. Salernum, one of the foremost European medical schools, began in the twelfth century to routinely conduct pig dissections to instruct its students about human anatomy. In the following century, Frederick II made a contribution to the education of the students by requiring all of them to attend at least one human dissection each year at the school. The importance of anatomical studies and human dissections became evident. Soon, other schools were making similar requirements, and medical practitioners' knowledge about the human body and its anatomy and physiology climbed to new heights. These studies, in turn, triggered more questions about specific areas, such as the circulatory, nervous, or respiratory systems. Among physicians, medical specialties became more commonplace. Among scientists, these initial anatomical and physiological studies opened the door to investigation into the intricate mechanisms at the foundation of human biology.

Animal science also continued to thrive. Led by the initial shift in philosophy by Albertus Magnus and enhanced by the many insights of noted philosophers and scientists in the thirteenth and fifteenth centuries, animal studies slowly began to transform into modern zoology. In the sixteenth century, Swiss scientist Konrad von Gesner (1516-1565) wrote his illustrated, five-volume work *Historia Animalium,* which included depictions of many animals never before seen by Europeans, while denouncing all fictitious varieties. The book is seen as a milestone in the development of zoology. In 1599, Italian naturalist Ulisse Aldrovandi (1522-1605) published his comprehensive bird-related encyclopedia *Ornithologia*. Other naturalists and scientists followed suit with general zoological studies, as well as investigations into specific species. With these and future contributions of famous and lesser-known scientists, modern zoology found its place as a respected field of science.

LESLIE A. MERTZ

Further Reading

Books

Byers, Paula K."St. Albertus Magnus." In *Encyclopedia of World Biography, second edition*. Detroit: Gale Research, 1998.

Ley, Willy. *Dawn of Zoology*. Prentice-Hall, Englewood Cliffs, 1968.

Magner, L. *A History of the Life Sciences*. New York: Marcel Dekker Inc., 1994.

Mayr, E. *The Growth of Biological Thought: Diversity, Evolution and Inheritance*. Cambridge, Mass.: The Belknap Press of Harvard University Press, 1982.

Periodicals

Nickel, Helmut."Presents to Princes: A Bestiary of Strange and Wondrous Beasts, Once Known, for a Time Forgotten, and Rediscovered." In *Metropolitan Museum Journal* 26 (1991): 129-38, New York.

The Art and Science of Falconry

Overview

Falconry is the practice of hunting with birds of prey such as falcons or hawks. There is evidence to suggest that falconry was practiced in Assyria (present-day Iraq and Turkey) as early as the eighth century B.C. It reached a peak in popularity in Europe during the Middle Ages. One of the main participants falconry was Frederick II (1194-1250), a king of Germany and Sicily crowned Holy Roman Emperor in 1220. Despite being one of the most politically powerful people in central Europe, Frederick found time to be an enthusiastic falconer and observer of birds. He authored a book on falconry titled *De Arte Venandi cum Avibus* (On the art of hunting with birds). This book was unusual in that it was based almost entirely on the author's own observations, rather than on the statements of other scholars.

Background

Frederick's book (often known simply as the Falcon Book) is much more than a hunting guide. It is divided into six parts, the first part of which is a general description of birds—not just falcons and hawks, but the hundreds of species with which Frederick was familiar. He gives detailed descriptions of their behavior, including feeding, breeding, and migration habits. He also discusses their anatomy and physiology—the structure and functions of their various parts—including their skeletal system, eyes, wing feathers, and internal organs.

In the second and third parts of the book, he moves more specifically to birds of prey and describes the capture and training of such birds. The remaining parts of the book discuss hunting with particular species of falcons. Frederick made much use of his observations of falcons in their natural surroundings as he cared for and trained his birds. For instance, he noted that young falcons in the wild were fed regurgitated meat by their parents. Therefore, he ordered that his young captured falcons should be fed finely chopped fresh meat—and not just any meat, such as that of a barnyard chicken, but that of wild birds that would be the normal prey of falcons. Thus, he attempted to raise his birds on food that would closely resemble what would have been fed to them by their parents.

Frederick noted that the sense falcons rely most heavily on is sight. (He concluded that falcons do not locate their prey by smell based on his observation that birds whose eyes had been sealed, or stitched closed, could not locate meat thrown near them.) He made use of this fact in training birds. By temporarily taking away their sight by sealing their eyes or by placing a leather hood over their heads, he was able to control the stimuli to which the birds were exposed. In this way, he was able to control their behavior.

As the initial part of training, Frederick suggests that the falconer should repeat a specific phrase while feeding the falcon. Eventually, the bird will learn to associate this sound with being fed. As a result, the phrase can be used to calm an agitated bird; it becomes less restless because it assumes it is going to be fed. This type of behavior modification, known as classical conditioning, became widely-known through the experiments of Ivan Pavlov (1849-1936) near the beginning of the twentieth century (about 700 years after the publication of the Falcon Book). In classical conditioning, an animal learns to associate one stimulus with another. By ringing a bell whenever a dog was shown food, Pavlov trained the dog to respond to a second stimulus (the sound of the bell) in the same way as it normally responded to another stimulus (the sight of food). In Frederick's case, the normal stimulus for the falcon was the taste of food and the second stimulus was the sound of specific spoken phrase. Both stimuli soon had the same response, a calming of the bird.

STATUS SYMBOLS OF THE MIDDLE AGES

During the Middle Ages, falconry became popular throughout Europe. The main purpose of hunting with falcons usually wasn't to obtain meat. Instead, most people participated in the sport simply for entertainment. However, falconry served another purpose as well: people used it as an excuse to show off to their neighbors. Falconry was an expensive hobby. Some species of falcons were worth more than their weight in gold. There were also other expenses involved, such as fresh meat for the birds, equipment, and training. In addition, nobles competed with one another to host elaborate falconry parties. These events were used to establish and maintain power. The grander the party and the more numerous the falcons, the richer and more important the host seemed.

Today, if people purchase a status symbol that is more than they can afford, the worst that might happen to them is that their new sports car will be repossessed and their credit ruined. In the Middle Ages, however, the consequences of owning a status symbol above your rank could be much more severe. *The Boke of St. Albans*, published in 1486, lists laws of ownership regarding falconry. These laws state the types of falcons a person could own depending on his or her rank. For instance, a king could own a gyrfalcon, an earl could own a peregrine, and a lady could own a merlin. According to the book, people caught owning a bird that was above their rank not only had their falcon repossessed—but also had their hands cut off.

STACEY R. MURRAY

Impact

Frederick's book made an important contribution to medieval science. He based his conclusions presented in the book on 20-30 years of his personal experiences. He refused to accept the ideas of other scholars without testing them for himself. This attitude toward science was not typical for a person of the Middle Ages.

In Frederick's time, relatively few books were published in Western Europe. (The printing press would not be invented until about 200 years after his death.) Those Europeans who produced books were primarily concerned with religion and spiritual matters and often had little use for the material world, including scientific study. Their own observations rarely appeared in what they wrote; instead, they tended to rely on the work of ancient philosophers such as Aristotle (384-322 B.C.). Frederick's Falcon Book, however, consists of nothing but the author's own observations of the material world; it represents a change in attitude that pointed toward the beginning of experimental science. Frederick can be seen as a predecessor of Albertus Magnus (1200-1280) and Roger Bacon (1220-1293) later in the thirteenth century. They also believed that science should be based on the observation and experience of nature.

Frederick, in fact, even went so far as to disagree with Aristotle on certain points. He states in his preface to the Falcon Book, "We discovered by hard-won experience that the deductions of Aristotle, whom we followed when they appealed to our reason, were not entirely to be relied upon." (When Frederick says "we," he means is referring to himself. It was common practice for monarchs to refer to themselves with plural pronouns.) He also states, "In his [Aristotle's] work, the *Liber Animalium* (History of animals), we find many quotations from other authors whose statements he did not verify and who, in their turn, were not speaking from experience. Entire conviction of the truth never follows mere hearsay."

Frederick, however, tried to verify everything about birds that he read or that was told to him by others. For example, he had heard of so-called barnacle geese that lived in northern Europe and supposedly hatched from barnacle shells attached to the rotting wood of ships. He had samples of such wood sent to his court to find out whether this was true. Based on his tests, he concluded that the geese did not in fact come from barnacles. Frederick proposed that the geese hatched from eggs like other birds, but made their nests in remote areas unfrequented by people. His hypothesis was later proven to be true, but the birds are still known as barnacle geese to this day.

Manuscript illustration of falcon hunt. *(Art Resource. Reproduced with permission.)*

Prior to the publication of the Falcon Book, other short works on falconry had appeared in Europe. However, none of the other authors were nearly as thorough or as knowledgeable as Frederick was. He attempted to bring the realm of falconry to an exact science. Those who read the book were taught not just about falconry or birds; they were also taught the art of observation as well as how to describe these observations to others in a clear manner. (Even though the book was written more than 700 years ago, its style is so matter of fact that it is still easily readable today.) This skill of observation could be applied to any area of science, not just ornithology (the study of birds).

The Falcon Book had a great impact on scientific thought because its author was not simply an unknown scholar applying such skills, but the Holy Roman Emperor—one of the most powerful figures in Europe. In fact, the book probably could not have been written by someone who did not have Frederick's political power and social status. These allowed him to send for birds from across Europe and Asia. Thus, he was able to observe firsthand a much wider variety of species than other naturalists would have had access to at that time. For example, Frederick was the first to realize that two falcons normally considered to be two separate species were really the same; their differences were due to the different climates in the different parts of the world in which they lived. Frederick also had access to the knowledge of falconers and other naturalists from across Europe and the Middle East. His court fostered scholars in the natural sciences, and he had the works of Aristotle and other ancient philosophers translated into Latin. His religious crusade to the city of Jerusalem in 1228 brought back other experienced falconers from Arabia and Syria.

The Falcon Book served as a model for other books that were soon written in the same manner and style, but on different topics. Frederick himself had great influence on the publication of another work called *Horse-Healing* by Jordanus Ruffus. It represents what many consider to be the first veterinary book produced in Western Europe. Frederick suggested that Ruffus write the book and served as one of Ruffus's sources of information, being an expert on horses himself. This book was also translated into many languages and also served as a template for similar works.

STACEY R. MURRAY

Further Reading

Books

Frederick II, Holy Roman Emperor. *The Art of Falconry, being the De Arte Venandi Cum Avibus*. Translated and edited by Casey A. Wood and F. Marjorie Fyfe. Boston: Charles T. Branford Co., 1943.

Kantorowicz, Ernst. *Frederick the Second*. London: Constable & Co. Ltd., 1931.

Madden, D.H. *A Chapter of Medieval History*. Port Washington, NY: Kennikat Press, 1924.

Van Cleve, Thomas Curtis. *The Emperor Frederick II of Hohenstaufen: Immutator Mundi*. Oxford: The Clarendon Press, 1972.

Other

"Falcons and Man—A History of Falconry." *PBS Online, 2000*. http://www.pbs.org/falconer/man/index.htm.

Biographical Sketches

Pietro d'Abano
c. 1250-c. 1318
Italian Physician and Philosopher

A professor of medicine at Padua, Italy, Pietro d'Abano (also called Peter of Abano) attempted a synthesis of Arab medicine, Greek philosophy, and the Catholic worldview that prevailed in the Europe of his day. He is remembered for his book *Conciliator differentiarium*, and for his efforts at making Padua one of the Western world's centers for medical study.

As his name indicates, Pietro was born in the town of Abano, near Padua. During his career, he traveled throughout France and Sardinia, and visited Constantinople. He also met the celebrated traveler Marco Polo (1254-1324), from whom he obtained information about Asia.

In addition to his work at Padua, Pietro practiced medicine in Paris. During his travels, he was said to have discovered and translated a lost work of Aristotle (384-322 B.C.), and this may have influenced his attempt—in the tradition of Ibn Sina (Avicenna; 980-1037) and others—to reconcile Greek thought with revealed religion in *Conciliator differentiarium*.

Such ideas were still considered dangerous in late medieval Europe; moreover, Pietro's dabblings in mathematics and astrology, and particularly his writings on magic, made him an object of suspicion. He was said to possess crystal vessels through which (to borrow a modern term) he "channeled" seven demons, each of whom gave him special knowledge about one of the seven liberal arts.

As time went on, rumors of Pietro's alleged abilities as a sorcerer became more and more outlandish. He was said to have the power to cause money to return to his purse after he had spent it, and when a neighbor forbade him to drink from a certain spring, he supposedly used his dark powers to divert the stream from the neighbor's property.

The latter claim was sufficient to bring him before the Inquisition, which tried and acquitted him. Later he was brought before the court on the same charges of sorcery, but before the second trial ended, he had died. He was declared guilty posthumously, and the inquisitors ordered that his bones be dug up and burned.

JUDSON KNIGHT

Taddeo Alderotti

c. 1223-1295

Italian Physician

Founder of a medical school in Bologna, Italy, Taddeo Alderotti was an early advocate of serious medical study and practice. It was because of his efforts that the city authorities extended to medical teachers and students the same legal status as that of their counterparts in law school.

Alderotti was born in Florence, perhaps in 1223, though estimates of his birth year vary from 1215 to 1233. In 1260, he began teaching medicine at Bologna, which during the preceding century had emerged as a center of learning for all of Europe. There Holy Roman emperor Frederick I Barbarossa (1123-1190) had established the first Western university in 1158, by which time the town had begun to develop a community of medical students.

Years later, Dante Alighieri (1265-1321) would describe Alderotti in his *Divine Comedy* as a "Hippocratist," or follower of Hippocrates (c. 460-c. 377 B.C.). The Greek "father of medicine" was indeed Alderotti's model, and like Hippocrates he sought causes for illness in science rather than religion—a revolutionary idea in the thirteenth century. He also reintroduced Hippocrates' practice of teaching medicine at the patient's bedside.

Among Alderotti's books was the *Consilia*, a series of case studies presented alongside medical opinions on each case. Also included was a record of the preventive measures applied by the physician, as well as both dietary and therapeutic treatments. Not only did Alderotti pioneer this type of medical literature, but he also wrote one of the first medical books in a modern language, the practical family physician's handbook *Sulla conservazione della salute*.

Much of Alderotti's work indicated that a reawakening was taking place in the European scientific community, whose curiosity had been held in check for many centuries; but it was equally clear that Alderotti was not simply taking up where the great physicians of the ancient world had left off. Forebears such as Galen (c. 130-c. 200) had been confronted with religious prohibitions against the dissection of cadavers, but by the late thirteenth century Christian laws proscribing such activities had been loosened. Alderotti's *Expositio in arduum aphorismorum Hippocratis volumen* contains descriptions of dissections and experiments in comparative anatomy performed by Alderotti and others.

After his death, Alderotti's legacy continued through students who became physicians and teachers, including the four Varignana brothers, Dino and Tommaso di Garbo, and Pietro Torrigiano Rustichelli—all ardent exponents of Galen. Alderotti also indirectly influenced Pietro de Tussignana (d. 1410) and Bavarius de Bavariis (d. c. 1480), court physician to Pope Nicholas V.

JUDSON KNIGHT

Arnau de Villanova

c. 1235-1311

Spanish Alchemist and Physician

Although Arnau de Villanova (sometimes referred to as Arnold of Villanova or Arnaldus Villanovanus) is remembered primarily as an alchemist, he was also an influential figure in the development of modern medicine. Not only was he one of the first to systematically use alcohol in treating certain diseases, but he also distinguished himself by relying more on observation than on a slavish obedience to the writings of Galen (c. 130-c. 200) and other ancient masters.

Born near Valencia, Arnau was of Catalan descent. During his time, much of Spain remained under Muslim control, though the Christians' Reconquista (reconquest) was gaining strength. Indeed, one of the key battles of the Reconquista had taken place in Arnau's hometown, where in 1094 El Cid (Rodrigo Díaz de Vivar, c. 1043-1099) had become the first Christian leader to gain a significant victory over Arab forces.

Educated by Dominican monks, Arnau went on to study medicine at Naples. He then went on to a highly successful medical practice, with kings, popes, and other dignitaries among his patients, and traveled widely throughout Spain, France, Italy, and North Africa. As a result of this exposure, he became fluent in a number of languages, including Arabic, Greek, and Latin—in other words, three of the four languages (other than Hebrew) in which virtually all Western scientific information was written.

Arnau wrote a number of works on alchemy, of which the most famous was *The Treasure of Treasures, Rosary of the Philosophers*. After successfully treating King Peter III of Aragon in 1285, he was awarded a chair at the University of Montpellier, in a part of southern France then under the control of Aragon. There he distinguished himself not only for his reliance on observation and his use of alcohol for treatments, but also for his dietetic prescriptions.

He also received a castle in Tarragona from Peter, and went on to become a favorite of Peter's son, James II. But while on a mission for James in 1299, Arnau was arrested on orders of the Holy Office, the authorities directing the Inquisition. The charge was heresy, and its basis was certain writings in which he had discussed the Antichrist. He spent more than a year in prison, but the intervention of Pope Boniface VIII and Philip IV (the Fair) of France gained his release in 1301.

Arnau lived another decade, during which time he met the alchemist Ramon Llull (c. 1235-1316) in Naples. He died on a voyage from Sicily to Avignon, the city to which the papacy had recently been moved. Arnau, who had been on his way to treat Pope Clement V, was buried in Genoa.

JUDSON KNIGHT

Gerard of Cremona

1114-1187

Italian Scholar and Translator

Gerard of Cremona is remembered not for any original contributions to scientific knowledge, but rather for his role as a translator. Thanks to Gerard, the *Canon of Medicine* by Ibn Sina (Avicenna; 980-1037), destined to become the most important work in the medical sciences of Europe for half a millennium, made its way into Latin. A secondary effect of Gerard's work was the debate concerning the role of reason and faith, sparked in part by Avicenna's work. This would in turn help bring about the growth in scientific curiosity that characterized the late medieval period.

Though some historians would later claim that Gerard came from Carmona in Spain, he was almost certainly from Cremona, in Italy. Nonetheless, he did spend most of his adult life in and around Toledo, Spain, and one of the principal sources concerning Gerard is the *Chronicle* by Francisco Pipino (fl. 1300), a Dominican friar in Toledo.

According to Pipino, Gerard traveled to his city in pursuit of scientific knowledge. Specifically, he was fascinated by the works of the ancient astronomer Ptolemy (c. 100-170), and desired to learn Arabic so that he could translate the latter's *Almagest*. (At that point Greek copies of the *Almagest* had disappeared, and it existed only in Arabic translations.)

Though Ptolemy would later be seen to have a damaging effect on Western scholarship, in particular with his adherence to a geocentric or Earth-centered model of the universe, it was necessary for Europeans to absorb the many useful aspects of Ptolemy's work before they could outgrow it. Thanks to Gerard, this seminal text appeared for the first time in Latin, and had a monumental impact on the physical sciences in Europe in the following years.

Gerard also translated Avicenna's *Canon*, destined to become European physicians' preeminent source of knowledge on medicine until the late seventeenth century. According to Pipino, he rendered a total of 76 Arabic texts into Latin, including works on dialectic, geometry, philosophy, physics, and other sciences. Thus Gerard, who died in 1187 and was buried at the Church of St. Lucy in Cremona, played an enormous role in bringing the highly advanced learning of the Arab world to the West. This also means that he had a greater impact on the rebirth of learning, which culminated with the Renaissance, than all but a handful of much more famous men.

JUDSON KNIGHT

Guy de Chauliac

1300-1368

French Physician and Surgeon

Guy de Chauliac has been called "the most eminent surgeon of the European Middle Ages." Other sources view him in the historical shadow of his French predecessor, Henri de Mondeville (1260-1320), who lived a half century earlier. In large part, Guy's fame is based on his publication of *Chirurgica magna,* a text on surgery that was in standard use until the seventeenth century. But he is also remembered because he was so successful in his practice of medicine and surgery, serving kings and popes in his long career. His fame as a practitioner, gained by his treatment of wounds, cataracts, ulcers, and fractures, was only surpassed by the enduring popularity of his publications on anatomy and surgery. He is also well-remembered for his willingness to remain in Avignon to treat patients during the Black Death rather than flee to the safety of the countryside.

Born in the French village of Chauliac, Guy initially trained as a cleric but received his medical education and surgical training in Toulouse, Montpellier, and Paris. He also studied anatomy at Bologna. He practiced surgery in Avignon, France, when that city was the residency of the papacy, serving popes Clement VI, Innocent VI, and Urban V.

By practicing in Avignon, then one of the intellectual and scholarship crossroads of Europe, Guy was fortunate in that he had access to translations of Greek and Arabic texts on surgery and medicine.

Guy de Chauliac was a controversial figure in his day because of his opinions about how wounds should be treated. His methods were considered "meddlesome" by his contemporaries. Guy, who shocked the medical world by stating that nature alone was not sufficient for wound healing, advocated widening, cleaning, and draining wounds rather than letting nature take its course, as was conventional wisdom at the time. He not only removed foreign objects from wounds, but also used purifying agents such as wine, turpentine, and brandy. Guy also advocated binding wounds closed with adhesive tape, sutures, or by cautery. He discussed wounds of different classes, such as "hollow" wounds, contused wounds, ulcerated wounds, and even bites.

During his lifetime two epidemics of bubonic plague, or Black Death, struck in Avignon, first in 1348 and then in 1360. The plague had decimated European cities and Avignon was not to be spared. Unlike many physicians who fled the cities, Guy remained in Avignon and treated patients. Although he contracted bubonic plague and nearly died, his written account of the Black Death, observed in scientific objectivity, gives historians one of the few firsthand, non-mythological accounts of its ravages and estimates of mortality.

Guy noted that the plague came in two types, each with slightly different symptoms. Some struck with the disease died in three days from one type and in five days from the second. By his direction, Pope Clement VI went into seclusion and escaped infection.

In 1363, toward the end of his career, Guy published *Chirurgica magna,* or "Grand Surgery," a collection of eight books. In them he reviewed the history of surgery and discussed surgery as a science and a part of medical practice, rather than just the tool of barbers and butchers, as was surgery's early status. He also published a chapter on anatomy and stressed that it should be learned and taught through hands-on dissection of the "recently dead" rather than taught through drawings, as was the practice for Henri de Mondeville.

Guy de Chauliac's "Grand Surgery" was not to be replaced until the writing of French surgeon Ambroise Paré (1510-1590) in the mid-1500s.

RANDOLPH FILLMORE

Hildegard von Bingen
1098-1179
German Medical Author, Composer, and Visionary Mystic

The achievements of the Benedictine nun Hildegard von Bingen were astonishing in their range and excellence, particularly for a woman of the twelfth century. During a time in which female activities were restricted, she became the first woman authorized by the Pope to write theological works, the only medieval woman to preach publicly, the author of the first known morality play, the most prolific medieval composer of liturgical plainchant, an artist who created unique illuminated manuscripts, the author of the first book written by a woman on herbal medicine, and one of the earliest scientific writers to discuss sexuality and gynecology from a female perspective.

Hildegard, born in 1098 in the rural Rhineland, was the tenth child of a noble family. At the age of eight, she was dedicated to the Church as a "tithe"—an offering of a tenth of one's wealth—at the convent attached to the Benedictine monastery at Disibodenberg, near Bingen. Here she became the student of the renowned teacher and recluse Jutta von Spanheim (1092-1136), receiving a Latin education based on holy scriptures and Church fathers.

When Jutta died, Hildegard was elected magistra or teacher. Five years later she began to write about the extraordinary visions she had experienced since childhood, which filled her

Manuscript illustration from Hildegard von Bingen's *Liber divinorum operum. (Gianni Dagli Orti/Corbis.Reproduced with permission.)*

with intuitive knowledge of spiritual mysteries. She sought counsel from Bernard, Abbot of Clairvaux (1090-1153), later named a saint, and from Pope Eugenius III (d. 1153), both of whom encouraged her to write and speak about her experiences. At age 50, Hildegard founded a new monastery at Bingen. In her late 60s she traveled to the great cathedrals of the area, preaching and advocating reform among the clergy.

In 1178, at age 80, she refused the order of Eugenius III to exhume the body of a friend buried in her monastery cemetery. The Pope claimed the man was not entitled to a Church burial, and placed the monastery under a decree of interdict, banning all religious activities there. Hildegard insisted the man had been absolved, and eventually the Pope withdrew the order and restored the monastery to full participation in the Church. Hildegard died in her sleep in 1179.

In her religious writings, Hildegard was concerned primarily with the divine, but in her scientific and medical works she focused on material, observable conditions. The exact sources of her knowledge are not known, but she had access to Latin texts as well as vernacular material, and profited from the long-standing Benedictine tradition of caring for the sick at the monastery's infirmary, most likely equipped with its own herbal garden. Between 1150 and 1160 Hildegard produced two medical texts. The first, *Causae et Curae* (Causes and cures), is a compact handbook on the etiology (cause), diagnosis, and treatment of diseases, and also contains chapters on human sexuality, psychology, and physiology. The detailed descriptions reflect a quality of scientific observation rare in that period.

Hildegard's second medical text, *Physica*, is an extensive pharmacopoeia (guide to healing agents) organized in nine books, cataloging the medicinal qualities and uses of plants, elements, trees, stones, fish, birds, quadrupeds, reptiles, and metals. *Physica* displays a thorough knowledge of what was then known about the natural world, and gives a reliable picture of how medicine was practiced at monastic centers.

In several ways, her texts depart from the twelfth-century norm. First, Hildegard was a nun writing about medicine. Only one other female medical writer, Trotula of Salerno, Italy (d. c. 1097), is known from the entire medieval period. Second, Hildegard did not merely copy selections from pre-existing works as a scribe might do, but rather sought to explain the reasons for health and disease in the broader context of religion and natural science. Although she saw the human fall from grace as the ultimate cause of disease, she did not advocate passive acceptance of suffering. She advanced the philosophy that practical skills and knowledge can help alleviate suffering and promote healing, thus justifying the energetic pursuit of knowledge and dignifying the practice of medicine.

Hildegard was much admired by her contemporaries, who called her "Sybil of the Rhine" and "Jewel of Bingen." During her incredibly productive career as writer, teacher, composer, reformer, and founder of monasteries, Hildegard was consulted by kings, emperors, popes, and other notable figures of her age. Much of her correspondence, music, mystical, and scientific writings survive. Largely forgotten for centuries, her compositions and writings have enjoyed a resurgence of interest in recent decades; her music is frequently performed and recorded, and scholarly studies of her religious and scientific writings are increasing. Although not canonized, Hildegard has been beatified and is frequently referred to as St. Hildegard.

DIANE K. HAWKINS

Hisdai ibn Shaprut

c. 915-c. 975

Jewish Spanish Physician

An influential figure in the court of Caliph 'Abd ar-Rahman III, Hisdai ibn Shaprut helped bring on a golden age of Jewish learning in Muslim Spain. Among his many specific contributions to medical scholarship was his translation of a pharmacological treatise by the Greek physician Dioscorides (c. 20-c. 90) into Arabic.

Hisdai, whose full name was Hisdai Abu Yusuf ben Isaac ben Ezra ibn Shaprut, had the good fortune to be born in a land controlled by Muslims rather than Christians. Jews had fared poorly in Spain prior to the Muslim invasion in 711, and few of their number living in Christian Europe had an opportunity to excel as scholars. In Muslim Spain, by contrast, it was possible that a Jew might not only engage in acclaimed intellectual pursuits, but also occupy a position of great political influence.

Such was the case with Hisdai, who became the caliph's court physician. (The Umayyads, escaping the Abbasid Caliphate's takeover in the Middle East in 750, had established a rival caliphate ruling Spain and Morocco.) In time Hisdai became a sort of untitled vizier (executive official), greatly valued for his linguistic versatility—he knew Hebrew, Arabic, Latin, and Greek—and for his sensitivity in diplomatic matters.

At one point, 'Abd ar-Rahman (891-961) used Hisdai to negotiate a treaty with the Byzantine emperor, who sent the manuscript of Dioscorides to Spain as a token of good will. By translating that text into Arabic, Hisdai made the knowledge it contained available throughout the Arab world, the culture that was making the greatest advances in scientific learning at the time.

Hisdai also acted as intermediary in a dispute between the Christian kingdoms of León and Navarre, and carried on a correspondence with the ruler of the Khazar Khanate, a Jewish kingdom in what is now Russia. After the death of 'Abd ar-Rahman in 961, he continued to serve the latter's son and successor, al-Hakam II.

Using his influence to advance the cause of his people, Hisdai gathered learned Jewish scholars around him such as the grammarian Dunash ben Labrat (c. 920-c. 990) and the lexicographer Menahem ben Saruq (c. 910-c. 970). He also greatly encouraged the study of the scriptures and the Talmud, a rabbinical commentary on the latter. Because of his influence, Spain became a recognized center of Jewish culture, rivaling other centers such as Iraq in importance.

JUDSON KNIGHT

Hunayn ibn Ishaq (Johannitius)
808-873
Arab Scholar and Physician

Hunayn ibn Ishaq, known in the West as Johannitius, is important primarily for his work as a translator: it was through his efforts that numerous writings from ancient Greece, which he translated into Arabic, were preserved. In this he played a role similar to that of his Western counterpart Gerard of Cremona (1114-1187), who three centuries later translated many of Hunayn's works back into a European language, Latin. Unlike Gerard, however, Hunayn wrote original works; furthermore, because he came earlier, his importance as a preserver of ancient knowledge is perhaps even more significant.

Though he was an Arab, the fact that Hunayn's family subscribed to Nestorianism, an Eastern variety of Christianity, perhaps gave him a closer psychological connection to Europe than he might have had otherwise. He studied at Baghdad, cultural center of the Arabic world, and later at Alexandria. In time he came under the employment of the Caliph al-Ma'mun (786-833), who established the House of Wisdom as a center for the translation of Greek texts.

One of the principal challenges facing the scholars at the House of Wisdom was the acquisition of manuscripts for translating. At one point al-Ma'mun sent a team to Byzantium to obtain texts, and it is likely that Hunayn, as the most knowledgeable scholar of Greek, took part in this expedition. Hunayn wrote about traveling throughout Mesopotamia, Syria, Palestine, and Egypt in search of a single manuscript—which he finally located in Damascus, though half was missing. This degree of determination would be notable even today, but in view of the difficulties associated with travel in the pre-modern era, it is almost beyond belief.

The works Hunayn did locate and render into Arabic were pieces of inestimable value: writings by Hippocrates (c. 460-c. 377 B.C.), Plato (428-348 B.C.), Aristotle (384-322 B.C.), Galen (c. 130-c. 200), and other ancient masters. Since many of the Greek originals have since been lost, the importance of Hunayn's translation work is hard to overestimate. In the short run, he influenced a revival of interest in the Greeks that helped to spark the prodigious scientific advancements of the medieval Arab world. Later these ideas would make their way to the West, in part through Gerard's translation of texts from Arabic to Latin, and this would facilitate the rebirth of scientific curiosity in Europe.

After al-Ma'mun's death, Hunayn continued to work for a series of caliphs at the House of Wisdom; then in 847, al-Mutawakkil appointed him chief court physician, a position he would hold for the remainder of his life. Hunayn also wrote works on ophthalmology, as well as an original introduction to Galen's *Ars parva*.

JUDSON KNIGHT

Ala ad-Din Abu al-'Ala 'Ali ibn Abi al-Haram al-Qurayshi ad-Dimashqi ibn an-Nafis
1210-1280
Arab Palestinian Physician

Ibn an-Nafis was a famous Arab physician and writer who contributed to the value of Arab medicine by helping preserve and systematize existing medical knowledge, as well as commenting on and explaining the ideas in these documents. He proposed the circulation of the blood 300 years before this was identified in the West by Michael Servetus (1511-1553), Realdo Colombo (1516?-1559?), and William Harvey (1578-1657).

Born at al-Qarashi near Damascus, Ibn an-Nafis studied medicine at the great medical college-cum-hospital in Damascus, founded by Nur al-Din Zangi. As with many physicians of his day, Ibn an-Nafis's interests were wide and varied. He was versed in logic, grammar, theology, literature, and law, in addition to medicine. He became a renowned scholar at the Shafis School of Jurisprudence as well as a great physician.

Ibn an-Nafis then moved to Cairo, where he served as principal of the famous Nasri Hospital and trained many famous medical specialists. He was appointed chief physician by the Mamluk sultan al-Bunduqdari, who reigned from 1260 to 1277, and served as his personal physician.

Like many Arab writers, Ibn an-Nafis was very thorough and systematic. His writings were numerous, and he joined the hosts of those who repeatedly copied manuscripts, preserving and organizing them. He was careful to preserve the spirit of Hippocrates (460-377 B.C.) and Ibn Sina (980-1037) and wrote commentaries on their work. But Ibn an-Nafis went further. He made original comments from his observations and departed from just slavish copying and organizing to critical thinking.

In addition to being a physician, he wrote on religion and law. However, it was in the field of medicine that he is most recognized. His book *Kitab al-Shamil* (Comprehensive book on the art of medicine), was written in his thirties. It consisted of 300 volumes of notes, of which he published only 80. Until 1952 this massive work was thought to be lost. It was subsequently rediscovered and catalogued in Cambridge University's collection of Islamic manuscripts. Earlier, a librarian had catalogued four manuscripts of this work without realizing who the author was. Another of Ibn an-Nafis's works, *Majiz al Qanun,* was vastly popular in his day and become the source of many subsequent commentaries. A number of early Arabic manuscript copies of Ibn an-Nafis's works are on display in Damascus.

The book, so far unpublished, had an important section on surgical technique that threw new light on Ibn an-Nafis as a surgeon. In it he defined three stages for each operation. First, there is the presentation where the diagnosis is made and the patient entrusts his life and body to the surgeon; second, the operation; and last, the preservation or post-operative care. The book gives detailed descriptions of duties of surgeons and relationships among patients, surgeons, and nurses. He also discusses decubitus or bed sores, posture, bodily movement, and manipulations of instruments. He illustrated his points with case histories.

However, Ibn an-Nafis is best known for his commentary on the anatomy of Avicenna, the *Sharh Tashrih al Qanun,* including one passage in which he describes the pulmonary blood circulation. Galen (130-200) had earlier proposed that circulation of the blood in the heart went through tiny holes in the walls of the septum. Ibn an-Nafis challenged Galen, asserting that the blood could not pass through the tough septum, but must first go to the lungs. He proposed that the lungs were made of several parts, the bronchi and branches of small arteries and veins connected to porous flesh. This discovery can be fixed at 1242. For years the descriptions fell into obscurity. It would be 300 years before others developed the idea. Servetus, Colombo, and Harvey are credited with the discovery of circulation. Some historians think these physicians may have had access to Ibn an-Nafis's work because several translations were made from the Arabic manuscripts into Latin around 1547.

Another important, but rarely mentioned, contribution by Ibn an-Nafis is his proposal that the nutrition of the heart is extracted from small vessels in the heart's wall. He was the first to postulate coronary circulation.

Ibn an-Nafis's *Sharh al Qanun* consists of four books, including commentaries on medicines and drugs, and diseases that are not specific to certain organs. He also wrote a commentary on Hippocrates' *Epidemics,* a book on ophthalmology, and a general reference for physicians.

Ibn an-Nafis was reputed to have recorded his own experiences, observations, and deductions, rather than using reference books. His Islamic religion and beliefs about mercy towards animals prevented him from doing experimental anatomy. If he had studied the anatomy of animals, he probably could have developed a very accurate description of the anatomy of circulation.

The date of Ibn an-Nafis's birth is disputed; while the accepted date is 1210, some place it at 1200 or 1213. Toward the end of his life he bequeathed his house and library to the newly founded Dar al Shifa, or House of Recovery, also called the Qulawum, or Mansuri Hospital, established in 1284. He died on December 17, 1280, of an unknown illness.

EVELYN B. KELLY

Abu'al-Walid Muhammad ibn Ahmad ibn Muhammad ibn Rushd (Averroës)

1126-1198

Spanish Physician, Philosopher, Astronomer and Jurist

Ibn Rushd, known to the West as Averroës, is famous for his commentaries on Aristotle

(384-322 B.C.), which were widely used as standard texts until the sixteenth century. This body of work earned him the appellation "the Commentator." He is also remembered for his medical treatise *Kulliyat* and philosophical work *Tahafut at-tahafut.*

Ibn Rushd was born in 1126 to a family of important jurists in Cordoba, Spain. While in Marrakesh, Morocco, to help with the reform of education (1153), he met the astronomer Abu Bakr ibn Tufayl (1105?-1184). Six years later, at Seville, Ibn Tufayl introduced Ibn Rushd to the Sultan's son, Abu Ya'qub Yusuf. As a result of their meeting, Ibn Rushd accepted the task of providing comprehensive commentaries on Aristotle's works. When Abu Ya'qub became caliph in 1169, Ibn Rushd was appointed *qadi* (religious judge) of Seville and then of Cordoba (1171). He was summoned to Marrakesh in 1182 to assume Ibn Tufayl's position as physician to Abu Ya'qub. After Abu Ya'qub's death in 1184, Ibn Rushd continued to enjoy privileged status under the new caliph, al-Mansur Ya'qub ibn Yusuf. When war with the Christian powers in the north became imminent in 1195, Ibn Rushd fell into disfavor because his views were thought subversive of Islamic orthodoxy. He was temporarily banished. After the victorious al-Mansur returned to Marrakesh, Ibn Rushd was summoned to his side and all edicts against him were canceled. He died on December 10, 1198, in Marrakesh.

Ibn Rushd is best known for his tripartite commentaries of Aristotle's work. The *Jami* was a brief summary of a text for beginning students. The *Talkhis* was a longer, intermediate analysis. The *Tafsir* was intended for advanced students and provided a comprehensive and original exegetical analysis. Traditional Arabic commentaries were heavily influenced by Neoplatonic accretions and thus distorted Aristotelian ideas. Ibn Rushed achieved a fair measure of success in restoring Aristotle's original meaning. His commentaries exerted a great influence on thirteenth-century Aristotelianism. Their clarity was such that, in the Latin West, they were pedagogically preferred to Aristotle's primary texts.

More than 20 works dealing with medicine are attributed to Ibn Rushd. These represent the denouement of Muslim medical thought, as Ibn Rushd criticized its rigid conformity to authority and tradition, calling for a renewed emphasis on empirical evidence.

His major medical treatise was *Kulliyat* (General medicine), which incorporated portions of Ibn Sina's (980-1037) *Qanun fi at-tibb* (Canon of medicine) as supplemented by his own original contributions. *Kulliyat* covers topics from organ anatomy and hygiene to the prevention, diagnosis, and treatment of diseases. Ibn Rushd asked his friend Ibn Zuhr to write a companion piece to *Kulliyat.* The two works were intended to serve as a medical textbook and were often printed together.

In astronomy, Ibn Rushd provided the most significant exposition and defense of Aristotelian cosmology. Aristotle had maintained that each planet was attached to a celestial orb, with the orbs nested within each other. According to Aristotle, these concentric spheres moved with a natural, uniform circular motion about a common center—the Universe's center. Ibn Rushd developed this view further, insisting that the motion had to be about a physical center—the earth. He attacked Ptolemy's use of epicycles and equants because they clearly violated these constraints. Thus, the Ptolemaic system could not be physically real.

Tahafut at-tahafut (The incoherence of the incoherence) is Ibn Rushd's most important philosophical work. It is a response to al-Ghazali's attack on the efficacy of rational inquiry in religious matters. Ibn Rushd rejected this position, arguing that natural reason is adequate for any intellectual investigation. Ibn Rushd also wrote important works on jurisprudence.

STEPHEN D. NORTON

Ibn Sina

980-1037

Arab Physician and Philosopher

Known in the West as Avicenna, the Arab thinker Ibn Sina was among the most influential figures in European philosophy and science during a period of half a millennium. As a philosopher, he played a highly significant role in affecting a synthesis between Greek science and the Muslim faith, an equation in which European thinkers would substitute Christianity for Islam with equally powerful results. As a physician, he dominated thought during the period from the late twelfth to the late seventeenth century, during which time his *Canon of Medicine* was the single most important medical text in all of Europe.

Born in Afshana in what is now Afghanistan, Ibn Sina (whose full name was Abu Ali al-Husayn ibn Abd-Allah ibn Sina) was raised in

Ibn Sina. *(New York Public Library Picture Collection. Reproduced with permission.)*

Bukhara, now part of Uzbekistan. He displayed an early talent as a student, and at the age of 10 had already read the entire Koran. He gained other useful knowledge from an Indian teacher who exposed him to Indian principles of mathematics, including the numeral zero, first used by Hindu mathematicians.

Encouraged by his family, who placed a high value on study, Ibn Sina began his formal education at age 16, and had soon mastered the available texts on medical theory, natural science, and metaphysics. He then supplemented his book learning by beginning a medical practice, in which he discovered a great deal more through empirical study.

Appointed as court physician to the sultan of Bukhara, Ibn Sina gained access to the latter's library, and by the age of 18 had consumed all its books. His study of logic led him to Aristotle (384-322 B.C.), whose writings initially upset him because he found himself unable to square the Greeks' pagan teachings with those of the Koran. One day, however, his reading of another Islamic scholar helped him unlock the seeming contradiction, and he was so overjoyed that he gave alms to the poor in gratitude.

When Bukhara's sultan died, Ibn Sina was forced to wander. At one point he earned a living lecturing on logic and astronomy at Jurjan near the Caspian Sea. Later he moved to Ray, near what is now the Iranian capital of Tehran, and there established a thriving medical practice. An attack by Turks forced him to relocate to Hamadan in western Iran, where he came under the protection of the emir Shams al-Daula and developed a following of devoted students.

With the death of the emir, Ibn Sina once again found his future uncertain, so he beseeched the ruler of nearby Isfahan for a governmental position. When the new emir of Hamadan learned of this, he had Ibn Sina imprisoned. Later, Ibn Sina gained his release and escaped to Isfahan, where he spent the remainder of his career in service to the emir, Ala al-Daula.

Ibn Sina wrote some 100 books on a variety of subjects, including an encyclopedia of nearly two dozen volumes. Most important to the scientific world were his *Kitab al Shifa* and *al-Qanun fi al Tibb*. The former was a philosophical encyclopedia informed by an Aristotelian worldview—though an Aristotelianism tempered with both Muslim and Platonic spirituality. Due to an uneven translation, the *Shifa* had a limited impact in the West, whereas the *Qanun*—known in Europe as the *Canon*—would have an influence too great to overestimate.

The *Canon* consisted of five books on medical theory, simple drugs, special pathology and therapeutics, general diseases, and pharmacopoeia. In it, Ibn Sina shows the influence not only of Aristotle—whose ideas pervade, even in the book's organizational structure—but of the ancient physicians Hippocrates (c. 460-c. 377 B.C.), Dioscorides (c. 20-c. 90), and Galen (c. 130-c. 200) as well. This synthesis of ancient and medieval ideas would have an enormous impact in Europe, and for this some of the credit must be given to Gerard of Cremona (c. 1114-1187) for his highly useful translation.

During his latter career in service to Ala al-Daula, Ibn Sina went on a number of military expeditions with the ruler, and thus had an opportunity make additions to the botany and zoology covered in the *Shifa*. Like many Muslims of his time, Ibn Sina owned slaves, one of whom turned against him when Ibn Sina was in his fifties. Hoping to steal his money, the slave put opium into Ibn Sina's food; but with his knowledge of medicine, he was able to treat himself and recover. The drug overdose weakened him, however, and in 1037 he had a relapse and died.

JUDSON KNIGHT

Pope John XXI (Peter of Spain)

c. 1215-1277

Spanish Physician and Pope

Perhaps the most famous of the medieval scientist-popes—men whose contributions to learning were as important as the fact that they held the most powerful throne in Europe—was Sylvester II, or Gerbert (945-1003). Equally noteworthy, however, was John XXI, who for most of his life was known as Peter of Spain. In addition to his *Treasury of Medicines for the Poor,* Peter wrote a work on optics, along with one of the medieval world's most influential textbooks on logic.

He was born Pedro Julião in Lisbon between 1210 and 1220, and though Portugal by then had established an identity separate from that of its larger neighbor, he would become known as Pedro Hispano, Petrus Hispanus, or Peter of Spain. During the late 1220s and early 1230s, Peter studied at the University of Paris, where his instructors included Albertus Magnus (c. 1200-1280). There he became intrigued by the writings of Aristotle (384-322 B.C.), particularly with regard to the natural sciences, and devoted himself to the study of medicine.

After earning his master's degree, in 1247 Peter was appointed professor of medicine at Siena, Italy, whose university was then newly founded. While at Siena, he wrote *Summulae logicales* (Small logical sums), destined to remain in wide use for the next three centuries. It may also have been during this period, before church affairs increasingly occupied his time, that Peter wrote *Liber de oculo* (Concerning the eye).

Beginning in 1261, Peter moved through the ranks of the church's upper echelons, and became a close associate of Teobaldo Visconti (1210-1276). Peter was serving as archdeacon of Vermuy in the Diocese of Braga, Portugal, when Teobaldo left for the Holy Land to accompany King Edward I of England on the Ninth Crusade. At this point the papacy had been vacant for three years, a situation that had a number of repercussions—most notably the fact that Marco Polo (1254-1324) and his father and uncle, in need of an audience with the pope, were forced to stall their departure on their now-celebrated journey. Teobaldo, who met the Polos in Palestine, shortly afterward received the startling news that *he* was to be the new pope. He rushed back to Italy, where as Pope Gregory X he appointed Peter his personal physician in 1272.

While serving Gregory in this capacity, Peter wrote his *Thesaurus pauperum* (Treasury of medicines for the poor.) The latter is, as its name indicates, a medical handbook for those who could not afford the care of a physician, and in time it would become a highly popular source of medical knowledge. The book paid special attention to herbal treatments, and despite his role as a priest, Peter even discussed plants a woman might use for contraception, including calamint, costus, pepper, rue, and sage. Elsewhere Peter discussed remedies for a wide variety of bodily ailments.

Though Peter was appointed Archbishop of Braga in the spring of 1273, Gregory soon had him consecrated as bishop. Gregory himself died in 1276, and was followed in quick succession by Innocent V and Adrian V. The latter reigned for only 39 days before dying, and a month later, the College of Cardinals elected Peter as Pope John XXI. (Actually, he was the twentieth pope to use the name John, but due to an error in Vatican record-keeping that went back to the tenth century, there was no John XX.)

As pope, John XXI dealt with a number of foreign-policy issues, including the ambitions of Charles of Anjou to control Italy, as well as conflicts between the English royal house and the papacy. Most intriguing, because of the potential consequences of these events, were his dealings with eastern monarchs. He received delegates from Abaga, Khan of Tartary, who beseeched the pope's assistance in a crusade against the Muslims and asked to have missionaries sent to his Central Asian khanate. Also, the pope received ambassadors from Byzantine emperor Michael Palaeologus, who along with the new patriarch of Constantinople, John Beccus, desired a rapprochement between the eastern and western churches; indeed, both Michael and his patriarch were willing to submit to the pope and the Roman Catholic Church.

But neither the conversion of the Tartars, nor the reunification of the Christian church, was to be: both undertakings were cut short by John's unexpected death. Eager to continue his scientific studies amid the fast-paced life of the papal residence, John (who like Gregory lived in Viterbo rather than Rome), had a special laboratory/apartment built on the back of his living quarters. He was working alone there when on May 14, 1277, the roof caved in on him, and he died six days later as a result of his injuries.

Soon after his death, John's prodigious medical learning made him a target for political enemies, who claimed that the late pope had dabbled in magical arts. His reputation suffered somewhat as a result, but has been rehabilitated in later centuries.

JUDSON KNIGHT

Guido Lanfranchi
1250-1306
Italian French Surgeon

Guido Lanfranchi is regarded as the founder of surgery in France. He was born in 1250 in Milan, Italy, and was educated by the famous physician William of Saliceto (1210-1277), who contributed to the renaissance in medical teaching at Salerno.

Lanfranchi (also known as Lanfranc of Milan, Lanfranc, or Lanfranco) had to leave Milan in a hurry in 1290 because he was on the wrong side of a dispute between the powerful Guelf and Ghibelline families. He escaped to Lyons, France, a city that was accepting of diverse ideas. Five years later he moved to Paris to become a professor, but this was cut short when it was discovered that he was married; professors were supposed to remain unmarried. However, Lanfranchi was not too unhappy about losing this job when he learned that teachers at the medical college were housed in the artist's center with straw covered floors, where the rats outnumbered the students. The independent college of Saint Come subsequently hired Lanfranchi and gave him a choice of more respectable quarters. Later, he did lecture at the university in Paris.

The study and practice of surgery as a learned art was just emerging, spreading to Bologna and, with the formation of the Confraternity of Saint Cosmos and Damian, moving into Paris during the thirteenth century. However, most of the surgery that was practiced at that time was still barbaric and done by untrained barber-surgeons.

Lanfranchi was convinced that medicine and surgery were inseparable; he firmly asserted that a good internist must know surgery and, likewise, a good surgeon must know medicine. He promoted the Hippocratic oath (the physician's pledge to do no harm) and attempted to make surgery as free from pain as possible.

Lanfranchi felt the need to document his experiences in a book. At first he penned the popular *Chirugia Parva,* ("Little book of surgery"). He later expanded it into his great work, the *Chirugia Magna,* ("Grand surgery"). Both books enjoyed widespread circulation and were much used. His beliefs were conservative, but also constructive and creative. For example, he was the first to use a silver tube in the windpipe to free objects that might cause suffocation. Both works were translated into French, Italian, Spanish, German, English, Dutch, and Hebrew.

His *Chirugia Magna* is divided into sections on general principles, anatomy, embryology, ulcers, fistulas, fractures, and luxations, baldness and skin diseases, phlebotomy and scarification, cautery, and diseases of various organs.

Among Lanfranchi's most important contributions was a chapter on brain injuries, in which he described symptoms and signs of concussion. He developed a system of percussion to determine skull injury. Tying a waxed string to a patient's tooth and holding it taut, he would pluck the string like a musical instrument and the tones produced would indicate skull injury.

He also referred to trephination, a procedure practiced by twelfth-century Arab surgeons, which involved boring a hole in the skull to treat various ailments. Lanfranchi alluded to this procedure in his writings and showed the instrument used to bore through the skull. He referred to "a trepane wich the brayn scalle schal be trepaned with."

Lanfranchi was one of the first to promote learned medicine in medieval Europe. During a time when surgery began to separate from the barbers as a profession, he was one of the first to write about the role and practice of the surgeon. He realized that surgeons must have specialized skills and suggested that exceptional mental and physical attributes were also essential for the task. He asserted that the surgeon must have a restrained and modest disposition, and wrote that "a surgian must have handes wel shaped, long smale fyngres and his body not quakying." Contemporary English translations of his writings emphasized the physical and mental qualities of the surgeon. Lanfranchi, a great humanist and person of high morals, also added that the surgeon must be versed in philosophy and logic and be able to read scriptures critically.

Still, Lanfranchi encountered well-established cultural hierarchies in medieval society that relegated the surgeon to the role of a menial subordinate who shared the duties of the pharmacist, dentist, and barber. This subordinate status is well illustrated in a famous woodcut by Johannes de Ketham in the *Fasciculus Medicinae* (1491). The picture shows a distinguished professor lecturing from his elevated "Chair of Medicine," while his surgical assistant, armed with an enormous butcher knife, performs the dissection on the operating floor down below.

Lanfranchi was not fond of his colleagues in Paris, whom he considered unqualified, illiterate, and clumsy mechanics. However, he was valued by his distinguished successors, Henri de Mondeville (1260-1320) and Guy de Chauliac (1300-1368). Lanfranchi later ended his brilliant career by becoming the personal physician of Philip the Fair.

EVELYN B. KELLY

Floridus Macer
fl. 1100s
French Herbalist and Physician

Not only are the dates of Floridus Macer's life uncertain, even his identity has been a subject of dispute. One of the few things known about him, in fact, is that he wrote *De viribus herbarum*, which became one of the most widely consulted texts on herbal medicines during the late medieval and early Renaissance periods.

A great number of historians link the pseudonym "Floridus Macer"—or more often, simply Macer—with the twelfth century French physician and bishop, Odo of Meung or Maung. Others, however, assert that he was in fact Marbode, bishop of Rennes. From these competing views, it is at least possible to discern the barest of outlines to Macer's biography: that he was French, a cleric (itself not a surprise, since most learned men of medieval Europe were), and that he lived some time during the twelfth or perhaps the thirteenth century.

Macer's fame rests entirely on *De viribus herbarum*, or "On the powers of herbs." Written in Latin hexameter verse, the book discusses the medicinal qualities of various plants; for instance, Macer recommended savin, juniper, and spearmint as contraceptives. Heavily influenced by ancient thinkers, including Hippocrates (c. 460-c. 377 B.C.), Pliny the Elder (A.D. 23-79), Dioscorides (c. 40-c. 90), and Galen (c. 130-c. 200), the book is a mixture of science and superstition. Many of the cures recommended by Macer were derived by experimentation and observation; others are pure witchcraft or magic. In any case, the use of rhymes not only distinguished the *Herbarum* from other texts; it also made it easy to teach, and thus the book became a favorite of instructors and students alike.

It appears that the name "Macer" is a reference to Aemilius Macer, a Roman herbalist mentioned in the writings of the ancient poet Ovid. Initially scholars attributed the *Herbarum* to this author, but already in the first known manuscripts of the work (which date from the thirteenth century), the book is described thus: "Of the power of herbs, the author being Odo, called Macer of the Flowers."

With each edition of the *Herbarum*, errors made in copying became compounded, but by 1373 a more or less definitive edition appeared. In 1477, the *Herbarum* became one of the first books ever printed, and later versions contained ever more discrepancies Nonetheless, the book continued to be a standard medical text until the eve of the Enlightenment.

JUDSON KNIGHT

Ali ibn Abbas al-Majusi
d. 994
Persian Physician

Ali ibn Abbas al-Majusi, known in Europe by his latinized name, Haly Abbas, was born in Ahvaz in the kingdom of Persia (now part of Iran). He was a highly influential physician during the tenth century. However, very little is known of his life other than the writings he left behind. He is most famous for authoring a work entitled *al-Maliki,* or *The Royal Book*, which remained the most important medical encyclopedia in the Arabic and European world for more than a century after its creation.

Al-Majusi apparently received his medical training from a private tutor. He also studied the works of ancient Greek physicians that had been translated into Arabic. He eventually went on to serve as court physician to King Adud ad-Dawlah (936-983) in the city of Baghdad (located in modern Iraq). In 981 this king founded the Adudi hospital, where al-Majusi worked.

In about 980 al-Majusi completed *The Royal Book*. It was widely used not only in Persia and other Arabic countries, but also in many parts of Europe after it was translated into Latin. *The Royal Book* is a collection of medical knowledge meant to be used as a reference for physicians. The first half deals with theories behind medical treatment. It covers such topics as anatomy (the structure of the body's parts) and physiology (the function of these parts). The second half of the book deals with medical treatments themselves, such as drugs and surgery. In fact, *The*

Royal Book was the first Arabic work to give detailed instructions regarding surgery.

For example, one operation al-Majusi describes is the treatment of an aneurysm—a bulge in a type of blood vessel called an artery that results from a weakening in the artery's wall. Al-Majusi states that surgery on large arteries should be avoided because of risk of death from blood loss. (Blood transfusions would not be widely used until the twentieth century.) For smaller arteries, however, al-Majusi advised physicians to cut open the patient's flesh to expose the blood vessel and then to tie it off at either end of the aneurysm with silk thread. A very similar procedure is used to treat aneurysms in small arteries to this day.

In the portion of the book dealing with medicines, al-Majusi states that the best way to determine the effects of a drug is to test it on healthy people as well as the sick and to keep careful records of the results. He offers a classification system for drugs based on their properties and also describes methods of preparing pills, syrups, powders, ointments, and so forth. Other chapters of the book discuss diet, exercise, and even bathing as they relate to health.

Much of the material in *The Royal Book* is based on the writings of Galen (130-200). Galen was an influential Greek physician, and by al-Majusi's lifetime more than 100 of his books had been translated into Arabic. Al-Majusi attempted to correct errors in Galen's works that had been revealed in the centuries since they had been written. Al-Majusi also wanted to arrange the information accumulated by Galen into a form that would be easy for physicians to use.

Al-Majusi's other main source was the Arabic physician ar-Razi (Rhazes; 865-923). Ar-Razi's most famous medical work was called the *Comprehensive Book.* Although al-Majusi clearly valued the *Comprehensive Book,* he criticized it for not being well organized and for being too long. Its length—more than 23 volumes—made it so expensive that almost no physician could afford to own a copy.

Al-Majusi's *Royal Book* solved these problems, as he organized and clarified ancient Greek and more recent Arabic medical knowledge into a single, more affordable book. However, his *Royal Book* was not entirely based on the work of others; al-Majusi also included his own observations. For instance, he stated that both arteries and veins carried blood. Most physicians of the time thought that veins carried blood, while arteries carried air. *The Royal Book* would also have a profound influence on Ibn Sina's (980-1037) *Canon of Medicine,* considered by many to be the most important medical book of the Middle Ages.

STACEY R. MURRAY

Henri de Mondeville
1260-1320
French Surgeon, Teacher, and Writer

Henri de Mondeville was one of the physicians and surgeons who established a great center for the teaching of medicine, surgery, and anatomy at the medieval university of Montpellier. Very little is known about the early life and education of Henri de Mondeville, who is remembered for his contributions to surgery and medicine. He was born in Mondeville, near Caen, or Emondeville, France about 1260. He died in Paris, about 1320. He is also known as Henry of Mandeville, Henricus de Amondavilla, Armandaville, Hermondavilla, Mondavilla, or Mandeville. According to clues found in his sur-

THE WEREWOLF DISEASE

The werewolf myth first appeared among the ancient Greeks and Romans and was well known throughout the Middle Ages. Thieves striking at night would sometimes wear wolf skins, knowing that the fear of werewolves would make their victims hand over money more quickly. Such behavior tended to reinforce tales of shape-shifting monsters.

Perhaps as a result of the myth's popularity, people with mental disorders who lived during this time were unusually susceptible to believing that they themselves were werewolves (even though in some places suspected werewolves were burned alive). In fact, one of the disorders al-Majusi discusses in *The Royal Book* is lycanthropy (which comes from the Greek words *lykos,* meaning "wolf," and *anthropos,* meaning "man"). Patients with this condition, he says, behave like dogs and lurk about graveyards at night. They may have yellowish skin, dark eyes, and bite marks on their legs. He considered lycanthropy to be incurable and classified it as a mental illness (rather than a supernatural one). Today, lycanthropy is still considered to be an actual, though very rare, disorder in which the patient believes he or she is a wolf or some other type of animal.

STACEY R. MURRAY

viving work and contemporary sources, Henri studied medicine and surgery at Montpellier, Paris, and Bologna in Italy. However, it is known that Henri studied theology and philosophy. He was apparently a cleric, and never married, but he never received a salary or grant as a clergyman. Because of his travels and his education, Henri de Mondeville can be considered a link between the Italian and French surgical and anatomical traditions of the thirteenth century.

In 1301, before embarking on his distinguished scholarly career, he served as surgeon to the armies of Philip the Fair. His skill as a surgeon was apparently well recognized and in demand, as demonstrated by the fact that he was in the service of royalty for the rest of his life. His royal patrons included Philip the Fair, Philip's brother Charles of Valois, and Louis X. Although Henri's association with royalty involved travel to various parts of England and France, he often complained that his service on behalf of the king did not provide sufficient financial compensation. By 1303 Henri was also teaching surgery and anatomy at Montpellier. Unfortunately, Henri felt that the demands of his royal patrons and the large numbers of patients and students who sought his attention interfered with the composition of his treatise on surgery.

In 1303 Guy de Chauliac, who is often called the father of French surgery, attended Henri's lectures. According to Guy de Chauliac (1300-1368), Henri had "demonstrated" anatomical matters with 13 anatomical illustrations, or charts of human anatomy. This approach was very innovative at the time and Mondeville is generally considered the first medical teacher to have lectured with the aid of illustrations. Some of the illustrations used in Henri's lectures have survived in the form of miniature copies. As might be expected, although the illustrations show some signs of a novel trend towards naturalization, they were not anatomically accurate.

In 1306, while lecturing in Paris, Henri began composing his *Cyrurgia*, which was to be a comprehensive textbook on surgery. Originally, Henri expected the *Cyrurgia* to encompass anatomy, the general treatment of wounds, special surgical pathology, injuries, fractures, poisons, and antidotes. However, his health had deteriorated significantly by about 1316 and he was unable to complete his planned treatise. Modern scholars have found about 20 surviving manuscript copies of Henri's *Cyrurgia*, or parts of it. Henri's writings and ideas were highly regarded by his contemporaries, but the Latin manuscripts for the *Surgery of Henri de Mondeville* were not translated and printed until the end of the nineteenth century. A German translation was printed in 1892 under the title *Die Chirurgie des Heinrich de Mondeville* and a French translation appeared in 1893.

With its clear and concise style, Henri's text reflects a practical, common sense approach to medicine and surgery, as well as considerable familiarity with classical and contemporary medical texts. The surviving manuscript versions of the *Cyrurgia* contain over 1,300 references to the works of some 60 different authors, including over 400 citations to the works of Galen (c. 130-c. 200). Clearly, Henri respected the work of the ancients, but he was willing to express his own opinions and did not regard even Galen as an infallible or final authority. Like Hugh of Lucca and Theodoric of Lucca, Henri consistently opposed deliberate efforts to make wounds suppurate, i.e. generate pus. An advocate of meticulous cleanliness, Henri urged surgeons to clean wounds without unnecessary probing. To make this approach more practical, he invented an instrument to extract arrows, advised surgeons to remove pieces of iron from the wounds by using a magnet, and developed improved needles and thread holders. His textbook was the first to insist that surgeons adopt two techniques that were essential to the development of modern surgery, that is, cleaning and suturing wounds. The primary goal of wound management, according to Henri, was to allow wounds to close and heal promptly, without infection. Therefore, Henri urged surgeons to keep instruments clean and rejected the use of irritant dressings.

Other intriguing aspects of Henri's work include his interest in the ancient belief in the healing power of light. The use of red light in the treatment of smallpox is primarily associated with the work of John of Gaddesden in 1510, but Henri had already used red light for this purpose. The importance of keeping the patient comfortable, cheerful, and hopeful was another significant aspect of Henri's teachings. Indeed, Henri thought hope was so important to recovery that the surgeon should even consider providing forged letters describing the death of the patient's enemy. The surgeon was urged to see that every aspect of the patient's regimen encouraged joy and happiness, including the use of music and jokes told by cheerful friends and relatives. Even when dealing with leprosy, a disease that Henri considered incurable, he advocated diligent, compassionate, and palliative treatment. No matter what the difficulty, the surgeon must banish anger, hatred, and sadness from the sickroom. Henri de Mondeville is regarded as the founder of the French surgical fraternity of the Collège de St. Côme and a major influence on late medieval medicine and surgery.

LOIS N. MAGNER

Mondino dei Liucci

1270-1326

Italian Anatomist

Mondino dei Liucci was the first documented person to dissect the human body in public. His *Anothomia* (*Anatomia*), written in 1316, became the most important textbook for dissection until Andreas Vesalius (1514-1564) wrote *De humani corporis fabrica.* Since *Anothomia* was the only book of its kind for many years, it outlived its expected "shelf life." As late as 1580, after both Berengario da Carpi (1460?-1530) and Vesalius had published improved anatomical works, Mondino's book still remained in use.

Mondino received his degree at Bologna in 1290 (1300 according to some sources). He came from a family of medical people; his father was an apothecary and his uncle a professor of medicine. Also known as Mondino de Luzzi and Mundinus, Mondino was a contemporary of Henri de Mondeville, whose name his is sometimes confused with.

Mondino's *Anothomia* was actually a "how-to" guide to dissection rather than a text on anatomy. It was divided into six parts and different from previous works because anatomy had never been the sole subject of a study. Prior to Mondino, anatomy was incorporated in surgical texts.

He described how to prepare the body and in which order to remove and examine organs. To dissect the most perishable organs first was an important consideration during a time when there was no refrigeration. Dissections were done infrequently, sometimes only once a year, and out of doors.

The other problem with preparing a handbook on anatomy was that in order to know the difference between usual, normal variation, and abnormal anatomy, one needed a great many specimens. Since bodies were only obtained after executions and were usually male, few comparative opportunities were available. As female criminals were less often executed and female anatomy remained misunderstood, the paradigm for understanding the human reproductive system was expressed in terms of male organs. The uterus was believed to have seven cells, a theological teaching rather than anatomic construct.

According to Singer, Mondino's language was convoluted and not very good. During a time when medical knowledge was borrowed from both Greek and Arabic sources, Mondino's nomenclature was confusing and made reading difficult. For example, the sacrum was referred to by several names that appear to be of Arabic origin (*alchatim*, *alhavius*, and *allannis*), but other bones of the pelvis, the hips, acetabulum, and the corpora quadrigemina (a part of the midbrain with four rounded eminences) were all referred to by the same word "anchoe." It is believed that Mondino introduced the term "matrix" for uterus.

He also used illustrations, an anomaly during a time when images of any kind were believed to be amateurish and fraudulent. The reason for this was related to theological discourse. Since graven images of God were forbidden in the Bible, and the body was a creation of God, any representation of it was considered blasphemy.

Unlike other pre-Vesalian anatomists, Mondino did not have access to the works of Galen (130-200). Subsequent anatomists criticized him for not adhering to Galenic tradition. But this was an advantage because Galen's observations were derived from animals, not humans. Galen's errors compounded as succeeding writers copied rote tradition as opposed to making their own direct observations.

Unlike his contemporaries, Mondino practiced his own form of direct observation. At the time it was customary for the learned professor to read from Galen (far from the odoriferous cadaver) while an *ostensore* pointed to the part and a dissector (a barber-surgeon) performed the cutting. A student or a group of students took turns holding up organs for all to see. Although the supervisor spoke Galen's words, he did not actually see what the student below was displaying. For example, the doctor-professor would read aloud that the liver has five lobes, but the student would hold up a four lobed liver (human livers have four lobes).

During Mondino's lifetime, no one was brave enough to ask why timeworn authority superceded what could be plainly seen, and refuted, by one's own eyes. It was believed that observation, because it came through the senses, could not be trusted, as the senses were thought to be base and degenerate; only spiritual matters or matters of God, not "man," could be trusted. In this way, Mondino was an iconoclast because he challenged long-established notions and wrote about what he observed firsthand.

LANA THOMPSON

Moses ben Maimon
1135-1204
Spanish Philosopher and Physician

Moses ben Maimon, also known as Maimonides (or by the acronym RaM-BaM), had a tremendous impact on Judaism and philosophy, as well as medicine. His most significant contribution, however, is arguably in the realm of Jewish law.

Maimonides was born to a prominent family in Cordoba, Spain. He spent much of his time studying with learned masters, including his father, Maimon. However, before turning 13, Maimonides was forced to live in the midst of war and persecution.

Cordoba, as part of Islamic Spain, enjoyed full religious freedom until a revolutionary Islamic sect known as the Almohads, or "the Unitarians," captured the city. At that time, the Maimons, along with many other Jewish families, were forced to practice their religion in secret while appearing as Muslims when in public. Maimonides continued his studies while he and his family remained in Cordoba.

Finally, after 11 years of practicing Judaism underground, the Maimons left Spain. The year was 1159 and they eventually settled in Fez, Morocco. Although this territory was also ruled by the Almohads, the Maimons felt that, because they were unknown there, it would be easier to maintain their Jewish faith undetected. This assumption proved to be untrue, as a mentor to Maimonides was arrested and executed after being found guilty of being a practicing Jew.

The family picked up and moved again, this time to Palestine. However, their stay did not last long. After only a few months they moved again to a town near Cairo, Egypt, where they were free to practice Judaism openly.

After finally finding a safe and comfortable home, Maimonides seemed to have found peace. It was only then that his personal life began to unravel. His father died shortly after the family's pilgrimage to Egypt. Then, Maimonides' younger brother, David, died in a shipwreck. It was the combination of these two deaths that left Maimonides as the head of the family. That meant that he was unable to continue his work as a rabbi because, at the time, that profession was considered a public service and was, therefore, an unpaid position. And so Maimonides turned to his medical knowledge and became a physician.

He did so well in his pursuits that he became the court physician to the Muslim sultan Saladin, all the while continuing his role as a leader in the Jewish community.

During his lifetime, Maimonides wrote several works that were very influential. His first, written in Arabic at the age of 16, was called *Millot ha-Higgayon,* or "Treatise on Logical Terminology." He also produced *Ma'amar ha'ibur,* or "Essay on the Calendar," in Arabic and Hebrew.

The first of his major works, which he began at age 23, was called *Kitab al-Siraj.* This was a commentary on the *Mishne,* a collection of Jewish law that dates back to the earliest times. It took Maimonides 10 years to complete this work.

One of his most notable pursuits was known as *Mishne Torah,* or "The Torah Reviewed." In this work, which he began writing in 1170, he explored Jewish law. It was made up of 14 books and took 10 years to complete.

Beginning in 1176, Maimonides spent the next 15 years on his subsequent work, *Dalalat al-ha'irin* (The guide for the perplexed), in Hebrew, *Moreh nevukhim.* This was his most daring effort, as it asserted three major views: God's will is not bound by nature, man cannot know God, and God is an intellectual entity.

Although he produced many other works during his lifetime, *Perplexed,* more than any other work by Maimonides, aroused opposition. His contemporaries saw his views as dangerous and heretical. However, his influence has stood the test of time, becoming a major part of religious philosophy for centuries to come.

AMY LEWIS MARQUIS

Abu Bakr ar-Razi
c. 865-c. 923
Persian Physician

Abu Bakr Muhammed ibn Zakariya ar-Razi (who was known as Rhazes in Europe) was an individual of sweeping intellect and broad interests. He is widely regarded as one of the finest physicians who ever lived. A prolific author, his writings on medical science influenced the practice of medicine throughout the West until at least the seventeenth century.

Ar-Razi was born near present-day Tehran, Iran, in what was then Persia. In his early education and experience, he showed interest and ability in a wide range of fields, studying and becoming accomplished in music, alchemy (and chemistry), philosophy, mathematics, and physics. But it was to the field of medicine that he chose to devote his greatest efforts, spending his life not only as a practicing physician, but also compiling accounts of medical treatments from a number of sources, testing them himself, adding his own innovations and improvements, and writing treatises and encyclopedias containing his results. He became widely known and respected. Early in the tenth century, he was chosen as physician to the court of Adhud Daulah, the ruler of Persia, and was asked to found and direct the new hospital in Baghdad.

Ar-Razi did creative work in a number of areas of medicine, especially obstetrics, pediatrics, ophthalmology, gynecology, and mental illness. His book *al-Judari wa al-Hasabah*, which dealt with the diseases of children, was the first published study of smallpox and chickenpox. It went through 40 editions and was still in use in 1866. He set up a special section in the Baghdad hospital for patients with mental illnesses and dealt at length in his writings with the effects that psychological factors have on physical wellness and illness. He also practiced an early form of psychotherapy.

He had modern ideas in other aspects as well. He believed that proper diet had an important effect on health. He tested the effects of experimental therapies on animals before administering them to humans. He used opium as an anesthetic during surgery and was the first to use alcohol for medicinal purposes. He gave the first description of the surgical procedure for the removal of cataracts. He also instituted a set of professional standards for the practice of medicine that emphasized the humane treatment of patients.

Ar-Razi was a prolific writer, producing more than two hundred books. Of these, the best known is *al-Hawi*, which is translated as "Comprehensive book." This work was a twenty-volume encyclopedia containing the details of Greek, Syrian, and Arabic medical knowledge, combined with his own practical experience. It was translated into Latin in 1279. Reprinted a number of times, it was still in use in the sixteenth century and was one of the works on which the development and practice of European medicine was based.

He also worked and published in other fields. He studied chemistry and chemical instrumentation and is credited with being the first to produce sulfuric acid. He codified the division of all things into the categories of animal, mineral, and vegetable, setting the stage for future divisions among the biological and physical sciences. He was also a philosopher of some renown, developing a system based on five eternal principles: creator, spirit, matter, space, and time. He regarded himself as a follower of Plato which put him somewhat at odds with the developing religion of Islam.

Ar-Razi was not only a preserver of ancient scientific and medical knowledge, he collected and tested it, added his own ideas and innovations, and passed it along in a better form for the use of his successors. He deserves to be recognized as one of the great physicians of history.

J. WILLIAM MONCRIEF

Abul Qasim Khalaf ibn al-Abbas al-Zahrawi

936-1013

Spanish Surgeon

Abul Qasim Khalaf ibn al-Abbas al-Zahrawi (latinized as Albucasis) was a Spanish Arab surgeon who made advances in the emerging art of surgery and wrote extensively on the surgeries that he performed. His reputation spread beyond Spain throughout the Muslim world; he was the personal physician to King al-Hakam II of Spain.

Al-Zahrawi was born in A.D. 936 in Zahra near Cordova, Spain. He is remembered for his 30-volume, 1,500-page medical encyclopedia called *Al-Tasrif li-man 'ajaza 'an al-to 'lif* which is translated as "The recourse of him who cannot compose (a medical work of his own)." In this publication he details what was known and what he had observed about the science of medicine. Three of the books cover in detail the subject of surgery. Some of the procedures detailed include surgery of the eye, ear, and throat, cauterization (applying heat to tissue usually to treat skin tumors or open abscesses), treatment for anal fistulas, and the need to drain blood from a chest wound. He also wrote on setting dislocated bones and fractures and the dissection of animals.

One of the procedures detailed was the removal of stones from the bladder. The surgeon was instructed to insert his finger into the rectum of the patient, move the stone down the neck of the bladder, and then make an incision in the wall of the rectum to remove the stone. An issue facing the patient and surgeon at this time was if the pain of the condition, in this case stones in the bladder, was greater or less than the pain of the surgical procedure. The risk of death from this surgery was also a concern.

Beyond surgical procedures, al-Zahrawi wrote on phlebotomy (the procedure of opening up a vein), obstetric procedures, midwifery, and child rearing. He was the first to give illustrations of medical and dental instruments, a number of which he designed, and 200 of which appear in the thirtieth volume of the *Al-Tasrif*. Three of the instruments that he invented were an instrument to examine the ear, an instrument to examine the urethra (the canal through which urine is discharged from the body), and an instrument used to remove objects from or insert objects into the throat. Al-Zahrawi's work continued to influence the practice of medicine for five centuries; he died in 1013.

MICHAEL T. YANCEY

Biographical Mentions

Abu Mansur Muwaffak ibn Ali Harasi
fl. 976

Muslim scholar who was the author of an important Persian pharmacological treatise *(Kitab al-Azhari al-lughawi)* that was composed about 970. A Latin translation survived as an epitome prepared in 1055. The Latin text was known as *Liber fundamentorum pharmacologiae*. The Latin manuscript was translated into German and published in 1893. The German edition *(Die pharmakologischen Grundsätze: Liber fundamentorum pharmacologiae des Abu Mansur Muwaffak bin Ali Harawi)* was reprinted in 1996, demonstrating the importance of this work for the history of Arabic medicine, botany, botanical medicine, materia medica, and pharmacology.

Albertus Magnus
c. 1200-c. 1280

German naturalist and philosopher who wrote highly detailed accounts of plants and animals in *De vegetabilibus et plantis* and *De animalibus*. He also had an impact on scientific study overall by promoting new investigation rather than the near-complete reliance on the work and thoughts of past scholars. The latter was common practice at the time. This contribution by Albertus Magnus is often cited as a turning point in the history of science.

al-Asmai
740-828

Arab philologist who wrote on zoology, anatomy, botany, and animal husbandry. Among his writings was the *Kitab al-insan,* concerning human anatomy. Al-Asmai is credited as the first significant Arab writer on zoology, botany, and animal husbandry, subjects destined to become highly popular among scientists under the Abbasid Caliphate (750-1258). During the ninth and tenth centuries, a number of Arab scholars consulted his work.

Bartholomeus Anglicus
fl. 1220s-1240s

English encyclopedist whose *De proprietatibus rerum* (1240) was the first important European reference work of the medieval period. The volume contained extensive information on the natural sciences, and included references to the work of Aristotle (384-322 B.C.), Hippocrates (c. 460-c. 377 B.C.), and other influential figures of the ancient world, as well as contemporary Arab, Jewish, and European scholars. Included in the book were extensive discussions of plant and animal life, as well as information on minerals. *De proprietatibus rerum* was later translated into a number of European languages, and appeared in more than 14 print editions before 1500.

Benedictus Crispus
fl. 681-c. 730

Italian scholar whose *Commentarium medicinale* contained extensive information on medicinal herbs. Benedictus, who served as archbishop of Milan, wrote the *Commentarium,* an elementary manual in verse form, in about 700. The book discusses a variety of plants, and their uses in treating illnesses.

Bertharius
fl. 857-884

Italian scholar whose writings are a principal source regarding medical treatments administered in medieval monasteries. Bertharius, who served as abbot at the renowned monastic center in Monte Cassino, wrote two treatises, *De innumeris remediorum utilitatibus* and *De innumeris morbis,* both concerning monastic medicine. Killed during an Arab invasion of the Italian peninsula in 856, he was later canonized.

Burgundio of Pisa
fl. c. 1150

Italian translator who rendered several texts by Galen into Latin. These translations, straight from Greek, represent a somewhat rare circumstance in the transmittal of classical knowledge to medieval Europe: more often than not, texts from ancient Greece first made their way to the Middle East, where they were translated into Arabic before eventually appearing as Latin texts in the West. Burgundio apparently traveled to Constantinople at one point, and on orders from Pope Eugenius III, also translated works by the theologian St. John Damascene.

Chen Chuan
fl. 700

Chinese physician who provided one of the first clinical descriptions of a key side effect relating to diabetes mellitus. The disease had first been identified in the West by Arataeus of Cappadocia (fl. A.D. 100s); however, Chen Chuan in c. 700 was one of the first scientists to note the sweetness of urine in patients suffering from that condition.

Constantine the African
c. 1020-1087

Moorish or Arab Italian scholar who pioneered in the translation of medical works from Arabic into Latin. Born either in Carthage (now in Tunisia) or in Sicily, then an Arab possession, Constantine spent most of his life in Italy. He studied at the University of Salerno, the first organized medical school in Europe, before entering Monte Cassino, the oldest and most important Western monastic center. Among the medical writers he translated were the Greek physicians Hippocrates (460-377 B.C.) and Galen (130-200); the Persian Ali ibn al-Abbas (Haly Abbas); and the Jewish physician Isaac Israeli.

Dunash ben Tamin
c. 900-c. 960

Jewish medical scholar and physician who wrote a mystical treatise on healing entitled *Sefer yetzira,* (The book of creation). Dunash lived in the city of Kairouan in Tunisia, a center of learning for the area that was particularly noted for its Jewish scholars, including Rabbi Jacob and Rabbi Nissim. Writings by Dunash include one of the first comparative studies of the Hebrew and Arabic languages.

Moses Farachi
fl. c. 1279

Italian translator who rendered a key text by ar-Razi (Rhazes; c. 865-c. 923) into Latin. Charles of Anjou, King of Sicily, commissioned Moses to translate a medical encyclopedia written by the Arab physician and philosopher. The translated encyclopedia, which in its Arabic form consisted of 10 volumes, appeared in Latin as *Liber continens* in 1279.

Jacoba Felicie
fl. 1300s

French healer who challenged gender hierarchy in medicine and the Statute of 1271. The faculty of the University of Paris brought charges against Felicie in 1322 for practicing medicine without a license. Despite her success with patients, they refused to grant her a license because she was a woman. The fact that she took pulses and studied urine was held against her because "medicine was a science transmitted from texts, not a craft to be learned empirically."

Gentile da Foligno
d. 1348

Italian physician and medical lecturer who was a pioneer in the study of hygiene. An instructor at the University of Perugia after it was established in 1308, Foligno wrote a commentary on the *Canon* of Ibn Sina (Avicenna; 980-1037). He also wrote an important treatise on hygiene in 1332, and in 1348 presented an essay on the Black Death then striking the city of Perugia—a plague that would take Foligno's own life. Few of his prophylactic suggestions or remedies resemble modern science: among his ideas was that the planets had caused the Plague, and that the drinking of potable gold would cure it.

Frederick II
1194-1250

German-Italian Ruler who contributed significantly to the foundation of European medicine and to medieval intellectual life. The grandson of Frederick I Barbarossa, he was elected Holy Roman Emperor in 1220, a position he held until his death. Frederick licensed the medical school at Salerno and decreed that all practicing physicians must have the approval of its faculty. Salerno was influential in the foundation of the medieval medical schools at Montpellier, Paris, Bologna, and Padua.

Jon Gardener
fl. 1400s

English gardener who wrote *The Feate of Gardening* (1440-50), the earliest account of principles of gardening written in English. The *Feate* lists over 100 plants and gives instructions for sowing and planting as well as tree grafting. It also provides advice on herb cultivation, useful particularly for apothecary gardens. While nothing is known of Gardener (whose name is also spelled Gardener, Gardiner, and Gardner), his document is a singular and practical text that provides a scientific base, rather than the usual folkloric one, for gardeners to build upon.

Giles of Rome
c. 1243-1316

Italian scholar who supported the doctrine of substance put forth by Thomas Aquinas (c. 1225-1274). An Augustinian hermit also known as Aegidus Romanus and Doctor Fundatissimus ("Best-Grounded Teacher"), Giles may have studied under Thomas in Paris between 1269 and 1272. Some years later, defending Thomistic substance in particular—and Scholasticism in general—against attacks from the higher clergy, he wrote *Theoremata de esse et essentia* (Essays on being and essence). He also produced commentaries on the *Organon*, writings concerning logic by Aristotle (384-322 B.C.).

Guarino Veronese
c. 1370-1460

Italian scholar and pioneer of Greek studies in Western Europe who translated numerous ancient works into Latin and edited several Roman writings. Sometimes known as Guarino da Verona, he studied at Constantinople, and returned to his homeland with a number of manuscripts. These he translated over the course of his career during which Guarino taught in various places around Italy. Among his works were translations of the geographer Strabo (c. 63 B.C.-A.D. 24) and the biographer Plutarch (c. 46-c. 119), and editions of the playwright Plautus (c. 254-184 B.C.), the historian Livy (59 B.C.-A.D. 17), and the scholar Pliny the Elder (A.D. 23-79).

Guglielmo da Saliceto
c. 1210-c. 1277

Italian physician and cleric, also known as William of Saliceto, widely regarded as the leading surgeon in Western Europe during the thirteenth century. Guglielmo practiced in Lombardy until 1270, when he took a position at the University of Bologna. He also served as city physician of Verona. In his *Cyrugia* (1275), he argued for a closer relationship between medicine and surgery—something that, while it seems obvious to modern people, was not clear at the time.

Hugh of Lucca
c.1160-c. 1257

Italian surgeon who helped to maintain the classical tradition of medicine and surgery during a period in which surgery reached its low point. Hugh of Lucca is primarily remembered for promoting the use of the soporific sponge as a surgical anesthetic and for his belief that wounds should be allowed to heal cleanly, without suppuration. Hugh's ideas and methods were preserved in the writings of his student, Theodoric, Bishop of Cervia (1210-1298).

Ibn al-Baitar
d. 1248

Arab botanist and pharmacist whose works included botanical and pharmaceutical encyclopedias. Ibn al-Baitar spent his early career in Spain before embarking in 1219 on an expedition across the North African coast, where he collected a number of herbs and medicinal plants. In 1224 he became chief herbalist for al-Kamil, governor of Egypt, whose conquest of Syria three years later made it possible for Ibn al-Baitar to collect plant specimens there as well. His *Kitab al-jami fi al-adwiya al-mufrada*, which remained in wide use until the late eighteenth century, discusses some 1,400 medicinal plants, more than 200 of which had not been previously identified. *Kitab al-mlughni fi al-adwiya al-mufrada*, his other major work, discusses a variety of drugs and their specific application to a variety of ailments.

Ibn Tufayl (Abubacer)
d. c. 1185

Arab physician and philosopher, widely regarded both for his medical work and for his novel *Hayy ibn Yaqzan*. The book's title is rendered in English as *Walk On*, and it presents Ibn Tufayl's mystical philosophy in a tale regarding a hermit who achieves enlightenment while living alone on a secluded island. The protagonist eventually comes to understand that the world is divided between a very few who understand spiritual truths through their unaided reason, a larger group who apprehend truth through spiritual symbols, and the great mass who simply accept the laws that emanate from those symbols.

Ibn Zuhr
1091-1161

Arab physician, known in Latin as Avenzoar, credited as the first parasitologist, who was the first to describe scabies. Ibn Zuhr, who practiced in Seville, emphasized observation and experimentation, and as a result of dissecting numerous corpses, was widely knowledgeable regarding anatomy. He gave the full clinical description of a tracheotomy operation, as well as of scabies, an itch caused by a mite. His most influential work, whose title in English is *The Book of Simplification Concerning Therapeutics and Diet,* was written at the request of Ibn Rushd (Averroës; 1126-1198), and remained in use among European scholars until late in the eighteenth century.

Isaac Israeli
832?-932?

Jewish Egyptian physician and philosopher who wrote on a number of subjects and greatly influenced European medicine. The dates of Isaac's life are uncertain—he could have been born as late as 855, and died as late as 955— but it is known that he lived to be about 100 years old. During his long career, he served the Fatimid caliphs in Egypt and wrote eight medical works on topics that included fevers, urine, pharmacology, ophthalmology, and various diseases and treatments. Translated by Constantine the African (c. 1020-1087), these writings would have a great impact in Western Europe. As a philosopher, Isaac is considered the father of Jewish Neoplatonism in the Middle Ages.

Ishaq ibn Hunayn
d. 910

Arab physician and translator who wrote several medical tracts. Son of Hunayn ibn Ishaq (808-873), he worked with his father on various translations. Among the original works by Ishaq was one whose title is translated into English as "The Salutary Treatise on Drugs for Forgetfulness."

al-Jahiz
776-868

Arab scholar whose works include one of the first and most significant encyclopedias of zoology produced in medieval Islam. Al-Jahiz produced more than 200 works, of which approximately 30 survive, and was one of the first Muslim scholars to write on scientific subjects in the language of laymen. His most famous work was the *Kitab al-hayawan* (Book of animals). Among the topics addressed in this seven-volume encyclopedia are animal communication and psychology, forms of social organization among ants and other creatures, and the impact of diet and climate on animal life.

John of Gaddesden
1280-1361

English surgeon who is perhaps best known as the model for Geoffrey Chaucer's "doctor of physick" character in *The Canterbury Tales.* Chaucer suggests that the doctor also relied on astrology and a knowledge of humors in his practice. John studied at Oxford and trained in medicine at the Montepellier school in Paris. He successfully treated King Edward II's son for smallpox. Writing in Latin, John wrote a medical text called *Rosa Anglica,* the English Rose.

Margery Kempe
1373?-1438?

English mystic and autobiographer whose dictated life story, *The Book of Margery Kempe,* is considered the first autobiography in English. A well-to-do matron who bore 14 children, Kempe started having visions and received the "gift of tears" shortly after her first pregnancy and began a series of pilgrimages that took her across Europe and throughout the Holy Land. Illiterate but opinionated, Kempe did not shy from theological argument and challenged clergy and commoner alike; but she also sought guidance, especially from the anchoress Julian of Norwich. Her disruptive behaviors and defiance of authority resulted in a trial for heresy, but she was found innocent. Following divine inspiration, Kempe dictated her *Book* as a record of her spiritual life. Idiosyncratic to a fault, the *Book* is a picaresque travelogue as well as a testament of faith.

Marbode of Rennes
1035-1135

French scholar who wrote on the medicinal uses of herbs and even gems. Bishop of Rennes (1061-81), Marbode wrote *Liber de lapidibus,* which may be the same work as his *De gemmarum.* In any case, both were on the spiritual and medicinal properties of precious stones. Regarding the diamond, for instance, Marbode wrote, "nocturnal spirits and bad dreams it repels.... Cures insanity.... For these purposes the stone should be set in silver, armored in gold, and fastened to the left arm." Elsewhere Marbode wrote that the herb artemis could cause an abortion in a pregnant woman.

Lapo Mazzei
fl. c. 1395

Italian lawyer and ancestor of physician and merchant Philip Mazzei (1730-1816). Lapo served as attorney for Santa Maria Nuova, a hospital in Florence established by the father of Beatrice Portinari, the heroine of Dante's *Divine Comedy.*

al-Muqtadir
fl. 908-932

Arab caliph who established a renowned hospital in Baghdad. By the time of al-Muqtadir's reign, the Abbasid Caliphate (750-1258) had begun its long, slow decline, and he was not a particularly notable leader. He did, however, take the important step of founding the Baghdad hospital in 918, and installing ar-Razi (Rhazes; c. 865-c. 923) as its director.

Nicholas of Salerno, also known as Nicolaus Salernitanus
fl. 1140

Italian physician who is traditionally considered the author of a famous medieval treatise on pharmacology and antidotes known as the *Antidotarium Nicolai.* This formulary is an important source of information about medieval pharmacy, materia medica, dentistry, and pharmacotherapy, and it was probably based on the anonymous *Antidotarius magnus* (composed between 1087 and 1100). That text was probably based on an older Salernitan manuscript by Constantine the African (Constantinus Africanus, 1020-1087). Nothing is actually known about the life of Nicolaus Salernitanus, who allegedly lived during the first half of the twelfth century. The name Nicolaus became associated with the *Antidotarium* and other traditional formularies after the twelfth century. The first printed edition of the *Antidotarium Nicolai* was published in Venice in 1471. A text containing a reprint of the first printing, along with a modern German translation, was published in 1976. The *Antidotarium* has also been translated into Dutch and Czech. Scholars continue to struggle with the dating and lineage of the work.

Nicola da Reggio
1280-1350

Italian physician who was involved in the recovery and translation of the works of Galen (130-200) into Latin. One of the texts translated by Nicola da Reggio (also known as Nikolaus of Rhegium) was Galen's *De temporibus morborum.* These translations stimulated interest in classical Greek medical theory and practice.

Notker Labeo
c. 950-1022

German scholar who translated numerous works from Latin into German. Widely regarded as an authority on a variety of subjects, including the natural sciences, Notker took the then unheard-of step of rendering scientific works and other ancient writings from Latin, the language of the educated few, into the vernacular. Among the ancient writers whose work he translated were Aristotle (384-322 B.C.) and Boethius (480-524). He also translated the *Bucolica,* a work on nature by Virgil (70-19 B.C.).

Tommaso Parentucelli, Pope Nicholas V
1397-1455

Italian scholar who, as Pope Nicholas V (r. 1447-1455), fostered learning in the sciences and arts by establishing the Vatican Library. The son of a physician, Tommaso rose through the ranks until, during the Council of Basel (1431-1449), he was elected to the church's leading position. After removing the antipope Felix V, he set about rebuilding much of Rome, including a renovation of the Vatican, and soon established the Vatican Library "for the common convenience of the learned." By the time of his death, the library contained some 1,200 manuscripts, one-third of which were in Greek, and most of which focused on classical learning. Within another quarter-century, the library contained more than 3,500 volumes, making it by far the largest in Europe, and securing Tommaso's lasting reputation as "the humanist pope."

Richard II
1367-1400

English king who in 1388 established the first sanitary laws in his country. Son of Edward the Black Prince and grandson of Edward III, whom he succeeded as king in 1377, Richard initially ruled through regents. During his reign, Wat Tyler's Rebellion (1381) was brutally suppressed in spite of efforts by the youthful king to end the trouble peaceably. In a 1399 struggle with Henry of Bolingbroke, the future Henry IV, he was deposed and placed in prison, where he died a year later. Richard's wife, Queen Anne (1366-1394), invented the sidesaddle.

Benedetto Rinio
fl. c. 1410

Italian herbalist who compiled one of the most well-known herbal encyclopedias of the era. Rinio's *Liber de simplicibus,* published in Venice in 1410, contained 440 illustrations by Venetian artist Andrea Amadio. The book discussed 450 domestic and 111 foreign varieties of herb, and included each plant's name in Latin, Greek, German, Arabic, several dialects of Italian, and Slavonic. Also included were notes on the best seasons for gathering a particular herb, and the parts of the plant most useful for medicinal purposes.

Roger II of Sicily
1095-1154

Norman king of Sicily who established the first compulsory medical examinations for European doctors. Roger, whose reign began in 1130, waged a successful war against the Byzantine Empire, created a civil service, and made his capital at Palermo a center of culture. In 1140, the year when he established the examination system, he granted the medical school at Salerno his official protection. Years later, in 1224, his grandson, Holy Roman Emperor Frederick II (1194-1250), put in place an even more rigorous examination process for physicians in Sicily.

Roger of Salerno
fl. 1100s

Italian surgeon and medical writer who was probably the greatest surgeon produced by the early school of Salerno. His treatise on surgery, *Practica chirurgiae*, which was probably composed about 1180, was one of the most important texts composed by a member of the School of Salerno. Roger taught and practiced at Parma before 1180, and, according to some authorities, Guido Aretino, one of his pupils, compiled the *Practica chirurgiae* from Roger's lecture notes and manuscripts about 1170. The *Practica chirurgiae* was considered a classic for at least three centuries and many of the surgical texts later associated with Salerno were probably based on Roger's own manuscripts and lectures. Roger recommended end-to-end sutures, mercurial inunction for chronic skin diseases, and seaweed for goiter. His writings supported the medieval belief that "laudable pus" (suppuration or pus formation) was an essential phase of wound healing. Roland of Parma, one of Roger's pupils, edited Roger's works and published the most important version of the *Practica chirurgiae* about 1230. Roger of Salerno has also been known as Roger of Palermo, Roger of Parma, Rogerius Salernitanus, Ruggiero Frugardi, and Roger Frugardi.

Roland of Parma
fl. 1200s

Italian surgeon and medical writer who edited the *Practica chirurgiae*, the great treatise on surgery that had been composed by his teacher, the distinguished twelfth-century surgeon Roger of Salerno. The edition of the *Practica chirurgiae* published by Roland about 1230 was the most important version of this treatise. This work was considered a classic for at least three centuries and many of the surgical texts later associated with Salerno were probably based on it. The manuscript has been preserved and was published in Rome in 1969 as the *Chirurgia Rogerii, per Rolandum Parmensem* (The surgery of Roger, edited by Roland of Parma, Codex Ambrosianus I, 18). Roland of Parma was an influential teacher and distinguished surgeon in his own right. (Parma, in northern Italy, was the site of an important medical school before the development of formal university medical education.) Roland later moved to Bologna, which soon surpassed Parma as a center of medical learning. Roland of Parma was also known as Rolando and Rolandino.

Sabur ibn Sahl
d. 860

Arab Christian physician who compiled a text that remained a dominant work in Islamic pharmacopeia for four centuries. Divided into 22 books, his *Aqrabadhin* (c. 850) presented a wide variety of antidotes for various illnesses.

Simon of Genoa
1270-1303

Italian physician and medical writer who compiled the manuscript that later became the first printed medical dictionary. His work on materia medica was highly respected and influential. Simon served as physician to Pope Nicholas IV. His *Synonyma medicinae, seu clavis sanationis* was published in 1471, but no complete copy of this work has survived. Simon of Genoa was also known as Simon Januensis or Genuensis.

Walahfrid Strabo
c. 808-849

Swabian scholar whose *Liber de cultura hortorum* describes a number of herbs and medicinal plants and their healing properties. A Benedictine monk, he served as tutor to Holy Roman Emperor Charles II (the Bald) in about 829, and wrote a number of theological works. He also revised the famous biography of Charlemagne (742-814) by the latter's contemporary Einhard (c. 770-840).

Ali Rabban al-Tabari
d. 870

Arab physician and teacher of ar-Razi (Rhazes; c. 865-923) whose *Firdous al-hikmat* (The paradise of wisdom), c. 850, established the foundations of Islamic medicine. The seven-volume encyclopedia, which was the first major work to bring together disparate branches of medical science, discusses the organs, diet, a wide variety of diseases, and numerous other topics.

Theodoric, Bishop of Cervia
1210-1298

Italian medical writer and surgeon whose writings helped to preserve the classical tradition of medicine and surgery and the ideas of his mentor, the surgeon Hugh of Lucca. Theodoric, like his mentor, is primarily remembered for using the soporific sponge as a surgical anesthetic, simple drugs for the treatment of wounds, and for the belief that wounds should be allowed to heal cleanly, without the formation of pus. Theodoric's Latin manuscripts have been translated into English under the title *The Surgery of Theodoric*.

Trotula
fl. c. 1097

One of the most famous physicians and teachers at Salerno, Italy, Europe's first medical school, Trotula headed the gynecology and obstetrics department and was a respected practitioner. Trotula was the earliest woman physician to write an influential medical work, *De Mulierum Passionibus* (Of the diseases of women). This book, traditionally attributed to her, may have been a later compilation of her work; it was regarded as an authoritative text in Italy up to the sixteenth century.

Bartolomeo da Varignana
fl. c. 1300-c. 1310

Italian physician and pioneer in forensic pathology. In Bologna in 1302, Varignana performed the first medico-legal autopsy— that is, an autopsy conducted for the purpose of obtaining information regarding a crime. He was later appointed to the faculty of the University of Perugia after the latter was founded in 1308.

Guido da Vigevano
fl. 1330s

Italian physician and inventor who became one of the first writers to include illustrations in a work on anatomy. In 1335, Guido presented Philip VI of France (r. 1328-50) with *Texaurus regis Franciae,* a guide both to health and military technology. Ten years later, he produced a medical manuscript that included depictions of human anatomy which by today's standards would be considered crude, but which were relatively detailed for their time. These drawings, from Guido's own hand, set a new standard for the use of illustrations to augment text in anatomical works, and proved a powerful force toward the reshaping of European studies in anatomy during the late medieval period.

Vincent of Beauvais
c. 1190-c. 1264

French encyclopedist who compiled one of the most wide-ranging reference works of the Middle Ages, with information on psychology, physiology, and other sciences. A Dominican who served in Paris, Vincent wrote a number of works, but none was on the order of his massive encyclopedia, *Speculum majus*. The work consists of 80 books divided into nearly 10,000 chapters, and addresses a variety of topics, including botany, zoology, psychology, physiology, and other scientific subjects, as well as history, philosophy, and theology.

William of Moerbeke
c. 1215-c. 1286

French scholar who translated numerous works by ancient Greek scientists, in many cases making the writings of thinkers such as Aristotle (384-322 B.C.) available for the first time to the scholars of medieval Western Europe. A member of the Dominican order, William was acquainted with such leading figures of the time as Thomas Aquinas (c. 1225-1274) and Witelo (c. 1230-1275), and served as consultant to Pope Gregory X during the Council of Lyons in 1274. In addition to Aristotle, he translated commentaries on the latter, as well as works by Hippocrates (c. 460-c. 377 B.C.), Ptolemy (c. 100-170), and the Neoplatonist Proclus (c. 410-485).

Yuhanna ibn Masawaih
924-1015

Assyrian physician and scholar who translated Greek medical works, mainly those of Galen and Hippocrates, into Arabic and Syriac. Amid the scholarly atmosphere of tenth-century Baghdad, Ibn Masawaih, a Christian given the Latin name of Mesue, or Masawaih the elder, joined a small group of Arab-speaking Christian and Muslim scholars whose translations of classical Greek texts gave rise to the Islamic revival of learned medicine. Ibn Masawaih also conducted original research, aided by his dissection of apes, and is credited with more than 30 original medical works, including *Daghal al-ain* (Disorders of the eye) and a pharmacological compendium.

Yuhanna ibn Sarabiyun (Serapion)
fl. 900s

Nestorian physician, known in Europe as Serapion of Alexandria, who authored *On Simple*

Medicines, a compendium that illustrated botanical cures and other medicinal practices. Serapion, a Christian who wrote in Syriac, worked among a group of Christian and Islamic scholars who translated classical Greek medical works into Arabic and Syriac, giving impetus to the rise of learned Islamic medicine in the tenth through twelfth centuries. Later, in the fifteenth century, *On Simple Medicines* was translated into Latin, along with other Arabic medical texts, providing a framework for the development of Western medicine in the late Middle Ages.

Bibliography of Primary Sources

Abano, Pietro d'. *Conciliator differentiarum.* A synthesis of Arab medicine, Greek philosophy, and the Catholic worldview that prevailed in the Europe of his day.

Albertus Magnus. *De Vegetabilibus* (c. 1250). This work became the most significant work on natural history written in western Europe during the Middle Ages.

Alderotti, Tadeo. *Consilia.* A series of case studies presented alongside medical opinions on each case. Also included was a record of the preventive measures applied by the physician, as well as both dietary and therapeutic treatments.

Alderotti, Tadeo. *Sulla conservazione della salute.* One of the first medical books in a modern language, a practical family physician's handbook.

Arnau de Villanova (Arnaldus Villanovanus). *The Treasure of Treasures, Rosary of the Philosophers.* The most famous of Arnau's works on alchemy.

Dunash ben Tamin. *Sefer yetzira* (The book of creation) (c. 940). A mystical treatise on healing by the Jewish medical scholar and physician.

Floridus Macer (pseudonym: Odo of Meung). *De viribus herbarum* (On the powers of herbs). One of the most widely consulted texts on herbal medicines during the late medieval and early Renaissance periods. Written in Latin verse and heavily influenced by ancient thinkers, the book is a mixture of science and superstition. Many of the cures recommended by Macer were derived by experimentation and observation, and others may have been intended as charms or incantations. In any case, the use of rhymes not only distinguished the *Herbarum* from other texts; it also made it easy to teach, making it a favorite of instructors and students alike.

Gerard of Cremona. Translated Ibn Sina's, or Avicenna's, *al-Qanun fi al tibb* (Canon of medicine). From the late twelfth to the late seventeenth century this was the single, most important medical text in all of Europe. It consisted of five books on medical theory, simple drugs, special pathology and therapeutics, general diseases, and pharmacopoeia. It was a synthesis of ancient and medieval ideas that had an enormous impact in Europe, and for this some of the credit must be given to Gerard of Cremona for his highly useful translation.

Guglielmo da Saliceto. *Cyrugia* (1275). Here Guglielmo argued for a closer relationship between medicine and surgery—something that seems obvious to modern people, but was not clear at the time.

Guy de Chauliac. *Chirurgia magna* (Grand surgery) (1363). A collection of eight books that reviewed the history and practice of surgery, arguing for its place in medical practice, rather than simply a tool of barbers and butchers. A chapter on anatomy stressed that it should be learned through hands-on dissection of the "recently dead" rather than taught through drawings.

Hildegard von Bingen. *Causae et Curae* (Causes and cures) (c. 1150-1160). A compact handbook on the etiology (causes), diagnosis, and treatment of diseases, it also contains chapters on human sexuality, psychology, and physiology. The detailed descriptions reflect a quality of scientific observation rare in that period.

Hildegard von Bingen. *Physica* (c. 1150-1160). An extensive pharmacopoeia (guide to healing agents) organized in nine books, cataloging the medicinal qualities and uses of plants, elements, trees, stones, fish, birds, quadrupeds, reptiles, and metals. The work displays a thorough knowledge of what was then known about the natural world, and gives a reliable picture of how medicine was practiced at monastic centers.

Ibn Rushd (Averroës). *Jami*, *Talkhis*, and *Tafsir*. A three-part commentary of Aristotle's work. The *Jami* was a brief summary of a text for beginning students. The *Talkhis* was a longer, intermediate analysis. The *Tafsir* was intended for advanced students and provided a comprehensive and original exegetical analysis. His commentaries exerted a great influence on thirteenth-century Aristotelianism. Their clarity was such that, in the Latin West, they were pedagogically preferred to Aristotle's primary texts.

Ibn Rushd (Averroës). *Kulliyat* (General medicine). Major medical treatise that incorporated portions of Ibn Sina's (980-1037) *Qanun fi at-tibb* (Canon of medicine) as supplemented by ibn Rushd's own original contributions. *Kulliyat* covers topics from organ anatomy and hygiene to the prevention, diagnosis, and treatment of diseases.

Ibn Rushd (Averroës). *Tahafut at-tahafut* (The incoherence of the incoherence). Ibn Rushd's most important philosophical work. It is a response to al-Ghazali's attack on the efficacy of rational inquiry in religious matters. Ibn Rushd rejected this position, arguing that natural reason is adequate for any intellectual investigation.

Ibn an-Nafis. *Kitab al-shamil* (Comprehensive book on the art of medicine). Based on 300 volumes of notes, of which he published only 80. Until 1952 this massive work was thought to be lost, but it was found and catalogued among Cambridge University Library Islamic manuscripts.

Ibn an-Nafis. *Majiz al-Qanun.* A summary of Ibn Sina's *Canon of Medicine* that was popular in his day. The book had an important section on surgery that defined three stages for each operation. The book also gives detailed descriptions of surgeons' duties and relationships between patients, surgeons, and nurses. Nafis also discusses bed sores, posture, bodily movement, and manipulations of instruments, using case histories to illustrate points.

Ibn an-Nafis. *Sharh Tashrih al-Qanun.* A commentary on the anatomy of Ibn Sina, in four books containing commentaries on medicines, drugs, and systemic diseases. In one famous passage the author describes pulmonary blood circulation, directly contradicting Galen. Nafis also proposed that the lungs were made of the bronchi and branches of small arteries and veins connected to porous flesh. Several translations were made from the Arabic manuscripts into Latin around 1547.

Ibn Massawaih. *Daghal al-ain* (Disorder of the eye) (c. 825). Ophthalmology flourished in Islamic medicine with works such as this one by Ibn Massawaih.

Ibn Sarabiyun, Yuhanna (Serapion of Alexandria). *On Simple Medicines* (900s). A compendium that illustrated botanical cures and other medicinal practices. In the fifteenth century, *On Simple Medicines* was translated into Latin, along with other Arabic medical texts, providing a framework for the development of Western medicine in the late Middle Ages.

Ibn Sina (Avicenna). *al-Qanun fi al tibb* (Canon of medicine). From the late twelfth to the late seventeenth century this was the single, most important medical text in all of Europe. It consisted of five books on medical theory, simple drugs, special pathology and therapeutics, general diseases, and pharmacopoeia. It was a synthesis of ancient and medieval ideas that had an enormous impact in Europe, and for this some of the credit must be given to Gerard of Cremona for his highly useful translation.

Ibn Sina (Avicenna). *Kitab al shifa* (Book of healing). A philosophical encyclopedia shaped by an Aristotelian worldview, tempered with both Muslim and Platonic spirituality. Due to an uneven translation, it had a limited impact in the West.

Lanfranchi, Guido. *Chirugia magna* (Grand surgery). Based on the author's popular *Chirugia parva* (Little book of surgery), this larger, more expansive version also enjoyed widespread circulation and was much used. The book is divided into sections on general principles, anatomy, embryology, ulcers, fistulas, fractures and luxations, baldness and skin diseases, phlebotomy and scarificaiton, cautery, and diseases of various organs. Like its predecessor, it was translated into French, Italian, Spanish, German, English, Dutch, and Hebrew.

Majusi, Ali ibn Abbas al- (Haly Abbas). *al-Maliki* (The royal book, c. 980). Medical encyclopedia that remained the most important in the Arabic and European world for more than a century after its creation. It was widely used not only in Persia and other Arabic countries, but also in many parts of Europe after it was translated into Latin. This work was the first Arabic work to give detailed instructions regarding surgery, and it clarified and edited the works of ar-Razi and Galen.

Marbode of Rennes. *Liber de lapidibus* and *De gemmarum* (c. 1070). These may be the same work. In any case, both were on the spiritual and medicinal properties of precious stones. Regarding the diamond, for instance, Marbode wrote, "nocturnal spirits and bad dreams it repels.... Cures insanity.... For these purposes the stone should be set in silver, armored in gold, and fastened to the left arm." Elsewhere Marbode wrote that the herb artemis could cause a miscarriage in a pregnant woman.

Mondeville, Henri de. *Cyrurgia* (Surgery). A comprehensive textbook on surgery encompassing anatomy, the general treatment of wounds, special surgical pathology, injuries, fractures, poisons and antidotes. Surviving manuscript versions contain over 1,300 references to the work of some 60 different authors, including over 400 citations to the works of Galen, although the work did not regard any of them as an infallible or final authority. His textbook was the first to insist that surgeons adopt two techniques that were essential to the development of modern surgery, that is, cleaning and suturing wounds.

Mondino dei Liucci. *Anothomia Mundini* (Anatomy) (1316). A guide to dissection rather than a text on anatomy, this was the most important textbook for dissection until Vesalius wrote the *Fabrica.* It was also the first book dedicated solely to the study of anatomy.

Moses ben Maimon. Known in the West as Maimonides. *Dalalat al-ha'irin* (The guide for the perplexed). Maimonides' most daring work. It asserted three major views: God's will is not bound by nature, man cannot know God, and God is an intellectual entity. Although he produced many other works during his lifetime, this work aroused the most opposition. His contemporaries saw his views as dangerous and heretical. However, his influence would stand the test of time, becoming a major part of religious philosophy for centuries to come.

Razi, ar- (Rhazes). *Al-judari wa al-hasabah* (A treatise on smallpox and measles). A landmark in the development of the concept of specific disease entities and the value of diagnostic precision. Translated into Latin, the book had a profound influence on the treatment of smallpox in Europe.

Razi, ar- (Rhazes). *Kitab al-hawi* (Comprehensive book). A 20-volume encyclopedia containing the details of Greek, Syrian, and Arabic medical knowledge, combined with the author's own practical experience. Translated into Latin in 1279 as *Liber continens*, it was reprinted a number of times, and was still in use in the sixteenth century. The book was fundamental to the development and practice of European medicine.

Robert of Chester. *Liber de Compositione Alchemiae* (The book of the composition of alchemy) (1144). This book was a translation of Jabir ibn Hayyan's (c. 721-c. 815) alchemical text *Kitab al-Kimya.* This work was the first Latin translation of an Arabic alchemical work to appear in western Europe.

Theodoric. *Chirurgia* (Surgery). Based on the teachings of Hugh of Lucca (1160-1257). The book featured a treatise on wound management that debunked the previous century's surgical opinion that the production of pus in a wound was desirable. Hugh and Theodoric felt that wounds should be immediately joined by cautery, as recommended by Islamic surgeons.

Zahrawi, al-. *Al-Tasrif li-man 'ajaza 'an al-to 'lif* (The recourse of him who cannot compose [a medical work of his own.]) A 30-volume, 1,500-page medical encyclopedia that discusses, among other topics, surgical techniques, setting dislocated bones, and the dissection of animals.

JOSH LAUER

Mathematics

Chronology

820 Al-Khwarizmi, an Arab mathematician, writes a mathematical text that introduces the word "algebra" (*al-jabr* in Arabic), as well as Indian numerals, including zero; henceforth these are mistakenly referred to as Arabic numerals.

c. 825 Al-Hajjaj ibn Yusuf makes the first Arabic translation of Euclid's *Elements*, which will be much more widely studied in the Islamic world than in the West for many centuries.

870 Thabit ibn Qurra, who translated most of the major Greek mathematicians into Arabic, publishes a treatise that becomes the first example of original mathematical work in the Middle East.

c. 900 Abu Kamil Shudja applies algebra to geometry; Albategnius applies spherical trigonometry to astronomy, and establishes several new theorems; and Qusta ben Luga al-Ba'labakki develops the use of tangents and cotangents.

c. 950 One of the first significant Latin translations of an Arabic text by al-Imrani, appears in the West.

c. 988 Arab mathematician al-Kuhi solves, and discusses the conditions of solvability for, equations higher than the second degree.

c. 1020 Al-Karki writes what is considered the greatest treatise on algebra produced by an Arab scholar, and also becomes the first to work with higher roots.

1175 Gerard of Cremona, an Italian scholar, translates Euclid's *Elements* and Ptolemy's *Almagest* into Latin for Western scholars.

1202 In *Liber Abaci*, a work that will prove instrumental in putting an end to the old Roman system of numerical notation, Leonardo Fibonacci explains the use of "Arabic" numerals.

1248 Chinese mathematician Li Yeh is the first to represent a negative number, drawing a cancellation mark across a numerical symbol.

c. 1340 Johannes de Lineriis of France becomes the first to designate fractions by placing the numerator over the denominator, with a horizontal line dividing the two.

c. 1400 Persian mathematician Al-Kashi is the first to use decimal fractions.

1436 Italian artist Leone Battista Alberti lays the groundwork in his *Trattato della Pittura* for a new science of perspective founded on rigorous geometrical principles; nine years later, his *De re aedificatoria* introduces the use of multiple mathematical considerations into architecture.

Overview: Mathematics 700-1449

During the classical period of mathematics in the ancient world, many fundamental branches of mathematics originated and were developed to a remarkable degree. However, due to various catastrophes, much of this ancient mathematical learning was lost, particularly in medieval Europe. The period between 700 and 1449 was a time of recovery for European mathematics, with many ancient texts restored, copied, and translated. However, it was also a period of innovation in many other parts of the world, and the mathematical knowledge of many cultures were exchanged. By chance, as well as design, European mathematics was to become the greatest benefactor of this process of recovery and melding, and social and cultural forces would give European mathematics the impetus to rapidly develop in the following centuries.

Mathematics has a long history, dating back at least as far as the Mesopotamian culture (in modern-day Iraq) over 4,000 years ago. Soon after, many other ancient cultures also began to invent addition, multiplication, and develop their mathematics. The Babylonians created a number system, sexagesimaal (base-60), developed a practical geometry, elementary algebra-like calculations, and fractions. The Egyptians used a decimal system (base-10), and developed a practical geometry that enabled them to construct and determine the areas of many simple figures, most famously demonstrated by the construction of the pyramids. Greek mathematics went beyond the practical elements of other ancient traditions and looked at abstract, philosophical, and mystical aspects of numbers. Greek scholars developed mathematics to levels unsurpassed for many centuries. The influence of Greek ideas was so dominant that many of the mistakes in their writings went unquestioned by later thinkers. The Roman Empire was primarily concerned with practical forms of mathematics, but preserved Greek writings. However, with the collapse of the Roman Empire in the fourth century much of the ancient mathematical knowledge of the Greeks was destroyed, lost, or scattered.

In other parts of the world mathematics was still actively studied. Chinese mathematics began to advance greatly from the seventh century on. Cubic and quadratic equations, as well as astronomical and applied mathematics, were significantly developed. In the thirteenth century, a number of mathematicians wrote important summaries of Chinese mathematics. Ch'in Chiu-Shao (1202?-1261?) wrote on the solution of numerical equations. Yang Hui (c. 1299) recorded work on arithmetical progressions, proportions, simultaneous linear equations, quadratic equations, and others areas. Chu Shih-Chieh (fl. 1280-1303) summarized many earlier discoveries in his writings. The Chinese did not study in isolation, absorbing many Hindu ideas and influencing Japanese mathematics. However, after the thirteenth century a number of political and social factors discouraged further mathematical development in this region.

On the Indian subcontinent there was also a long tradition of mathematical study. Mathematics was mainly studied as an aid to astronomy, and there were also mystical and poetical elements. There is evidence of a strong Greek influence on early Hindu mathematics, but the stress was on arithmetic, rather than geometry. Hindu innovation and development appears to have peaked before 700, but there were a number of important books and individuals in the following centuries. The writings of mathematicians such as Mahavira (c. 850) and Sridhara (c. 850) display the wide range and complexity of Hindu mathematics, including fractions, finding squares and cubes, and the calculation of areas. In the twelfth century Atscharja Bhaskara's (1114-1185) work was only a little more advanced than earlier texts, and represents the last great summary of Hindu mathematics before the Muslim conquest of India. While Hindu mathematics waned, the influence of their developments spread far and wide. Most famously, the numeric system we use today is derived from Indian mathematics, as is the concept of the zero.

In the years between 700 and 1449, the most dynamic and flourishing mathematical tradition was that of the Arab world. The expansion of the Islamic faith, by conquest and conversion, went as far west as Spain, and as far east as India. Arab rulers encouraged specialists of all types, and their courts became centers of learning and research in medicine, astronomy, mathematics, and other disciplines. Arab scholars were ideally placed at the crossroads of many other mathematical cultures, and they absorbed ideas from the Babylonians, the Greeks, the Hindus, and others. Muhammad ibn-Musa al-Khwarizmi

(780?-850?) helped popularize Hindu mathematical concepts such as the zero, fractions, and the Hindu numeral system. His writing was so influential that when Hindu numerals were introduced to Europe, by way of translations of his work, they were called Arabic numbers. Al-Khwarizmi also gave us the word *algebra,* and his early work in this field was carried on by many later court mathematicians, such as al-Karkhi (c. 1020). The influence of Greek mathematics led to developments in the fields of geometry, astronomy, and trigonometry by scholars such as Abu al-Wafa (940-998).

The city of Baghdad was the earliest center of Arab mathematical study, but with the expansion of the Muslim world other important centers sprang up in Egypt, Morocco, and Spain. In Egypt, important work on the volumes of paraboloids was done. Moroccan mathematicians excelled in the fields of conics and astronomy. But by far the biggest Arab center of learning outside Baghdad was in Cordova, Spain. Schools and libraries were founded there in the tenth century, and Spain became the main point of contact for Arab and European scholars. However, by the end of the fifteenth century the Spanish territories had been lost, and political and social concerns in the Arab world led to a decline in mathematical research.

In Europe little mathematical knowledge remained after the fall of Rome. Over the centuries more texts were lost, to disasters and the decay of time. European mathematicians came to rely on a small collection of Latin translations. Translators, such as Boethius (480-524), simplified the ancient Greek texts, leaving out difficult material, which often included explanatory figures, numbers, and calculations, making the texts hard to follow. In the Middles Ages mathematical studies included arithmetic, music, geometry, and astronomy, but were not popular subjects. However, as the economic and political stability of Europe improved, European society developed a new need for mathematics. The lost ancient mathematical texts were reintroduced by contact with the Arab world. However, many of these recovered Greek writings suffered from the multiple translation process they had undergone over the centuries, from Greek, to Arabic, and finally to Latin. Yet slowly mathematical knowledge caught up with that of the ancient Greeks.

New innovations were also introduced from the East, though not without resistance. Gerbert of Aurillac (946?-1003) attempted to introduce the abacus and the Hindu-Arabic numeral system. Later attempts to popularize the Hindu-Arabic numeral system by Adelard of Bath (1090-1150) and Leonardo of Pisa (Fibonacci) (1175-1250) also met with resistance, but eventually they replaced Roman numerals due to the ease and quickness of calculation they afforded.

New fields of study were created in Europe, such as the study of observed motion (kinematics). A number of fourteenth-century scholars used mathematical ideas to help describe physical actions. Thomas Bradwardine (1295-1349) wrote on accelerated motion, instantaneous velocity, and force, and his work was expanded upon by Nicole of Oresme (1323-1382), and others. From this simple beginning, the eventual integration of physics and mathematics would become a cornerstone of modern scientific knowledge.

European trade began to grow rapidly late in the Middle Ages, producing a need for more numerate clerks and administrators. New innovations, such as the plus (+) and minus (-) signs were introduced to make bookkeeping easier. Mathematics was also pursued for pleasure. Recreational mathematics provided intellectual entertainment, and helped popularize the study of numbers. Even chess was considered a mathematical pursuit, and one worthy of the noblest gentleman.

The years immediately after 1450 saw Europeans rapidly develop many areas of learning, including mathematics. The Renaissance, as this period has come to be known, is considered by many the beginning of the modern period of world history, with advances in science and technology that enabled European domination of much of the globe. However, without the recovery of ancient knowledge that occurred in the years between 700 and 1449, the spectacular developments of the Renaissance could not have occurred. Through contact with other cultures, and especially the absorption of Arab ideas and innovations, European learning in fields such as mathematics was able to go beyond the work of ancient scholars. New fields of study unknown to the Greeks were opened, leading to such developments as the calculus of Isaac Newton (1642-1727) and Gottfried Leibniz (1646-1716), which would revolutionize both mathematics and science.

DAVID TULLOCH

Islamic Mathematics in the Medieval Period

Overview

During the medieval period Islamic mathematicians enjoyed a dynamic and vibrant profession that, contrary to many popular teachings, made significant contributions to their field that continue to affect the way mathematics is practiced today. They did not simply preserve the glories of Greek mathematics and transfer some concepts from Hindu mathematicians to Europe. Rather, they developed sophisticated systems of algebra, introduced many now-standard mathematical notations, and helped mathematics move away from the largely geometrical formulations of the Greeks to a more symbolic and abstract structure that is much closer to the manner in which mathematics is practiced today.

Background

The first great flowerings of mathematics occurred in Babylonia, Egypt, and Greece. With the passing of time, these cultures either vanished or became assimilated into the Roman Empire. In particular, the Greek tradition in mathematics helped establish the form of European and Roman mathematics for several centuries.

To the east, China, Persia, and India were also developing their own mathematical traditions, largely independent of one another and wholly independent of Europe. With the fall of Rome in the fifth century, progress in mathematics largely stopped in Europe, although Asian cultures remained unaffected.

In the seventh century, in the Middle East, Muhammad founded the Islamic religion. Within a few decades, Islam had spread throughout the Arabian peninsula and, within a century, it became one of the dominant religions in the Middle East. Although followers of Islam did not originally express much of an interest in study, this quickly changed and they set about copying, translating, and adding to the mathematical knowledge of Greece, India, Persia, and (when possible) China.

It was long thought that Islamic scholars did little more than simply translate Greek texts, holding them until Europe emerged from its intellectual interregnum during the Middle Ages. In reality, Islamic mathematicians did far more than this. Specifically, they invented the algebra that most learn in school today, made significant advances in the field of trigonometry, and helped form a synthesis of mathematical ideas, fusing the best of Greek mathematics with important Hindu and Persian concepts to create a mathematical structure that was far grander than what they had inherited.

For several centuries, the Middle East, specifically Baghdad, was the world's center for mathematics. Islamic scholars such as al-Khwarizmi (780?-850?), Thabit ibn Qurra (836?-901), Abu al-Wafa (940-998), al-Kharki, al-Biruni (973-1048), Omar Khayyam (1048?-1131), and others literally reshaped the entire field, benefiting all of mathematics and much of science.

Impact

The most succinct way to describe the impact of Islamic mathematicians is to note that they completely changed the "flavor" of mathematics during their dominance in the field. One example of this is the change from the largely geometric formulations of the Greeks to the largely symbolic formulations that we use today. In other words, most of the problems addressed by Greek mathematicians were solved by describing them mathematically and then trying to find a geometric solution, such as the famous solution to the Pythagorean Theorem, which shows a triangle with a square drawn on each of the triangle's sides. In addition, the introduction of numerals, borrowed from Hindu culture, greatly facilitated the performing of calculations (try to imagine solving a problem that goes "first you have to find the product of twenty-one and fifteen" instead of simply reading $21 \times 15 = ?$). Islamic contributions to trigonometry also served to meld the Greek and Hindu approaches, and their contributions to number theory were equally impressive.

Of prime importance was the development of algebra (originally *al-jabr*) by the great Arab mathematician al-Khwarizmi (770-840). Simply put, algebra is a way of letting symbols represent numbers or other concepts, and manipulating those symbols according to a fixed and logical set of rules. Most learn aspects of al-Khwarizmi's algebra in school as they learn, for example, to solve an equation to find the value of x. Other algebraic problems include finding the "roots" of an equation, that is, where a particular line or curve will intersect the x-axis on a piece of graph paper. In general, any problem in which symbols are manipulated in order to arrive at a final answer is a type of algebra problem. Extracting a square root is yet another example of an algebraic problem.

This was a crucial step in the advancement of mathematics because it freed mathematicians from either performing interminable calculations for every problem solved or from having to try to construct a geometrical representation of the problem and trying to solve that. Although the Greeks brought geometrical mathematics to surprising levels of sophistication, this methodology is inherently limited because one can only solve problems that can be diagrammed in a way that lends itself to solution. The majority of mathematical problems are not amenable to solutions in this manner and, in any event, it is an indirect way to work in most cases. By abstracting the problem and using symbols to represent the important variables and concepts, it is possible to carry the problem to an analytical (or "symbolic") solution and, as the final step, to "plug in" actual numbers to find a numerical solution to the problem. This is the way that most scientific and engineering calculations are performed, and it is a powerful conceptual tool to be able to bring to bear on any problem.

The introduction of the Hindu numbering system was hardly less important to the practice of mathematics. By using a single symbol for each of the 10 numerals and using zero as a placeholder, mathematical calculations suddenly became almost simple. For example, consider the following problem, written in both Roman and Hindu notation:

CXXVI multiplied by XII = ?

$126 \times 12 = 252 + 1260 = 1512$

In both cases, of course, the final answer is the same. However, solving this problem using Roman numerals is almost impossible; better to use an abacus or some other multiplication aid. In addition to aiding in computation, Hindu numerals were just easier and less ambiguous to read. One might erroneously read XIII as XII, for example, or might write XI as IX, errors far less likely to occur when each numeral is unique.

Islamic mathematicians seemed almost giddy with the possibilities raised by these two concepts. Throughout the mathematics of this era we find impossibly long tables filled with numbers, each the result of painstaking calculations. Some mathematicians actually completed these calculations in base 60 as well as in base 10, and one table contained nearly a half million entries. Although, by today's standards, this sort of work seems pointless (not to mention mind-numbingly dull), it was important to undertake at that time. Future mathematicians were able to use these tables of numbers in much the same way that, until the advent of cheap and powerful computers and calculators, statisticians used tables of statistical data, scientists used tables of logarithms and trigonometric functions, and so forth. By devoting their lives to calculating and completing these tables, Islamic mathematicians did a great service to their colleagues and to those who were to follow.

Islamic contributions to number theory are far from being the only other contribution to mathematicians during this intellectual golden age. Indeed, entire books have been written on this one subject. However, it is fair to say that Islamic contributions to number theory during the medieval period were also both important and impressive.

In some respects, it is difficult to distinguish between pure number theory at the level practiced by medieval Islamic mathematicians and the algebra they introduced. Simply put, number theory is the study of the properties of numbers. For example, the search for more digits in the number *p*, the search for a pattern among these digits, or the proof that no such pattern exists are examples of number theory. In a more practical area, number theory also deals with developing methods of extracting the roots of numbers (such as square or cube roots), or with finding pairs of numbers that meet certain criteria.

Number theory is often thought of as mathematics at its purest because, for the most part, discoveries in number theory have little practical utility and are sought instead for their mathematical and conceptual beauty. Although Islamic mathematicians did not originate this area of inquiry, they certainly pursued it vigorously, discovering many properties of prime numbers and several interesting and important theorems that would not be proven for several centuries, and asking questions that helped bring to number theory a degree of mathematical and intellectual rigor.

In general, it can be claimed that Islamic mathematics represented a sort of middle road between the mathematics of Greece and that of India, with a particularly Islamic flair. Islamic mathematicians translated virtually every surviving Greek text on mathematics and they were certainly aware of the Greek discoveries and formulations of problems. In fact, the earliest Muslim text describing algebra describes problems that could only have been translated from the Greeks. To this, they added the numbering system used by mathematicians in the East, as well as some Hindu mathematical techniques. However, their approach was more analytical than that of the Greeks and less mystical than that of the Hindus.

The golden age of Islamic mathematics lasted for only a few centuries, from about the eighth century through the twelfth century. There were several subsequent resurgences, and Islamic mathematics did not go into a decline until about the fifteenth century, with the death of al-Kashi (1380?-1429), the last great Islamic mathematician of this era. By that time, however, Europe was recovering from its long intellectual stagnation, ready to accept back the mathematics it had ignored for so long.

P. ANDREW KARAM

Further Reading

Al-Daffa, A.A. *The Muslim Contribution to Mathematics.* Atlantic Highlands, NJ: Humanities Press, 1977.

Boyer, Carl, and Uta Merzbach. *A History of Mathematics.* New York: John Wiley and Sons, 1991.

Maor, Eli. *Trigonometric Delights.* Princeton, NJ: Princeton University Press, 1998.

Rashid, R. *The Development of Arabic Mathematics.* Dordrecht: Kluwer Academic, 1994.

Developments in Chinese Mathematics

Overview

Chinese mathematical progress during the medieval period reached its pinnacle between 1000-1300, during which time there was both innovation and discovery, and an increasingly systematic organization of the great traditions of Chinese math. Advances were made in solving problems for remainders, solving numerical equations, negative numbers, "magic squares" (vertical and horizontal arrangements of numbers that resulted in interesting properties among the columns), extractions of roots and calculations of congruences, and polynomials. Much of the mathematical effort took the form of proofs for existing problems and commentaries on mathematical histories and the problems presented in those histories. Despite the diversity of mathematical topics and challenges undertaken by scholars during this time, genuine mathematical progress was relatively limited, and theoretical mathematics all but unknown, other than in the form of diversions such as magic squares. The emphasis throughout was upon the pragmatic uses of math, particularly in surveying, astronomy, finance, and the keeping of accurate calendars.

Background

The great classic of Chinese mathematical literature, *Nine Chapters on Mathematical Procedures,* dating from the first century A.D., gathered within its contents the bulk of Chinese mathematical knowledge of the time. Its centrality to Chinese mathematics served not only as a foundation for further mathematical development, but also as a tool for preserving fundamental mathematical knowledge, a base upon which mathematical education was built. While the book addressed the writing of numbers, its primary emphasis was on the use of physical aids to computation, especially counting rods, sticks of bamboo that symbolically represented numbers and, more importantly, the place value each number held in a computation. Counting rods enabled the Chinese to undertake and solve many complex problems, including those involving fractions, division, solutions for area and volume, extraction of roots, and calculations involving π.

For ten centuries much of the focus of Chinese math was directed at commenting upon and annotating the *Nine Chapters.* New explanations and proofs of solutions further enhanced the importance of the *Nine Chapters,* with the production of commentaries being among the major activities of Chinese mathematicians. Although many of the commentaries and proofs served to extend and improve the operations described in the *Nine Chapters,* it was not until some time after A.D. 1000 that a systematic attempt was made to completely revise the text.

Impact

By 1050, commentators, particularly Jia Xuan, were working to simplify the extraction of both square and cube roots. Evidently Jia Xuan achieved some success in this quest, though this is surmised only from subsequent commentaries—the mathematician's own work does not survive.

During these centuries, the ability to perform rapid calculation using counting rods or similar devices, such as a "checkerboard" whose grid represented numerical positions, was highly prized by China's rulers. Indeed, much of the emphasis of Chinese mathematics during the period was focused on calculation and practical applications of the formulae in the *Nine Chapters*. Measurement, accurate calendars and date calculations, and financial math were the primary applications of computational ability.

So highly prized were mathematical abilities that by the eleventh century an "Office of Mathematics" was established by the government. While the purpose of the office was to employ mathematics for governmental uses, a side effect was the further codification of Chinese mathematical knowledge in a new book, called the *Ten Classics of Mathematics*. This instructional text accepted the essence of the *Nine Chapters* and extended its effectiveness through new solutions, proofs, and methods.

By the twelfth century Chinese mathematicians were beginning to range further into new territory. Advances were made in geometry, particularly the development of methods of incorporating negative numbers into mathematical coefficients, an advance vital to the development and evolution of equations.

Equations themselves were increasingly the focus of Chinese mathematical effort. The publication in 1247 of *Mathematical Treatise in Nine Sections* (a title obviously intended to honor the original text) saw mathematician Ch'in Chiu-shao (1202-1261) address the nature of algebraic equations, their properties, as well as geometric and astronomical properties, including the accurate handling of remainders in complex calculations. Much of this work anticipated mathematical advances that would not be made in Europe for several centuries.

At almost exactly the same time, in China's northern section (Ch'in Chiu-shao lived in southern China), another mathematician, Li Yeh (1192-1279) published *Sea-Mirror of Circle Measurements* (1248), a geometry text important for its contribution to the development of polynomials (constant numbers multiplied by variables in algebraic equations.) Working from the constants and their coefficients, Li Yeh showed how to solve for the indeterminate power, which he referred to as the "celestial unknown."

The more unknowns that could be solved, the more complex the problems that could be tackled. Chu Shih-chieh (1280?-1303) wrote *Precious Mirror of the Four Elements* (also known as *Jade Mirror of the Four Unknowns*) in which he showed how to solve polynomials, including multiple unknowns.

By the beginning of the fourteenth century increasing attention was being paid to the role of negative numbers, which when written were either depicted in a different color from the positive numbers, or were denoted by having a diagonal line drawn through them (or through one numeral if the number was larger than 9.) The attention paid to negatives, though, was not further pursued; some scholars argue that the Chinese sense of order and placement made the idea of negatives unappealing. Nor did the Chinese see large practical applications for negative values. The concept was identified, but allowed to languish.

Exerting a large appeal, on the other hand, was a mathematical construction that had little practical value. This was the "magic square," which, according to Chinese legend, had been discovered by the emperor Yu the Great around 2000 B.C.

The "magical" quality of the squares was found in the fact that with the proper arrangement of numbers in vertical columns, the sums of those columns would be reflected in the square's consequent horizontal and diagonal columns.

During the thirteenth century Yang Hui (1238?-1298?) devoted much effort to the construction of highly complex magic squares and magic circles, all of which demonstrate remarkable properties of constants.

More importantly, though, Yang Hui collected much important mathematical knowledge in a book that included a major section on the importance of systematic mathematical education, rather than the traditional path of simple memorization—a step toward understanding the importance of mathematics theory as well as the practical contributions of mathematics.

In addition, he presented the first clear explication of decimal notation, and anticipated Pascal's Triangle, a geometrical exercise that would be perfected (and named for) French mathematician and philosopher Blaise Pascal (1623-1662).

Yang Hui's achievements were in many ways the final flowering of the great age of Chinese mathematics. Following his work, most Chinese mathematical texts were recapitulations of earlier works. Once contact with the West was established in the centuries to come, Chinese mathematics texts were among the first books to be translated, and affected the far livelier development of mathematics in Europe. By the late 1600s the current was reversed, and Western mathematics would begin to flow into China, absorbing, and ultimately subsuming Chinese traditions and techniques.

The great and ongoing impact of Chinese mathematics is found in the sense of importance that the Chinese themselves ascribed to mathematics. While their culture neither encouraged—nor needed—the exploration of mathematics as a pure or theoretical endeavor, the culture did elevate mathematicians to a higher social rank and position within the bureaucracy, reflecting Chinese understanding of the practical value of mathematics in government and finance. The passage of Chinese mathematical ideas into the West would serve to accelerate Western mathematical development even as Chinese mathematics remained relatively stagnant.

KEITH FERRELL

Further Reading

Dunham, William. *Journey through Genius: The Great Theorems Of Mathematics.* New York: John Wiley & Sons, 1990.

Gernet, Jacques. *A History of Chinese Civilization.* Cambridge: Cambridge University Press, 1996.

Katz, Victor, *A History of Mathematics: An Introduction.* 2nd ed. New York: Addison-Wesley, 1998.

Needham, Joseph. *Science and Civilization in China: Mathematics and the Sciences of the Heavens and the Earth,* Vol. 3. Cambridge: Cambridge University Press, 1959.

Swtz, Frank J., ed. *From Five Fingers to Infinity: A Journey through the History of Mathematics.* Chicago and Lasalle: Open Court, 1994.

Mathematics in Medieval India

Overview

Indian mathematicians developed some of the most important concepts in mathematics, including place-value numeration and zero. By developing new techniques in arithmetic, algebra, and trigonometry, medieval Indian mathematicians helped make modern science and technology possible. Their innovations were brought to the West when treatises by Muslim scholars were translated into Latin.

Background

The years from A.D. 320 to about 500 were critical in the development of Indian civilization. In the north, under the Gupta dynasty, Sanskrit culture thrived, great universities were founded, and the arts and sciences flourished. In the south, where Hindu and Buddhist dynasties reigned, merchants seeking new trade opportunities started colonies and spread Indian culture throughout surrounding regions, especially Southeast Asia.

During the Gupta period, the observatory at Ujjain in central India was the heart of mathematical scholarship, and many mathematical techniques were developed to meet the needs of astronomers. The astronomical text the *Surya Siddhanta,* written by an unknown author some time around A.D. 400, contains the first known tabulation of the sine function. Indian mathematicians also developed the concept of zero, the base-10 decimal numeration system, and the number symbols, or *numerals*, we use today.

The entirety of Indian mathematics were compiled by the mathematician Aryabhata (476-550) in a collection of verses called *Aryabhatiya* in 449. The book describes both mathematics and astronomy, covering spherical trigonometry, arithmetic, algebra and plane trigonometry. Aryabhata calculated π to four decimal places, computed the length of the year almost exactly, and recognized that the Earth was a rotating sphere.

The famous astronomer Brahmagupta (598-670) wrote important works on mathematics and astronomy, including *Brahma-sphuta-siddhanta* (The opening of the universe) and *Khandakhadyaka.* He studied solar and lunar eclipses, as well as the motions and positions of the planets. Unlike Aryabhata, Brahmagupta believed that the Earth was stationary, but he, too, calculated the length of the year with remarkable precision.

Impact

The numeration system developed in India facilitated further advances in mathematics. Earlier ways of writing numbers, such as Roman numerals, used symbols to represent individual quantities, and these were added to determine the value. For example, X was the symbol for 10, and XXX was the symbol for 30, and 50 was L. Numbers expressed this way can be lengthy: 1,988 is MCMLXXXVIII. More to the point, there is no convenient way to do computations with them. People who used Roman numerals and other similar systems did their calculations with counting aids such as the *abacus.*

In contrast, Hindu arithmetic used number symbols that went only from 1 to 9, and instead of using more symbols for higher numbers, they introduced a place-value system for multipliers of 10. Each place had an individual name: *dasan* meant the tens place, *sata* meant the hundreds place, and so on. To express the number 235, the Hindus would write "2 sata, 3 dasan, 5". Seven hundred and eight would be "7 sata, 8".

Toward the end of the Gupta period, Indian mathematicians found a way to eliminate the place names while keeping the advantages of the place-value system. They used a symbol called *sunya,* or "empty" to designate a place with no value in it. This is equivalent to the symbol we call zero. With this they could write 708 for "7 sata, 8," and easily distinguish it from "7 dasan, 8," or 78. The physical alignment of tens, hundreds, etc. in columns resulted in the development of new arithmetic techniques for working with numbers.

About 800, the Hindu mathematician Mahavira demonstrated that zero was not simply a placeholder, but had an actual numerical value. His tenth-century successor Sridhara further recognized that the zero was as meaningful a number as any of the others. Without the zero, modern mathematics, and therefore most of modern science, would have been impossible.

The twelfth century mathematician Bhaskara (often called Bhaskara the Learned) was, like many of his predecessors (such as Brahmagupta), head of the Ujjain observatory and a gifted astronomer. His two mathematical works, *Lilivati* (The graceful) and *Bijaganita* (Seed counting) from the series *Siddhantasiromani*, were the first to expound systematically the use of the decimal system, based on powers of 10. He compiled many problems with which earlier mathematicians had struggled, and presented solutions.

Bhaskara was starting to understand the special nature of dividing by zero, as he specifically noted that 3/0 is infinite. He was, however, unable to generalize this to any number divided by zero. He enumerated the convention of *signs* in multiplication and division: two positives or two negatives divided or multiplied yields a positive result, and a positive and a negative divided or multiplied gives a negative result.

In algebra, Bhaskara built on the work of Aryabhata and Brahmagupta. He used letters to represent unknowns, as we do in algebra today. Bhaskara developed new methods for solving *quadratic equations,* that is, equations containing at least one variable raised to the second power (x^2). He studied regular polygons with up to 384 sides, in order to calculate increasingly precise approximations of π.

One of the first Muslim mathematicians to write about Indian techniques was Muhammad ibn Musa al-Khwarizmi, a teacher in the mathematical school at Baghdad. His book *Al-Khwarizmi Concerning the Hindu Art of Reckoning* was translated into Latin as *Algoritmi, de numero Indorum.* The Latinization of his name from "al-Khwarizmi" to "Algoritmi" eventually became our word for a mathematical procedure, *algorithm.* When his book on elementary mathematics *Kitab al-jabr wa al-muqabalah* (The book of integration and equation) was translated into Latin in the twelfth century, the term *al-jabr* became *algebra.*

Indian mathematical techniques were disseminated in the West through texts such as these. They were first brought to Moorish Spain, where they then spread to the rest of Europe. However, they did not come into common use there until the digit symbols were standardized after the invention of the movable-type printing press in the mid-1400s.

During the thousand years that followed the Gupta dynasty, successive waves of invaders poured through India. First came the Huns, who descended from central Asia beginning around 450, finally conquering the Gupta empire 50 years later. Muslims arrived from Arabia in the 700s, and from Afghanistan and Persia at the turn of the millennium. Muslim sultans ruled from Delhi between 1206 and 1526, helping to disseminate Indian mathematical advances throughout the Islamic world. Because Indian symbols were introduced to the West by Muslim mathematicians, they came to be known as "Arabic" numerals. Today scholars generally refer to our way of expressing numbers as the *Hindu-Arabic numeration system*.

SHERRI CHASIN CALVO

Further Reading

Bose, D.M., S.N. Sen, and B.V. Subbarayappa (eds.). *A Concise History of Science in India.* New Delhi: Indian National Science Academy, 1971.

Ibn Lablan, Kushyar. *Principles of Hindu Reckoning.* Madison, WI: University of Wisconsin Press, 1966.

Murthy, T. S. Bhanu. *A Modern Introduction to Ancient Indian Mathematics.* New Delhi: Wiley Eastern Ltd., 1992.

Rao, S. Balaachandra. *Indian Mathematics and Astronomy.* Bangalore, Jnana Deep Publications, 1994.

Sarasvati, T. A. *Geometry in Ancient and Medieval India.* Delhi: Indological Publishing, 1979.

Schulberg, Lucille. *Historic India.* New York: Time-Life Books, 1968.

Srinivasiengar, C. N. *The History of Ancient Indian Mathematics.* Calcutta: World Press Private Ltd., 1967.

Swetz, Frank J. *From Five Fingers to Infinity. A Journey through the History of Mathematics.* Chicago: Open Court, 1994.

The Return of Mathematics to Europe

Overview

Early medieval mathematics was based on only a few classical texts, since the majority of ancient mathematical knowledge had been lost after the fall of Rome. Slowly, over many centuries, these texts were reintroduced to Europe through contact with Arab mathematicians, who had preserved and extended classical learning. Social and economic changes in Europe created a demand for a newly sophisticated mathematical learning.

Background

Mathematics flourished in the Greek world from 600 B.C. to A.D. 300 in what has been called the Golden Age of Mathematics. The rise of the Roman Empire saw mathematical philosophy take a back seat to practical methods, since the Romans, in general, preferred language studies to abstract mathematics, but Greek learning was still preserved and studied.

With the fall of Rome and the collapse of the empire in the fourth century, however, many ancient mathematical works were lost or destroyed. Europe endured a period of anarchy and political fragmentation. Trade became localized, and towns and cities shrank in importance and size. These economic and political changes all reduced the role of mathematics in society.

What little knowledge from the ancient world remained was preserved in the Byzantine Empire (roughly modern-day Turkey), or in monasteries scattered across Europe where they were stored and copied over the centuries. While the Byzantine Empire used little of this knowledge themselves, they shared it with neighboring Arab lands; in this way, much of Greek learning was translated into Arabic.

In order to preserve these ancient texts over many centuries as parchment aged and crumbled, frequent copies had to be made. In Europe monks tended to concentrate their efforts on theological and philosophical texts, not mathematical or scientific ones; as a result, many great works literally crumbled into dust. Scholars came to rely on a small selection of Latin texts, many compiled into large encyclopedias in the fifth and sixth centuries by scholars such as Boethius (480-524). These large collections simplified complicated concepts for the European audience, which often meant deleting mathematical figures and calculations. In addition, medieval writers frequently did not use numbers, so even books on technical subjects such as glassmaking and jewelry tended to contain nonspecific quantities, such as "a medium-sized piece" or "a bit more."

This kind of simplification was necessary because education in the Middle Ages contained almost no higher mathematics. Although arithmetic was taught as part of the seven liberal arts—the *quadrivium* (arithmetic, music, geometry, and astronomy) and the *trivium,* (grammar, rhetoric, and dialectic)—it was merely the *theory* of numbers, not the calculation of problems we associate with the subject today. Medieval mathematicians focused on basic properties, such as odd and even numbers, ratios, proportions, and the harmony of numbers. Addition, subtraction, multiplication, and division were separate subjects, collectively called *computus,* which were rarely taught.

Furthermore, the chaotic political and economic situation in Europe, combined with the preference for religious and philosophical texts, severely limited the scope and application of mathematical ideas in medieval society. Mathematical computation was restricted to that needed for the small-scale trade of the era. Over time, however, the economic and political stability of Europe began to improve, and mathematics slowly revived to meet the needs of the changing society.

Impact

One of the major factors in medieval life was the church. The Christian tradition was ambivalent toward numbers. Parts of the Bible seemed to support mathematics, such as the use of apocalyptic numbers in Daniel and Revelation. However, there were also some sections that appeared hostile. An abbot in 1130 stopped his men counting their provisions to see if they would survive an impending crisis, as this would suggest they did not trust God to see them safely through. He referred them to the biblical story of King David, who was punished for counting the people in his kingdom.

However, two religious trends helped spur the return of mathematics: the revival of numeric apocalyptic prophecy, and the need to calculate the correct date for Easter. Calendar reform became a major religious issue in the fifth century, and many popes recruited the best mathematical minds from across Europe to ensure that Christ's resurrection would be celebrated on the correct date.

The reign of Charlemagne (742-814) produced a short-lived revival of numerical knowledge, as well as increased trade and localized political stability. Irish monks, whose libraries of ancient texts were well-preserved, brought their knowledge to Charlemagne's court, and scholars from across Europe flocked there to study mathematics. However, Viking invasions ended the calm, and few texts survived to influence later scholars. One important lasting innovation did come from the so-called Carolingian Renaissance: the foundation of cathedral and monastery schools across Europe, some of which evolved into universities.

Through contact with the Arab world, Europeans rediscovered the mathematical heritage of the ancients, and added a few discoveries of their own. However, it was an uneven and frustratingly slow process. Gerbert of Aurillac (946-1003) attempted to introduce a number of Eastern innovations into European mathematics in the tenth century, but with limited success. Gerbert studied in southern Spain, then in Islamic lands. He learned Arabic, and was particularly impressed with the abacus and the Hindu-Arabic numeral system, the basis of the numerals we use today. Gerbert brought these innovations back to Italy.

By contrast, the European abacus (or counting board) that was in use at the time had severe limitations, and was much slower than the string-bead abacus that can still be seen in use in the Middle East today. By its very nature the abacus discouraged mathematical writing, as computations were confined to a particular time and space, and the transitory steps of calculation were not recorded, only the result. Very little information on the medieval abacus remains today, and only a few examples survive.

Hindu-Arabic numerals did not prove popular, despite later attempts by individuals such as such as Adelard of Bath (1090-1150) and Leonardo of Pisa, also known as Fibonacci, (1175-1250) to popularize them. The Hindu-Arabic system included the number zero, which caused some philosophical problems for many Europeans. It appeared to represent a mystical quantity, entirely abstract and somewhat frightening. Some condemned the zero as heretical. Europeans preferred to use Roman numerals until the sixteenth century, although strange blends of both systems did occur. The maker of one calendar in 1430 wrote the length of a year as "ccc and sixty days and 5 and sex odde howres." Even more confusing was an inscription of IVOII to represent the 1502. Europeans were also slow to recognize the usefulness of mathematical symbols, like +, −, and =.

In the twelfth century the Crusades accelerated exposure to the Arab world. Many important trade and intellectual contacts were made, and more lost Greek mathematical texts were found and translated into Latin. These works, however, often suffered from multiple translations from Greek to Arabic to Latin, and many errors had crept in. It was also a large task, and it took a handful of dedicated translators until the middle of the fifteenth century to complete it. Copies were few, and extremely expensive, as they had to be painstakingly handwritten. Only with the late-fifteenth century printing revolution was the continued survival of the ancient mathematical texts ensured, with cheap, plentiful copies.

Individuals attempted and failed to introduce mathematical innovations because mathematics had little relevance to European life. However, as the political, economic, and social structures developed a new need for mathematics, mathematical learning began to emerge. The rise of commerce fueled a need for more numerate clerks and scribes in the business sector. There was a corresponding rise in the importance of trading centers, and strategic urban towns began to grow rapidly. Towns became the focal points of the training and employment of the newly numerate. The relative peace and stability of Europe of the twelfth century also led to rise in political administration, and a new interest in counting everything from money to soldiers. Prosperous times helped fund new buildings, and architecture absorbed the ancient Greek mathematical models, giving rise to the splendor of medieval cathedrals.

Mathematics began to invade all areas of medieval life. Clocks began to divide the day into regular intervals. Alchemists used numbers for the supposed mystical properties they contained. By the end of the thirteenth century mathematics was reentering medieval life apace. New translations recovered much that had been lost a thousand years before, and new ideas were discovered in Arabic and Hindu scholarship. The abacus helped revive the art of calculation, and the teaching of mathematics was being demanded by many sectors of medieval society.

However, the growth of the economy and of urban centers was dramatically interrupted in the fourteenth century, with wars and the Black Death killing as much as one third of the European population, and stunting the intellectual revival. Yet the impulses that had begun the revival of mathematics still remained, and with a new period of relative peace and stability from the mid-fifteenth century mathematics once again flourished. Mathematics spread into more fields, such as the application of geometry in painting to produce perspective. The development of printing ensured that ancient learning could no longer be lost through lack of copies. The mathematicians of the Renaissance would later help forge the way for the eventual marriage of mathematics and science; one of the fundamental characteristics of the modern age.

DAVID TULLOCH

Further Reading

Burton, David M. *The History of Mathematics: An Introduction*. Allyn and Bacon, Inc., Newton, Massachusetts, 1985.

Cooke, Roger. *The History of Mathematics: A Brief Course*. John Wiley and Sons, Inc, New York, 1997.

Crosby, Alfred W. *The Measure of Reality, Quantification and Western Society, 1250-1600*. Cambridge University Press, Cambridge, 1997.

Murray, Alexander. *Reason and Society in the Middle Ages*. Clarendon Press, Oxford, 1978.

Wagner, David L. (ed.). *The Seven Liberal Arts in the Middle Ages*. Indiana University Press, Bloomington, 1983.

Development of Algebra during the Middle Ages

Overview

During the Middle Ages, while the intellectuals of Christian Europe concerned themselves with theology and the common people with subsistence agriculture, a vibrant scientific and mathematical culture developed in the Islamic world. Among its achievements was the development of algebra, which would be reintroduced into Western mathematics through the Latin translation of a book, the *al-jabr*, by the ninth-century Persian astronomer and mathematician al-Khwarizmi.

Background

In its modern sense, algebra is the branch of mathematics concerned with finding the values of unknown quantities defined by the equations that they satisfy. Problems of an algebraic type are recognizable in the surviving mathematical writings of the Egyptians and Babylonians. Ancient Greek mathematics included the development of some algebraic concepts, but mainly in connection with geometry. The Greek mathematicians and philosophers were uncomfortable with the existence of irrational numbers, numbers such as the square root of two; irrational numbers cannot be expressed as the ratio of two whole numbers and are almost unavoidable in algebraic problems. One of the later Greeks to write on algebraic topics was Diophantus (c. A.D. 210-c. 290), who was affiliated with the famous library at Alexandria about the year 250. Diophantus was the first to use symbols to introduce unknown quantities that then allowed equations to be written.

By the time of the fall of the Roman Empire in the fifth century A.D., the study of mathematics had made considerable progress in both China and India. In India it was put to use in the service of astronomy and Hindu religious practices. A number of Indian scholars, notably the astronomer and mathematician Brahmagupta (598-c. 665) made contributions to algebra and arithmetic in the treatment of astronomical problems. Most of these are known from the *Brahamasphuta Siddhanta*, which includes a discussion of algebraic equations that goes beyond the work of Diophantus, and a treatment of arithmetic using the system of ten numerals, including zero, that is essentially that used today.

In 632 the prophet Muhammad established an Islamic state centered in Mecca. Following his death in the same year, his followers began to conquer lands to the east and west. By 750 Arab armies had established control of a region extending from southern Spain through North Africa, Asia Minor, and into India. Initially hostile to other cultures, the Islamic rulers soon began to welcome the scholars of many different lands within their realm. By 766 an Arabic translation of a major Hindu mathematical work, most likely the *Siddhanta* of Brahmagupta, had been prepared. At Baghdad, the Caliph al-Ma'mun (786-833) established a "House of Wisdom" modeled on the earlier Greek academy at Alexandria. Al-Ma'mun encouraged the translation of the mathematical works in other languages into Arabic. One of society's members would be the mathematician and astronomer al-Khwarizmi (c. 780-c. 850).

Our word "algebra" comes from al-Khwarizmi's book *Kitab al-jabr wa al-muqabala,* or *The Compendious Book on Calculation by Completing and Balancing.* In this book (referred to as the *al-jabr*) al-Khwarizmi described the art of finding the value of an unknown quantity in equations by rearranging terms. The book had three parts, only one of them dealing with algebra as we know it, the remainder dealing with measurement and with some legal questions concerning inheritances. Most remarkably, the discussion of algebra is presented entirely without symbols; even the numbers are spelled out in words. Thus the quantity $(x/3 + 1)$ would be described by al-Khwarizmi as "a thing, take a third of it and add a unit."

Although al-Khwarizmi, unlike the ancient Greeks, was untroubled by irrational numbers, he was unwilling to accord negative numbers the same status as positive ones. Thus he stated that there are six fundamental forms of equations involving powers no greater than the square of a single unknown quantity. The modern student of algebra would recognize each of these as a special case of the general quadratic equation, $ax^2 + bx + c = 0$, with a, b, or c being possibly zero or negative. Al-Khwarizmi's fundamental forms all have positive and nonzero coefficients.

For each of the fundamental forms, al-Khwarizmi explained how a solution may be obtained by a sequence of mathematical operations. He gave rules for solving each of them, using the ordinary arithmetic operations of addition, subtraction, multiplication, and division, plus the taking of square roots. The "al-jabr" of the book's title actually refers to the process of "restoration" or "completion" by which negative quantities appearing in a problem are eliminated to obtain one of the standard forms. "Al-muqabala" refers to the combining of terms involving the same power of the unknown, again to obtain one of the six standard forms.

The *al-jabr* suggests that al-Khwarizmi was familiar with both the Hindu and Greek traditions in mathematics. Hindu influence is even more apparent in his astronomical work, but the symbol-free presentation of the *al-jabr* is also more consistent with Sanskrit mathematical texts than with Greek. On the other hand, al-Khwarizmi's use of geometrical figures to illustrate equations suggests a familiarity with Euclid's *Elements*. While the *al-jabr* included no symbols, the Hindu system of numerals would be described by him in a second book now known only in Latin translation as *Algoritmi de numero Indorum*. It is from the first word of this title, meaning "al-Khwarizmi's," that we obtain the modern word "algorithm," meaning a systematic method of obtaining a mathematical result.

Among other Persian scholars to be concerned with algebra was the Persian poet and astronomer Abu'l-Fath Umar ibn Ibrahim al-Khayyami (1048-1131), better known as Omar Khayyam. Khayyam's poetry was translated into English in the nineteenth century, but his mathematical work remained unknown in the West until the 1930s. By 1079 Khayyam had written a manuscript on cubic equations that gave the solution of every type of cubic equations with positive real number solutions. He also discovered a relation between the coefficients generated by the expansion of the binomial $(a + b)^n$ that would be independently discovered by the

French mathematician Blaise Pascal (1623-1662) six centuries later.

Impact

The survival and circulation of manuscripts before the invention of the printing press is an uncertain matter. Al-Khwarizmi's *al-jabr* is in many respects less advanced than earlier works. It is possible that the practical and elementary nature of the presentation was instrumental in its survival while other texts have been lost. Scholars thus are uncertain about how much of the *al-jabr* was original and how much may have been taken from earlier Arab or Hindu sources. Nonetheless, it is through this book that the study of algebra was reintroduced into Europe during the Renaissance. The *al-jabr* was translated into Latin in 1145 as the *Liber algebrae et almucabala* by Robert of Chester (fl. c. 1141-1150), an English scholar living in Islamic Spain.

By the time of the Renaissance, mathematics in Christian western Europe was in a far less advanced state than in the Islamic world. The Byzantine Empire had, however, provided a refuge for Greek-speaking scholars. In 1543, after Turkish forces overran its capital, Constantinople, Byzantine scholars began to seek refuge in Italy, where rich and powerful families like the Medici were collecting manuscripts and providing financial support for scholars. The appearance of Johannes Gutenberg's (1390?-1468) printing press would make mathematical ideas far more widely available. Over 200 new books on mathematics appeared in Italy before 1500.

In 1545 a book entitled *Ars Magna* or *The Great Art* by the Italian mathematician Girolamo Cardano (1501-1576) appeared. This work incorporated significant new results, including the solution of the cubic and quartic equations. Cardano used letters to represent known quantities, but unlike Diophantus he stopped short of using other letters for the unknown quantities. The next major step forward in algebra occurred with the work of French lawyer and writer François Viète (1540-1620), who introduced the modern practice of using letters to represent the unknown as well as known quantities. With Viète's new notation, it became easier to think of solving an algebraic equation as finding the values of x for which a definite function of the variable x would equal zero. This set the stage for the study of functions themselves and the study of transformations of functions caused by introducing new variables, ideas important in modern algebra, trigonometry, and calculus.

Algorithms to obtain various results have been known, of course, throughout the history of mathematics. One of the most famous is the algorithm attributed to the Greek mathematician Euclid (fl. c. 300 B.C.) for finding the greatest common factor of two integers. The various recipes given by al-Khwarizmi for transforming equations into one of the six standard forms and for solving each of them are also examples of algorithms. The notion of algorithm would gain increased attention in the twentieth century. Algorithms are, after all, rules for the manipulation of symbols, and nineteenth-century mathematics had become increasingly concerned with the formalization of mathematics—that is, translating the statements of mathematics into purely symbolic form. In 1900 the great German mathematician David Hilbert (1862-1943) proposed that an algorithm might be found that could produce a solution to any mathematical problem expressed as a string of symbols. A young British mathematician, Alan Turing (1912-1954), was prompted by this proposal to examine the notion of algorithm, particularly in arithmetic, far more closely, and ultimately to prove that no such algorithm could exist. In his work, however, Turing first described the operation of a simple symbol processing machine, now called a "Turing machine," that could be built from electronic components and that has evolved into the digital computer of the present day.

DONALD R. FRANCESCHETTI

Further Reading

Bell, Eric Temple. *Development of Mathematics*. New York: McGraw-Hill, 1945.

Boyer, Carl B. *A History of Mathematics*. New York: Wiley, 1968.

Grattan-Guiness, Ivor. *The Rainbow of Mathematics: A History of the Mathematical Sciences*. New York: Norton, 1997.

Kline, Morris. *Mathematical Thought from Ancient to Modern Times*. New York: Oxford University Press, 1972.

Toomber, G. J. "Al-Khwarizmi, Abu Jafar Muhammad Ibn Musa." In *The Dictionary of Scientific Biography*, vol. 7. New York: Scribner's, 1973: 358-65.

The Use of Hindu-Arabic Numerals Aids Mathematicians and Stimulates Commerce

Overview

The modern system of notation, using ten different numerals including a zero and using position to denote value, appears to be the invention of Hindu mathematicians and astronomers, reaching its present form by the seventh century. The system became known in western Europe through the works of Islamic commentators whose works were translated into Latin. The Hindu-Arabic numerals, as they are now known, greatly facilitated arithmetic computations, particularly multiplication and division. They also allowed more rapid calculation of the mathematical tables needed for surveying, navigation, and the keeping of financial records and thus contributed to the extensive exploration and the growth of capitalism that characterized the Renaissance.

Background

One of the essential requirements for any type of mathematics is a means of representing quantities. At first, tokens might be used—pebbles or small clay objects with one pebble representing each sheep in a herd, for example. Eventually, tokens of different shapes might be used to represent certain multiples, say five or ten sheep instead of one. With the invention of writing, it made more sense to use marks pressed into clay or made on paper or papyrus to keep track of one's possessions. The Babylonians, Egyptians, Indians, and Mayans all had developed elaborate systems to represent quantities by the first century A.D. The Babylonians developed a sophisticated number system based on the number 60, using it in commerce and for astronomy and astrology. By the last century B.C., this system included a symbol for zero, which was used as a placeholder in expressing quantities.

Of the several number systems, those that had the greatest effect on the development of mathematics in Europe of the Middle Ages were the Roman, the Chinese, and the Indian or Hindu, transmitted to the Western world by the Arabs and now known as Hindu-Arabic numerals. In the Roman system, still occasionally used today, letters of the alphabet were used to represent units and multiples of five or ten. In the Roman system the year 2004 can be written quite compactly as MMIV, with a hint of positional notation in that the "I" appearing before the "V" means that the one it represents is to be subtracted from the five represented by the "V." Roman numerals were adequate for record keeping and could be added and subtracted easily, but were far more cumbersome in multiplication and division and certainly not suited to the needs of modern science or commerce.

The Chinese system was not a decimal system, based on the number 10, but a centesimal system, based on separate symbols for the whole numbers between 1 and 9 and for multiples of 10 between 10 and 90. By alternating the pairs of symbols, the Chinese were able to represent numbers of any size. Because the Chinese number symbols were composed of single strokes, it was possible to represent them by short sticks, and to do arithmetic by moving sticks about according to preset rules. This led in the Middle Ages to the use of counting boards by merchants to do simple arithmetic. The Chinese were also responsible for the abacus, an arrangement of beads on wires that facilitated the ordinary operations of addition, subtraction, and multiplication.

The number system we use today, based on the numerals 0-9 and using position to denote different powers of 10, originated in India. Mathematical thinking in India dates back to at least 800 B.C. Number symbols first appear in the third century B.C., including among many alternatives the so-called Brahmi symbols, which include separate symbols for the numerals 1 through 9 and the multiples of 10 from 10 to 90. The Brahmi figures gradually evolved into the "1,2,3..." of today. By the year 600, they had come to predominate and to include a symbol for zero and for the use of positional notation. The *Brahamasphuta Siddhanta*, a treatise on astronomy written by the astronomer and mathematician Brahmagupta (598-c. 665) includes a treatment of arithmetic using the system of ten numerals, including zero, along with rules for fractions, the computation of interest, and rules for using negative numbers. Interestingly, Brahmagupta appeared to treat the fraction 0/0 as equal to zero, and avoided the question of dividing other numbers by zero.

Beginning in A.D. 632 Arab armies expanding from the Arabian peninsula established an Islamic empire that would stretch as far eastward as India and as far west as Spain. In 755 it split into two kingdoms, one with its capital at Baghdad. There, Hindu scientists and mathematicians found themselves welcome, despite their different religious beliefs. In Baghdad they could meet the descendants of the Greek scholars who had fled to Persia, bringing their mathematical interests, after the Emperor Justinian closed Plato's academy in A.D. 529. By 766 some Hindu mathematical work had been translated into Arabic. At Baghdad the Caliph al-Ma'mun (786-833) established a "House of Wisdom" modeled on the earlier Greek academy at Alexandria, with a library and observatory.

One of the scholars at the House of Wisdom was al-Khwarizmi (c. 780-c. 850). Al-Khwarizmi's book on algebra, popularly known as the *al-jabr*, was translated into Latin in 1145 by Robert of Chester (fl. c. 1141-1150), an English scholar living in Islamic Spain. Al-Khwarizmi would also become known to the Western world for a book known only in Latin translation as the *Algoritmi de numero Indorum*, or *Al-Khwarizmi on the Hindu Method of Calculation*, in which he explains the Hindu number system and how it can be used in arithmetic calculations. It is from the title of this book that we obtain the word "algorithm" for any systematic method of calculation.

Among the readers of this book were Leonardo of Pisa (c. 1170- c. 1250), also known as Leonardo Fibonacci, an Italian who traveled throughout northern Africa and became familiar with the Arab system of numbers and methods of calculation. In 1202 he wrote the *Liber Abaci* or *Book of Calculations*, in which he described the Arabic system of numbers. Although the Hindu-Arabic system of numbers was not entirely unknown in Europe, it was Fibonacci's book that led to its widespread adoption in commerce and record keeping.

Impact

The most powerful aspect of the Hindu-Arabic system is the existence of a separate numeral for zero that can serve both as a placeholder and as a symbol for "none." The zero also appears, apparently independently, in the Chinese number system and in the system developed by the Mayans of Central America.

The use of Hindu-Arabic numerals and positional notation has become so prevalent in the modern world that it is difficult to imagine what mathematics would be like without it. On the practical side, banking, accounting, and capitalism in general could hardly be possible without a system that allowed the expression of large numbers in compact form and the easy calculation of interest. While merchants could perform the required calculations for a purchase or sale using the abacus or a counting board, the new method was faster and left a permanent record. The sort of record keeping required by bankers who accepted funds on deposit, and then lent or invested the money, required a lasting and compact notation, which the Hindu-Arabic system provided.

The more efficient system of arithmetic greatly facilitated the calculation of mathematical and astronomical tables that were used in surveying, construction, navigation, and the casting of astrological horoscopes. Prior to the twentieth century and the development of the electronic computer, all such tables had to be calculated by hand by specially trained human "computers."

The decimal system, based on ten distinct digits, is of course a reflection of human anatomy with its ten fingers. One disadvantage to the system is that the number ten is divisible only by two and five. From time to time individuals have advocated a duodecimal system based on the number twelve, which is divisible by two, three, four, and six. King Charles XII of Sweden, who reigned from 1697 to 1718, even contemplated introducing such a system in his kingdom but did not proceed. For electronic computer applications, in which numbers are stored and represented by "two-state" devices, each a sort of switch that can be on or off, the binary system in which numbers are represented by ones and zeros is ideal. While this system uses only two digits, the principle of number construction and the algorithms for the arithmetic operations parallel those for the decimal system based on Hindu-Arabic numerals.

DONALD R. FRANCESCHETTI

Further Reading

Bell, Eric Temple. *Development of Mathematics*. New York: McGraw-Hill, 1945.

Boyer, Carl B. *A History of Mathematics*. New York: Wiley, 1968.

Cajori, Florian. *A History of Elementary Mathematics*. New York: Macmillan, 1930.

Grattan-Guiness, Ivor. *The Rainbow of Mathematics: A History of the Mathematical Sciences*. New York: Norton, 1997.

Kline, Morris. *Mathematical Thought from Ancient to Modern Times*. New York: Oxford University Press, 1972.

The Rediscovery of Euclid's *Elements*

Overview

The principal Greek compendium of geometry, Euclid's *Elements*, was translated into Arabic in the ninth century. Muslim mathematicians were then able to combine geometry with the arithmetic and algebra they learned from the Hindus and develop new advances of their own. In the twelfth century the work was translated into Latin, making it more accessible to European scholars. The *Elements* was still regularly used as a textbook of geometry until about 1900.

Background

Euclid was a Greek mathematician who lived in Alexandria, Egypt, around 300 B.C., and founded the first school of mathematics there. He wrote many books, some of which have been lost. However his 13-volume treatise on geometry, called the *Elements*, was among the most important mathematical texts in history.

The *Elements* was a synopsis of the work of many mathematicians, including Hippocrates of Chios, Eudoxus, and Theaeteus. It was unique for its logical exposition of contemporary knowledge and techniques in geometry, with *postulates* that were self-evidently true and proofs of *theorems* derived from the postulates. It also included sections on algebra and number theory. For example, it includes a proof that the number of *primes*, which are divisible only by themselves and 1, are infinite.

Alexandria, along with the rest of Egypt, was conquered by the Romans in 31 B.C. The Roman Empire was heir to much of the Greek intellectual tradition. However, after the fall of the Roman Empire in the fifth century A.D., the task of preserving classical achievements in mathematics and science fell mainly to the Islamic world.

Impact

Islam expanded rapidly after the death of the prophet Muhammad in A.D. 632, eventually extending from Spain well into central Asia. Arabian warriors conquered Alexandria in 642. Muslim scholars who now had access to the ancient centers of learning took advantage of their opportunity to study and disseminate the classical texts. Many of these were received in a peace agreement with the Byzantine emperor.

Caliph al-Ma'mun, who ruled from Baghdad between 813 and 833, caused a translation and study center to be built there, along with an astronomical observatory. In this institution, called the House of Wisdom, both Muslim and Christian scholars were employed to translate texts from the Greek and Syriac languages into Arabic. Many of the translators of mathematical texts were themselves well known mathematicians. The enterprise was financed by the caliph and his successors, as well as other wealthy Muslim intellectuals.

Translations of the *Elements* were made by al-Hajjaj ibn Matar, Ishaq ibn Hunayn, and Tabit ibn Qurra. Euclid's *Data*, *Optics*, *On Division*, and his astronomical text *Phaenomena* were translated. Other important translations included Ptolemy's seminal work on astronomy, the *Almagest*, Archimedes's *Sphere and Cylinder* and *Measurement of the Circle*, Diophantus's *Arithmetica*, the *Sphaerica of Menelaus*, and almost all the works of Apollonius of Perga, known as the Great Geometer. In addition to the classical material, works written in Sanskrit by the great Hindu mathematicians were also translated. The translations disseminated from the Baghdad center stimulated independent mathematical research in the Islamic world for the next 600 years.

The Muslim mathematicians who studied at the House of Wisdom in the ninth century included Muhammad ibn Musa al-Khwarizmi, whose works brought Hindu arithmetic and algebra to the West. Al-Khwarizmi adopted the *axiomatic* (proof-oriented) presentation Euclid employed in his geometrical texts to elucidate these new mathematical disciplines.

With both the geometry of the Greeks and the advanced arithmetic and algebraic techniques of the Hindus, the Islamic world had a much more complete set of tools for dealing with mathematical problems. So, for example, solid geometry problems that could not be solved with a ruler and compass could be represented algebraically. Conversely, geometric curves could be used to represent algebraic equations. The brilliant Persian poet, astronomer and mathematician Omar Khayyam (c. 1050- c. 1123) found geometric solutions to cubic equations.

One of the lines of research that was vigorously pursued by Muslim mathematicians was the investigation of Euclid's *parallel postulate*.

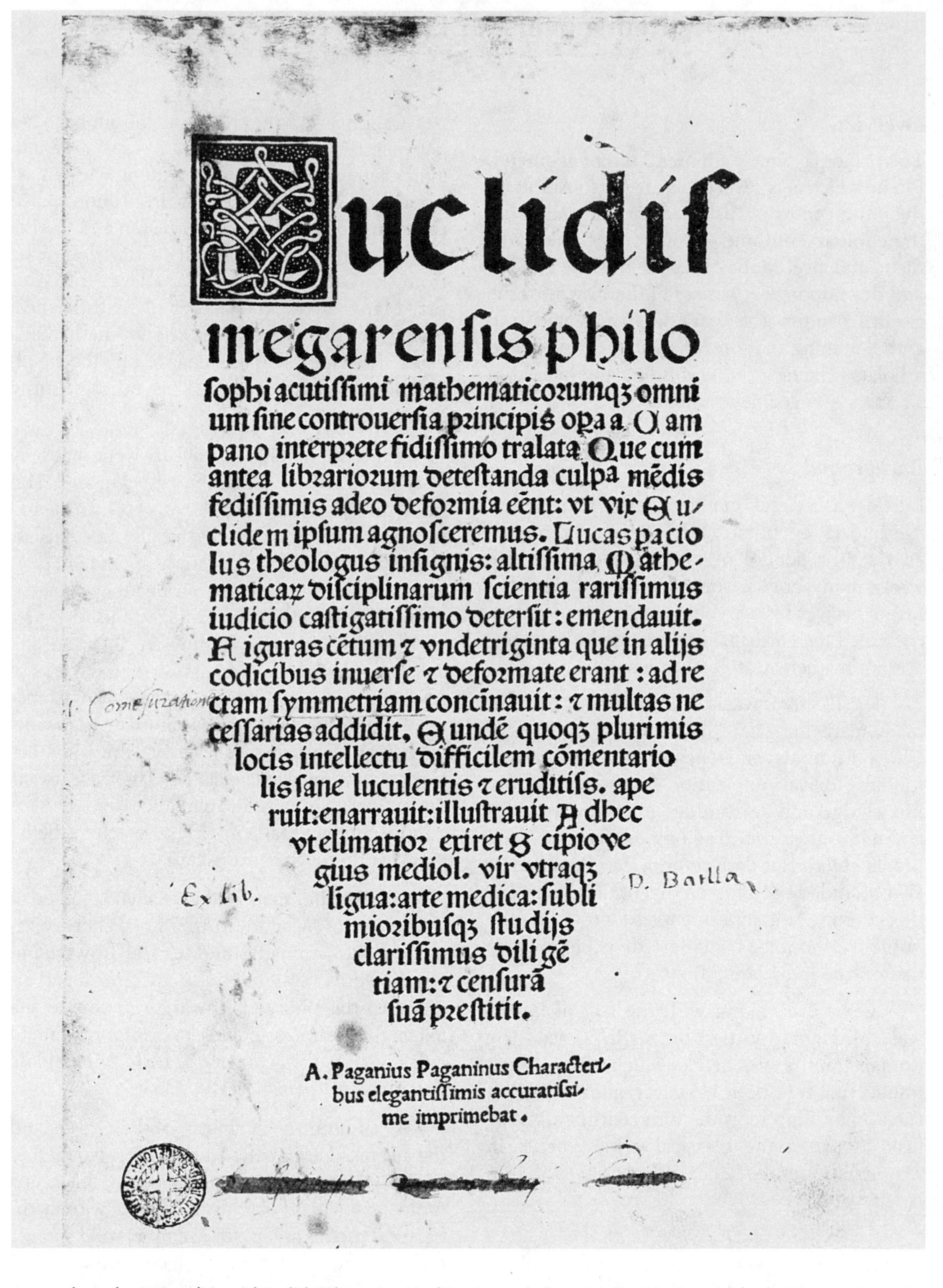

Euclidis
megarensis philo
sophi acutissimi mathematicorumq; omni
um sine controuersia principis opa a Cam
pano interprete fidissimo tralata. Que cum
antea librariorum detestanda culpa mēdis
fedissimis adeo deformia eēnt: vt vix Eu
clidem ipsum agnosceremus. Lucas pacio
lus theologus insignis: altissima Mathe
maticarum disciplinarum scientia rarissimus
iudicio castigatissimo detersit: emendauit.
Figuras cētum & vndetriginta que in aliis
codicibus inuerse & deformate erant: ad re
ctam symmetriam concinnauit: & multas ne
cessarias addidit. Eundē quoq; plurimis
locis intellectu difficilem cōmentario
lis sane luculentis & eruditis. ape
ruit: enarrauit: illustrauit. Adhec
vt elimatior exiret Scipio ve
gius mediol. vir vtraq;
ligua: arte medica: subli
mioribusq; studiis
clarissimus dili gē
tiam: & censurā
suā prestitit.

A. Paganius Paganinus Characteri
bus elegantissimis accuratissi
me imprimebat.

A page from the Latin edition of Euclid's *Elements.* *(Archivo Iconografico, S.A./Corbis. Reproduced with permission.)*

The *Elements* was based on theorems derived from five "common notions," or postulates. The first four are simple, stating that:

(1) Any two points can be joined with a straight line

(2) A finite straight line of any length can be drawn upon a straight line.

(3) A circle may be described with any center and radius.

(4) All right angles are equal.

The fifth, known as the parallel postulate, is considerably more complex. It states, "If a straight line meets two other straight lines, so as to make the two interior angles on one side of it together less than two right angles, the other straight lines will meet if produced on that side on which the angles are less than two right angles."

Later, a simpler wording, also known as *Playfair's Axiom*, was devised. It states, "Through a point on a given line, there passes not more than one parallel to the line."

Many mathematicians disagreed that this was a "common notion," and judged it insufficiently self-evident to be termed a postulate. Even Euclid proved his first 28 propositions without reference to the parallel postulate, as if reluctant to use it himself. The mathematicians who followed him attempted to prove the parallel postulate as a theorem, using only postulates 1 through 4, the 28 propositions proved using them exclusively, and the definitions given in Euclid's work. If they could do that, they would prove in effect that the entire geometry was based on the first four postulates alone.

A frequently encountered problem was *petitio principii*, or *circular reasoning*. This is the attempt to prove a statement by using another assertion which, when considered carefully, turns out to be equivalent to, or a consequence of, the one being proved. Of course, reputable scholars do not do this intentionally. Circular reasoning is generally a consequence of the fact that mathematical statements can be formulated in quite different ways, just as Euclid's formulation of the parallel postulate sounds nothing like Playfair's Axiom. Sometimes a proof would be accepted for many years before the circular reasoning was discovered.

Omar Khayyam approached the problem of proving the parallel postulate using a four-sided figure, or *quadrilateral*, with two equal sides perpendicular to the base. He realized that he could accomplish the proof by demonstrating that the other two angles in the figure were also right angles. Although he never succeeded in this effort, from then on the parallel postulate was generally discussed in terms of this quadrilateral.

The *Elements* was first translated into Latin in the 1100s, making it more readily available to European scholars. Advances in European understanding of geometry followed, paving the way for thirteenth-century advances in geometrical optics. The German monk Theodoric of Freiberg studied the reflection and refraction, or bending, of light in spherical droplets, and first understood how rainbows are formed. Euclid's work was also brought to India by the Muslims, returning the mathematical boost the Hindus had given them with their arithmetic and algebraic techniques.

The *Elements* was commonly used as a geometry textbook until the dawn of the twentieth century. Today mathematicians recognize two main classes of geometries, classical *Euclidean* geometry in which the parallel postulate is assumed to be true, and *non-Euclidean* geometries, such as spherical geometry, in which it does not exist.

SHERRI CHASIN CALVO

Further Reading

Artmann, B. Euclid. *The Creation of Mathematics*. New York: Springer-Verlag, 1999.

Berggren, J. L. *Episodes in the Mathematics of Medieval Islam*. New York: Springer-Verlag, 1986.

Heath, T. L. *The 13 Books of Euclid's Elements*. New York: Dover, 1956.

Knorr, W. R. *The Ancient Tradition of Geometric Problems*. New York: Dover, 1993.

Rashed, Roshdi. *Entre arithmétique et algèbre: Recherches sur l'histoire des mathematiques arabes*. Société d'édition Paris: Les Belles Lettres, 1984.

Rosenthal, F. *The Classical Heritage in Islam*. Berkeley and Los Angeles: University of California Press, 1973.

Arab Contributions to Trigonometry

Overview

Trigonometry is one of the most practical branches of mathematics, finding uses in engineering, physics, chemistry, surveying, and virtually every other science and applied science. It is also one of the oldest branches of applied mathematics; practical problems in crude trigonometry have been dated to Egypt in about 1850 B.C., and the ancient Greeks developed more sophisticated trigonometry about 2,000 years later. Since that time, trigonometry has played a crucial role in many branches of mathematics and science, and is indispensable to our understanding of science and technical disciplines today.

Background

The earliest mention of a problem relating to trigonometry is in an Egyptian papyrus dated to about 1850 B.C. Although the concepts used are not stated in conventional trigonometric terms, it is obvious from the context that a form of "proto-trigonometry" existed at this time and was used to help ensure the pyramids were constructed according to the architect's specifications. However, it is almost certain the Egyptians did not place their calculations in a mathematical context that would allow them to draw any other conclusions from their results—the math involved was only applied to construction projects.

The next milestone in the development of trigonometry as a true mathematical discipline was reached by the Babylonians when they divided the circle into 360 equal divisions, or degrees. They did this because a year in their calendar had 360 days, so each day represented a degree. Since the Babylonians used a base-60 number system (as opposed to our base-10 system), 360 degrees was a tidy "fit" into their existing mathematics. The Babylonians also invented the *gnomon*, a device for measuring the angular distance of stars or planets above the horizon, which was similar to the protractor.

It is interesting to note how deeply ingrained the Babylonian numbering system is today: our hours have 60 minutes of 60 seconds each, we continue to use circles with 360 degrees, and our maps use 60 minutes of arc to a degree and 60 arc seconds to an arc minute. Clocks, maps, and protractors throughout the world are based on this system, even though a decimal (base-10) system would be easier to use.

The Greeks were first to elevate trigonometry to the level of an independent branch of mathematics. Greek trigonometers such as Pythagorus, Euclid, and Aristarchus advanced trigonometric theory and also championed new practical uses. Perhaps the most ambitious of these uses were Erastosthenes's calculation of the circumference of Earth and Hipparchus's determination of the distance of the Moon from Earth. In both cases, the final results were surprisingly close to current accepted values, in spite of the crude instruments used at the time.

In India, the Hindus made further advances during and after the fifth century. These advances included the construction of some early trigonometric tables and, more important, the invention of a new numbering system that made calculating much simpler. Hindu mathematicians based their version of trigonometry on variants of the sine function. The Hindu system led not only to the sine function, but to the cosine, tangent, and other familiar trigonometric functions we use today.

During their centuries of contact with the Greeks and Hindus, Arabic mathematicians adopted many of their mathematical discoveries. Among prominent Arabic mathematicians who helped translate Hindu mathematical texts or introduced Hindu mathematics to the Arabs were al-Battani (c. 850-929), Abu al-Wafa (940-998), and al-Biruni (973-?). Al-Battani adapted Greek trigonometry and astronomical observations to make them more useful. Al-Biruni was among the first to use the sine function in astronomy and geography, and Abu al-Wafa helped apply spherical trigonometry to astronomy, among other important contributions.

Impact

Arab mathematicians and scientists of the Middle Ages did more than translate Greek texts into Arabic, they translated *specific* Greek texts to use as reference materials for their own research in these areas. The Arab world lay between two other intellectual powerhouses—India and Greece. Arab scientists were exposed to the rich mathematical tradition of their own culture and, to this, they added the best of both Greek and Hindu mathematics and science. They were then able to synthesize these elements into a new way of looking at mathematics, as well as putting their mathematics to work on practical problems.

Abu al-Wafa made several significant contributions to the mathematics of the day. He made the first recorded mention of negative numbers in a book he wrote in the latter half of the tenth century. Today we take negative numbers for granted, but a thousand years ago, negative numbers were not widely accepted because they did not make intuitive sense to the people of that time. For example, we can all visualize having an apple, but how do you visualize having a "negative" apple? What does it look like? How do you count it? People of Abu al-Wafa's day were not used to thinking in these terms, and many simply refused to. Abu al-Wafa described negative numbers in monetary terms, referring to them as a "debt." This description of negative numbers could be grasped intuitively and was instrumental in bringing negative numbers into mainstream mathematics.

Abu al-Wafa's construction of tables of sines was also important. Having tables of sines may seem mundane because today we have calculators that instantly calculate all the trigonometric functions. To use trigonometric functions in calculations 1,000 years ago, one had to know their values, and these came either from hand calculation or from tables that had been laboriously calculated by hand and distributed. When he decided to calculate the value of the sine function for all angles at 15' (¼-degree) increments, Abu al-Wafa committed himself to an arduous and mind-numbingly repetitive task that required not only a great deal of commitment but also an almost unimaginable attention to detail. However, his work made these tables available for future generations of mathematicians who used his tables or their derivatives for centuries.

Abu al-Wafa was also first to introduce the concept of the tangent, the secant, and the cosecant to Arab mathematics. These functions, all derivatives of the sine function, are extremely useful in many areas of study, including physics, engineering, architecture, and surveying. The tangent had been described by Hindu mathematicians, but Abu al-Wafa showed how all the concepts could be used in mathematical calculations. By introducing these functions, Abu al-Wafa helped to increase the value of trigonometry by creating concepts that expanded its range.

If Abu al-Wafa had only translated some obscure texts into Arabic and generated some interesting functions, he might have passed into history without further notice. However, Abu al-Wafa and other Arab scholars helped to blend mathematical concepts from two distinct mathematical traditions into a synthesis which was much more important than either of its parts. Arab mathematicians took the geometric trigonometry (trigonometric identities derived from geometric drawings) of the Greeks, and added the mathematical sophistication and superior numbering system of Hindu mathematics, to create a trigonometry that very much resembles that of today. By doing this, they helped to create one of the most useful branches of mathematics.

Since the time of Abu al-Wafa, trigonometry has become essential to virtually all the sciences and applied sciences. Consider these examples:

1) The motion of rotating objects is usually described in terms using trigonometric functions. Engineering designs that involve rotating pieces (including wheels, camshafts, gears, motors, and fans), depend on trigonometry.

2) The motion of objects moving cyclically, such as pendulums, bridges swaying in a strong wind, and buildings oscillating after an earthquake is described using trigonometric functions. In fact, any periodic or oscillatory action can be described using trigonometry, including electrical equipment that uses oscillating electric fields, electronics, and the orbits of spacecraft.

3) Trigonometric functions also underlie much of physics, including quantum physics. These functions help describe phenomena ranging from probability functions that describe electron orbitals around an atom to the rotation of galaxies.

The technology, sciences, and mathematics upon which industrialized societies depend are based on trigonometry. Abu al-Wafa and his fellow Arab mathematicians and scientists could not have imagined how their work would someday be applied, but their discoveries are an indispensable part of our modern society.

P. ANDREW KARAM

Further Reading

Books

Boyer, Carl, and Merzbach, Uta. *A History of Mathematics*. John Wiley and Sons, 1991.

Maor, Eli. *Trigonometric Delights*. Princeton University Press, 1998.

Internet Sites

The MacTutor History of Mathematics Archive. http://www-history.mcs.st-andrews.ac.uk/history/index.html.

Omar Khayyam and the Solution of Cubic Equations

Overview

Omar Khayyam (c. 1048-1131), also known as Umar al-Khayyam, was a Persian poet, scientist and mathematician. Khayyam's greatest work in mathematics was his enumeration of the various types of cubic equations and his solutions of each type. Khayyam's work on cubic equations was a synthesis of Greek geometry, Babylonian and Hindu arithmetic, and Islamic algebra. His work greatly influenced future Islamic mathematicians, and through them the mathematicians of Renaissance Europe.

Background

Omar Khayyam was born in what is now Iran. He is best known in the West as a great Persian poet and philosopher. Khayyam's poetry (called the *Rubaiyat of Omar Khayyam*), written in the form of quatrains, was translated and adapted by Edward FitzGerald in the nineteenth century. These poems were admired as shining examples of Eastern culture. Although it is questionable just how much of the poetry attributed to Omar Khayyam was really written by Umar himself, his standing in the West has been made primarily based on his reputation as a poet.

Omar Khayyam's work as a mathematician was well known in the eastern world. However, it was not until the middle of the nineteenth century that Omar Khayyam the mathematician was "discovered" by the West, when Franz Woepke published *L'algèbre d'Umar al-Khayyami*. And it was not until 1931 that the mathematics of Khayyam was made available to the English-speaking world with David S. Kasir's translation of Woepke's book.

Omar Khayyam spent the majority of his life in various royal courts serving local rulers as court astronomer/astrologer. In this capacity, one of his primary duties was to create accurate calendars, an important work in medieval Islam needed to find the correct times for religious observances. It was often difficult for Khayyam to continue his work, as rulers changed and court intrigue caused him to fall out of favor with each ruler's successor.

Omar Khayyam's most important contribution to mathematics was his work involving cubic equations. A cubic equation is an equation whose highest degree variable is three, for instance $x^3 + 3x^2 - 2x + 5 = 0$. Although Khayyam did not contribute significant original methods of solution, he did write one of the first treatises that enumerated the different types of cubic equations and attempted to find the general solution of each different type of cubic equation.

In order to understand how Omar Khayyam solved cubic equations, we must first understand what mathematics was like in the eleventh and twelfth centuries. The mathematicians of the Islamic Empire were the intellectual descendants of Greek mathematicians from the last five centuries before Christ, and the Greeks had given the Islamic mathematicians geometry. Islamic mathematicians of the Middle Ages were also greatly influenced by the ancient numerical mathematics of the Babylonians and the more recent contact with the mathematics of Hindu India. The new science of algebra was being developed during this time. Even with the development of algebra, credited to al-Khwarizmi (c. 780-850), the ancient science of geometry continued to play a central role in Islamic mathematics.

In his work, *On the Sphere and the Cylinder*, the Greek mathematician Archimedes proposed and solved the problem, "to cut a given sphere by a plane so that the volumes of the segments are to one another in a given ratio." Since Archimedes stated that *volumes* of segments are related and volume is a three-dimensional measurement, this is the geometric equivalent of the algebraic solution of a cubic equation. Manaechmus, another Greek mathematician, showed how conic sections (a conic section is made by cutting a cone with a plane to produce geometric figures such as parabolas and ellipses) could be used to solve a similar geometric problem that is the equivalent to the algebra problem $x^3 = a^2b$.

The question of solving cubic equations by the use of conic sections was one that interested many Islamic mathematicians of the Middle Ages. These solutions involved a mixture of the geometric techniques inherited from the Greeks and the new algebraic techniques developed in the Islamic Empire. Mathematicians such as Ibn al-Haytham, al-Biruni, al-Mahani, and al-Khazin all contributed solutions to certain types of cubic equations. But it was Omar Khayyam who wrote the first treatise that set out the complete theory of cubic equations.

Omar Khayyam's book known as *Algebra* (c. 1078) contains all that is known of his work on cubic equations, as well as his work on quadratic equations. (The book's full title is *Hisāb al-jabr w'al-muqābala*, which can be translated as, "The calculation of reduction and restoration." The word *al-jabr* in the title is an Arabic word that means "to restore", and is the origin for the word "algebra.") A quadratic equation is one whose degree is two, such as $x^2 + 3x - 5$. However, Khayyam's work on quadratic equations was not as original as his work on cubic equations.

Omar Khayyam's work on cubic equations involved exhaustive evaluations of all the different forms of a cubic equation. For instance, he considered $x^3 + bx = a$ and $x^3 + a = bx$ to be different types of equations with distinct methods of solution. All told, Omar Khayyam provided solutions to more than a dozen different forms of cubic equations. One example of Khayyam's method for solving these equations is his solution of the equation $x^3 + bx = a$ by finding the intersection of a circle and a parabola. The intersection of these two curves is the solution of the equation. Recall that a cubic equation must have three solutions. In this case, the other two solutions are imaginary numbers, a concept that was not accepted until many centuries later; therefore, Khayyam found the only real solution to the problem. He also ignored any negative solutions that might occur.

The problems solved by Khayyam and his Islamic contemporaries were often given in terms of what we call today, "word problems." For instance, the following problem yields a cubic equation that must be solved:

> *Divide ten into two parts so that the sum of the squares of both parts plus the quotient obtained by dividing the greater by the smaller is equal to seventy-two.*

Omar Khayyam solved the resulting cubic equation by finding the intersection of a circle and a hyperbola.

Impact

If an algebra student today were to look at the solutions of cubic equations found in Omar Khayyam's *Algebra*, they would likely not find much that looked familiar. Islamic algebra did not employ the symbols we use today in modern algebra. Problems and their solutions were written out in words and illustrated using geometric constructions. For instance, the problem we would write as $x^3 + bx = a$ would read as follows: "a cube and roots equal to numbers." It would actually be four centuries after Omar Khayyam before European mathematicians, particularly the Italian known as Tartaglia, developed algebraic methods and symbols similar to our modern usage.

Why then do we consider Omar Khayyam's work, and the work of many other Islamic mathematicians, as contributions to algebra? Although Omar did not use symbols and did rely on geometric constructions, his work was fundamentally algebra. This is because he sought to find solutions by assuming the solution was known (this is what we use variables for today) and proceeding to solve the problem. This procedure is exactly the same as was used by Cardano in the sixteenth century to find solutions to cubic equations using symbolic algebra much as we would use today.

Renaissance Europe owed much to these Islamic mathematicians. Translations and improvements on Greek mathematics, made by medieval Islamic mathematicians, kept alive the advanced theories and techniques that provided a foundation for Western mathematics. The West is also indebted to Islamic mathematicians for adopting and transmitting our modern number system from India. This system, which introduced the simple yet important concept of zero, has come to be called the Hindu-Arabic number system.

It is possible that Khayyam's mathematical work influenced Renaissance European mathematicians directly. Historians continue to find previously unknown connections between medieval Islamic science and scientists of Renaissance Europe. In particular, European mathematicians were influenced by an Islamic mathematician by the name of al-Tusi (sometimes spelled at-Tusi), who lived a century later than Omar Khayyam. It is known that al-Tusi was in turn influenced by the work of Omar Khayyam, although it is not clear if Renaissance Europe knew of his work.

The importance of Omar Khayyam was that he gathered together all that was known about cubic equations, as well as adding some new ideas of his own. His work formed a bridge between the strictly geometric methods of the Greeks and what would become the modern algebraic methods for solving such problems. This bridge was strengthened by Islamic mathematicians who came after Omar Khayyam. The mathematicians and scientists of Renaissance Europe were indebted to their Islamic predecessors, and these Renaissance scientists helped to form the foundation upon which the Scientific Revolution was built.

TODD TIMMONS

Further Reading

Berggren, J. L. *Episodes in the Mathematics of Medieval Islam*. New York: Springer-Verlag, 1986.

Kasir, D. S. *The Algebra of Omar Khayyam*. New York, 1931.

Katz, Victor J. *A History of Mathematics*. Reading, MA: Addison-Wesley, 1998.

Medieval Kinematics

Overview

Medieval European mathematicians inherited from the works of Aristotle (384-322 B.C.) a tradition of kinematics that was more qualitative than quantitative. Without proceeding far in the direction of experiment, the medieval scholars of France and England could still use arguments that suggested new conclusions about how to describe motion. In the tradition of Aristotle, these arguments had to apply to all sorts of change. In the hands of subsequent physicists the arguments proved the basis for laying a foundation of a quantitative physics devoted to the study of motion based on experiment as well as theory.

Background

Kinematics is the study of motion. By the end of the twelfth century the works of Aristotle on the subject were being introduced into Western Europe, frequently via translations from Arabic and Hebrew intermediaries. The transmission and preservation of the texts of Aristotle had a value in their own right, but there were clearly issues about motion that Aristotle did not resolve. The works of Aristotle made it difficult to isolate questions simply related to the movement of objects. Aristotle had looked at motion as simply another form of change, and consequently his arguments about motion had to apply equally well to changes of color or even to objects coming into existence and ceasing to be. One of the results was that Aristotle could not take a quantitative approach to issues of motion because it was unclear how one could apply numerical values to other forms of change.

One of the additional difficulties for the study of motion in the Greek tradition was the Euclidean background for the geometry that would be used to describe the motion mathematically. Euclid (330?-260? B.C.) had devoted a good deal of space in his *Elements* to discussion of different sorts of ratios. The treatment that was available during the Middle Ages did not allow for the possibility of taking a ratio between different sorts of quantities (like distance and time). Since fundamental notions like velocity are now expressed as a ratio of distance and time, there was not a mathematical basis for expressing the ideas of motion as they were developed subsequently.

New approaches to kinematics were formulated primarily in the universities of Paris and Oxford, which had only recently come into existence. In the thirteenth century there were arguments within the Church about the relative value of scientific and religious studies, and those who followed the scientific path had to worry about the dangers of accusations of heresy. Fortunately Albertus Magnus (1206-1280) gave the imprimatur of the Church to a variety of scientific pursuits and steered clear of doctrinal issues. In the aftermath of his leadership, scholarship within the university could look at issues discussed in Aristotle without having to treat his view as enshrined by the Church.

At Oxford, the group of mathematicians associated with Merton College looked at laws governing "local motion," as the movement of physical objects in space was called to distinguish it from other sorts of change. Aristotle had explained motion in terms of objects having natural places and seeking to return to those places when they had been moved out of them. It is perhaps characteristic that one of the chief accomplishments of the Merton school of physicists was coming up with a suitable definition for "uniform speed" and "uniformly accelerated motion." Uniform motion involved going through equal distances in equal time intervals. The slightly more complicated notion of uniform acceleration involved adding equal increments of velocity in equal time intervals. On this basis they were able to derive a relationship between the distance traveled by an object and the speed at which it was traveling. All of these notions had to be expressed in words, since there was no notation for representing the quantities in an algebraic equation.

Even more impressive was the work of Nicole Oresme (1320?-1382), a French mathematician whose work anticipated many of the developments of the next few centuries. In particular, Oresme came up with the idea of representing motion by a picture, which resembled the graphs made familiar a few centuries later by the work of René Descartes (1596-1650). Oresme looked at the general shapes of motions as represented on paper and argued that differences in the shapes of the motions indicated that there were different laws of motion being obeyed. As with the Merton school, Oresme was saddled with the Aristotelian and Euclidean traditions that made him try to apply his diagrams to changes other than motions. While the members of the Merton school had run into logical problems about the possibilities of motion, Oresme worked out his results in Euclidean terms without worrying so much about the philosophical issues that they raised.

Impact

Perhaps the most immediate impact of the work of the Merton school in England and Oresme in France was the realization that there were many issues about motion that were not settled by Aristotle. Even though they did not verify their results through experimentation and they trusted a fundamental Aristotelian view of motion, they were careful to clarify some points that Aristotle had left obscure and to produce quantitative results. In an era inclined to take Aristotle's texts as definitive, this was an important contribution to justifying continued work in areas of science that Aristotle had tackled.

In physics the work of Oresme had a great influence on Galileo (1564-1642). It is certainly true that Galileo was in revolt against many of the ideas about science taken for granted in the educational system of his time. On the other hand, Galileo also recognized the value of Oresme's use of graphical representations for motion and built on them in his own works for his fellow scientists and the general public. Galileo's use of experimentation enabled scientists to apply to a wide variety of situations the mathematical calculations of three centuries before and to confirm that the numerical values obtained applied to objects in the physical world. With Galileo, the importance of Oresme's work was easier to recognize, as Galileo had not felt obliged to look at the broader issues of changes other than motion.

Within mathematics the works of the Merton school and of Oresme had more of an influence through the coordinate geometry of Descartes. In looking back at the texts of Oresme it is easy to see diagrams that resemble graphs with coordinate axes. As a result, there is a temptation to claim that Oresme must have invented analytic geometry long before Descartes. The problem is that Oresme does not use points as representations of two coordinates, but instead was concerned just with the general shape of the picture. The same approach has been popular with certain branches of mathematics in the twentieth century, such like catastrophe theory. Descartes's work, by contrast, used graphs in a more strictly quantitative way and laid the foundations for work that went well beyond what Oresme could have envisaged.

Most of the effects of the work of the Merton school and of Oresme did not become visible until several centuries later. One of the reasons was the limited access to texts in the days before printing rendered them more generally available. It is also the case that some of the scientific innovation in the centuries that immediately followed Oresme took place in Italy and Germany, where works from France and England were not so well known. Even though all the medieval physicists in Western Europe wrote in Latin, the transmission of texts from one part of the world to another was a painfully slow process. As a consequence, much of the work that was done in the first few centuries after the revival of Aristotle in Western Europe had to be rediscovered (either independently or when the texts became available) some centuries later.

Physics (whose very name has a Greek root referring to "nature") had received its initial formulation in the works of Aristotle. Because of self-imposed limitations in his work, Aristotle could not even ask some of the questions that are fundamental in the study of motion. The work of the Merton school and of Oresme provided clear formulations of some of the questions and even started on the path to quantitative answers. With the development of mathematical notation and the introduction of experimentation over the next few centuries, the work in medieval kinematics could serve as the basis for a revolution in the study of motions of bodies on the Earth and in the skies.

THOMAS DRUCKER

Further Reading

Clagett, Marshall. *The Science of Mechanics in the Middle Ages.* Madison: University of Wisconsin Press, 1959.

Dijksterhuis, E.J. *The Mechanization of the World Picture.* Princeton, NJ: Princeton University Press, 1986.

Grant, Edward. *The Foundations of Modern Science in the Middle Ages.* Cambridge: Cambridge University Press, 1996.

Pedersen, Olaf. *Early Physics and Astronomy: A Historical Introduction.* Cambridge: Cambridge University Press, 1993.

Combinatorics in the Middle Ages

Overview

Combinatorics is concerned with defining a finite or discrete mathematical system, and then solving problems relating to the selection and arrangement of numbers or items within that system. A typical problem in combinatorics is to determine the number of possible configurations of a particular type, such as the values that could occur when rolling a pair of dice. Combinatorics has many applications in probability as well as in algebra and geometry.

Background

Combinatorics is an ancient branch of mathematics. In the Rhind Papyrus of Egypt, one of the oldest mathematical texts in existence, the 79th problem asks, "There are seven houses; in each house there are seven cats; each cat kills seven mice; each mouse has eaten seven grains of barley; each grain would have produced seven hekat. What is the sum of all the enumerated things?"

The Rhind Papyrus was copied by a scribe named Ahmes in about 1550 B.C. from a text that was about 300 years old at that time. A similar problem was preserved in the English nursery rhyme that begins, "When I was going to St. Ives, I met a man with seven wives."

One of the first recorded combinatorial constructs was the *magic square.* The magic square is an array of numbers in which the rows, columns, and diagonals all add up to the same sum. Magic squares appear in the Chinese book the *I Ching,* which dates from the twelfth century B.C.

Impact

Medieval scholars were fascinated by magic squares. Like the Chinese, Arab mathematicians who worked on them, including al-Buni, invested them with mystical significance. However, when the constructs were introduced to the West, in a text written by the Greek Byzantine scholar Manuel Moschopoulos in 1300, they were presented as pure mathematics.

Jewish mathematicians of the Middle Ages, including Rabbi Abraham ben Meir ibn Ezra (1092-1167), who lived in Muslim Spain, and Levi ben Gershom (1288-1344) of France, were particularly interested in combinatorial problems. The Hebrew mathematical tradition arose in the area of modern Israel around the eighth century, influenced by contacts with Islamic scholars and the innovations they brought from the Hindus, and influencing Arab mathematicians in turn. It spread with Jewish merchants and immigrant populations into Arab, Persian, Turkish, and Christian territories. Together, Jews and Arabs helped to bring the Hindu number system (what we often call "Arabic numerals") into Europe. At the time, the West was still laboring under the awkward Roman numeral system, which made arithmetic calculations difficult.

An important feature of Hebrew mathematics was its early understanding of variation in such values as the size of bricks and the positions of boundaries. Counting the number of possibilities in a situation, and weighing their likelihood, was also important in deciding legal questions in accordance with the complicated Talmudic code. This led to a tendency to apply combinatorics to statistical and probabilistic problems.

While it is often difficult to determine who originated specific mathematical methods, it may be established that the techniques are known in a particular community when they appear in the writings of scholars there. In India, the Jaina school of mathematics, established in the first century B.C., included a subject called

Vikalpa, or permutations and combinations, in its course of study.

The Indian mathematician Bhaskara (1114-1185?) included a number of algebraic and combinatorial problems in his book *Lilivati* ("The Graceful," said to be named for his daughter). Often the problems were couched as diversions, such as riddles about the number of geese in a flock or swans in a lake. One was to figure out the number of combinations of stressed and unstressed syllables that are possible in a six-syllable verse.

An important algebraic application of combinatorics, for which rules were given in *Lilivati,* is the computation of the *binomial coefficients.* The binomial coefficients are those of the individual algebraic terms when the expression $(a + b)^n$ is multiplied out. For example, $(a + b)^2$ is equal to $a^2 + 2ab + b^2$. The coefficients in this expression are 1, 2, 1.

Binomial coefficients are often written in a triangular array called "Pascal's Triangle." The construct was named for the famous French mathematician and scientist Blaise Pascal (1623-1662), who developed many of the founding principles of probability theory. Its peak is a 1, because $(a + b)^0$ is equal to 1. The next row has two 1s, because $(a + b)^1 = a + b$; both coefficients are equal to 1. Each succeeding row is made up of the next set of coefficients, derived from writing 1s on each side, and computing each inner coefficient by adding the two numbers directly above it.

Despite the name, Pascal was not the first to use a triangular figure to compute binomial coefficients. The same figure appears in an algebra text by Ibn Yahya al-Maghribi al-Samawal (1130?-1180?), a Jew (and later convert to Islam) in Baghdad, who, in turn, attributed it to al-Karaji (953-1029?). It was known to the Persian philosopher Nasir ad-Din al-Tusi (1201-1274) by 1265, and also had been developed in China, probably well before its appearance in the work of Chu Shih-chieh (1270?-1330?) in 1303.

Combinatorial problems have often been associated with games and puzzles. The Italian mathematician Leonardo Fibonacci (1170?-1250?) is most famous for devising a series in which each number is the sum of the previous two: 1, 1, 2, 3, 5, 8, . . . However, he is also credited with the "Rabbit Problem" of combinatorics. This puzzle requires the computation of the number of descendents of a pair of rabbits. According to the problem, the initial pair is left in an enclosure to produce a pair of offspring a month, which are assumed to become similarly fertile when they are two months old. In 1256 the Islamic mathematician Ibn Kallikan constructed a puzzle in which a chess board had 1 grain of wheat on the first square, two on the second, four on the third, eight on the fourth, and so on.

It was centuries before combinatorial techniques were widely used in the West. Pascal and his contemporary Pierre de Fermat (1601-1665) employed combinatorics to develop probability theory. The Swiss mathematician Leonhard Euler (1707-1783) solved the famous Königsberg Bridge problem by proving it had no solution. The river town of Königsberg stretched across two islands and the opposite banks of the river, and seven bridges connected these four landmasses. The object of the problem was to cross all seven bridges without having to cross any of them more than once, starting and ending from the same point. Euler showed that for such a journey to be possible, each landmass would have to have an even number of bridges connected to it. Similar recreational problems in combinatorics were circulated during the nineteenth century.

Around 1900, combinatorial theory contributed to graph theory, symbolic logic, and topology. The first comprehensive textbooks in the field, such as Eugen Netto's (1848-1919) *Lehrbuch der Combinatorik,* were written at about this time. Combinatorics became more prominent in mathematics in the twentieth century, with the development of computing technologies. Combinatorial problems arose in the areas of computer design, information theory, and coding. Graph theory has applications in designing and analyzing transportation and communication networks and other systems. However, the field still exists as a collection of discrete problems and methods, and mathematicians continue to search for a theory or set of principles to unify them.

SHERRI CHASIN CALVO

Further Reading

Books

Cooke, Roger. *The History of Mathematics: A Brief Course.* New York: John Wiley and Sons, 1997.

Grattan-Guiness, Ivor. *The Norton History of the Mathematical Sciences.* New York: W.W. Norton, 1997.

Ifrah, Georges. *The Universal History of Numbers.*New York: John Wiley and Sons, 2000.

McLeish, John. *Number: The History of Numbers and How They Shape Our Lives.* New York: Fawcett Columbine, 1991.

Periodical Articles

Biggs, N.L. "The Roots of Combinatorics." *Historia Mathematica* 6, No. 2 (1979): 109-36.

The Muqarnas: A Key Component of Islamic Architecture

Overview

The *muqarnas,* a Muslim variety of stalactite vault, is a primary characteristic of Islamic architecture. Developed during the mid-tenth century in both northeastern Iran and central North Africa, two ends of the sprawling expanse that constituted the Dar al-Islam, the muqarnas, with its honeycomb texture, became a common feature in palaces and temples. Indeed, it became a prominent feature in nearly all Islamic structures from the eleventh century on, persisting as a key element of the Islamic architectural vernacular until modern times.

The muqarnas is a form that embodies the ideals of Islamic civilization: its physical form, characterized by fluidity and replication, is based as much on Islamic theological principles as it is on the more mundane principles of structural engineering.

Background

There are four main attributes of the muqarnas that distinguish its appearance. First, it is three-dimensional, thereby providing volume in built structures. Second, the degree of this volume is variable. As a result, this variability allowed architects to implement the muqarnas as a purely architectonic intended to provide support to a structure, or as an ornamental device. As a third characteristic, the muqarnas knows no logical or mathematical boundaries. None of its elements are a finite unit of composition; as a result, there are no logical or mathematical boundaries limiting the scale of the composition of a muqarnas. In that sense, perhaps it is useful to liken the tiered honeycomb patterns of the muqarnas to a figure such as the fractal. The fractal, the shape of which is determined by a simple algebraic equation, also knows no limitation and can conceivably stretch to infinity. Likewise, the complexity of the muqarnas is limited only by the skill of the architect and the builder. The fourth characteristic of the muqarnas is that, because of its variable volume, a three-dimensional unit can easily be transformed into a two-dimensional figure.

Impact

The muqarnas is one of the key components of Islamic architecture. It is as much a component of the vernacular of Islamic architecture as the Ionic column is of Greco-Roman architecture. Likewise, the muqarnas became a standard architectural feature at a moment when the Islamic architectural style exerted its greatest influence. As a result, medieval buildings from Cordoba, Spain, to Damascus, Syria, exhibit the intricate lattice work of the muqarnas.

The Dar al-Islam was the preeminent power structure of the Muslim world from about the eighth century until the thirteenth century. The great influence of the muqarnas in the built environment of cities from the Atlantic to China is testament to the political might that the Islamic juggernaut exerted over the medieval world. Indeed, the muqarnas was developed as the Islamic power structure developed. After the prophet Mohammed's death in 632, the Islamic power structure expanded rapidly. Indeed, in 732, only 100 years after Mohammed's death, the Arabs had extended from the desert to central France, where they suffered a decisive defeat at Poitiers. An Islamic culture, evident in art, literature, and architecture, developed as an Islamic empire became reality.

Islamic art and architecture are characterized by their reliance on decoration and repetition. The role of decoration in Islamic art helps to characterize it. This extends to architecture as well. The extent to which a form has an architectural and decorative function helps to determine its overall value in an Islamic context to a degree unknown to Western aesthetics. For Western art and architecture, form and function are separate categories, and a building must not necessarily be both functional and aesthetically pleasing.

For Islamic art and architecture, however, these two components are synonymous.

The style of decoration used by Islamic artists and architects resulted in buildings and objects overlaid with decorative elements. The buildings themselves were often structural cores that were covered, coated, and enveloped with different materials. The bricks of a muqarnas, for instance, may have been covered with wood or tiles that added additional depth and complexity to an already-complex core.

While Westerners often hold the mistaken belief that Islamic art is limited to two dimensions, Islamic art is capable of transforming the two-dimensional into the three-dimensional through the repetition and interlacing of patterns. Interlaced designs create the illusion of different planes. Furthermore, the textured surfaces of building materials such as stucco and brick multiply the possibilities available to artists and architects. Indeed, the masterly control that Islamic designers exhibit over geometric patterns enables them to manipulate two-dimensional patterns in order to produce three-dimensional, optical effects. The complexity of the muqarnas, for instance, allows designers to balance positive and negative areas of space, thereby creating buildings that are functionally and aesthetically whole.

Indeed, in the Islamic world, the role of the architect-engineer was closely associated with the mathematician. In the West, on the other hand, the role of the architect came into existence through a process whereby designer-builders slowly distinguished themselves from the ranks of builders and craftsmen. The Islamic architect Ahmad ibn Muhammad al-Hasib, for instance, was referred to as al-hasib, which means "the mathematician." Issues of geometric complexity were the province of the Islamic architect.

These fundamental components of Islamic decoration are not without philosophical and theological significance, however. Indeed, Muslim philosophical concepts are embodied in their architectural forms. The relations between these forms are determined by what we may refer to as a spiritual mathematics. Islamic spirituality is linked with the variety of qualitative mathematics developed by Pythagoras (580?-500? B.C.). For Pythagoras, numbers were the ultimate elements of the universe. In other words, numbers were divine: for a Pythagorean, an understanding of relations between numbers can only result in an understanding of the nature of the universe.

Islamic spirituality, then, sees a direct link between the intellect and the spirit. These relations are, essentially, transcribed into spatial metaphors. The intellectual and the spiritual are ways of comprehending, respectively, the corporeal and invisible worlds that overlap. The mathematical structure of Islamic art and architecture reflects the point at which the intellectual flows into the spiritual. As such, repetition serves a conceptual function. It is intended to provide a space where tensions are resolved: despite the intricate complexity of the muqarnas, its form is also determined by constant rules dictating the relationships between its components. Consider the definition of the muqarnas given by the fifteenth century Timurid mathematician al-Kashi (1380?-1429):

> *The muqarnas is a ceiling like a staircase with facets and a flat roof. Every facet intersects the adjacent one at either a right angle, or half a right angle, or their sum, or another combination of these two. The two facets can be thought of as standing on a plane parallel to the horizon. Above them is built either a flat surface, not parallel to the horizon, or two surfaces, either flat or curved, that constitute their roof. Both facets together with their roof are called one cell. Adjacent cells, which have their bases on one and the same surface parallel to the horizon, are called one tier.*

Al-Kashi uses geometric terms, indicating that the muqarnas is the space in which a particular mathematical problem is enacted. In this problem, the possible number of permutations is determined by variations on the 90° angle. As al-Kashi indicates, the angle at which two facets intersect can be any conceivable angle, as long as it corresponds to "a right angle, or half a right angle, or their sum, or another combination of these two." The type of combination permissible is not limited: only the elements that may be used in such a combination are predetermined. These components are variations of the right angle. The rules that guide the design of the muqarnas remind us of the architectural function of the device. The muqarnas is, in essence, little more than a squinch, or an interior corner support. Corners are made of 90° angles. Corners often need additional support because that is where stress is concentrated. The muqarnas reminds us, over and over again, that it supports weight directly. Indeed, the more ornate it is, the more weight it can hold. Each additional tier creates a greater surface area of the structure, adding increased support. In terms of Western architecture, this seems paradoxical. For the Westerner, form and function are divorced: frequently one is sacrificed at the expense of the other. Elements such as the muqarnas, however, reveal that in Islamic architecture practical applications and aesthetic considerations frequently overlap.

The elaborate possibilities available to the designer of the muqarnas also serve to obscure its physical function. Furthermore, its appearance is meant to call to mind divine associations for the viewer as well. The stalactite structure raises the eye upward. In this sense, the large precise forms closest to the viewer congeal into a single form higher up in the vault. This can represent earthly reflection of supernatural archetypes: the earthly forms and angles with which we are familiar in the material world are imperfect approximations of unknowable heavenly forms. Conversely, the downward movement of the muqarnas implies the descent of the heavenly towards the earth and the encasement of the divine in material forms.

This connection between mathematics and theology is not specific to Islamic art and architecture, however. The mathematical focus of these disciplines reflects the mathematical nature of the Muslim religion. Indeed, the Koran, the Muslim holy text, is noted for its numerological significance. The Koran features a bewildering mathematical structure; Muslim scholars seek to decode the numerical significance in its letters and words. In this sense, intellectual mastery of mathematic subtlety leads to spiritual illumination. Architectural forms reveal the structure of the divine because, in the Pythagorean sense, architecture is one of the most godly of all arts. Like music, architecture is an art based on numbers, and the possibilities offered by numerical proportion come closest to emulating the divine proportions of God's physical creation. While words often speak plainly, for mystics, numbers reveal divine proportion.

DEAN SWINFORD

Further Reading

Creswell, K.A.C. *A Short Account of Early Muslim Architecture.* Baltimore: Penguin Books, 1958.

Hillenbrand, Robert. *Islamic Art and Architecture.* New York: Thames and Hudson, 1999.

Michell, George, ed. *Architecture of the Islamic World: Its History and Social Meaning.* London: Thames and Hudson, 1978.

Nasr, Seyyed Hossein. *Islamic Art and Spirituality.* Albany: State University of New York Press, 1987.

Recreational Mathematics in the Middle Ages

Overview

Mathematics in the Middle Ages was largely concerned with commenting on traditional texts, most notably the *Elements* of Euclid (fl. 300 B.C.). This work supplied both the vocabulary in which much mathematics was done and the kind of problems to whose solution mathematical effort was devoted. After the fall of Rome in 476, it took many centuries even for a complete text of Euclid to be available to mathematicians in western Europe. As a result, much of the mathematics that was being done could be called "philosophy of mathematics" and involved an investigation of the properties of the numbers that were the building blocks of mathematics in Euclid and elsewhere. Amid all this work (which resembles a kind of number mysticism), there were some techniques that could be incorporated in mathematics when it was less involved with issues of theology.

Background

Geometry was the form of investigation that the most sophisticated mathematical work of the Greeks took. Arithmetic was more or less built into the system of geometry that Euclid developed, although he did not describe the system of calculation in detail. In the geometrical work of Euclid and his successors, the mathematical ideas were given a firm foundation by means of explicit definitions, axioms, and a sequence of theorems whose logical relations to one another were set out equally explicitly. Arithmetic was done by means of representing numbers of lengths, although this did have some peculiar consequences. Thus, when a number (represented by a length) was multiplied by a number (represented by a length), the product would be represented by an area. This would scarcely have made for an easy system of calculation.

By contrast, there were problems and speculations about arithmetic stretching back to Diophantus of Alexandria, about whose life little is known. He gave methods of solution to what are now called "word problems," but then there were no other sorts of problems available to students of arithmetic in Greek mathematics. There was an even longer tradition of number mysticism going back to the earliest commentators on the Bible. It seemed difficult to make sense of some of the numbers used in books of the Bible (from Genesis to Revelation), and theologians applied themselves to understanding how the numbers fit together. Issues ranged from seeking explanation for why human beings were created on the sixth day of creation to figuring out how the Israelites in Egypt could have increased so rapidly over four generations. The kind of mathematics being trotted out to solve such problems was not impressive by Greek standards, but in the absence of Greek mathematical texts it was the best available. It was not until the eleventh century that Greek mathematical texts began to be available in translation in western Europe, after they had been handed down through the mathematicians of Islamic civilization in the Middle East and Spain.

Typical of the sort of arithmetic investigation popular among the scholarly community during the Middle Ages was the work of Rabbi ben Ezra (1092-1167). A Jew working within Arabic civilization in Spain, he wrote about the positive whole numbers in a variety of settings and tended to take an attitude in which religion was mixed with observations of nature. In paragraphs about numbers like six or seven, he would bring in Biblical references as well as referring back to the kind of number speculation associated with the disciple of Pythagoras (c. 570-c. 500 B.C.) In this tradition the numbers had properties associated with them (in the same way that people did) like bravery or masculinity. Despite the rather far-fetched nature of some of these speculations, ben Ezra also worked in the area that would nowadays be called combinatorics. This involves the number of ways of arranging objects and making selections from them. It seemed as though his work was connected with astrological issues, keeping track of the number of ways that the planets could be grouped. Astrology had a more distinguished place in medieval science than it has had in more recent times.

Perhaps the best known of the medieval investigations of combinatorics was the work of Ramon Llull (c. 1232-1316). Just as with ben Ezra, Llull was not interested in the abstract study of mathematics and patterns for their own sake. Instead, he was strongly interested in trying to set out the truths of the Christian religion in a way that would make them convincing to the followers of Islam. Llull was from Catalonia in northeastern Spain, and the question of dealing with the Muslim enclaves still left in Spain at the time was not just a matter for idle speculation. Llull sought to put the attributes of God in a list and then to present the possible combinations of those attributes as facts of Christian belief. He wrote of manufacturing ways of enumerating these combinations and came up with different means of representing them throughout his life. Although the motivation for his efforts was religious, once he created the diagrams for representing the attributes of God, he subordinated theology to combinatorics.

Impact

The work on arithmetic and speculations about the whole numbers continued under the heading of religion and philosophy after they had been given up on as mathematics. Book VII of the *Elements* of Euclid had sounded something like a philosophical foundation for the whole numbers, and arithmetical reflections of religious ideas seemed not to fit in so badly. When Book V became available to western Europe, its advantages for doing mathematics enabled the arguments about Book VII to be relegated to a backwater. The mathematical traditions that were imported from the Arabic scientific world offered more substantial problems on which to work. Numerology did not long remain in the curriculum of mathematical studies at universities and earlier instruction.

The Pythagorean speculations about number did not entirely disappear from discussions in the philosophy of mathematics. The Indian mathematician Srinivasa Ramanujan (1887-1920) was said to have an intimate, firsthand acquaintance with many integers, which explained his ability to come up with generalizations about their property. There is a mystical strain to some of the attitudes of Kurt Gödel (1906-1978) in which one can hear an echo of medieval attitudes towards the integers. Within mathematics itself such speculation had vanished, as there were too many interesting problems that could be couched in more concrete terms.

The individual who followed most closely in the footsteps of Ramon Llull was the German polymath Gottfried Wilhelm Leibniz (1646-1716). Leibniz had a detailed acquaintance with the mathematics of his day and independently created the calculus (along with Isaac Newton), to name just one of the major contributions he made to mathematical progress. Combinatorics had been the study of some earlier work in the seventeenth century, most of it not concerned with the work of Llull.

Leibniz, by contrast, was trying to deal with the problem of communication between Christian denominations as well as between Christians and non-Christians. He harked back to the approach that Llull had used in search of a universal language in which all the truths of religion could be framed. His hope was to create a language in which statements could not only be given a comprehensible form, but their truth or falsity could be settled by a mechanical procedure. He wrote a good deal about the benefits of such a language but did not accomplish much by way of constructing it. Leibniz's view of logic and the form of sentences led him to the sort of enumeration of combinations that was built into Llull's work.

By the start of the twentieth century, mathematical logic had begun to supply a more concrete realization of Leibniz's vision. Technical developments in logic, however, had led to a more complex situation than anything found in Llull. It seems hard to believe the claims that Llull's work had anything to do with the development of computer science, since the structures for the computer scientist would be unrecognizable to Llull.

Recreational mathematics has frequently supplied topics for investigation outside the mainstream of contemporary mathematics and ended up providing new directions for mathematical research. In the Middle Ages the mathematical language and texts available were sufficiently impoverished that the recreational byways did not lead to anything near the boundaries of current developments. When too much of the mathematics being done is recreational, there is not enough mathematical technique to be called upon to be applied to the solution of recreational problems.

THOMAS DRUCKER

Further Reading

Clagett, Marshall. *Mathematics and its Application to Science and Natural Philosophy in the Middle Ages*. Cambridge: Cambridge University Press, 1987.

Katz, Victor J. *A History of Mathematics: An Introduction*. New York: HarperCollins, 1993.

Rabinovitch, Nachum L. *Probability and Statistical Inference in Ancient and Medieval Jewish Literature*. Toronto: University of Toronto Press, 1973.

Biographical Sketches

Abu al-Wafa

940-998

Persian Mathematician and Astronomer

Abu al-Wafa introduced the trigonometric functions of the secant and cosecant, and developed a method for computing sine tables. As an astronomer, he worked in a Baghdad observatory, where he created the first wall quadrant, a device used in observing the movement of heavenly bodies. He also wrote and translated a number of works.

His full name was Muhammad ibn Muhammad ibn Yahya ibn Isma'il ibn al'Abbas Abu al-Wafa al-Buzajani, and he is known variously to history as Abu al-Wafa, Abu al-Wafa al-Buzajani, Abu'l Wefa, and Abul Wefa. The details of his early life are unknown, and the first definitive date from his career is 959, when he began his work at the Baghdad observatory.

Though he is rightly credited for his invention of the wall quadrant, Abul Wefa did not—as some historians later claimed—discover the variation or inequality of the moon's motion. His work on lunar theory required the development of new trigonometric methods, and thus it was that he calculated tangent and cotangent tables. He also created the secant and cosecant functions, and employed a new means of calculating sine tables, as an aid to astronomical observation. In addition, he proved the generality of the sine theorem for spherical triangles.

Abu al-Wafa also made an invaluable contribution to scientific knowledge by translating ancient works of Euclid (c. 325-c. 250 B.C.) and Diophantus of Alexandria (c. 200-c. 284) into Arabic. These writings would eventually influence scholarship in Europe, where many of the Greek originals had been lost. Likewise some of Abu al-Wafa's commentaries on Euclid, Diaphanous, and al-Khwarizmi (c. 780-c. 850) have disappeared. In the realm of applied mathematics, he wrote works whose titles have been translated as *Book on What Is Necessary from the Science of Arithmetic for Scribes and Businessmen* and *Book on What Is Necessary from Geometric Construction for the Artisan.*

JUDSON KNIGHT

Adelard of Bath

1075-1160

English Scholar

The contributions made by Adelard of Bath to mathematical knowledge were primarily in the area of scholarship, translation, and historical writing rather than original theory. Nonetheless, his achievements were crucial to the growth of learning in Western Europe, and included the first translation of Euclid's (c. 325-c. 250 B.C.) *Elements,* which became the first important geometry textbook in the West, and al-Khwarizmi's (c. 780-c. 850) *Tables,* which introduced Arab astronomy to Western thinkers. In presenting ideas derived from the East, Adelard brought two other concepts of staggering importance to the attention of Western mathematicians: trigonometry and Hindu-Arabic numbers.

Though he is known as an Englishman, Adelard spent much of his career outside his home country. In France, he studied at Tours and from 1100 taught at Laon. He later spent seven years traveling, first visiting the medical school at Salerno in southern Italy, a highly significant institution that served as a precursor to the idea of a university. He also visited Sicily, which had long been held by Arabs (though by then it was under Norman control), and it is likely that he gained his knowledge of Arabic while there. Later years found Adelard in Cilicia, now part of Turkey, from whence he traveled through Syria and Palestine. By 1130 he was back in Bath.

Early in his career, Adelard wrote a study of arithmetic based on the work of Boethius (480-524), as well as a philosophical work in 1116 that shows a heavy influence from Plato (427-347 B.C.). With his exposure to the East he gained access to new ideas—some of which were not new at all; rather, these were concepts from ancient Greece which had been lost to Western Europe, but were recovered by Arab thinkers. Thus he became influenced by the philosophy of Aristotle (384-322 B.C.), who was much more highly regarded than Plato by Arab scientists, and undertook his translations of Euclid.

With regard to the latter, three versions of the *Elements* have been attributed to Adelard. The first appears to have been drawn from an Arabic translation attributed to al-Hajjaj (661-714). By contrast, the wording of the second is so different from that of the first that historians assume it was based on another, unknown, Arabic version of Euclid. The second version is considered less thorough than the first, and did not acquire the status of his predecessor; however, a third translation—which may or may not have been the work of Adelard—was later quoted extensively, by Roger Bacon (1213-1292) and others.

In his work on al-Khwarizmi, it became necessary for Adelard to use symbols hitherto unknown in the West: the numerical characters developed in India, adapted by the Arabs (hence the erroneous name "Arabic numerals"), and still in use throughout the world today. Thanks to Adelard, Western scientists were also introduced to the concept of astronomical tables, and other works by him exposed them to trigonometry, as well as the abacus and the astrolabe.

JUDSON KNIGHT

Leon Battista Alberti

1404-1472

Italian Architect and Writer

Like many key figures in the history of mathematics during the Middle Ages, Leon Battista Alberti was not a professional mathematician, and the advances he made in that discipline were in service to another—in his case, architecture. Nonetheless, his contributions were of great importance and included the first general study on the laws of perspective (*Della pittura,* 1435) and a book on cryptography that includes the first use of a frequency table. Alberti also worked with Toscanelli dal Pozzo (1397-1482), who later provided the maps used by Christopher Columbus (1451-1506) for his now-famous first voyage on a project involving geometrical mapping.

Born in Genoa on February 14, 1404, Alberti was the illegitimate son of Lorenzo Alberti, a prominent Florentine who had been banished from his city three years before. He attended school at Padua, then enrolled at the University of Bologna in 1421. There he earned a degree in canon law in about 1428, and around this time received an appointment as prior of San Martino in Tuscany. Alberti would hold that position for the remainder of his life, but he also served the church in a number of other capacities.

Alberti wrote a large number of works on a vast array of subjects, from architecture to the family to morality to law. These began to appear in the 1430s, and his most important writings—those on the arts—began after he was appointed to the court of Pope Eugenius IV in 1434. The following year saw the publication of the Latin *De pictura,* which appeared in a more popular Italian version as *Della pittura* in 1436. The latter marked the first mathematical explanation of the theories of one-point linear perspective developed by the architect Filippo Brunelleschi (1377-1446).

The 1430s and 1440s saw Alberti in a number of locations around Italy, serving a papal court that was often forced to relocate temporarily due to unrest in Rome. By 1443, the pope had returned to his home, and it was during this time that Alberti began his seminal Latin study of architecture, *De re aedificatoria.* The latter, which drew on the ideas of the Roman writer Vitruvius (c. A.D.1), would have an enormous impact on European architects in the two centuries that followed.

Alberti's work as an architect dates primarily to the period beginning in 1447, when he designed the Rucellai Palace in Florence. There followed numerous commissions throughout Italy, though the grandest of these—a vast building plan for Rome commissioned by Pope Nicholas V in 1450—was never realized.

In 1471 Alberti, who had played a key role in reviving Italians' interest in their architectural past, served as a guide to the Roman ruins for a distinguished party that included Lorenzo de' Medici (1449-1492). Alberti died in Rome soon afterward, most likely in early April 1472.

JUDSON KNIGHT

Bhaskara

1114-1185

Indian Mathematician and Astronomer

Bhaskara, one of the greatest medieval Indian scholars, pioneered learning in a number of areas, most notably in his approximations of π. Director of the astronomical observatory at Ujjain, he was at the center of scientific activities in the India of his time, and his work in number systems and equations represented a level of understanding far beyond that of contemporary Europeans.

Bhaskara is known variously as Bhaskara II, Atscharja Bhaskara, Bhaskaracharya, and Bhaskara the Learned. His role in Ujjain is significant, because that city—located in the central part of the subcontinent—was a focal point of learning for Hindu India. During this time, Muslim forces occupied what is now western India, Pakistan, and Afghanistan, but they were unable to penetrate deeper into India, and thus the scientists at Ujjain were able to continue their studies largely undisturbed.

Not only did Bhaskara know and understand the uses of the numeral 0, his demonstration that there are two solutions to the equation $x^2 = 9$ shows an understanding of negative numbers. He worked on what later came to be known as Pell's equation, or $x^2 = 1 + py^2$, solving the latter for $p = 8, 11, 32, 61,$ and 67. This resulted in some very large numbers; thus where $p = 61$, $x = 1{,}776{,}319{,}049$, and $y = 22{,}615{,}390$.

Bhaskara's best-known work is the *Siddhanta Siromani,* (Head Jewel of Accuracy), a series of books on mathematics and astronomy. The first two books, *Lilavati* (The beautiful) and *Bijaganita* Seed counting) deal largely with arithmetic and algebra. Although some of their content builds on the work of earlier mathematicians, Bhaskara also demonstrates the first consistent use of the decimal system, and began the practice of using letters to represent variables, a technique used to this day. He also presented several approximations for π and solved quadratic equations.

In addition to the last two books of the *Siddhantasiromani,* which deal with the heavens, Bhaskara also wrote *Karanakutuhala* (Calculation of astronomical wonders), a significant text on the motions of the planets and the numerical techniques used to study them. Not content merely with the ethereal realms of numbers and stars, in about 1150 Bhaskara also made one of the first descriptions of a machine for perpetual motion.

JUDSON KNIGHT

al-Biruni

973-c. 1050

Arab Mathematician, Astronomer, and Geographer

Al-Biruni, sometimes referred to as Abu Rayhan al-Biruni, conducted a great deal of original mathematical and astronomical work, most notably when he collaborated with Abu al-Wafa (940-998) on determining the longitudinal difference between Baghdad and his city of Kath in what is now Uzbekistan. He also served an important function as a transmitter of ideas from India, where he spent part of his career, to the Arab world. In all, al-Biruni published well over 100 works.

Born near Kath in the town of Khwarizm on September 4, 973, al-Biruni gained an early education in astronomy and mathematics from Abu Nasr Mansur, whose family ruled the region at that time. As a youngster, he made astronomical observations using a meridian ring.

His world at Kath was soon shattered when in 995 the ruler of nearby Jurjaniya invaded. Al-Biruni most likely fled to the town of Rayy. There he met the astronomer al-Khujandi, with whom he discussed the latter's observations made using a mural sextant. Later al-Biruni would write on this subject in his *Tahdid.*

By 997, al-Biruni was back in Kath, where he observed a lunar eclipse. As it happened, Abu al-Wafa had observed the same eclipse from Baghdad, and by comparing the difference in time between the two readings, the two scientists were able to determine the longitudinal difference between the two cities. Not only was this a remarkable example of scientific collaboration in an era that lacked electronic communication or easy travel, the findings of al-Biruni and Abu al-Wafa represented one of the few instances in which accurate longitudinal calculations were made in the premodern period.

During the years that followed, al-Biruni traveled extensively in Muslim central Asia, collecting information he would use in his monumental history of the region. The latter, known in English simply as the *Chronology,* appeared in 1000, and is one of the most useful sources available for information on Iranian calendars and local history.

Al-Biruni became a leading member of the court at Jurjaniya beginning in 1004, and conducted many of his astronomical observations during the period that followed. But in 1017, he was confronted with the same sort of political turmoil that had marred his youth: in that year, Sultan Mahmud of Ghazni (r. 997-1030), brother-in-law to the shah of Jurjaniya, invaded and kidnapped al-Biruni.

The scientist made the best of his captivity, and after Mahmud sent him to Kabul in 1018, he began making astronomical observations there. Mahmud also provided him with his introduction to India, a land the sultan invaded in 1022 and again in 1026. As a result, al-Biruni had the opportunity to learn Sanskrit and collected information on Hindu astronomy and astrology.

It was during this period that al-Biruni wrote many of his works, including *On Shadows* (c. 1021), *Tahdid* (1025), and *On Chords* (1027), which along with other writings provided an invaluable history of both Hindu and Muslim advances in astronomy during the period from the eighth to tenth centuries. In later years he wrote on subjects that included specific gravity, gemology, pharmacology, and—in the *Tahfim*—astrology. He died in about 1050 in Afghanistan.

JUDSON KNIGHT

Thomas Bradwardine

c. 1290-1349

English Mathematician

Known as "the Profound Scholar," Thomas Bradwardine wrote a wide array of mathematical and scientific works on subjects ranging from geometry to the physics of motion. He developed mathematical formulae for physical laws that served as a precursor to later efforts at quantitative measurement involving physical processes.

Bradwardine may have been born in Chichester, England, and scholars give various dates for his birth. He entered Merton College in Oxford in 1321, earning his M.A. in 1323 and his B.Th. before 1333. During his 14 years at Merton—he left in 1335—he wrote the majority of his works on mathematics, logic, physics, and philosophy.

In *De proportionibus velocitatum in motibus* (1328), Bradwardine analyzed principles put forth by Aristotle (384-322 B.C.) concerning the physics of bodies in motion. His contemporary Jean Buridan (1300-1358) would later critique the Greek philosopher's erroneous assertion that force must be applied as long as a body stays in motion, and Bradwardine also seemed to take issue with Aristotle.

Taking Aristotle's claim that velocity is equal to the force acting on a body divided by its resistance, Bradwardine showed mathematically that this is impossible. His position, however, was more that of a believer who attempts to root out unbelief than that of a critic. Hence he "resolved" Aristotle's contradiction by maintaining that there is a correspondence between an arithmetic increase in velocity and a geometric increase in the original force-resistance ratio. This incorrect conclusion would hold sway for more than a century. Nonetheless, his application of mathematical formulae to physical laws would also spur advances in scientific study.

With *Speculative Geometry,* Bradwardine became the first mathematician to make use of "star polygons," a form that would later interest Johannes Kepler (1571-1630). *Speculative Arithmetic* draws from the writings of Boethius (480-524), and *On the Continuum* discusses atomic theory as interpreted by Aristotle. Bradwardine's works of logic include *On Insolubles* (concerning logical problems such as "I am telling a lie") and *On "It Begins" and "It Ceases".* In the latter, he maintained that there can be no final instant to an interval of time, because the end marks the first moment in which that interval has ceased to exist.

Appointed canon of Lincoln in 1333, Bradwardine went on to become chancellor of St. Paul's Cathedral in London four years later. He soon became chaplain to King Edward III (r. 1327-77), and in this capacity was an eyewitness to several early battles of the Hundred Years' War (1337-1453), travelling to France with the invading force that won major victories at Crécy on August 26, 1346, and at Calais the following month.

In 1348 Bradwardine was appointed Archbishop of Canterbury, but Edward inexplicably rescinded the appointment. He was elected a second time on June 4, 1349 and this time the king offered no resistance. Bradwardine was consecrated on July 10 at Avignon, but shortly afterward fell victim to the disastrous Black Death then sweeping Europe.

JUDSON KNIGHT

Campanus of Novara

1220-1296

Italian Mathematician

Campanus of Novara, also known as Johannes Campanus and Giovanni Compano, produced a translation of Euclid's (c. 325-c. 250 B.C.) *Elements* that was destined to remain the definitive version among European scholars for two centuries. He also wrote on astronomy, providing the first European description of a planetarium.

As his title suggests, Campanus came from the town of Novara in Italy, where he was born in 1220. He served as chaplain to three popes beginning with Urban IV (r. 1261-64), and was working in the service of Boniface VIII (r. 1294-1303) when he died.

In 1260, Campanus produced his translation of Euclid. Original Greek texts of the latter's definitive work had long since been lost, but it had been preserved by Arabic scholars. Adelard of Bath (1075-1160) had been the first to translate the Greek mathematician back into a European language—Latin, used by scholars in the West at the time—but Campanus's version superseded even the translations produced by Adelard. When the first printed editions of the *Elementa geometriae* appeared in 1482, they used the translation of Campanus, which would remain dominant for the next two centuries.

In the realm of astronomy, Campanus wrote *Theorica planetarum,* in which he provided detailed observations on the construction of a planetarium. In addition, the book discussed the longitude of the planets and a geometrical description of the planetary model used.

In its data on planets, the *Theorica* showed the influence of the ancient Greek astronomer Ptolemy (c. 100-170), whose work had a powerful and sometimes unfortunate impact on late medieval science. Also evident was the influence of the *Toledan Tables,* edited in 1080 by the Arab astronomer al-Zarqali (Arzachel; 1028-1087) from computations made by al-Khwarizmi (c. 780-c. 850), al-Battani (c. 858-929), and al-Zarqali himself. Campanus's original contribution included determinations of the time for each body's retrograde motion, instructions for using the astronomical tables, and calculations of planets' size and distance from the Earth.

Campanus's *Computus major* concerned the properties of time and calendrical computations. His other works included *Tractatus de sphaera, De computo ecclesiastico,* and *Calendarium.*

JUDSON KNIGHT

Ch'in Chiu-shao

c. 1202- c. 1261

Chinese Mathematician

The work of Ch'in Chiu-shao represents the culmination of Chinese studies in indeterminate analysis. His principal writing was *Shu-shu chiu chang* (1247), variously translated as *Mathematical Treatise in Nine Sections* and *The Nine Sections of Mathematics.*

According to the modern system for spellings of Chinese names, Ch'in's would be rendered as

Qin Jiushao. As was typical of scholars in his country, he adopted a literary name, in his case Hao Tao-ku. He was born in the town of P'u-chou, Szechuan Province, and in 1219 joined the army, serving as captain in a volunteer unit who put down a local rebellion against the Sung Dynasty. Afterward he went on to a number of positions within the Sung state, becoming governor of Ch'iung-chou Province, now part of Hainan.

In 1247, Ch'in published his *Nine Sections,* which discussed a range of subjects that included simultaneous integer congruencies, linear simultaneous equations, algebraic equations, the areas of geometrical figures, and what became known as the Chinese Remainder Theorem. As was typical of Chinese mathematical texts, the book was heavily concerned with practical applications crucial to the administration of a Confucian bureaucracy, including matters relating to calendars and finances.

Though it remained unpublished except in manuscript form for many years and has never been fully translated, the *Nine Sections* would remain an influential text in Chinese mathematics for the next four centuries. Later Johann Karl Friedrich Gauss (1777-1855) and other mathematicians would delve into the questions of congruencies investigated by Ch'in.

In addition to his work in mathematics, Ch'in was recognized for his knowledge of the arts including poetry, music, and architecture, as well as his martial talents which included fencing, archery, and riding. In 1258 he lost his position as governor amid charges of corruption and died about three years later in Mei-chou, Kwangtung Province.

JUDSON KNIGHT

Chu Shih-chieh

fl. c. 1280-1303

Chinese Mathematician

The career of Chu Shih-chieh advanced studies in arithmetic and geometric series, as well as finite differences. He produced two notable written works, both of which were destined to have an impact on mathematical studies in East Asia for centuries.

Chu (Zhu Shiejie in the modern spelling) was also known by the literary name of Han-ch'ing. He appears to have come from the area of Yen'shan, near Beijing, and spent much of his career as a teacher travelling throughout China. As his reputation spread, more students came to him requesting instruction.

The first of his notable works, which appeared in 1299, was *Suan-hsueh ch'i-meng,* or *Introduction to Mathematical Studies.* As its name indicates, this was a book written for novices. The manuscript disappeared from China some time after his death, and was only recovered in the nineteenth century; in the meantime, it had spread to Japan and Korea, where it came into wide use as a textbook beginning in the 1400s.

By contrast to the earlier work, *Ssu-yuan yu-chien* (Precious mirror of the four elements 1303;) was a book to challenge the thinking of mathematical scholars. The four elements referred to in the title were four variables in a single algebraic equation, which Chu expressed using what he called the "method of the celestial element."

Chu's transformation method for solving equations, which he developed up to the degree of 14, would not be equaled by European mathematicians until the nineteenth century, with the Ruffini-Horner procedure of Paolo Ruffini (1765-1822) and William George Horner (1786-1837). The book also discussed what came to be known as the arithmetic or Pascal triangle, actually discovered earlier by other Chinese mathematicians.

Chu's era was one of those periods of turmoil that have traditionally punctuated Chinese history. In 1279, the Sung Dynasty had been overthrown by the Mongols, whose Yüan Dynasty was destined to last fewer than 90 years before it too was replaced by the Ming, China's last native-born ruling house. Given the instability in the country at that time, it is perhaps no surprise that it would be many years before Chinese mathematicians made new advances.

JUDSON KNIGHT

Leonardo Pisano Fibonacci

c. 1170-c. 1250

Italian Mathematician

A towering figure in the history of mathematics—to say nothing of mathematical studies during the medieval period—Leonardo Pisano Fibonacci is credited with the introduction of the Hindu-Arabic numeral system to Europe. He is also remembered for the Fibonacci sequence, whose recursive quality has made it a continued subject of fascination for mathematicians.

The dates of Fibonacci's life are uncertain, and his name itself presents something of a challenge. He was born in Pisa, and thus is sometimes known as Leonard of Pisa. His father was William or Guillermo Bonacci, and the name Fibonacci appears to be a contraction meaning family of Bonacci, or son of Bonacci. Thus the mathematician became known to history by his given name, that of his city, and the contraction: Leonardo Pisano Fibonacci.

Pisa at that time was a great trading center, and when Fibonacci was a boy, his father received an appointment as director of a warehouse in the North African port of Bugia. There Fibonacci trained under a Moorish instructor, who used the system of numerals 0 through 9 developed centuries before in India and later adopted throughout the Arab world.

Though other European scholars aware of Eastern advances in numbering had attempted to introduce so-called "Arabic numerals" to the West, European mathematics—including business arithmetic—was still conducted using the cumbersome old Roman symbols I, V, X, L, C, D, and M. The latter system had no expression for zero, nor was there a concept of place value, and this meant that large numbers could only be calculated using an abacus. This in turn meant that it was impossible to provide written verification of a result, and thus mathematical progress was severely limited.

After leaving Burga and travelling to a number of destinations, including Egypt, Syria, Constantinople, Greece, Sicily, and France, Fibonacci returned to Italy. There in 1202 he produced *Liber abaci,* or *Book of Calculations,* in which he introduced Europeans to the wondrous numerical system he had learned in the East. The book explained the rudiments of reading and using Hindu-Arabic numerals, and went on to a discussion of fractions, squares, and cube roots before addressing more challenging applications in geometry and algebra. Fibonacci, himself trained in business mathematics, also included chapters on the practical uses of the numeral system.

The publication of *Practicae geometriae* in 1220, which examined a number of algebraic, geometric, and trigonometric questions, added to Fibonacci's growing reputation. Soon he attracted the attention of Holy Roman emperor Frederick II (1194-1250), a renowned patron of the sciences, who in 1225 visited Pisa and held a mathematical competition to test Fibonacci's talents.

Johannes of Palermo, a mathematician working for the emperor, presented three questions to

Leonardo Fibonacci. *(Corbis. Reproduced with permission.)*

Fibonacci and his challengers. The first involved a second-degree problem and the second a third-degree or cubic equation. The last was a mere first-degree problem, but it involved a complex riddle concerning three men dividing an unspecified sum of money unequally. Fibonacci solved all three equations and his competitors withdrew without providing a single solution.

Later in 1225, Fibonacci wrote *Liber quadratorum,* a work he dedicated to the emperor. The book discussed a number of theorems involving indeterminate analysis, and recounted the problems put before him in the earlier competition. In 1228, Fibonacci presented a revised edition of *Liber abaci,* which he dedicated to his friend Michael Scot (c. 1175-c. 1235), an astrologer in Frederick's court.

The revised *Liber abaci* contained the famous "Fibonacci sequence" problem, which concerned a pair of rabbits who produce offspring at the rate of one pair a month, beginning in the second month. Given the proposition that each pair will reproduce at the same rate, and no rabbits will die, the problem addressed the question of how many rabbits would be produced at the end of a year. The answer was the sequence 1, 1, 2, 3, 5, 8, 13, 21, and so on. This is a recursive series, that is, one in which the relationship between successive terms can be expressed by means of a formula.

Fibonacci, who enjoyed great acclaim during his time and whose reputation has only grown in subsequent centuries, died in about 1250, during a war between Pisa and Genoa.

JUDSON KNIGHT

al-Kashi

1380-1429

Persian Mathematician and Astronomer

Al-Kashi made the first notable use of decimal fractions, and it would be two centuries before any mathematician improved on his calculation of π to 16 places. During his career, he worked with a number of prominent figures in the mathematical and scientific communities of Muslim central Asia, in particular Ulugh Beg (1393-1449) and Qadi Zada al-Rumi (1364-1436).

Known sometimes as Jamshid al-Kashi, the mathematician's full name was Ghiyath al-Din Jamshid Mas'ud al-Kashi. He was born in the desert town of Kashan, near the eastern tip of the central Iranian mountain range. During his early years, the region was terrorized by the invasions of Timur (or Tamurlane) (1336-1405), and the young al-Kashi lived in poverty. With the death of the conqueror and the accession of his son Shah Rokh, however, the life of the people in general—and of al-Kashi in particular—improved dramatically.

Al-Kashi made observations of a lunar eclipse in 1406, and in the following year produced an astronomical work, *Sullam al-sama,* whose title has been translated as *The Stairway of Heaven.* In 1410-1411, he wrote *Compendium of the Science of Astronomy,* which he dedicated to a member of the Timurid ruling house. Among the latter was Ulugh Beg, ruler of Samarkand and son of Shah Rokh.

Eager to win favor with Ulugh Beg, al-Kashi dedicated his *Khaqani zij* to the prince, who in 1420 founded a university in Samarkand and invited al-Kashi to join its faculty. Four years later, Ulugh Beg had an observatory built in the city. At Samarkand, al-Kashi worked with a number of other scholars, but his letters expressed admiration only for the work of Ulugh Beg and Qadi Zada.

In *Treatise on the Circumference* (1424), al-Kashi made his famous calculation of π. This improved greatly on the most accurate figure to date, obtained by Chinese astronomers in the fifth century, and would not be eclipsed until Ludolph van Ceulen (1540-1610) calculated the number to 20 decimal places some 200 years later.

The purpose of *The Key to Arithmetic* (1427) was to provide scholars at Samarkand with the mathematical instruction necessary for their studies in astronomy. In the book, al-Kashi developed a number of interesting geometric ideas, including a means of measuring a complex shape called a *muqarna* used to hide edges and joints in major public buildings. This book and others made use of decimal fractions, and though al-Kashi was the first to gain notoriety for their use, they had actually been developed earlier.

The Treatise on the Chord and Sine, unfinished at the time of al-Kashi's death and later completed by Qadi Zada, presented highly accurate sine calculations and touched on the idea of cubic equations. Al-Kashi died in Samarkand on June 22, 1429.

JUDSON KNIGHT

Omar Khayyam

1048-1131

Persian Mathematician, Astronomer, and Poet

Omar Khayyam occupies an unusual place among mathematicians, in that he is widely known—but for his poetry rather than for his achievements in mathematics. Two of the most famous lines in literature come from the 1859 translation of his *Rubaiyat* by Edward Fitzgerald: "A loaf of bread, a jug of wine, and thou..." and "The moving finger writes, and having writ, moves on...." Yet his work as a poet was but a small part of a career that included advancement of methods for solutions to algebraic equations and a highly accurate calculation of the year's length.

Ghiyath al-Din Abu'l-Fath Umar ibn Ibrahim al-Nisaburi al-Khayyami was born on May 18, 1048, in the Persian city of Nishapur. *Al-Khayyam* means tent-maker, the profession of his father. By the age of 25, Khayyam had already produced books on arithmetic, algebra, and music, and after moving to Samarkand (now in Uzbekistan) in 1070, he wrote his most significant algebraic work, *Risala fi'l-barahin ala masa'il al-jabr wa'l-muqabala* (Treatise on demonstration of problems of algebra).

During Khayyam's life, the region was under the control of the Seljuk Turks, and though their tempestuous rule created considerable upheaval, it also provided Khayyam with valuable opportunities for scientific study. In 1073, he was invited by the Seljuk prince (later sultan) Malik

Omar Khayyam. *(Corbis. Reproduced with permission.)*

Shah to establish an observatory at the new capital in the Persian city of Esfahan. While working at the observatory, Khayyam made an amazingly accurate calculation of the year's length: 365.24219858156 days.

Khayyam advocated reforms to the calendar, but the death of Malik Shah in 1092 left him without a patron, and the planned reform was suspended. For a time he also found himself under censure from Islamic purists, who judged his habit of scientific inquiry as a sign that he lacked the proper attitude of faith and reverence. Over time, however, his fortunes improved, and after 1118 he worked in the city of Merv (now in Turkmenistan) for Malik Shah's son Sanjar.

In his later mathematical studies, Khayyam confronted a difficult algebraic problem that led him to solve the cubic equation $x^3 + 200x = 20x^2 + 2000$. He observed that a more accurate solution would require the application of conic sections, and that it could thus be solved using only ruler and compass. This proved to be true, though it would be three-quarters of a millennium before any mathematician could put the idea to work.

The *Treatise on Demonstration of Problems of Algebra* approached the subject of cubic equations, and Khayyam became the first mathematician to propose a general theory for their solution. This, too, would have to wait many years, however, until the time of Sciopone dal Ferro (1465-1526) and others who successfully tackled the problem. Another prescient work was *Commentaries on the Difficult Postulates of Euclid's Book,* in which Khayyam prefigured aspects of non-Euclidean geometry.

On December 4, 1131, Khayyam died in the city of his birth, Nishapur.

JUDSON KNIGHT

al-Khwarizmi

c. 780-c. 850

Arab Mathematician and Astronomer

Although al-Khwarizmi was an early Arab proponent of the use of Hindu numerals—which were eventually adopted so widely throughout the Middle East that they came to be known as *Arabic numerals*—his advocacy in this area had its greatest impact when his books were translated for mathematicians in Western Europe. He used the term *al-jabr,* or restoration, for a method he applied in solving equations, and in the West this became *algebra.*

His name in full was Abu Ja'far Muhammad ibn Musa al-Khwarizmi, and he probably came from what is now Iraq. When Caliph al-Ma'-mum (r. 813-833) established his Dar al-Hikma or House of Wisdom, a center of scholarship in Baghdad, al-Khwarizmi was among the scholars invited to join this early think tank.

Al-Khwarizmi's first important mathematical text was *al-Kitab al-mukhtasar fihisab al-jab wa'l-muqa-bala,* also written as *Hisab al-jabr w'al muqabala* (The compendious book on calculation by completing and balancing). Probably written in the period 825-830, the book addressed matters of practical mathematics, offering solutions to problems involving unknown quantities. It was here that he introduced his method of restoration or completion, or separating the variable from other figures: thus $2x - 4 = x$ would become $x = 4$.

In his second major mathematical work, whose title is translated as *Treatise on Calculation with the Hindu Numerals,* al-Khwarizmi provided a guide for the use of the numerals 0-9. Up to this time, the Arabs had generally used an alphabetic numbering system, which like the Roman numerals used by Europeans lacked the valuable concept of place value. With the Hindu system, as al-Khwarizmi showed, it was possible (for instance) to know at a glance that 238 represented two units of 100, three units of 10, and eight units of 1.

Though al-Khwarizmi's book gained only limited use among Arab mathematicians, it would have an enormous impact in Western Europe, where it was translated into Latin during the early part of the twelfth century. In the meantime, Hindu numerals had taken hold in the Arab world, and by then Europeans had come to believe the system had developed there rather than in India.

Al-Khwarizmi also wrote books on geography, in which he provided latitudinal and longitudinal calculations for a variety of locations around the Middle East. His astronomical works amplified concepts developed earlier by Indian scientists, and in two other books he described the uses of the astrolabe for making astronomical measurements. In addition to these, al-Khwarizmi produced a commentary on the Jewish calendar, a history of the Islamic world during the early ninth century, and other books.

JUDSON KNIGHT

Levi ben Gershom
1288-1344
French-Jewish Mathematician and Astronomer

One of the few Jewish scholars of any discipline who came to prominence in Christian Europe, Levi ben Gershom dealt with problems in arithmetical operations and trigonometry. As an astronomer, he observed several eclipses and developed Jacob's staff, a mechanism for measuring the angular distance between heavenly bodies.

Known variously as Gersonides, Leo de Bagnolas, Leo Hebraeus, and Ralbag, Levi ben Gershom was born in the French town of Bagnols in 1288. He first came to prominence in 1321, with the publication of *Sefer ha Mispar,* or *Book of Numbers*. Among the subjects involving arithmetical operations discussed in the book was that of root extractions.

In 1342, Levi produced a work translated as *On Sines, Chords, and Arcs.* An investigation of trigonometry, *On Sines* presented five sine tables and examined the sine theorem for plane triangles. It is evident that Levi, while maintaining his faith (he wrote several commentaries on the Jewish scriptures), enjoyed a level of support from the Christian community unusual for a Jew in medieval Europe. Thus in 1343, the Bishop of Meaux commissioned him to write *The Harmony of Numbers,* a commentary on the first five books of Euclid's (c. 325-c. 250 B.C.) *Elements.*

On October 3, 1335, Levi observed a lunar eclipse, as well as an eclipse of the sun in 1337. He created a geometrical model to describe lunar movement and made a number of celestial observations using the camera obscura. The latter, a windowless room in which small quantities of light were allowed to enter, was a precursor to the modern camera and Levi was one of the first scientists to make use of it.

For his astronomical studies, Levi developed the Jacob's staff, which was about 4.5 feet (1.37 m) long and 1 inch (2.5 cm) wide, with a number of perforated tablets made to be slid along the staff. Each of these tablets represented a fraction of the staff's length and by using varying placements he was able to measure the distance between celestial objects, as well as their altitude and diameter.

In his *Sefer ha-hekkesh ha-yashar* (1319), Levi challenged cosmological ideas handed down by the Greek astronomer Ptolemy (c. 100-170). Many of the latter's assertions—for instance, his insistence that the earth is the center of the universe—were indeed spurious. Levi, however, made a few mistakes of his own, maintaining for instance that the glow of the Milky Way is a reflection of the light produced by the sun.

JUDSON KNIGHT

Li Yeh
1192-1279
Chinese Mathematician

In 1248, Li Yeh developed a new system of notation for designating negative numbers, using a cancellation mark drawn across the numerical symbol. He was also notable for the application of what he dubbed the "celestial unknown" in polynomial equations.

Li Yeh, who also went by the name Li Chih, lived during the period of the Sung Dynasty (960-1279), which up until 1127 had controlled all of China. In that year, however, an attack by the nomadic Juchen people of the north—hitherto putative allies of the Sung—forced the government to retreat southward. With its capital at Hangchow, perhaps the largest city in the world at its time, China enjoyed a second flowering during the period of the Southern Sung.

Despite the fact that it was occupied by "barbarian" invaders, mathematical scholarship continued in northern China, where Li Yeh made his home. He appears to have been serving as governor over the region of Chun Chou when he published his highly regarded *T'se-yuan hai-ching* or *Ceyuan haijing* (Sea mirror of circle measurements). The latter, which included his use of the symbol for negative numbers, presented 170 problems involving right triangles inscribed in circles or vice versa.

In both the *Sea Mirror* and *Yigu yanduan* (New steps in computation), Li Yeh applied his method of the "celestial unknown" (what modern mathematicians would call a variable), using polynomials in solving equations. It is possible that he had the manuscript of *New Steps* destroyed, however: according to one story—perhaps apocryphal— he ordered his son to burn all his books other than the *Sea Mirror.*

By 1234, the Mongols had replaced the Juchen in northern China, as they would replace the Sung in the south in 1279, a year that marked the establishment of the Yüan (Mongol) Dynasty. During his later years, Li Yeh served as director of an astronomical bureau under the Mongol emperor Kublai Khan (1215-1294), who is reported to have admired his scholarship greatly.

JUDSON KNIGHT

Nicole d'Oresme

1323-1382

French Mathematician

Nicole d'Oresme, sometimes referred to as Nicole Oresme or Nicholas of Oresme, pioneered the use of fractional exponents, and developed a type of coordinate geometry centuries before René Descartes (1596-1650). He discussed the Earth's rotation as well, and further distinguished himself with his writings on economics and his service to France's King Charles V.

Born in Normandy, Nicole studied theology and later enrolled in the College of Navarre at the University of Paris. He served as master of the college from 1356-1361, and in 1370 became royal chaplain in the court of Charles V. Much of his most important mathematical work dates from the 1360s.

In one book, published in about 1360, Nicole made the first use of fractional exponents, or fractional powers. (The notation in use today for expressing this idea, however, had not yet been developed.) Elsewhere he discussed a logical relationship between the calculating and graphing of values, thus paving the way for his own rudimentary form of coordinate geometry. In particular, Nicole suggested that a graph could be plotted for a variable magnitude whose value was a function—to use the terminology employed by modern mathematicians—of another. Descartes may indeed have been influenced by Oresme's work, which was later printed and saw numerous reprints.

Questiones super libros Aristotelis de anima, a commentary on Aristotle (384-322 B.C.) written in about 1370, discussed the Greek philosopher's ideas about the motion of the earth—ideas which by the time of the Renaissance would be proven wrong. To a degree Nicole anticipated these later critiques, maintaining initially that the earth is not stationary, and that a daily rotation would be possible. In the end, however, he went back on his own position, and embraced that of Aristotle. The *Questiones* also examined the nature of light, particularly its speed and reflective qualities.

Charles commissioned Nicole to prepare translations, from Latin to French, of several other works by Aristotle, an important step in the spread of knowledge from the scholars-only world of Latin to the vernacular tongues of Europe. Nicole also wrote an attack on astrology, *Du divinacions* (On divination), in which he maintained that events with alleged astrological causes can be explained by scientific phenomena instead. Perhaps his most famous work of any kind was *De moneta* (On money), written between 1355 and 1360, in which he maintained that a ruler has an obligation not to debase (reduce the precious-metal content) the currency of his people.

In 1377, Nicole was appointed bishop of Lisieux. He died in that town five years later, on July 11, 1382.

JUDSON KNIGHT

Georg von Peuerbach

1423-1461

Austrian Mathematician and Astronomer

Author of a work containing a table of sines, Georg von Peuerbach advanced mathematics in general with his use of Hindu-Arabic numerals in his sine tables. His student was the German astronomer Regiomontanus (1436-1476), who completed Peuerbach's monumental work on Ptolemy (c. 100-170), *Epitome in Almagestum Ptolemaei,* after his teacher's death.

Peuerbach's family name, like many in the Middle Ages, was derived from his hometown of Peuerbach, upper Austria, where he was born on May 30, 1423. He undertook his studies at Vienna and graduated in 1446. Seven years later, he earned a master's degree, then spent the year 1453-1454 travelling through Germany, France, and Italy as an astronomy lecturer. In 1454, he received an appointment as court astronomer to King Ladislas of Hungary.

In 1456, Peuerbach observed what came to be known as Halley's comet and recorded his observations. A year later, on September 3, 1457, he and Regiomontanus observed a lunar eclipse from a site near Vienna and also recorded it. Among Peuerbach's early astronomical works was *Tabulae ecclipsium,* which contained tables of his eclipse calculations. He later published additional tables and developed several astronomical instruments for making observations, as well as a large star globe.

Theoricae nova planetarum discussed the epicycle theory of the planets first presented by Ptolemy, and included Peuerbach's assertion that the planets revolve in transparent but solid spheres. Despite this erroneous notion, he was forward-thinking in his suggestion that the planets' movement is governed by the sun—an early step toward the refutation of the geocentric (Earth-centered) cosmology propounded by Ptolemy.

Nonetheless, Peuerbach remained a committed devotee of the ancient Greek mathematician right up to the time of his own death in Vienna on April 8, 1461. In 1462 or 1464, Regiomontanus completed the *Epitome,* which was finally published in 1474.

JUDSON KNIGHT

Johannes de Sacrobosco

1195-1256

English Mathematician and Astronomer

Johannes de Sacrobosco, also known as John of Holywood or John of Halifax, wrote a number of influential works in mathematics and astronomy. Like his contemporary Leonardo Pisano Fibonacci (c. 1170-c. 1250), he helped to popularize Hindu-Arabic numerals, and his *Algorismus* (c. 1230) became a widely used textbook in European universities.

Sacrobosco was born in the town of Holywood, Yorkshire, England, and received his education at Oxford. He became canon of the Order of St. Augustine at the Holywood monastery, then in 1220 continued his studies in Paris. A year later, he was appointed a teacher, and subsequently a professor of mathematics, at that city's university.

During the 1220s and 1230s, Sacrobosco produced a number of important texts. Divided into 11 chapters, the *Algorismus* addresses a range of subjects, including not only arithmetic functions but square and cube roots. Most notable, however, was Sacrobosco's use of Hindu-Arabic numerals, just then beginning to gain acceptance in the wake of Fibonacci's *Liber abaci* (1202).

Sacrabosco's *Tractatus de sphaera* (1220) had a similarly popularizing effect for the work of the Greek astronomer Ptolemy (c. 100-170). Among the topics discussed in the book are Ptolemy's ideas on planets and eclipses, as well as the position of the Earth in a universe that Sacrobosco believed to be spherical. In 1472, the *Tractatus* would be the first printed work of astronomy, and despite its many errors, it continued to be widely used until the 1600s.

In 1232, Sacrobosco wrote *De anni ratione*, which concerned the measurement of time. In addition to discussing the day, week, month, year, lunar phases, and ecclesiastical calendar, Sacrobosco used the book to make an early case for calendar reform. The Julian calendar, he noted, was 10 days in error, and he advocated a method of correction by omitting a single day every 288 years. In 1582, the Western system of dates would indeed be corrected with the adoption of the Gregorian calendar, though the system used was different from that promoted by Sacrobosco.

Other works by Sacrobosco included *Tractatus de quadrante,* in which he discussed the uses of the quadrant in astronomy. He died in Paris in 1256.

JUDSON KNIGHT

Thabit ibn Qurra

836-901

Syrian Mathematician

Not only was his writing on amicable or friendly numbers the first notable example of original mathematical work in the Middle East, but Thabit ibn Qurra distinguished himself with his translations of major writings by the mathematicians of Greek antiquity. He also translated works of astronomy, adding his own ideas to Ptolemy's concepts of the spheres, and discussed mechanics.

Al-Sabi Thabit ibn Qurra al-Harrani, born in the town of Harran in what is now Turkey, was not an Arab or a Muslim. Rather, his people spoke Syriac and were members of the Sabian sect. The Sabians worshipped the stars, and thus Thabit grew up in a situation particularly suited to a future mathematician with an interest in astronomy. In addition, the Sabians were strongly influenced by Greek culture, and Thabit became proficient both in Greek and Arabic.

A member of a wealthy family who inherited a handsome fortune, Thabit left his hometown to study mathematics in Baghdad. There he revised a translation of Euclid's (c. 325-c. 250 B.C.) *Elements* by Hunayn ibn Ishaq (808-873). Though two earlier translations had been made by al-Hajjaj (661-714) in addition to Hunayn's, Thabit's would become the definitive one.

Thabit translated a number of important works, and performed numerous original studies as well. His writings on ratios helped pave the way for the generalization of numbers, and many of his ideas may be seen as precursors to such later developments as number theory, spherical trigonometry, analytic geometry, integral calculus, and non-Euclidean geometry. In his *Book on the Determination of Amicable Numbers,* he discussed formulae for determining perfect and amicable numbers, and was reputedly the first to discover the amicable pair 17296, 18416.

Other writings of Thabit's include a treatise on the application of the Pythagorean theorem to an arbitrary triangle, as well as an astronomical study, *Concerning the Motion of the Eighth Sphere,* in which he stated incorrectly that there is an oscillation in the motion of the equinoxes. His *Kitab fi'l-qarastun* (The book of the balance beam) was a well-known work on mechanics. Thabit also wrote on philosophy and other subjects. He died on February 18, 901 in Baghdad.

JUDSON KNIGHT

Nasir al-Din al-Tusi

1201-1274

Arab Mathematician and Astronomer

In 1259, Nasir al-Din al-Tusi persuaded the Mongol conqueror Hulagu Khan (c. 1217-1265) to establish an observatory at Maragheh in what is now Azerbaijan. The Maragheh observatory became a center of learning, producing information such as the tables contained in al-Tusi's *Zij-i ilkhani.* Al-Tusi was not merely concerned with mathematics in its applied form, however; in fact, he is often given credit as the first thinker to use pure trigonometry.

Al-Tusi, also known as Muhaqqiq-i Tusi, Kwaja-yi Tusi, and Khwaja Nasir, was born in the town of Tus in northeastern Iran. His father was a jurist in the Shi'ite Islamic sect, and al-Tusi received a strong religious education supplemented with teachings, provided by his uncle, in logic, physics, and mathematics. As a teenager he studied philosophy, medicine, and mathematics in the Persian city of Nishapur.

From at least the time of al-Tusi's early teen years, his homeland had been in a state of turmoil, threatened from without by the prospect of Mongol invasion and from within by the Assassins. The latter, a radical political-religious group and perhaps the world's first formally organized terrorist organization, invited al-Tusi to join their numbers as a scholar under their protection. Whether this "invitation" was in fact a command is not known, nor is it clear whether al-Tusi, who threw in his lot with the Assassins after 1220—when he would have been only about 20 years old—did so voluntarily.

Under the protection of the Assassins in their various strongholds, al-Tusi continued his mathematical and scientific scholarship until Hulagu's forces stormed the Assassins' redoubt in 1256. Again, it is not clear whether al-Tusi betrayed his allies, or simply regarded Hulagu as someone liberating him from kidnappers—though in fact he had been with the Assassins for more than three decades by then. In any case, the Mongols slaughtered the Assassins, and al-Tusi suddenly found himself under the care of Hulagu, who greatly admired his scholarship.

When the dust of the invasion settled and Hulagu established his capital at Maragheh, al-Tusi suggested the idea of the observatory, which met with the new khan's hearty approval. It took three years to construct and prepare the observatory, during which time the Mongols brought in Chinese astronomers to assist their Persian counterparts. The completed observatory included a large copper wall quadrant, an azimuth quadrant, and other instruments designed by al-Tusi, and a library containing books on all manner of scientific pursuits. In fact the observatory was far more than simply a place for observing stars, and became a center for inquiry in mathematics, various sciences, and philosophy.

Another product of the Maragheh observatory was al-Tusi's *Zij-i ilkhani*, or "Ilkhanic Tables." (Hulagu had given himself the title "il-Khan.") Written in Persian and translated into

Arabic later, the book contained 12 years' worth of observations on the planets. In another astronomical work, al-Tusi developed a mathematical principle that came to be known as the "Tusi-couple," which enabled him to describe motion in uniform terms, regardless of whether that motion was linear or circular.

This in turn led to new hypotheses on the movement of planets, which some scholars consider the most important contribution to the understanding of the planetary system between the time of Ptolemy (c. 100-170) and that of Copernicus (1473-1543). It is possible that the latter may in fact have modeled his own use of a similar method on the Tusi couples.

Al-Tusi's *Treatise on the Quadrilateral* was perhaps the first mathematical work in history to treat trigonometry as a discipline in its own right, rather than as a mere application of astronomy. He also wrote on *n*th roots of an integer, and provided commentaries or revisions to Arabic versions of works by a variety of ancient Greek mathematical and scientific figures. Other writings by al-Tusi addressed logic, medicine, minerals, and ethics. He died on June 26, 1274, near Baghdad.

JUDSON KNIGHT

Yang Hui

c. 1238-c. 1298

Chinese Mathematician

Though he did not invent the Pascal triangle, Yang Hui did provide an early discussion concerning its use. He also wrote several texts, at least one of which proved influential in East Asian mathematics for centuries to come. Some scholars consider Yang Hui's career the apex of Chinese algebraic studies in the medieval period.

Almost nothing is known about the life of Yang Hui, except for the fact that he served as an official under the southern Sung. The Sung Dynasty, established in 960, had controlled all of China until 1127, when an attack by "barbarian" peoples from the north forced them to move their government southward. Yang Hui's contemporary and fellow mathematician Li Yeh (1192-1279), for instance, lived in northern China, which was controlled first by Juchen nomads and later by Mongols.

During this same time, southern China flourished, and despite the humiliating retreat that had forced the Sung to vacate, the Southern Sung period proved to be one of the cultural high points of Chinese history. With a population of 1.5 million, the capital city of Hangchow was probably the largest city in the world, and its size is particularly remarkable considering the fact that only the largest European urban centers had even as many as 100,000 inhabitants.

Throughout Sung China, the arts and sciences flourished, and though the Sung would lose their power to the Mongols during Yang Hui's lifetime, even the invaders did their best to permit the southern Chinese to go on as before. As for Yang Hui, he had served as a government official under the Sung, and it is quite possible that he continued to do so with their usurpers. (Unless, like many Chinese, he left government service as a means of protesting his country's invasion by foreigners whom the Chinese considered their inferiors.)

Yang Hui is credited with mathematical studies that appeared in 1261, 1275, and 1299. The first of these applied decimal fractions—centuries before these would be adopted in the West—and made use of the triangular configuration of binomial equations usually known today as Pascal's triangle. In many parts of East Asia, the latter is known as Yang Hui's triangle, though in fact it seems to have been in use long before Yang Hui, with Chinese references to it dating as far back as c. 1100.

The 1275 book was *Cheng chu tong bian ben mo*, (The beginning and end of variations in multiplication and division). The latter is particularly valuable for the insights it provides historians of mathematics regarding the curriculum used by Chinese mathematical students at that time. Furthermore, Yang Hui's syllabus, included in the book, shows him to have been a highly competent teacher who emphasized genuine learning rather than memorization—not always the case in medieval schools, East or West.

Another volume, *Suan-hsiao chi-meng*, appeared after Yang Hui's death in about 1298. The book was destined to be well-received in Japan, where it served as an impetus to mathematical inquiry for half a millennium.

JUDSON KNIGHT

Biographical Mentions

Abbo of Fleury
c. 947-1004

French scholar who commented on ancient computation tables in his *De numero et pondere super calculum victorii* (c. 1000). A monk at the abbey in Fleury, Abbo taught in England before returning to Fleury to become abbot in 988. He also served as a diplomat in the service of France's King Robert II. In addition to his mathematical work, Abbo wrote on astronomy, philosophy, and grammar.

Abu Nasr Mansur
970-1036

Arab mathematician and astronomer known for his commentary on the *Spherics* by Menelaus. Mansur came from the Banu Iraq family, who ruled Khwarazm near the Aral Sea, but in 995 lost power; thereafter, he served in the courts of other rulers. A student of Abul Wefa, Mansur was later teacher to al-Biruni and wrote as many as 25 works, of which seven were on mathematics. His work on the *Spherics* is particularly noteworthy because the original Greek manuscript has been lost. In addition, Mansur contributed to the development of trigonometry by discovering the sine law $a/\sin A = b/\sin B = c/\sin C$.

Abu Utman
fl. c. 914

Arab mathematician and physician who translated part of Euclid's (c. 325-c. 250 B.C.) *Elements*. Abu's translation, of the tenth book in the classic mathematical work, appeared in 914.

al-Adli
fl. c. 800

Arab chess master who wrote the first known work on the game, introducing ideas that would later influence recreational mathematics. Al-Adli's book, which appeared in about 800, is lost to history—as, indeed, is its title. It is known, however, that in it he discussed a number of chess problems called *mansubat*. Among these was one called "the Dilaram mate," concerning a legendary chess player who gambled his wife on a game of chess. Just when he was about to lose, Dilaram's wife whispered to him that if he sacrificed both of his rooks, he could checkmate his opponent in just a few moves—which he did.

Alcuin
735-804

English educator and cleric, also known as Alcuin of York, who, in answering Charlemagne's call for a revival of learning, was instrumental in bringing about the Carolingian renaissance and revision of the Roman Catholic liturgy. As minister of education for Charlemagne's Frankish empire, Alcuin established cathedral schools and scriptoria, which studied, copied, and preserved ancient Christian and pagan manuscripts, including those of the Greek mathematicians. The cursive script he developed for these scriptoria, called Carolingian miniscule, regularized European handwriting and made it more readable. He also wrote lesson texts on arithmetic, astronomy, and geometry and commentaries on religion and education.

Aryabhata II
fl. c. 950

Indian mathematician whose interests included recreational mathematics and astronomy. The "II" in his name refers to Aryabhata (476-c. 550), a much more well-known Hindu mathematician and astronomer after whom Aryabhata II modeled his work. Among the writings attributed to Aryabhata II is the *Mahasiddhanta,* in which he discusses "casting out nines," a concept in recreational mathematics.

al-Baghdadi
c. 980-1037

Arab mathematician who wrote an important work on arithmetic and number theory. Raised in Baghdad, al-Baghdadi moved to Nishapur in what is now Iran; eventually, however, political turmoil forced him to flee to the smaller town of Asfirayin. There al-Baghdadi, who came from a wealthy family, taught for free in the local mosque. Though he wrote primarily on theology, he produced two mathematical works, the more significant of which was *al-Takmila fi'l-hisab.* In this work, he examined differing arithmetic systems, including counting on the fingers, the sexagesimal system, and Hindu numerals, which he favored. The work also addressed the subjects of irrational numbers and business arithmetic.

Ibn al-Banna
1256-1321

Arab mathematician and astronomer who was the first to treat a fraction as a ratio of two numbers, and the first to use the term *almanac* (*al-manakh,* or "weather," in Arabic) for a book of astronomical data. Al-Banna, who spent most of his career in Morocco, wrote some 82 books on a number of subjects, including an algebraic treatise and an introduction to the *Elements* of Euclid (c. 325-c. 250 B.C.). His most important mathematical writings were *Talkhis amal al-hisab* (Summary of arithmetical operations), and *Raf al-hijab,* a commentary on the *Talkhis.*

Banu Musa: Jafar Muhammad ibn Musa, Ahmad ibn Musa, and al-Hasan ibn Musa
fl. 9th century

Arab mathematicians and astronomers who were among the first to study ancient Greek mathematical texts, and who thus helped lay the foundations for Arab mathematical study. Jafar concerned himself primarily with geometry and astronomy, Ahmad with mechanics, and al-Hasan with geometry, but their careers were so closely mingled that it is difficult to separate them. The brothers, collectively referred to as the Banu Musa, were all involved in the House of Wisdom, the "think tank" established by the Abbasid caliph al-Ma'mun. Their most important work, whose title is translated as *The Book of the Measurement of Plane and Spherical Figures,* examines two texts by Archimedes.

Filippo Brunelleschi
1377-1466

Florentine architect and artist traditionally believed to have played an important role in the development, and some say the invention, of mathematical perspective. While Brunelleschi was certainly the first architect and artist to study and systematically apply mathematical perspective in his works, recent research assigns him a more modest role in its development. He also invented a technique for raising domes without scaffolding.

Eilmer
980-after 1066

English monk who attempted to fly off the tower of Malmesbury Abbey with a set of wings attached to his arms and feet. It so happened that the great historian William of Malmesbury witnessed the event as a boy, leaving an account of it in his *Gesta regum anglorum.* Apparently Eilmer panicked after flying some 600 ft (183 m) and suddenly plummeted to earth, breaking both of his legs. In spite of this, he wanted to make a second flight, but the abbot forbade it. According to William, Eilmer also had the distinction of twice seeing the comet known today as Halley's comet. As a nine-year-old boy in 989, Eilmer believed the comet to be a portent of dire happenings, and indeed it was soon followed by a Danish attack in England. The second time Eilmer saw the comet, in 1066, the Normans invaded, bringing an end to the Anglo-Saxon world he had known.

Rabbi ben Ezra
1092-1167

Spanish Jewish scholar who wrote on a number of mathematical topics, introducing Europeans to concepts that originated in the Arab world. Ezra spent the first five decades of his life peacefully, but after 1140 political turmoil forced him to wander throughout Christian Europe. In his latter years, when he lived in Italy, he wrote his most important works, among them treatises whose titles are rendered in English as *Book of the Unit* and *Book of the Number*. The first concerned the Hindu-Arabic numerals 1 through 9, and the second the decimal system. Ezra discussed the concept of zero, which he called *galgal,* meaning circle. In spite of the attention directed toward his work, it would still be several centuries before the Hindu-Arabic system gained widespread acceptance in Europe.

Gerard of Cremona
c. 1114-1187

The most prolific translator from Arabic to Latin in the medieval period. He went to Spain to learn Arabic, and—either by himself or in collaboration—translated over seventy major works. He also corrected the work of earlier translators, adding in missing passages and making corrections. He may have written some original works. His translations had a massive influence for hundreds of years, contributing to the revival of ancient knowledge, and the European absorption of Arabic ideas.

Gerbert of Aurillac
c. 945-1003

French mathematician and astronomer who ascended to the papacy as Sylvester II. Having spent time in Spain, he took part in the transmission of Arabic science to the west, although his own work in mathematics was not original. He tended to list theorems without explaining how they were proved. In astronomy he used models rather than depending on textbooks and is credited with one of the earliest mechanical clocks.

Gregory of Rimini
d. 1358

Italian scholar and logician. Gregory worked as a professor at the Sorbonne, where he achieved acclaim as a professor and earned a number of distinguished titles from his students, including *Doctor acutus*, *Lucerna splendens*, and *Doctor authenticus*. A leading nominalist, Gregory was, along with William of Ockham (c. 1290-1349), a prominent figure in the debate over universals.

al-Hajjaj
fl. 786-d. 833

Arab scholar who was first to translate Euclid's *Elements,* and one of the first to translate the *Almagest* of Ptolemy, from Greek into Arabic. Al-Hajjaj actually translated the *Elements* twice, first for the caliph Harun al-Rashid (766-809), and later for another caliph, al-Ma'mun (786-833). As for the *Almagest,* the European name for this work comes from al-Hajjaj's translation of the title, *Kitab al-mijisti.*

Abu Ali al-Hasan ibn al-Hasan al-Haytham
965-c. 1040

Arabic mathematician and physicist known to the West largely for his work in geometry under the name Alhazen. The most famous problem he solved dealt with locating the point on a surface from which light at a point of origin will be reflected to another specific point. He wrote about difficulties in the standard text of geometry inherited from the Greeks and applied geometrical methods to analyzing problems in optics. His work on the area of crescent-shaped figures (lunes) went beyond what was available in Greek texts.

Heriger of Lobbes
c. 925-1007

Burgundian scholar and author of a work on arithmetic. Through persistence and ardent devotion to scholarship, Heriger became an unusually well educated man for tenth century Europe. He wrote a number of books, among them *Regulae de numerorum abaci rationibus*, an arithmetical work published many centuries later along with writings of his more famous contemporary Gerbert (a.k.a. Pope Sylvester II, 945-1003).

Johannes Hispalensis
fl. 1130s-1140s

Spanish scholar who translated several mathematical works, and added new terms to the vocabulary of mathematicians. A converted Jew, Johannes, sometimes called John of Luna or John of Seville, worked with the archdeacon of Segovia, Domengo Gondisalvi, on translating the works of ancient Greek and medieval Arab mathematicians and scientists. In about 1134, he translated a version of Ptolemy's (c. 100-170) *Almagest,* thereby influencing the exposure of that seminal work to European scholars. He was the first to use the term *numerus minuendus* (number to be diminished), which eventually became the English *minuend,* and he was one of the first to use the word *fractiones*.

Hunayn ibn Ishaq
808-873

Arab scholar and physician who translated numerous Greek writings—notably those of Plato (427-347 B.C.) and Aristotle (384-322 B.C.)—on mathematics and science. A member of the House of Wisdom founded by the caliph al-Ma'mun (786-833), Hunayn probably took part in an expedition to purchase manuscripts from Byzantium. Among those with whom he worked in the House of Wisdom were al-Khwarizmi (780-c. 850), al-Kindi (c. 801-873), al-Hajjaj (fl. 786-833), Thabit ibn Qura (826-901), and Jafar Muhammad ibn Musa (800-c. 873).

Ibrahim ibn Sinan
908-946

Arab mathematician and grandson of Thabit ibn Qurra (836-901). Ibrahim's most significant work involved the quadrature of the parabola, and he is credited with improving on the ideas of Archimedes (c. 287-212 B.C.) with his method of integration. He also made contributions in the field of mathematical philosophy. Seven of Ibrahim's treatises, including *On the Drawing of Three Conic Sections*, survive.

Ahmed ibn Yusuf
835-912

Arab mathematician who wrote works on ratio, proportion, and arcs. Born in Baghdad and raised in Damascus, Ahmed spent much of his career in Egypt, where he served as private secretary to the powerful ruler Ahmad ibn Tulun. His book on ratio, later translated into Latin by Gerard of Cremona (1114-1187), influenced Leonardo Pisano Fibonacci (c. 1170-c. 1250). Ahmed also wrote a work translated as *On Similar Arcs,* which developed ideas introduced by Euclid (c. 325-c. 250 B.C.), as well as treatises on the astrolabe and the *Centiloquium* of Ptolemy (c. 100- 170).

al-Imrani
fl. 900s

Arab mathematician. Among al-Imrani's writings were a treatise on astrology, and a commentary on the algebraic writings of Abu Kamil (c. 850-c. 930). His students included astrologer al-Qabisi (d. 967).

Jabir ibn Aflah
1100-1160

Arab mathematician and astronomer whose *Islah al-majisti* (Correction of the *Almagest*) proved to be a strong influence on scholars throughout the Mediterranean and Europe. A native of Seville,

Aflah knew Moses ben Maimon (Maimonides, 1135-1204). He developed a theorem in spherical trigonometry that bears his name, and created an instrument called the torquentum for making transformations between spherical coordinates. His critique of Ptolemy's (c. 100-170) *Almagest* exerted an impact on a number of thinkers, among them Ibn Rushd (Averroës, 1126-1198).

Jacob ben Tibbon
1236-1312

French Jewish mathematician and astronomer who translated several key works into Hebrew, and wrote a number of mathematical and astronomical treatises. Among the texts Jacob translated from Arabic were Euclid's (c. 325-c. 250 B.C.) *Elements* and Ptolemy's (c. 100-170) *Almagest*. In a treatise whose English title is *Jacob's Quadrant*, he explained the workings of a quadrant he had invented, and in his *Luhot* (Tables) he offered astronomical tables for the ascensions of stars over Paris. Dante Alighieri (1265-1321) mentioned the latter book in his *Divine Comedy*. Among the scientists influenced by Jacob's theories was Nicolas Copernicus (1473-1543).

al-Jawhari
c. 800-c. 860

Arab mathematician and astronomer who wrote on Euclid's (c. 325- c. 250 B.C.) *Elements* and became the first to attempt a proof of the parallel postulate. Born in Baghdad, al-Jawhari was a member of the prestigious House of Wisdom, an institution of scholars established by the caliph al-Ma'mun (r. 813-833). In his *Commentary on Euclid's Elements,* al-Jawhari presented some 50 propositions in addition to those offered by Euclid, and attempted (though unsuccessfully) to prove the parallel postulate. As an astronomer, al-Jawhari made observations both from Baghdad and Damascus.

al-Jayyani
989-after 1079

Arab mathematician and astronomer who expounded on Book V of Euclid's (c. 325-c. 250 B.C.) *Elements* and wrote the first treatise on spherical trigonometry. Al-Jayyani spent much of his career in the town of Jaén, capital of the Moorish principality called Jayyan; hence his name. In his *On Ratio,* he discussed what he identified as four types of geometrical magnitude outlined by Euclid—line, surface, angle, and solid—and added to these a fifth, number. *The Book of Unknown Arcs of a Sphere* approached the subject of spherical trigonometry, and provided formulas for spherical triangles. Al-Jayyani's writings would have a profound impact on European mathematics in general, and on the work of Regiomontanus (1436-1476) in particular.

Jia Xien
fl. c. 1100

Chinese mathematician sometimes credited with the development of what came to be known as the Pascal triangle. This triangular configuration of binomial equations was named after Blaise Pascal (1623-1662), who actually developed it centuries after his Chinese counterparts. Though Jia Xien, among others, is credited with discovering the idea, it is likely that it was already known in his time.

al-Karaji
953-c. 1029

Arab mathematician whose *Al-Fakhri* was an important work in the development of algebra. Al-Karaji, sometimes known as al-Karkhi, has been credited with developing the first theory of algebraic calculus. *Al-Fakhri,* which shows the influence of Diophantus (c. 200-c. 284), defines and gives rules for a number of monomials; examines "composite quantities" (sums of monomials); and makes use of mathematical induction. The book also includes a binomial theorem akin to that used in the Pascal triangle.

al-Khalili
1320?-1380?

Arab astronomer and mathematician who compiled an extensive series of tables necessary for Muslim prayer rituals. Working at the Umayyad Mosque in Damascus, al-Khalili was responsible for providing a means of timekeeping in order to regulate the hours of the five daily prayers that are a part of the Islamic faith. This required extensive knowledge of geometry, and his calculation of the exact direction of the Muslim holy city of Mecca—toward which all Muslims pray—forced him to confront challenging problems in spherical trigonometry.

al-Khazin, or al-Kazin
d. after 961

Also known as Abu Jafar Muhammad ibn al-Hasan al-Khurasani, or Abu Jafar al-Khazin, his writings may be the work of two separate people of the same name. Al-Khazin wrote on number theory and astronomy, but only two books survive intact. The astronomical work includes details of the astrolabe and other astronomical instruments. The mathematical work is a commentary on Book 10 of Euclid's *Elements*, and contains work on spherical trigonometry related to planetary motion. He was also an astrologer, and theorized a clever model of the solar system.

Abu Mahmud Hamid ibn al-Khidr al-Khujandi
940?-1000

Central Asian astronomer and mathematician who calculated the latitude for the city of Rayy, near modern Tehran, Iran, and developed a relatively accurate sextant. Born in what is now Tajikistan, al-Khujandi was a Mongol, but spent most of his career in Arab lands controlled by the Buyid dynasty, who seized control of the Abbasid capital at Baghdad in 940. Working at the observatory in Rayy, al-Khujandi developed a large mural sextant that indicated seconds, providing a level of accuracy beyond that of most existing instruments. He also calculated the obliquity of the ecliptic with less than perfect accuracy, and is sometimes incorrectly credited with discovering the sine theorem.

Kushyar ibn Labban
fl. 900s

Persian astronomer and mathematician. Like many astronomers, Kushyar approached mathematics as an aid to his work, and took a particular interest in trigonometry, studying tangent functions and assembling new astronomical tables. Perhaps Kushyar's most famous work is *Kitab fi usul hisab al-hind* (translated in 1965 as *Principles of Hindu Reckoning*), in which he examined arithmetic as practiced in India. He also wrote on astrology.

Levensita
fl. 718

Indian astronomer and mathematician who translated Hindu mathematical works into Chinese. These included texts describing angle measurement, as well as a table of sines for angles from 0° to 90°. He also made an unsuccessful attempt to introduce Hindu decimal numbers into general use in China.

Johannes de Lineriis
fl. 1320s

French astronomer, sometimes known as Jean de Lignières, whose *Canons of the Tables* became a widely used book. The *Canons*, a collection of astronomical tables, was adopted as part of the required educational texts for second-year students at the University of Bologna in 1344.

Muhyi l'din al-Maghribi
1220-1283?

Arab astronomer and mathematician who calculated the approximate value for the sine of 1°. Born in Spain, al-Maghribi spent his most fruitful years working with al-Tusi at the observatory established by their patron, the Mongol conqueror Hulagu Khan, at the city of Maragheh in what is now Azerbaijan. There al-Maghribi made a number of observations on the Sun and on the mathematical means required to calculate solar eccentricity and apogee. His calculation of the value of the sine for 1° appears in his *Treatise on the Calculation of Sines,* and would prove the most accurate until the work of Qadi Zada.

al-Mahani, also known as Abu Abd Allah Muhammad ibn Isa
c. 820-c.880

We know little about his life, and much of his writing is now lost. Only through the works of later authors quoting al-Mahani do we know he wrote on astronomy, geometry, and arithmetic. He worked on the problem of calculating the volumes of a sphere cut by a plane, and other difficult equations posed by Archimedes. His surviving works include commentaries on Euclid's *Elements*, the calculation of ratios, and other mathematical problems.

Mahavira
d. 850

We know nothing about him except that he wrote during the reign of Amoghavarsa, monarch of Karnataka and Maharastra, between 814/5 and 880. His sole surviving work, the *Ganitasarasangraha*, is a mathematical text in nine parts that discusses terminology, arithmetical operations, fractions, the calculation of areas, calculations relating to shadows and excavations, and other mathematical operations.

Manuel Moschopoulos
fl. c. 1300

Byzantine scholar who wrote a work on the subject of magic squares. A magic square is a square containing integers arranged in such a fashion that the sum of the numbers for each row, column, and main diagonal (and often in some or all of the other diagonals) is the same. The concept had first appeared in the writings of Arab mathematicians more than three centuries before, but Manuel was perhaps the first European scholar to comment on the topic.

Abu l'Hasan Ali ibn Ahmad al-Nasawi
1010?-1075?

Arab mathematician who wrote a summary of Euclid's *Elements*. Al-Nasawi, who worked variously for a prince in the Buyid dynasty and for a Shi'ite leader in Baghdad, composed numerous short works. The best known of these is his summary of the *Elements*, which he claimed to have written in order to provide those studying Ptolemy's *Almagest* with the necessary background in geometry. Al-Nasawi's work was largely forgotten until its rediscovery in Europe in 1863.

Abu'l Abbas al-Fadl ibn Hatim al-Nayrizi
865?-922?

Persian mathematician and astronomer who wrote a commentary on Euclid's *Elements* and made early use of the trigonometric tangent function. Little is known of al-Nayrizi's life, but his name suggests that he came from the town of Nayriz in what is now central Iran and he probably served the Abbasid caliphs in Baghdad. In addition to his writings on Euclid and Ptolemy's *Almagest* and *Tetrabiblos*, al-Nayrizi discussed the problem of calculating the exact direction of Mecca, an important part of the Muslim daily prayer ritual. In so doing, he used the tangent function, though he was not the first to develop that concept.

Maximos Planudes
1260?-1310?

Byzantine monk who reconstructed a number of the maps described in Ptolemy's *Geographia*. The maps in the original manuscript were incomplete, since some had only been explained by Ptolemy without accompanying illustrations. Planudes undertook the drawing of the additional maps, using Ptolemy's text as a guide, for the Byzantine emperor Andronicus III.

Plato of Tivoli
d. c. 1134

Known only through his surviving works, he translated a number of important mathematical and astronomical texts from Arabic and Hebrew to Latin. He worked with other translators, especially Savasorda (Abraham bar Hiyya ha-Nasi). His versions of Ptolemy and other Greek writers proved very popular, and many copies of his translations still survive across Europe. His work helped introduce many new areas to European mathematics, such as the solutions to quadratic equations, and advanced trigonometry.

Abu Sahl Waijan ibn Rustam al-Quhi
940?-1000?

Persian mathematician and astronomer who was a leader in the application of Greek geometric methodology among Muslim scholars. Al-Quhi (sometimes spelled al-Kuhi), served the Buyid dynasty, whose caliphs ordered him to make a number of astronomical observations. For this purpose, he directed the building of an observatory in Baghdad, at which he served as director following its opening in 988. He discussed geometric problems involving quadratic or cubic equations, and described a conic compass, or one with a leg of variable length, for drawing conic sections.

Qusta ibn Luqa al-Ba'labakki
fl. 860-900

Arab translator and scientist who was instrumental in transmitting Greek ideas to the Muslim world, and thus later to the West. As an original thinker, Qusta discussed the uses of tangents and cotangents in trigonometry and wrote on medicine, astronomy, logic, and natural science. His most important contribution, however, was in his translation of works by Aristotle, Diophantus, Aristarchus, and, in particular, the *Mechanics* by Hero of Alexandria.

Richard of Wallingford
c. 1292-1336

The son of a blacksmith, he eventually became the abbot of St. Albans in Britain. He studied at Oxford, became a Benedictine monk, and later studied theology and philosophy. He wrote on trigonometry, and on the theory, construction, and use of an instrument called the Albion (all-by-one) which could help calculate planetary positions. He also built and wrote about an astronomical clock, which is the earliest clock for which we have detailed descriptions.

Robertus Anglicus
fl. 1200s

English scholar who produced a commentary on *De sphaera* by thirteenth-century English mathematician Sacrobosco. In Robertus's text, entitled *Tractatus quadrantis vetus*, he discussed a demonstration of the force of gravity using a millstone. Robertus also wrote *Canons for the Astrolabe*, later printed in 1478.

Shams al-Din al-Samarqandi
1250?-1310?

Central Asian mathematician and astronomer who wrote a short study of Euclid's propositions in which he applied the work of numerous earlier mathematicians of the Muslim world. Little is known about al-Samarqandi, though his name indicates that he came from Samarqand in what is now Uzbekistan. His most important work was his 20-page paper on Euclid in which he referred to writings by Omar Khayyam, al-Tusi, and others. Qadi Zada later referred to al-Samarqandi's paper on Euclid in his own work. Al-Samarqandi also produced a number of other treatises on astronomy and philosophy, as well as a star catalogue for 1276-77.

al-Samaw'al
d. c. 1180

The son of Jewish immigrants, from his educated parents he learned medicine and science, and was self-taught in mathematics. At eighteen he had read all the mathematics he could find, and began writing his own. His surviving mathematical work brings together all the known algebraic rules of his age. A highly successful physician, he traveled widely in this capacity, and wrote on various medical topics. A treatise on sexology, and an anti-astrological work still survive.

Savasorda, also known as Abraham bar Hiyya ha-Nasi and Abraham Judaeus
c. 1065-c. 1136

He wrote a large text in Hebrew on practical geometry, which contains the earliest use of algebra in Europe including the solution to one form of quadratic equation. He included many practical elements in his writing, and was an active translator of Arabic mathematical works into Latin. He appears to have collaborated with Plato of Tivoli on several translations. Both his translated and original works had a major influence on later European thinkers.

Shao Yung
1011-1077

Chinese philosopher whose ideas later influenced Gottfried Wilhelm Leibniz (1646-1716) in his development of a binary arithmetical system—that is, one based on just two digits. Shao, who began as a Taoist and went on to embrace Confucianism and become one of the leading proponents of the neo-Confucianist school, first took an interest in Confucianism after studying the *I Ching,* or, (Book of changes). This book also influenced him to undertake numerological studies, and Shao Yung became convinced that the number 4 played a unifying role in existence. His idea that there is an underlying pattern for all that is—a pattern existent as much in the human mind as in the world of perceived experience—had an enormous impact on Confucian idealism.

Shuja, or Shudja, also known as al-Hasib al-Misri, or Abu Kamil Shuja'ibn Aslam ibn Muhammad ibn Shuja
c. 850-c. 930

He is remembered for his work in the field of algebra, but virtually nothing is known of his life. He added practical elements to mathematics; for example, he used algebraic equations to solve problems of inheritance. He used Greek and Babylonian ideas in his work, and introduced powers greater than two to Arab mathematics. His work had a large influence on Leonardo of Pisa (Fibonacci) and other later European mathematicians.

Abu Said Ahmad ibn Muhammad al-Sijzi
945?-1020?

Persian mathematician and astrologer who considered a number of challenging geometric problems. Notable among his works is *Book of the Measurement of Spheres by Spheres,* in which he presented 12 theorems involving a large sphere containing between one and three smaller spheres. Some historians maintain than in this work al-Sijzi approached the idea of a four-dimensional sphere, though it is likely that this is the result of his own misunderstanding of the factors he was considering. He did, however, approach the topic suggested in the title of another work, *Treatise on How to Imagine the Two Lines Which Approach but Do Not Meet When They Are Produced Indefinitely.*

Sridhara
d. c. 850

An Indian mathematician who wrote at least three mathematical works. Only two of these survive, the *Patiganita* and the *Patiganitasara* (or *Trisatika*). The *Patiganita* covers the mathematical operations of addition, subtraction, multiplication, division, finding squares, cubes and their roots, fractions, proportions, and some geometrical principles, but the last section is lost. The *Trisatika* is a brief summary of the *Patiganita.* His lost work was on algebra.

Sripati
1019-1066

Indian mathematician, astronomer, and astrologer who applied mathematical techniques to problems in astronomy, such as the study of spheres. His *Siddhantasekhara,* though a work of astronomy, contains chapters on arithmetic, algebra, and spheres. In the *Dhikotidakarana* (1039) he discussed solar and lunar eclipses, and in *Dhruvamanasa* (1056) he examined the problem of calculating planetary longitudes, eclipses, and planetary transits.

Sharaf al-Din al-Tusi
1135?-1213

Persian mathematician and astronomer who helped lay the foundations for algebraic geometry. In his *Treatise on Equations,* al-Tusi examined a number of equations, including cubics, and appears to have applied an early version of what would much later come to be known as the Ruffini-Horner method. In fact, this method had been developed earlier by Arab mathematicians, but al-Tusi's writing contains the first extant example of it.

Ulugh Beg
1394-1449

Central Asian ruler, astronomer, and mathematician who, in addition to gathering a number of leading scientific minds around him, made important contributions to trigonometry. A grandson of the Mongol conqueror Tamerlane, Ulugh Beg ruled the city of Samarqand, where he established a school that included Qadi Zada, al-Kashi, and others. His achievements included producing tables for sines and tangents that were correct to eight decimal places and helping to create the *Zij-i sultani,* a star catalogue regarded as a standard work until the seventeenth century. Little interested in ruling, Ulugh Beg was later usurped by his son, who had him put to death.

Abu Abdallah Yaish ibn Ibrahim al-Umawi
1400?-1489

Spanish Muslim mathematician whose *Marasim al-intisab fi'ilm al-hisab* is the earliest extant arithmetical treatise by a Muslim in Spain. One interesting characteristic of the text is the fact that al-Umawi apparently wrote it for use by mathematicians in the eastern Islamic world—al-Umawi later migrated to Damascus—at a time when the achievements of eastern Islamic mathematicians generally surpassed those of their counterparts in Spain and Morocco. *Marasim* examines topics such as addition, multiplication, arithmetic series, and geometric series.

Abu'l Hasan Ahmad ibn Ibrahim al-Uqlidisi
920?-980?

Arab mathematician whose *Kitab al-fusul fi al-hisab al-Hindi* (952-53) is the earliest known Arab work discussing the Hindu number system. Though he was born and died in Damascus, al-Uqlidisi spent much of his career traveling and may have learned about Hindu mathematics in India. In his treatise he explained Hindu numerals, translated problems used by earlier mathematicians into the terms of the new system, and answered questions that might be raised by students confronting the system for the first time.

Bibliography of Primary Sources

Alberti, Leon Battista. *Della pittura* (1436). The first mathematical explanation of the theories of one-point linear perspective developed by the architect Filippo Brunelleschi. With this work, a new science of perspective was founded on rigorous geometrical principles.

Alberti, Leon Battista. *De re aedificatoria* (1452). A seminal study of architecture that drew on the ideas of the Roman writer Vitruvius and had an enormous impact on European architects in the centuries that followed.

Battani, al-. *Kitab al-Zij*. Astronomical handbook that introduced new trigonometric methods for performing astronomical computations. The book consists of a star catalog and trigonometric tables, as well as solar, lunar, and planetary tables together with canons for their use. The treatise was written from a Ptolemaic perspective. However, al-Battani turned a critical eye toward Ptolemy's (second century A.D.) practical results, and designed improved instruments to collect more accurate data. These observations revealed errors in Ptolemy's *Almagest* and allowed al-Battani to provide corrected values for many of the main parameters of planetary motion. The *Zij* was translated in the twelfth century and again in the fifteenth, and it proved influential in the development of European astronomy.

Bhaskara. *Siddhanta Siromani.* (Head jewel of accuracy). A series of four books in verse. The first two, *Lilavati* (The beautiful) and *Bijaganita* (Seed counting) deal largely with arithmetic and algebra. The last two books, *Grahaganita* (Mathematics of the planets), and *Goladhyaya* (Chapter on the celestial globe), discuss astronomy.

Bhaskara. *Karanakutuhala* (Calculation of astronomical wonders). A significant text on the motions of the planets and the numerical techniques used to study them.

Biruni, al-. *Al-Qanun al-Mas'udi* (The Mas'udi canon). A major work on astronomy.

Biruni, al-. *Athar al-baqiyah* (Chronology of ancient nations) (1000). One of the most useful sources available for information on Iranian calendars and local history.

Biruni, al-. *Kitab as-Saydalah.* A treatise on drugs used in medicine.

Biruni, al-. *On Chords* (1027). Along with other writings, this text provided an invaluable history of both Hindu and Muslim advances in astronomy during the period from the eighth to tenth centuries.

Bradwardine, Thomas. *De proportionibus velocitatum in motibus* (1328). Analyzes principles put forth by Aristotle concerning the physics of bodies in motion.

Campanus of Novara. *Computus major.* A treatise on the properties of time and calendrical computations.

Campanus of Novara. *Theorica planetarum* (Theory of the planets). Provided detailed observations on the construction of a planetarium. In addition, the book discussed the longitude of the planets and a geometrical description of the planetary model used.

Ch'in Chiu-shao. *Shu-shu chiu chang* (Mathematical treatise in nine sections) (1247). An influential book that explored a range of subjects, including simultaneous integer congruencies, linear simultaneous equations, algebraic equations, the areas of geometrical figures, and what became known as the Chinese Remainder Theorem. The book was heavily concerned with practical applications crucial to the administration of a Confucian bureaucracy. Later Johann Karl Friedrich Gauss (1777-1855) and other mathematicians would delve into the questions of congruencies investigated by Ch'in.

Chu Shih-chieh. *Suan-hsueh ch'i-meng,* (Introduction to mathematical studies) (1299). As its name indicates, this was a book written for novices. The manuscript disappeared from China some time after the author's death, and was only recovered in the nineteenth century; in the meantime, it had spread to Japan and Korea, where it came into wide use as a textbook beginning in the 1400s.

Chu Shih-chieh. *Su-yuan yu-chien* (Precious mirror of the four elements) (1303). A book written to challenge the thinking of mathematical scholars. The four elements referred to in the title were four variables in a single algebraic equation, which Chu expressed using what he called the "method of the celestial element."

Fibonacci, Leonardo. (Leonardo of Pisa). *Liber abaci* (Book of calculations) (1202). This book introduced Europeans to the wondrous numerical system of the East, explained the rudiments of reading and using Hindu-Arabic numerals, and went on to a discussion of fractions, squares, and cube roots before addressing more challenging applications in geometry and algebra. Fibonacci, himself trained in business mathematics, also included chapters on the practical uses of the numeral system. This work would prove instrumental in putting an end to the old Roman system of numerical notation.

Fibonacci, Leonardo. (Leonardo of Pisa). *Liber quadratorum.* An examination of number of theorems involving indeterminate analysis.

Fibonacci, Leonardo. (Leonardo of Pisa). *Practicae geometriae* (1220). A discussion of algebra, geometry, and trigonometry.

Gerard of Cremona. Translated Euclid's *Elements* (trans. 1175), and Ptolemy's *Almagest* (trans. 1175). These translations, from Arabic back into Latin, helped European scientists rediscover these crucial works.

Ibn Yusuf, al-Hajjai. Translated Euclid's *Elements* (trans. c. 825). This was the first Arabic translation of Euclid's seminal work; thanks in large part to this translation, it would be much more widely studied in the Islamic world than in the West for many centuries.

Jordanus Nemorarius. *Planisphaerium.* A treatise on mathematical astronomy.

Jordanus Nemorarius. *Tractatus de Sphaera.* The second of Jordanus's two tracts on mathematical astronomy.

Kashi, al-. *Treatise on the Circumference* (1424). In this work Kashi made his famous calculation of π, which improved greatly on the most accurate figure to date, obtained by Chinese astronomers in the fifth century.

Kashi, al-. *The Key to Arithmetic* (1427). Written to provide scholars at Samarkand with the mathematical instruction necessary for their studies in astronomy, al-Kashi developed a number of interesting geometric ideas, including a means of measuring a complex shape called a *muqarna* used to hide edges and joints in major public buildings.

Kashi, al-. *The Treatise on the Chord and Sine.* Unfinished at the time of al-Kashi's death and later completed by Qadi Zada, this work presented highly accurate sine calculations and touched on the idea of cubic equations.

Khwarizmi, Muhammad ibn Musa, al-. *Al-Khwarizmi Concerning the Hindu Art of Reckoning.* Translated into Latin as *Algoritmi, de numero Indorum,* from which we get the word algorithm to describe the mathematical procedure discussed in this work.

Khwarizmi, Muhammad ibn Musa, al-. *al-Kitab al-mukhtasar fi hisab al-jabr wa'l-muqabalah* (The compendious book on calculation by completing and balancing) (Probably 825-830). Addressed matters of practical mathematics, offering solutions to problems involving unknown quantities. It was here that al-Khwarizmi introduced his method of restoration or completion, or separating the variable from other figures. When translated into Latin in the twelfth century, the term *al-jabr* became *algebra.*

Khwarizmi, Muhammad ibn Musa al-. *Treatise on Calculation with the Hindu Numerals.* In this major mathematical work, al-Khwarizmi provided a guide for the use of the numerals 0-9.

Levi ben Gershom. *De numeris harmonicis* (The harmony of numbers) (1343). A commentary on the first five books of Euclid's *Elements,* commissioned by the Bishop of Meaux.

Levi ben Gershom. *De sinibus, chordis et arcubus* (On sines, chords, and arcs) (1342). An investigation of trigonometry, the book included five sine tables and examined the sine theorem for plane triangles.

Levi ben Gershom. *Sefer ha-hekkesh ha-yashar* (Book of proper analogy) (1319), Levi challenged cosmological ideas handed down by the Greek astronomer Ptolemy (fl. 100s). Many of the latter's assertions—for instance, his insistence that the earth is the center of the universe—were indeed spurious. Levi, however, made a few mistakes of his own, maintaining for instance that the glow of the Milky Way is a reflection of the light produced by the sun.

Levi ben Gershom. *Sefer ha Mispar* (Book of numbers) (1321). A discussion of arithmetical operations, including root extractions.

Li Yeh. *T'se-yuan hai-ching* (Sea mirror of circle measurements). A book of 170 problems involving right triangles inscribed in circles or vice versa; included the author's use of a symbol for negative numbers.

Li Yeh. *Yigu yanduan* (New steps in computation). Here the author applied his method of variables, using polynomials in solving equations.

Omar Khayyam. *Commentaries on the Difficult Postulates of Euclid's Book*. Khayyam analyzes aspects of non-Euclidean geometry.

Omar Khayyam. *Risala fi'l-barahin ala masa'il al-jabr wa'l-muqabala* (Treatise on demonstration of problems of algebra). In this book Khayyam discussed the subject of cubic equations, and became the first mathematician to propose a general theory for their solution. He was the first to demonstrate that a cubic equation could have two roots.

Omar Khayyam. *Rubaiyat* (Quatrains). Translated by Edward Fitzgerald in 1859 and published as *The Rubaiyat of Omar Khayyam*, these metaphysical poems are the most famous of Khayyam's works.

Oresme, Nicole d'. *De moneta*, (On money). Written between 1355 and 1360, Oresme maintained that a ruler has an obligation not to debase his kingdom's currency.

Oresme, Nicole d'. *Livre du divinacions* (On divination). Here Oresme maintained that events with alleged astrological causes can be explained by scientific phenomena instead.

Oresme, Nicole d'. *Questiones super libros Aristotelis de anima* (1370). This commentary on Aristotle discussed ideas about the motion of the earth and also examined the nature of light, particularly its speed and reflective qualities.

Robert of Chester. *Algebra* (1145). Latin translation of Muhammad ibn Musa al-Khwarizmi's (c. 800-c. 847) *al-Jabr wa'l-Muqabalah*. This translation introduced western Europe to algebra. Al-Khwarizmi's was the primary influence on the development of European algebra, determining its rhetorical form and much of its specialized vocabulary. The modern term *algorithm* is derived from the translation's first line. The translation also contains the first use of *sinus* (from which the term "sine" is derived) in its modern trigonometric sense. (Gerard of Cremona [1114-1187] produced another translation in 1150).

Sacrobosco, Johannes de. *Algorismus* (1230). A widely used textbook in European universities, the book addresses a range of subjects, including not only arithmetic functions but square and cube roots. Most notable, however, was Sacrobosco's use of Hindu-Arabic numerals, just then beginning to gain acceptance.

Sacrobosco, Johannes de. *De anni ratione* (1232). A treatise on the measurement of time. In addition to discussing the day, week, month, year, lunar phases, and ecclesiastical calendar, Sacrobosco used the book to make an early case for calendar reform.

Sacrobosco, Johannes de. *Tractatus de quadrante*. A discussion on the uses of the quadrant in astronomy.

Sacrobosco, Johannes de. *Tractatus de sphaera* (1220). Popularized the work of the Greek astronomer Ptolemy. Among the topics discussed in the book are Ptolemy's ideas on planets and eclipses, as well as the position of the Earth in a universe that Sacrobosco believed to be spherical. In 1472, the *Tractatus* would be the first printed work of astronomy, and despite its many errors, it continued to be widely used until the 1600s.

Thabit ibn Qurra. *Book on the Determination of Amicable Numbers*. A discussion of the formulae for determining perfect and amicable numbers, reputedly the first to list the amicable pair 17296, 18416.

Thabit ibn Qurra. *Elements*. This revised translation of Euclid would become the definitive text.

Tseng Kung-Liang. *Wu Ching Tsung Yao* (c. 1044). Famous Chinese military text describing the construction of one of the earliest compasses, primarily used by soldiers on land for navigation.

Yang Hui. *Cheng chu tong bian ben mo* (The beginning and end of variations in multiplication and division) (1275). Valuable for the insights it provides regarding the curriculum used by Chinese mathematical students at that time. Furthermore, Yang Hui's syllabus, included in the book, shows him to have been a highly competent teacher who emphasized genuine learning rather than memorization

JOSH LAUER

Physical Sciences

Chronology

c. 775 Alchemist Abu Musa Jabir ibn Hayyan, known as the "father of Arab chemistry," discovers the principal salts of arsenic, sulphur, and mercury.

c. 850-c. 950 First serious challenges to Ptolemy: Irish-born philosopher John Scotus Erigena suggests that the planets revolve around the Sun (c. 850); Albategnius clarifies a number of fine points in Ptolemy (900); and Abd al-Rahman al-Sufi revives Ptolemy's catalog of fixed stars, preparing an accurate map of the sky that will remain in use for centuries.

914 The writings of al-Mas'udi, with their discussion of evaporation and the causes of ocean salinity, constitute the beginnings of scientific geography.

c. 1000 Arab physicist Ibn al-Haytham (also known as Alhazen) argues against the prevailing belief that the eye sends out a light that reflects off of objects; instead, he correctly posits that light comes from an external source such as the Sun.

c. 1025 Ibn Sina (also known as Avicenna) writes an alchemical treatise that later becomes a standard text in medieval Europe; in it, he argues that conversion of ordinary metals into gold or silver is not possible.

1090 Chinese court astronomer Su-sung builds one of the first timekeeping devices that relies neither on sunlight, sand nor other such unreliable contrivances: a huge astronomical clock operated by a water wheel.

1137 Arab astronomer and mathematician Abu Ja'far offers a general description of the laws of gravity.

c. 1175 In China, Chu Hsi states that fossils were once living organisms.

1260 German scholar Albertus Magnus writes *De Rebus Metallicus et Mineralibus,* in which he compiles a list of some 100 minerals; the work retains its authoritative status for several centuries.

c. 1275 English scholar Roger Bacon builds the first magnifying glass to be used for scientific purposes, and advocates the use of chemicals in the practice of medicine.

c. 1280-c. 1350 French scholar Peter Olivi develops the theory of impetus, which contradicts Aristotle by correctly maintaining that a moving body will continue in motion even when the propelling force is removed; 70 years later, Jean Buridan and Albert of Saxony independently corroborate this position.

c. 1300-c. 1325 Two very different scholars both advance scientific thinking: John Duns Scotus distinguishes between causal laws and empirical generalizations, laying the groundwork for the scientific method; and William of Ockham formulates "Ockham's razor," which states that when two theories equally fit all observed facts, the one requiring the fewest or simplest assumptions is preferable.

1335 Richard of Wallingford, an English scholar who made the first known studies of tides for the purpose of preventing floods, builds a mechanical clock that indicates high and low tide.

1440 Nicholas of Cusa, a German cardinal, invents the hygrometer for measuring moisture and the bathometer for measuring depths in water.

Overview: Physical Sciences 700-1449

The year 700 was a turning point in history because it marked the beginning of the medieval era. In the west, small European kingdoms began to unite into larger ones. In the Middle East Islam asserted itself politically and soon became a dominant cultural force. In both Christendom and the Dar-al-Islam the knowledge of ancient civilizations provided a foundation for the long period of growth toward modern western civilization, yet the continuity of learning in these two spheres was quite different.

In marked contrast, Far Eastern civilization was based on centuries of uninterrupted learning. Chinese physical science had begun as early as 2000 B.C., with observational astronomy and a 12-month calendar. Chinese optics was as ancient as that of the Greeks. Alchemy, too, had its roots in Asia in the second century A.D.

Things were also different in the West, where barbarian invasions and the collapse of Roman civilization led to the Dark Ages and the loss of ancient scientific knowledge. In the Middle East Islamic culture became benefited from a tradition that encouraged the pursuit of knowledge and the assimilation of knowledge from earlier cultures.

Medieval Islamic Physical Science

The legacy of ancient knowledge, held in trust by eastern Christian and Persian scholars and augmented by Indian scholars, was the focus of a concerted Islamic effort, beginning in the eighth century, to translate the great corpus of Greek philosophy, science, and mathematics into Arabic. Aristotle (384-322 B.C.), Pythagoras (c. 580-500 B.C.), Plato (c. 428-c. 348 B.C.), Claudius Ptolemy (c. 85-c. 165), and others were diligently studied, demonstrating the importance of knowledge, science, and philosophy in the Islamic world. This effort reached a high point in the early ninth century under the Abbasid caliph al-Mamun.

Arabic astronomers incorporated Indian, Persian, and Near Eastern ideas, along with the central synthesis of Greek thought on the cosmic realm. Among the important contributions they made to the sciences were the first large-instrument observatory (Maraghah in Persia) and the development of accurate astronomical instruments, including the astrolabe, which they used to catalog old stars and discover new ones.

In physics, Islamic scientists learned how to measure an object's specific weight (or weight density) and to use the balance. They also tried to correct Aristotle's theory of projectile motion, which led them to consider ideas on impetus and momentum. Their most important contribution may have been innovative experimental techniques in the study of optics, pioneered by ibn al-Haytham, also known as Alhazen (965-1038).

In the Islamic world, the alchemy of Alexandrian Greece and China were studied, producing the famous Islamic alchemist Abu Musa Jabir ibn Hayyan (c. 721-815). In the next century experimentation helped refine the acid-base theory in the science of chemistry, and Muhammad ibn Zakariyya ar-Razi, also known as Rhazes (c. 865-c. 930), an alchemist who was also the greatest physician in the Islamic world, divided all matter into organic and inorganic categories. Interestingly, the term "alchemy," whose name is derived from the Arabic *al-*

kimiya, was little known in the West until Arabic works were translated in the eleventh century.

Applied physical science was given full attention as well. Islamic thinkers studied geography, minerals, and geology. Ibn Sina, also known as Avicenna (980-1037), produced the first descriptions of Malayan minerals. The close contemporary and equally polymathic Abu Rayhan al-Biruni (973-1048) wrote extensively on physical geology, geography, the problem of determining the circumference of the earth, and his study of India, which resulted in the first correct description of the sedimentary nature of the Ganges basin.

Medieval European Physical Science

Early European scientific thought was hindered by Christian doctrinal distrust of surviving Greek science, which centered on astrology and the physical traditions from Aristotle, neither of which agreed with Scripture. St. Augustine was the first to retreat from this stance by stating that, when dealing with the physical world, experiment should count more than faith-based interpretation.

About the early tenth century, the first Arabic translations and commentaries appeared in Europe, translated into Latin by a dedicated generation of scholars: Domengo Gondisalvi (c. 1134), John of Luna (c. 1134), and others. By the twelfth century these Islamic contributions, along with original Islamic works (such as Ibn Haytham's *Book of Optics*), fueled much interest in Western science. Unfortunately, these Islamic hand-me-downs were often paraphrases and summaries of Greek thought rather than works in their entirety. Producing complete Latin translations of original Greek texts became the goal of another generation of translators, particularly Gerard of Cremona (c. 1114-1187) and his group at Toledo, followed by Michael Scot (c. 1230), and later William of Moerbeke (c. 1255-1278). Their efforts spurred critical appraisal of past scientific thought, and encouraged independent contributions to physical science.

In the West, little was known about Aristotle, although his works greatly influenced Islamic science, until about 1115, when Arabic paraphrases, commentaries, and partial translations made some of his works available. By the late twelfth and early thirteenth centuries Aristotle's scientific thought had gained much attention in England and France—and ecclesiastical resistance to them only increased curiosity and study. By the 1240s Aristotle had become the foundation of the formal scholastic framework used in the analysis of science and other ideas. Following a concerted effort, all of Aristotle's works had been translated from Greek into Latin by about 1278.

The Aristotelian foundation was built upon by Robert Grosseteste (c. 1168-1253) (who influenced the English Oxford Franciscans), Albertus Magnus (c. 1193-1280), and scholars at the universities of Paris and Toulouse. Soon, however, a more critical appraisal of Aristotle and his ideas surfaced. By the late thirteenth century a formalized school of logic called *nominalism* emerged. It first appeared in commentaries by John Duns Scotus (c. 1270-1308) and is best known from the work of William of Ockham (c. 1285-1349).

Aristotle's astronomical theory centered on a "Prime Mover" of the universe, which for many medieval European thinkers represented God. This was particularly true for Thomas Aquinas (1225-1274), who used the concept of a Prime Mover in the first of his logical proofs of God's existence. Ockham, on the other hand, by applying strictly nominalist principles, denied that the Prime Mover concept proved the existence of God.

Other Scientific Achievements

The era's heightened interest in science produced further seminal observations and conclusions. In 1269 Peter Peregrinus of Maricourt (fl. 1200s) provided the first experimental study on the magnetization of iron and magnetic poles. Mathematician Jordanus de Nemore (c. 1255) advanced static physics by defining the law of straight-lever equilibrium.

Studies on the reflection and refraction of light, especially rainbow optics, marked the work of Oxford's Grosseteste and the Oxford Franciscans, followed by those of the Pole Witelo (c. 1230-c. 1275), and culminated with Theodoric of Freiberg's (c. 1250-1311) comprehensive rainbow theory. The Oxford and Paris nominalists, particularly Thomas Bradwardine (c. 1290-1349), Jean Buridan (c. 1270-c.1358), and Nicole Oresme (c. 1320-1382), established concepts in the physics of motion, or dynamics.

The scholars at Oxford and Paris also delved into more speculative areas of physical science, such as the possibility of the vacuum (which was judged reasonable though nature seemed to abhor it). Another was the motion of the earth. Though Aristotle's physics and Earth sciences shaped medieval thought, his astronomy had been supplanted by Ptolemy's (fl. 139-161), despite criticism of its confusing epicenters and epicycles to explain the apparent retrograde movements of the planets. In either case, the earth was considered to be at rest, yet other incessant motions were proposed relating to changing center of gravity by Albert of Saxony (c. 1316-1390), Buridan's student at Paris. This opened the way for conjecture possibility by Oresme (1377) that the earth could rotate, and that its territory obeyed a gravity attraction to the earth's center.

Conclusion

Science during this period was undeniably in its infancy and hampered by its ties to ideological, rather than objective, standards. Despite this, the rediscovery of ancient writings and the exposure of European civilization to Eastern, particularly Islamic, achievements, set the stage for the flowering of scientific knowledge that would occur in the Renaissance.

WILLIAM J. MCPEAK

Science in Premodern China

Overview

The study of science in China during the period from 700 to 1449 must be placed within the larger framework of science in *premodern* China, thus encompassing preceding eras as well. The reason is that, unlike Western history, that of China is not as conveniently divided into "ancient" and "medieval" periods. Moreover, much of the scientific work in China during the age viewed by Westerners as a middle period was in fact on a continuum with that which had occurred in ancient times.

Premodern China saw enormous advances in all realms of science, from medicine to technology, that gave the Chinese enormous advantages over their counterparts in the Western world and, in many cases, the Middle East as well. As with the eras of scientific development in China, the varieties of science practiced in the "Middle Kingdom" must also be seen on a continuum, because Chinese scholars rarely made a distinction, for instance, between "pure" and "applied" science. It is useful, then, to view the various disciplines not so much as discrete entities, but as interlocking circles, an interdependent array of pursuits with astronomy and alchemy—forerunner of modern chemistry—near the center.

Background

Science flourished throughout many phases of ancient Chinese history, and achievements of early Chinese scientists covered a number of areas. At least as early as 1700 B.C., China had entered the Bronze Age, and archaeologists have found Iron Age tools that date back to about 1000 B.C. It appears that by the fifth century B.C., the Chinese had passed on to the next technological stage, making steel from iron.

Around 1500 B.C., Chinese astronomers created a calendar that took into account the phases of the Moon and the length of time it took for Earth to revolve around the Sun—that is, a year. They also recorded eclipses as well as a nova, a star that suddenly grows extremely bright before fading. Their observations of events in space make it possible for historians to be absolutely certain about dates in Chinese history after 840 B.C.

The historic period in China began with the Shang Dynasty (1766-1027 B.C.), which built its power using advanced forms of warfare technology. A Shang chariot squadron consisted of five horse-drawn wagons, each with a driver, an archer, and a soldier bearing a battle axe. In the Shang religion, one can see the elements of an early interest in chemistry. They practiced forms of divination—the study of physical material in order to find omens of the future—using tortoise shells and the bones of cattle or water buffalo. Priests would polish the outsides of the shells or bones, then dig out holes on the inside to make them easier to crack. After this, they applied heat to the underside. As cracks appeared on the top, they would study these cracks much as a palm-reader observes the lines in a person's hand for "omens" concerning his or her future.

Shang farms grew a wide array of crops, and it appears that Shang farmers employed an early form of crop rotation. Mulberry trees on farms in Shang China yielded a product for which China would become famous: silk. Another plant grown on Chinese farms was hemp, important both for ropes and for the narcotic qualities of the plant, which the Chinese used in medicinal treatments. The Shang also developed ice cream—made by mixing milk with soft rice, and then chilling this mixture in snow from the high mountains—in about 2000 B.C.

Shang rulers became increasingly oppressive until they were replaced by the Chou Dynasty (1027-246 B.C.), which maintained power longer than any system in the history of the world. Despite its longevity, the Chou Dynasty was fraught with political upheaval, yet scholarship flourished during the era as well. In 613 B.C., the Chinese made the first recorded sighting of what came to be known as Halley's Comet, which passes by Earth approximately every 76 years. By the fourth century B.C., astronomers in China had compiled a chart showing various stars' locations.

The Chou era also produced China's greatest philosophers, most notably Confucius (551-479 B.C.) and Lao-tzu (500s B.C.). Aside from Confucianism and Taoism, one important school of thought to emerge during this time was Naturalism, which maintained that for every force in one direction, it was right and necessary that there should be a force in the opposite direction. On the one side, there was an active, masculine quality called *yang;* on the other, an inactive, feminine quality known as *yin.* The combination of yin and yang, it was believed, produced everything that existed—a concept that would become central to Chinese thought in general, and Chinese science in particular.

Chou rule dissolved into anarchy, leading to the rise of a short-lived but highly significant dynasty, the Ch'in (221-207 B.C.), which gave China the name by which it is known to the rest of the world. The Ch'in ruler Shih Huang-ti (259-210 B.C.) united much of the country for the first time, established China as an empire—it would remain such until 1912—and began the building of the Great Wall. The autocratic Ch'in Shih-huang-ti also built a vast nationwide system of roads and canals, and standardized weights and measures, currency, the written language, and even the size of vehicle axles.

After the Ch'in, which quickly lost power following the death of its founder, the Han Dynasty took power, and would rule for most of the period from 207 B.C. to A.D. 220. Under WuTi (r. 141-87 B.C.), China established the rudiments of the Confucian civil-service system that would remain in place until modern times; issued banknotes; created a state monopoly over industries such as salt-making and iron production; and made its first contacts with the civilizations of the West.

The dissolution of Han rule led to successive periods of crisis and stability that lasted until the founding of the Sui Dynasty (589-618). As had happened before in China, however, political problems did not necessarily impede scientific progress. Thus this era saw developments in the study of medicine, the first use of coal for heat, the first appearance of kites, and the writing of the first encyclopedias. As for the Sui, their rule resembled that of the Ch'in: highly autocratic, short-lived, and historically significant. Also like the Ch'in, the Sui built enormous public works projects, most notably the Grand Canal, a waterway of some 1,000 miles (about 1,600 kilometers) that connected the Yangtze and Yellow rivers.

Impact

During the period that coincided with medieval times in Europe, Chinese scholarship experienced high points under two native-controlled dynasties, both distinguished by reforms and highly efficient governments. The first of these was the T'ang (618-907), which greatly extended the canal network put in place by the Sui, thus aiding the transport of goods from north to south in a land where most major rivers flowed eastward. As the economy of T'ang China thrived, new goods such as tea from Southeast Asia made their appearance. China in turn exported a variety of items, including silk and printed materials.

The latter was an outgrowth of two outstanding Chinese innovations: paper—first developed around A.D. 100 by Tsai-lung (c. 48-118) and not discovered in Europe until the fourteenth century—and block printing, a process whereby a printer carved out characters on a piece of wood. Its invention was attributed to Buddhist monks of the seventh century, who needed copies of sacred texts faster than they could be produced by hand-copying. The world's first printed text was a Buddhist scroll, later discovered in Korea and probably printed in China between 704 and 751. The ink for these early printed documents came from the black substance secreted by burning wood and oil in lamps; later, when this innovation passed to the West, it would incorrectly be called "India ink."

The T'ang Chinese were eager to import knowledge of astronomy and mathematics from

India. Indeed, this interest in outside knowledge was a hallmark of the early T'ang, who were more open to "foreign" ideas than most Chinese ruling houses—including the later T'ang. With the growing acceptance of Buddhism that had begun in about the third century A.D., numerous ideas arrived from India, including advances in number theory and the more systematic study of the skies practiced by Indian astronomers at the great center in Ujjain.

Chinese studies in astronomy, however, retained a quasi-religious flavor. As was typical of many premodern societies, there was sometimes little distinction between astronomy and astrology, and all work was heavily controlled by the state. An edict of the emperor in 840, for instance, stated that, "If we hear of any intercourse between the astronomical officials or their subordinates and officials of other government departments or miscellaneous common people, it will be regarded as a violation of security regulations which should be strictly adhered to." Nonetheless, it was still possible to make significant advances in such an environment, as the building of a water-driven astronomical clock in 721 attests. This was the creation of the Buddhist monk I-hsing and the military engineer Liang Ling-tsan, and it was the first known instrument to use an escapement, a device essential for regulating a clock's movements.

A humiliating military defeat by the forces of the Abbasid caliphate at Talas marked the watershed of the T'ang Dynasty, which steadily dwindled over the next 15 decades until China fell into a period of semi-anarchy identified by Chinese historians as "Five Dynasties and Ten Kingdoms." Finally, in 960 the establishment of the Sung Dynasty, which would last until 1279, brought order to the country.

Astronomical study flourished under the Sung Dynasty, which saw the building of an even more impressive astronomical clock than the 721 T'ang model. In 1077 Su Sung, a diplomatic envoy to the court of a "barbarian" ruler to the north, was embarrassed to discover that his host's calendar was more accurate than that of his own emperor. Therefore he requested and received permission from the emperor to build what he called a "heavenly clockwork," an astronomical clock powered by a water wheel.

Science historian Daniel J. Boorstin described Su Sung's impressive creation as "a five-story pagoda-like structure. On the topmost platform, reached by a separate outside staircase, was a huge bronze power-driven armillary sphere within which there rotated automatically a celestial globe.... Every quarter-hour the whole structure reverberated to bells and gongs, the splashing of water, the creaking of giant wheels, the marching of manikins." It was altogether a far more accurate time-keeping device than anything that would appear in Europe for many centuries, but after a new emperor took the throne in 1094, his court declared the previous ruler's calendars to be inaccurate—a custom of Chinese rulers. Thus the clock was allowed to fall into disrepair.

Sung astronomers also advanced Chinese methods of time-keeping by replacing the 10-day week with a 7-day version—based on the four phases of the Moon—in about 1000. Along with their counterparts in Arab lands, astronomers of this era witnessed a supernova, an exploding star, in 1006, and again in 1054. The first of these remained visible for a year, and the second—the first such event for which relatively detailed records exist—was described by Chinese and Japanese astronomers as the sudden appearance of an extremely bright light in the constellation known by Europeans as Taurus. Much later, during the Ming era, Chinese astronomers in 1433 would observe a comet also noted by the Italian Paolo Toscanelli (1397-1482).

Though the Sung were highly competent administrators, they lacked savvy in foreign affairs, and a disastrous alliance with the nomadic Juchen people of Manchuria forced them to move their capital to southern China in 1127. Nonetheless, the Southern Sung period was a time of tremendous scientific and technological advancement.

Notable during this period, for instance, were the geological studies of Chu Hsi, who in his *Chu Tsi Shu Chieh Yao* (c. 1175) suggested that fossils were once living organisms. Chinese astronomers in about 1270 built a torquentum, an instrument for making transformations between spherical coordinates. An improvement on the armillary sphere, it was the first such device to use an equatorial mounting.

Porcelain-making also reached its height during this era. Though porcelain's invention had been attributed to the semi-legendary figure Tao Yue (c. 608-c. 676), who combined the "white clay" or kaolin of the Yangtze River banks with other types of clay, it is likely that the craft actually began to take shape as early as the third century A.D. Not until the eighteenth century would Europeans develop means of making porcelain.

On the high seas, the magnetic compass made navigation at sea much easier. Land compasses, which used a naturally magnetic piece of lodestone, had been used in China as early as the fourth century B.C., but in the period between 850 and 1050, the needle compasses had been developed for use at sea. These took into account the shift in Earth's magnetic field, or magnetic declination. Though a description of a magnetic compass first appeared in European writings in about 1190, it would not be until the fifteenth century that Europeans became aware of the magnetic declination.

Along with the compass, improvements in shipbuilding enabled the Sung to send ships (junks) on merchant voyages. The larger Sung junks could hold up to 600 sailors, along with cargo. Around this time—to judge from a church carving that dates to 1180—Europeans began to use a ship's rudder, but this invention, too, had long existed in China. As early as the first century A.D., Chinese shipbuilders had used rudders in place of the more cumbersome steering oars.

Tea and cotton emerged as major exports during the Southern Sung era, while a newly developed rice strain, along with advanced agricultural techniques, enhanced the yield from China's farming lands. China also exported a variety of manufactured goods, including books and porcelain, while steel production and mining grew dramatically. Banks and paper money—one of Sung China's most notable contributions—also made their appearance.

The development of movable-type printing aided the spread of information: instead of carving out a block of wood, a printer assembled precast pieces of clay type to ink out written messages. This, one of the most significant developments in Chinese technology, is attributed to the alchemist and blacksmith Pi Sheng, who in c. 1045 cut a series of characters into clay cubes on an iron frame, making a solid block of type. Ultimately, however, the peculiarities of the Chinese language would encourage the use of block printing over movable type. It is easy enough to store and use blocks when a language has a 26-letter alphabet, as English does; but Chinese has some 30,000 characters or symbols, meaning that printing by movable type was extremely slow. Thus movable-type printing would have its greatest impact in Europe, where it was introduced by Johannes Gutenberg (c. 1395-1468) in about 1450.

Great strides in science and technology continued right up to the end of the Sung era, making the end of the Sung Dynasty all the more of a tragedy. The Sung created pumps for lifting water, and experimented with water power as a means of operating silk looms and cotton mills. The final years of Sung rule saw the development of one of history's most significant inventions, one that would completely alter the world of the future: the use of gunpowder for warfare.

Gunpowder, actually invented about 950, is counted along with paper, printing, and the magnetic compass as one of the four most significant discoveries made by the premodern Chinese. Here the connection between alchemy, chemistry, and technology in Chinese science is most clear. Alchemy had existed in China from ancient times, long before European alchemists began attempting to turn base metals into gold, but in China it was more closely tied to another elusive goal: the quest for immortality. Thus it naturally allied itself with medical study, which in China—as witnessed by the development of acupuncture in ancient times—often tended to venture into fields that are not fully encompassed by scientific understanding.

Long before the Sung era, Chinese alchemists had been experimenting with the properties of saltpeter and sulfur as a means of creating an elixir for immortality, and between 350 and 650 they produced numerous chemical concoctions involving these substances. By about 850, however, when a Taoist text warned against the dangers of mixing these substances, it had become apparent that the elixir was more likely to end life than to prolong it. Around 1040, Tseng kung-liang, building on the saltpeter-and-sulfur experiments of previous Chinese chemists, created the world's first formula for gunpowder. His was merely combustible, however, and not explosive; nonetheless, its military applications were clear, and it quickly gained use in flame-throwers. By the thirteenth century, Sung chemists had learned how to make explosive gunpowder.

As for firearms themselves, archaeological digs in Manchuria have uncovered a small gun, dating from c. 1288. The first European depiction of a gun—one that fired arrows instead of metal balls—dates from four decades later, in 1327, and it is likely that elements of this technology had arrived in Europe from China. During the last days of the Sung Dynasty, scientists were experimenting with rockets, which they used in warfare against the Mongols.

Impressive as these new forms of military technology were, however, they did not prevent the onslaught of the Mongols, who established China's first foreign ruling house, the Yüan (1264-1368). Under the rule of Kublai Khan (1215-1294), contact between East and West increased dramatically, leading to visits by outsiders such as Marco Polo (1254-1324), who brought back to Europe information about Chinese advances such as gunpowder, paper money, the compass, kites, and even playing cards.

Yet the Chinese chafed at rule by "barbarians," and were only too happy to see the overthrow of the Yüan Dynasty by the Ming (1368-1644), China's last native-born ruling house. The Ming emperor Yung-lo (r. 1403-24) sent a number of naval expeditions under Admiral Cheng Ho (c. 1371-c. 1433) to India, Ceylon, Yemen, and even Africa. These ships brought with them such Chinese luxuries as silks and porcelains, and returned bearing exotic animals, spices, and varieties of tropical wood. Centuries later, when archaeologists unearthed the ruins of Zimbabwe in Africa, they found broken pieces of Ming porcelain.

The naval expeditions were costly, and this fact brought them to an end; also expensive was the establishment of a vast palace complex. In 1421, when Yung-lo moved the capital from Nanjing in the interior to Beijing, he built a palace 5 miles (8 km) in circumference. Containing some 2,000 rooms where more than 10,000 servants attended the imperial family, it was not so much a palace as a city: hence its name, "Forbidden City," meaning that only the emperor and the people directly around him were allowed to enter. Built to illustrate the boundless extent of Ming power, the Forbidden City became—aside from the Great Wall—the best-known symbol of China in the eyes of the world.

These ventures, along with the restoration of the Grand Canal (which had fallen into disrepair under the Mongols), placed heavy burdens on the treasury and weakened the power of the Ming. So too did attacks by Chinese, Korean, and Japanese pirates on Ming merchant vessels, not to mention the appearance of European traders who were often pirates themselves. The Ming Dynasty would continue for several centuries before falling to Manchurian invaders, but like many Chinese ruling houses that preceded it, it experienced a long, slow period of decline.

Likewise, China itself, home to many of the world's great advances in science and technology, began to lag in comparison to the once-technologically inferior civilizations of the West. Just as Europe was awakening from the long sleep of the Dark Ages, China began to turn inward, and Chinese scholars and political leaders placed increasing emphasis on tradition rather than progress, order rather than experimentation. In the nineteenth century, the decaying Chinese monarchy would pay heavily for its reaction against change, and the country was beleaguered by outsiders, including not only Europeans and Americans, but Japanese. Only in the late twentieth century did China again begin to emerge as a formidable technological and scientific power.

JUDSON KNIGHT

Further Reading

Books

Boorstin, Daniel J. *The Discoverers.* New York City: Random House, 1983.

Needham, Joseph. *Science and Civilization in China* (4 volumes). New York City: Cambridge University Press, 1962-65.

Needham, Joseph. *Science in Traditional China: A Comparative Perspective.* Cambridge, Massachusetts: Harvard University Press, 1981.

Needham, Joseph; Wang Ling; and D. J. Price. *Heavenly Clockwork.* New York City: Cambridge University Press, 1960.

Schafer, Edward H. *Ancient China.* New York City: Time-Life Books, 1967.

Temple, Robert. *The Genius of China: 3,000 Years of Science, Discovery, and Invention.* New York City: Simon & Schuster, 1986.

Thorwald, Jürgen. *Science and Secrets of Early Medicine.* New York City: Harcourt, Brace, 1963.

Williams, Suzanne. *Made in China: Ideas and Inventions from Ancient China,* illustrated by Andrea Fong. Berkeley, California: Pacific View Press, 1996.

Internet Sites

"Chinese Inventions." http://www.hyperhistory.com/online_n2/connections_n2/chinese_inventions.html (October 22, 2000).

"Who Invented It? Chinese Inventions." http://www.asksia.org/frclasrm/lessplan/1000019.htm (October 22, 2000).

Greek Texts are Translated into Arabic

Overview

Greek was the language of philosophy, and therefore of science, in the Mediterranean world from the time of the Greek city states through the period of late antiquity. In the seventh century A.D., however, a new world power emerged. The rise of the Islamic Empire brought Muslim culture to North Africa, Spain, Persia, and India. During this period of expansion, Arabs encountered Greek philosophy for the first time, and a systematic effort to translate Greek works received royal support and encouragement. The wide variety and large number of Greek texts that were translated proved to be of lasting significance.

Background

The first dynasty of the Islamic Empire, the Umayyads (661-750), was more concerned with increasing the size of its empire and achieving political and military stability than with translating Greek scientific texts. In the middle of the eighth century the Abbasid dynasty (750-1258), led by the second caliph al-Mansur (709 or 714-775), moved the capital from Damascus to Baghdad, and began to offer royal support for the translation of Greek (as well as Persian and Indian) works of science and philosophy. Al-Mansur's grandson, al-Ma'mun (786-833), however, usually receives the most credit from historians for beginning the systematic translation that lasted until the end of the tenth century.

Al-Ma'mun established the *Bayt al-Hikma,* or House of Wisdom, in which documents from ancient Greece and contemporary Byzantium were translated into Arabic. While the ultimate significance of such an institution is hard to establish historically, it is clear that a precedent was set by the early Abbasids for the patronage of translations. Over the next century and a half, a huge corpus of Arabic translations were created.

A number of factors made the translation movement under the Abbasids successful. Besides a secure empire, which provided the funds necessary to support the project, the move to Baghdad was fortuitous. The area was home to two groups of schismatic Christians, the Monophysites and Nestorians, who had been forced out of the Byzantine Empire. Both of these groups had long been translating Greek works into Syriac. This provided both a body of texts and a group of professional translators ready to perform the service for their new Islamic rulers. Also significant was a basic cultural attitude shared by many of the Islamic faith: that foreign knowledge could be valuable, that truth in some form could be found outside of the revealed religion. The Abbasids saw that Greek works contained important and valuable knowledge, and this gave implicit approval for translation of ancient scientific texts. Finally, because the technology of papermaking had been brought from China into the Islamic Empire, the Muslims could produce books faster and cheaper than their European counterparts.

These circumstances, in combination with the support and encouragement of the ruling family, allowed a large number of Greek works to be translated into Arabic over the course of two centuries. Nearly the whole corpus of Aristotle's (384-322 B.C.) works—as well as a number of works falsely attributed to him—from metaphysics and natural philosophy to logic and ethics were translated, as were Plato's (c. 428-348 B.C.) and later Neoplatonic works, numerous scientific works in astronomy and astrology (Ptolemy, 127-145), mathematics (Euclid, fl. c. 300 B.C.; Apollonius, 262-190 B.C.; and Archimedes, c. 287-212 B.C.), and medicine (Hippocrates, c. 460-377 B.C., and Galen, c. 130-200), as well as some works of Indian and Persian origin. All contributed yet more depth to the Arabic appropriation of Greek science and philosophy.

Impact

The influence that Arabic translations of Greek science and philosophy had on the Arabic world can hardly be underestimated, although the "translation" process often worked both ways. Islamic scientists and their patrons chose which texts to translate, emphasizing and utilizing the portions of the texts that served their own purposes, just as they produced original works that best served their needs.

At the same time, however, Greek science and philosophy, as transmitted through the translation movement, had a huge impact on the culture of the Islamic Empire. The Abbasids actively encouraged the work, thereby setting an example for others. For two centuries patrons from a broad portion of society supported translations, paying translators and scientists and building libraries. In addition, large-scale institutions were created, such as hospitals and astronomical observatories. The texts produced were incorporated into the educational system, expanding the number of persons exposed to the material and incorporating such knowledge into an even broader cultural setting. Many Muslims appeared to value scientific knowledge for its own sake, as a way of understanding the world through the use of scientific theory and investigation.

Translations were often made with the expectation of practical benefit. Mathematics, for example, helped administer both the large imperial bureaucracy and private financial enterprises. Astronomy improved timekeeping and instrument making. Astrology, too, offered practical benefits; even if few believe in astrological prediction today, its rationality given the ancient understanding of the natural world made it attractive. Greek medicine offered a coherent, if often inaccurate, system through which doctors could preserve health and heal injury and disease. The translation of technological works led to the development of mechanical devices and advances in irrigation systems.

Islamic scientists also wrote commentaries on translated Greek scientific texts, thereby increasing their understanding of the world. Muslim astronomers developed a genre of advanced mathematical astronomical treatise that dealt with all aspects of astronomy, including planetary theory, observational techniques, and instrument-making. In some cases, alternative cosmological theories were created in response to perceived failures of Greek systems (such as the problematic relationship between Aristotelian physics and Ptolemaic astronomy). Optics was greatly advanced by scientists like al-Kindi (800-873) and Ibn al-Haytham (965-1040). In the field of mathematics, Islamic scientists propagated the use of decimal-place arithmetic and made important advances in algebra and trigonometry (which properly belonged to the field of astronomy during that time period). In the field of medicine, compilations of medical theory were produced in large numbers.

BAGHDAD'S HOUSE OF WISDOM

In the year 832, the caliph al-Ma'mun established the House of Wisdom in Baghdad. Its purpose was to combine the accumulated knowledge of Europe and the Arab world, making this knowledge available for study and helping preserve it for future generations of Islamic scholars. One of the most famous Islamic scientists who worked in the House of Wisdom was the astronomer and mathematician al-Khwarizmi, who wrote and translated numerous works there. Drawing on the work of ancient Greek and Hindu mathematicians and scientists, his works included an introduction of Hindu mathematics, the first formal book on algebra, and many astronomical works. The concept of knowledge centers like the House of Wisdom would eventually evolve into libraries, scientific organizations, and think tanks.

P. ANDREW KARAM

Ancient texts also provided a language in which to discuss scientific, philosophical, and religious questions. Greek had been used as a platform for philosophical issues since the time of the Greek city-states in the fourth century B.C., and had remained the language of sophisticated philosophy and science through the Hellenistic period (323-30 B.C.) the Roman Empire (27 B.C.-c. fifth century A.D.) in the West, and the Byzantine Empire (c. fifth century to 1463) in the East. With the spread of Christianity after the fourth century A.D., Greek became the language of theological debates. When Islamic intellectual culture tapped this rich vein of Greek language and tradition, it found a wealth of concepts and vocabulary that turned towards its own questions and debates. Greek philosophy had long dealt with issues that Islam raised, such as the nature of the divine and human, man's place in society, the role of law, and so on. Centuries of prior thought on these issues provided valuable assistance.

The translation movement ended around the end of the tenth century, possibly because the movement died out when Islamic scientists and their patrons decided enough material was available, or perhaps because it simply went out of fashion. Regardless of the reasons, the end of the translation movement ended neither Islamic science nor its influence. (A vital tradition of original scientific work continued at least into the thirteenth century.) Arguably the greatest beneficiary of Islamic science and the Greek scientific tradition it preserved was Western Europe. Beginning in the twelfth century, Western translators and scientists began to translate Greek and Arabic scientific works into Latin, and eventually began to compose their own original treatises, just as the Arabic translators of earlier centuries had done.

MATT DOWD

Further Reading

Dhanani, Alnoor. *The History of Science and Religion in the Western Tradition: An Encyclopedia.* Edited by Gary B. Ferngren, et. al. New York: Garland Publishing.

Endress, Gerhard. "The Circle of al-Kindi: Early Arabic Translations from the Greek and the Rise of Islamic Philosophy." In *The Ancient Tradition in Christian and Islamic Hellenism: Studies on the Transmission of Greek Philosophy and Sciences.* Edited by Gerhard Endress and Remke Kruk. Leiden: Research School CNWS, 1997.

Fakhry, Majid. *Philosophy, Dogma and the Impact of Greek Thought in Islam.* Brookfield, VT: Variorum, 1994.

Gutas, Dimitri. *Greek Thought, Arabic Culture: The Greco-Arabic Translation Movement in Baghdad and Early Abbasid Society (2nd–10th centuries).* New York: Routledge, 1998.

Hill, Donald R. "Science and Technology in 9th-Century Baghdad." In *Science in Western and Eastern Civilization in Carolingian Times.* Edited by Paul Leo Butzer and Dietrich Lohrmann. Boston: Birkhäuser Verlag, 1993.

Leary, De Lacy. *How Greek Science Passed to the Arabs.* London: Routledge and Kegan Paul, Limited, 1948.

Micheau, Françoise. "The Scientific Institutions in the Medieval Near East." In *Encyclopedia of the History of Arabic Science,* vol. 3. Edited by Roshdi Rashed. New York: Routledge, 1996.

Peters, F. *Aristotle and the Arabs: the Aristotelian Tradition in Islam.* New York: New York University Press, 1968.

Rosenthal, Franz. *The Classical Heritage in Islam.* Translated by Emile and Jenny Marmorstein. New York: Routledge, 1992.

Sabra, A. I. "The Appropriation and Subsequent Naturalization of Greek Science in Medieval Islam: A Preliminary Statement." In *History of Science,* 25 (1987): 223-243. Reprinted in *Tradition, Transmission, Transformation,* edited by F. Jamil Ragep and Sally P. Ragep, with Steven Livesey. New York: E. J. Brill, 1996.

Shayegan, Yegane. "The Transmission of Greek Philosophy to the Islamic World." In *History of Islamic Philosophy,* edited by Seyyed Hossein Nasr and Oliver Leaman. New York: Routledge, 1996.

Young, M. J. L., J. D. Latham, and R. B. Serjeant, eds. *Religion, Learning and Science in the Abbasid Period.* Cambridge: Cambridge University Press, 1990.

The Transmission of Arabic Science to Europe

Overview

After centuries of struggling to preserve the most basic elements of scholarship and literacy, in the tenth century European scholars became aware of the vast storehouse of knowledge held in the Islamic world. Between the eleventh and thirteenth centuries, much of this Arabic knowledge, including earlier Greek works of science, medicine, and philosophy, was translated into Latin and transmitted to European centers of learning.

Background

Beginning with the Crusades, European scholars learned of the advanced state of Arabic scholarship and the impressive collections of Greek works held within Islamic lands. Works by Aristotle (384-322 B.C.), Hippocrates (460?-377? B.C.), and Ptolemy (second century A.D.), unknown in Europe for centuries, had been carefully preserved, studied, and enhanced by Islamic scholars. In particular, Islamic studies of astronomy, mathematics, and medicine far exceeded anything known in Europe at the time.

In the eleventh century, Europe was on the verge of a cultural and economic revival. Important among the many causes of the revival were the end of Viking and Magyar invasions and the growth of strong monarchies. The resulting stability led to a growth of commerce, increased affluence, rapid population increase, and the birth of cities. Urbanization provided a centralization of wealth that encouraged the growth of schools and intellectual culture. Economic and social stability allowed money and time to be spent on leisure activities such as scholarship. The new urban schools, out of which eventually grew colleges and universities, sought to recover and master the Latin and Greek classics, and therefore provided a market and eager audience for translations of the newly discovered Arabic works.

Impact

In the eleventh through thirteenth centuries there were three main geographical areas in which contact between the Islamic world and the Latin world allowed for the transmission of knowledge from one culture to the other: Spain, southern Italy and Sicily, and the area encompassing the Holy Land. Spain was by far the most important of the three for its role in the direct transmission of Arabic knowledge into Latin Christendom. While some copies of Arabic works were brought to Europe by the crusaders, Italian traders, and ambassadors, the most important role played by Italy and the Middle East was to awaken European scholars to the intellectual riches of the Islamic empires. Stories told by crusaders and traders filled European scholars with wonder and pointed them in the direction of the Islamic world.

The place to go was Spain. Most of Spain had been under Islamic rule since the eighth century. For several centuries Muslims, Christians, and Jews coexisted peacefully under Islamic rule, and Arabic scholarship flourished in the eleventh and twelfth centuries under the Umayyad dynasty. Bilingual and multilingual Spanish scholars facilitated the translation of Arabic works into Hebrew and Latin. However, it was not only native Spaniards who produced translations, but foreign scholars as well who came to Spain, learned Arabic, and took their translations back to their homelands. As early as 967 the scholar Gerbert crossed the Pyrenees from France into Spain to study Arabic mathematics. What began as a trickle turned into a flood as the Christian reconquest of Spain during the eleventh and twelfth centuries allowed Arabic centers of culture and libraries of Arabic books to come into Christian hands. Toledo, the cultural center of Spain, fell to the Christians in 1085 and its intellectual riches attracted scholars from as far away as Wales and Scandinavia.

The greatest of all the translators was Gerard of Cremona (1114?-1187). Around 1140 he traveled from northern Italy to Spain in search of Ptolemy's *Almagest,* which he had learned about but had been unable to locate elsewhere. He found a copy in Toledo and learned Arabic in order to translate it into Latin. While there, he became aware of numerous Arabic texts on many other subjects and he devoted the next 30 or 40 years to translating this corpus into Latin. He produced an astonishing number of books, between 70 and 80, including over a dozen astronomical works, 17 treatises on mathematics and optics, many works of natural philosophy, and 24 medical works. Among these translations were many great and important works, such as Euclid's (330?-260? B.C.) *Elements,* al-Khwarismi's (780?-850?) *Algebra,* Aristotle's *Physics, On the Heavens,* and *On Generation and Corruption,* and Ibn Sina's (980-1037) *Canon of Medicine.* Perhaps most impressive, though, was the skill with which Gerard rendered these works into Latin. Often translators resorted to literal word-for-word replacement from Arabic into Latin, which resulted in nonsensical sentences and mangled meanings. Gerard, however, had such a good command of the languages and a clear understanding of the subject matter that he was able to produce translations that were true to the original meaning and nuances of the Arabic works.

While Italy was much less important than Spain in the translation activity from Arabic into Latin, its role in the accumulation of knowledge was not insignificant. Southern Italy and Sicily were important both for the translation activity of Constantine the African in Salerno in the eleventh century and especially for the translations directly into Latin of Greek works during the twelfth and thirteenth centuries. There had always been Greek-speaking communities in Italy and strong ties with the Byzantine Empire. Libraries of Greek works were rediscovered and translators such as James of Venice (c. 1140) and William of Moerbeke (c. 1270) attempted to provide European scholars with new or revised Latin translations of Aristotle, Plato (427?-347 B.C.), Archimedes (287?-212 B.C.), and Euclid from the Greek.

The primary motivation behind the translation effort was utility. Astronomical and medical works were sought out and translated first. Medical treatises had an obvious value, and Ibn Sina's *Canon of Medicine* was the most complete, scholarly compilation of medical knowledge to be found anywhere during the Middle Ages. Astronomy and its stepsister astrology were also very useful in the medieval world. Astrology was used in medicine, helping physicians and healers to determine the best time to perform cures and what general combination of humors (known as a complexion) a patient was likely to have based on astrological data. Astronomy was essential for calendar keeping and the prediction of celestial events. To fully understand and utilize complex astronomical and astrological works such as the *Almagest,* scholars also needed to translate and learn Greco-Arabic mathematical treatises. Moreover, medicine and astronomy both rested on certain philosophical underpinnings found in Aristotle and other Greek metaphysicians. Thus, translators who sought medical and astronomical works also found themselves delving into the natural philosophy and metaphysics of the Greeks and their Arabic commentators. At the core was Aristotle, and in Aristotle European scholars found a powerful system of logic and philosophy that could be utilized in any branch of scholarship.

By the middle of the thirteenth century, the flood of translation had slowed again to a trickle, as most of the Greek and Arabic philosophical and scientific works were by then available in Latin at the various European centers of learning. Throughout the next century and a half gaps in the translations were filled and the new learning spread to the farthest reaches of Latin Christendom, where it was incorporated into, or inspired, new educational institutions. It was at these universities and schools that the final phase of assimilation occurred, as the influx of Greco-Arabic knowledge became absorbed and institutionalized in Latin Christian theology, thought, and scholarship.

REBECCA BROOKFIELD KINRAIDE

Further Reading

Benson, Robert L., and Giles Constable, eds. *Renaissance and Renewal in the Twelfth Century.* Cambridge: Harvard University Press, 1982.

Burnett, Charles S.F. "Translation and Translators, Western European." In Vol. 12 of *Dictionary of the Middle Ages,* edited by Joseph R. Strayer, 136-42. New York: Scribner, 1982-1989.

Lindberg, David C. *The Beginnings of Western Science.* Chicago: University of Chicago Press, 1992.

Lindberg, David C. "The Transmission of Greek and Arabic Learning to the West." In *Science in the Middle Ages,* edited by David C. Lindberg, 52-90. Chicago: University of Chicago Press, 1978.

The Invention and Advance of Scientific Instruments

Overview

The study of the heavens produced the earliest surviving scientific instruments, and the need for accurate astronomical sightings and calculations provided the stimulus for technological and theoretical innovation that would allow mechanization and development in many other fields. However, the history of scientific instruments is patchy at best, as most devices were constructed of cheap materials such as wood, or even paper, and consequently have not survived. Only the most permanent structures or the most grand and expensive, made from metals or stone, have survived.

Background

Some of the earliest scientific instruments were markings on rocks that showed the position of the sunrise on a certain day of the year. Later, astronomical constructions were made that ranged from simple sundials to complex structures such as Stonehenge. In ancient Mesopotamia mudbrick buildings were designed specifically for observation of the stars and planets. The Egyptian pyramids were built with perfectly aligned north-south and east-west positions. Such accurate construction was achieved using a star-sighting tool called a merkhet, which enabled the direction of the north star to be found with a plumb line. Water clocks were used by the Egyptians and the Babylonians as far back as 1500 B.C. Such clocks were used for ceremonial purposes, and in the Greek and Roman legal systems for timing the speeches of those in court.

The Greeks used the study of mathematics to improve on earlier tools. Ptolemy (second century A.D.) modified the sundial so that it marked out equal hours throughout the year by angling the pointer (or gnomon) parallel to the earth's axis of rotation. The Greeks also introduced new tools in astronomy, from sophisticated spheres that demonstrated the motion of the heavens to small handheld devices.

However, the collapse of the Roman Empire saw much of the knowledge of the ancients destroyed, lost, or scattered. Little of the ancient Greek learning was preserved in Europe, and only a fraction was translated into simple Latin texts. The medieval European audience was not interested in the complex mathematics and mechanical sophistication of the ancient writings, and so the complexity of ancient astronomy and mathematics was lost.

Impact

Elsewhere, however, there was a growing demand for more complex learning, especially in the fields of mathematics and astronomy. Chinese, Hindu, and Arab scholars actively sought out new knowledge, resulting in cross-cultural exchanges of ideas. Arabic scholars were especially well located, with access to the knowledge of the Babylonians, Indians, Chinese, and Greeks. They procured texts by any means they could, the most important example being the acquisition of Ptolemy's *Algamest* in a peace treaty with the Byzantines. This single work contained sophisticated geometry and descriptions of a number of astronomical instruments.

Arab astronomers and timekeepers modified the instruments they found and created many new ones. The sundial was modified along the lines suggested by Ptolemy, and this method and its geometry perfected. Eventually this information was translated into Latin, and the sundial became the simplest and most popular method of timekeeping, at least when the sun was shining.

Less susceptible to interruption were water clocks, which used the steady release of water to display the hours. In India and China many complex and ornamental designs were constructed, some of immense scale. In 1086 a Chinese diplomat and administrator, Su Sung (c. 1086), began work on a giant astronomical clock powered by water. It reproduced the movements of the Sun, Moon, and Earth, and weighed many tons, occupying a tower over 40 ft (12.2 m) high. Su Sung's astronomical clock represents the best and worst of Chinese technology. The clock was a magnificent wonder, yet did not function well over extended periods. In Chinese science the pretence of accuracy was more important than mechanical precision itself. Astronomy was extremely important in Chinese life, so important that political considerations overrode scientific ones. Chinese official astronomers contrived their results, and relied on old texts, rather than actual observations using their instruments. Much of Chinese learning was destroyed in political purges, and their mighty instruments and clocks decayed and ceased to function.

Islamic water clocks were common in large towns beginning in the ninth century. In 1050 a large water clock was constructed in Toledo, Spain, which told the hours of day and night and the phases of the Moon. Larger and more sophisticated models were later developed.

European water clocks may have been copied from either Arabic models or from the Roman tradition. The earliest date from around the tenth century. By the end of the twelfth century there was a guild of clockmakers established in Cologne, and water clocks became more common. However, it is the development of the mechanical clock for which Europe is famous. Some mechanical developments were already obvious in complex water clocks throughout the world. In Europe, however, there emerged a conscious drive to develop a weight-driven clock during the middle of the thirteenth century. Richard of Wallingford (1291?-1336?) appears to have been the first to solve the problems of keeping regular time with a weight-driven device, constructing a working piece in 1327. Shortly after, Giovanni Dondi (1318-1389) constructed a sophisticated astronomical clock using a similar mechanism. Mechanical clocks quickly spread across Europe, and most major towns built a civic clock out of local pride and a desire to regulate the hours of the day.

Astronomical clocks were based on a device known as an armillary sphere. This was an instrument for representing the motion of the stars, made from a number of rings that could be rotated around an axis. On the rings were drawn various bright stars, and these could be sighted relative to a fixed horizon. Such instruments were described in the writings of Ptolemy and other Greek astronomers. Islamic astronomers developed complex and accurate spheres, and used them for both demonstration purposes and as an aid to observation. A related instrument was the equatorium, which allowed the user to calculate the position of a planet for a given time. The earliest written description of an equatorium dates from the late eleventh century. European universities introduced them as teaching aids in the twelfth century.

By far the most common and versatile astronomical instrument of the period was the astrolabe, a two-dimensional representation of the night sky. The origins of the instrument are now lost, but an anecdote from the medieval period playfully suggests that, while riding on a donkey one day, Ptolemy dropped an armillary sphere, which the donkey stood on, thereby creating the first astrolabe. Essentially the astrolabe was a flat model of the universe that could be held in the hand, although some Arab instruments were made very large in order to improve their accuracy. An astrolabe consists of a movable framework with markings representing various bright stars, and a fixed plate that acts as the horizon for the observers latitude, combined with pointers and a scale on the back. The astrolabe could be used to show the time, or when a star would rise, set, or be an its highest point in the night sky, as well as a number of other uses.

Astrolabes were originally only useful for a specific latitude, but this was solved by making many discs for different latitudes. Later, a more intricate solution was found by al-Zarqali (1028-1087), whereby a single disc could represent any given latitude. Arab astronomers made many other modifications and expanded the usefulness of the astrolabe. For example, they were able to make it capable of telling the time of day as a function of solar latitude. These daytime markings are found on all early European astrolabes, even though the method used does not give accurate results for northern latitudes. This reveals the reliance Europeans had on Arab technology, slavishly copying every detail, even when unnecessary.

Slowly, however, European instrument-makers improved their skills and developed new innovations. The mariner's astrolabe was introduced by Portuguese sailors in the fifteenth century in order to help them explore and map the African coast with some safety and accuracy. This device was a simplified version of the astronomer's astrolabe, and was used to calculate latitude by observations of the Sun at local noon, or by sighting bright stars. Such astrolabes were still in use in the eighteenth century.

Another common astronomical instrument was the quadrant (or quarter-circle), which calculated the angle of a star or the Sun. Islamic astronomers improved upon the design of ancient quadrants, making them bigger and more versatile.

Collections of instruments were gathered together in observatories in China and the Arab world. The most famous observatory of the period was at the Arab town of Maragha (in modern Iran), which flourished in the thirteenth century. It contained large and accurate instruments, and attracted many Muslim, Christian, Jewish, and Chinese astronomers.

The measurements made at observatories were used to improve existing knowledge of the heavens, to calculate astronomical tables listing the positions of planets, the lunar phases, when eclipses would occur, and other events. The most popular of these tables were also some of the earliest, calculated by al-Battani (858-929). His astronomical tables were adopted by Europeans and used until the middle of the fifteenth century. Many tables, almanacs, and calendars were written by Islamic, Chinese, Indian, and European scholars, and often these scientific writings became embroiled in political arguments. European astronomers of the period were almost exclusively concerned with calendar reform, and the debates over the "correct" dates had a strong religious tone, especially in the computation of when Easter should be celebrated.

The development and innovation in timekeeping and astronomical instruments had wide implications in many other areas. Navigation benefited directly from both astronomical and timekeeping devices, enabling the European explorers of the fifteenth century to estimate their position with some degree of accuracy. Mechanical developments in other fields also developed from the pioneering work of early instrument-makers. From the ingenious automata, designed for the entertainment of rich patrons, to the practical developments in later centuries of complex engineering tools in industry, the beginnings of precision devices and machinery can be found in the efforts of early clockmakers and astronomers to make their instruments more accurate and versatile.

DAVID TULLOCH

Further Reading

Hill, Donald. *A History of Engineering in Classical and Medieval Times.* London and Sydney: Croom Helm, 1984.

Landes, David. *Revolution in Time.* Cambridge: Harvard University Press, 1983.

Ronan, Colin A. *The Shorter Science and Civilisation in China: 1.* Abridged by Joseph Needham. Cambridge: Cambridge University Press, 1978.

Medieval Religion, Science, and Astronomy

During the European Dark Ages there was no coherent system of scientific or philosophical thought. Throughout Western civilization, theological doctrine and dogma replaced the rational and logical inquiry of the ancient Greek scholars. During the thirteenth and fourteenth centuries, however, the rediscovery of Aristotle's (384-322 B.C.) philosophy, as preserved by Arabic scholars, renewed interest in the development of logic and scientific inquiry. The critical writings of St. Thomas Aquinas (1227-1274), Roger Bacon (c. 1214-1292) and William Ockham (also spelled Occam, c. 1285-c. 1349) regarding Aristotelian ideas ultimately laid the intellectual foundations for the seventeenth century scientific revolution by de-emphasizing the primacy of understanding based upon scriptural revelation or authority.

Background

Although the origins of astronomy and cosmology (the study of the origin, structure, and evolution of the universe) predate the human written record, by the height of ancient Greek civilization the cause of natural phenomena was attributed to the collective whim of a pantheon of gods. Although monotheistic in the same sense as ancient Judaism, out of this pantheism (a theology that includes multiple gods) arose the idea that there was an infinite being, Plato's (c. 428-c. 347 B.C.), "The One," and Aristotle's "Prime Mover." Aristotle's influence over astronomy and cosmology was to extend for nearly two millennia and, as a set of philosophical and scientific explanations of the universe, Aristotle's assertions ultimately became integral to the tightly interwoven fabric of philosophy, science, and theology that came to dominate the late medieval intellectual landscape.

In his work *De caelo*, Aristotle discussed the motion of bodies and the structure of matter, and set forth a model of a finite geocentric universe (an Earth-centered spherical universe). Based on his observations that celestial bodies seemed to move in circular paths and recurring patterns across the night sky, Aristotle argued that the cosmos was not infinite. In contrast to a Platonic distaste for actual observation, however, Aristotle also argued for a spherical Earth based upon his observation of the shadows of the Earth on the Moon and of the progressive disappearance of ships over the horizon. The simplicity and reproducibility of Aristotle's arguments allowed a degree of secularization in the study of natural phenomena. This secularization allowed Aristotle to advance the concept of an internal or external "mover" as something that puts a body into motion from a state of absolute rest. Accordingly, once put in motion, bodies required the continued intervention of a "mover" to continue their motion. Intuitively, Aristotle argued that there a must be a "prime mover" of all things who continued to direct movement, especially the observed motion of celestial bodies.

Aristotle proposed that the cosmos was composed of concentric, crystalline spheres to which the celestial objects were attached and upon which they moved. The outermost sphere was the domain of the prime mover who caused the outermost sphere to rotate at constant angular velocity. This motion of the outer sphere imparted movement to inner spheres that resulted in the observable movements of planets and stars. Based on Aristotelian concepts, in A.D. 150 the Greek astronomer Ptolemy developed a model of an Earth-centered cosmos composed of concentric crystalline spheres that was to dominate the Western intellectual tradition until challenged by the Polish astronomer Nicolaus Copernicus (1473-1543). Ptolemy's model, as published in his *Almagest*, featured epicycles (movement of the planets about smaller circles imposed on greater circles) that allowed practical adjustments of planetary motion to account for retrograde motion (the apparent temporary reversal of the motion of planets on the celestial sphere later explained by planets overtaking one another in their respective orbits) and varying planetary brightness.

Aristotle's logic and Ptolemy's model presented a well-developed—seemingly intellectually invulnerable—rational explanation of the cosmos that left little room for faith. The fall of the Roman Empire, however, plunged Western civilization into the Dark Ages during which most of Greek philosophy and reasoning was lost in Europe.

Impact

The Western rediscovery of Aristotle's philosophy in the twelfth century brought the theologically dominated medieval world into sharp conflict with Aristotelian logic as contained in the *Book of Causes* and other works of Arabic scholars. Although the original author of the *Book of Causes* is unknown (some scholars argue that it originated in ninth or tenth century Baghdad, others that it was written in twelfth century Spain), the book elaborated upon Aristotelian concepts by asserting that from each cause there results a certain order in its effects. In contrast to medieval beliefs regarding miracles, the *Book of Causes* argued that God could not do anything contrary to the order he had already established. Aristotle's concept of a prime mover as a being that itself remains unmoved, unchanging, and impersonal was also incompatible with the Christian concept of a God who regularly intervened in the affairs of humans through miracles. In addition, Aristotle's argument that the universe was circular and eternal contrasted with the Christian doctrine of creation.

Most early medieval scholars rejected the eternity of the universe as philosophically absurd. Some however, made tenuous connections to Aristotle's prime mover by asserting that God was ultimately the cause of all phenomena and that God worked through natural mechanisms.

As part of a broader effort to harmonize logic with religious revelation (i.e., Greek logic and rationality with Christian monotheism), Thomas Aquinas attempted to harmonize the work of Aristotle with medieval theological doctrine. Aquinas stated his own purposes in his *Commentary on the Book of Causes* as an attempt to "delineate the first causes of things" because the "ultimate happiness possible in this life must lie in the consideration of first causes, because what little we can know about them is worthier of devotion and nobler than all that we can know about lower things."

For Aquinas, reason and revelation were both legitimate paths to knowledge and truth and therefore both philosophic and scientific inquiry could be used to prove theological concepts. In his work *Summa Theologica*, Aquinas set forth a comprehensive defense of Christian doctrine. Aquinas claimed five proofs for the existence of God based on motion, efficient causes, necessity and contingency, goodness, and teleology in nature.

Aquinas's first "proof" of the existence of God was, in fact, based on the Aristotelian concept of a prime mover. Aquinas argued that the stars needed a continuing influence or source of motion and that they must therefore have had a "prime" or "first mover" who Aquinas identified as the Christian God. In accordance with the assignment of the role of prime mover in Aristotelian cosmology to the Christian God, the outermost celestial sphere became identified with the Christian heaven.

The classical notion of causality is succinctly expressed in Aquinas's *Commentary on Aristotle's Physics* in which Aquinas achieved a synthesis of Greek pagan religion with Christian theology and, in so doing, challenged the traditional reasoning that asserted that understanding came only from belief. With the synthesis of Aquinas, this philosophical process was reversed so that belief came from, or was strengthened by, a deeper understanding of nature.

For Aquinas there was an absolute and direct correlation between natural phenomena and divine revelation. The writings of Aquinas can be viewed within the larger context of an effort to characterize God as both the creator and continuing guiding force behind physical phenomena. Accordingly, descriptions of nature could be fully explained by logical inquiry and the human mind—as created by God—could be relied upon to produce more reliable descriptions of nature. Another important contribution to the renewed interest in scientific thought resulted from Roger Bacon's work in optics, upon which Bacon based assertions of the usefulness of natural knowledge as means to support religious belief.

Aquinas's synthesis was not, however, to go unchallenged. Early in the fourteenth century William Ockham and others began concerted studies into the causes of motion. Ockham soon rebelled against Aristotelian physics by advancing the impetus theory of movement. The impetus theory rejected the need for a mover to be in contact with a body in order for a body to continue its motion. According to Ockham, bodies continued their motion due to their own impetus, a forerunner of modern concepts regarding momentum.

Ockham cautioned against too great a reliance on logical inquiry. For Ockham, the movements of bodies and celestial objects could be described in terms of the properties of matter. More importantly, the work of Ockham showed the fallibility of reliance on intuitively obvious or logical explanations. Moreover, in contrast to the Aristotelian assertion that thoughts and ideas were a type of real and objective matter, Ockham asserted that reality was based upon tangibility in space and time, not upon metaphysical logic.

In his arguments, Ockham coined a maxim which has since become known as Ockham's razor (*entia non sunt multiplicanda praeter necessitatem*) that asserts that the best scientific theory, all other factors being equal, is the explanation or theory which requires the fewest new starting assumptions. This principle of parsimony (simple explanation) was used by Ockham to refute the need for the numerous entity-laden and complex models that had been devised to explain reality. Ockham's razor is best expressed in his assertion, "*Nulla pluralitas est ponenda nisi per rationem vel experientiam vel auctoritatem illius, qui non potest falli nec errare, potest convinci*" (No plurality should be assumed unless it can be proved by reason, experience, or by infallible authority) and remains a well-accepted means of choosing between two theories regarding the same physical phenomena.

Many centuries away from the empiricism that was to fuel the scientific revolution, Ockham and other nominalist philosophers ultimately fell back upon the maxim that faith was a more reliable guide than reason. Regardless, the writings of Aquinas, Bacon, and Ockham greatly contributed to a reestablishment of reason and logic as a path of knowledge and ultimately de-emphasized the role of understanding based upon mystical scriptural revelation.

K. LEE LERNER

Further Reading

Aquinas, St. Thomas. *Commentary on Aristotle's Physics.* Translated by Blackwell, R., J. Spath, and W. Thirlkel. South Bend, IN.: Dumb Ox Bks., 1998.

Aquinas, St. Thomas. *Summa Theologica.* Allen, TX.: Christian Classics, Inc., 1998.

Burtt, Edwin A. *The Metaphysical Foundations of Modern Physical Science.* Atlantic Highlands, N.J.: Humanities Pr. International, Inc., 1982.

Gilson, Etienne. *The Spirit of Medieval Philosophy.* Notre Dame, IN.: Univ. of Notre Dame Pr., 1991.

Hetherington, Norriss S. *Enclyclopedia of Cosmology: Historical, Philosophical, and Scientific Foundations of Modern Cosmology.* New York: Garland Publishing, Inc., 1993.

Peter Peregrinus Initiates the Scientific Study of Magnets

Overview

The earliest experimental study of magnetism can be found in a letter written by Petrus Peregrinus in 1269. Peregrinus was the first individual to describe the existence of two magnetic poles in each magnet, to describe the attraction between unlike poles, and to explain the creation of new poles when a magnet is broken in two. A designer of instruments, Peregrinus also described improvements in the magnetic compass, which made it far more useful for navigation on the high seas. Roger Bacon, a Franciscan friar teaching at the universities of Oxford and Paris, popularized the experiments of Peregrinus, including studies now lost. The letter about magnets was copied numerous times and widely circulated. Later it would stimulate the researches of William Gilbert, an English physician whose treatise on magnets would initiate the modern study of electricity and magnetism.

Background

The basic phenomena of magnetism were known to the ancient Greeks. The philosopher Thales (624-546 B.C.) was familiar with lodestone, a naturally occurring magnetic rock, and felt it necessary to attribute to it a "soul" because it was able to cause motion. The Greek medical writer Galen (A.D. 130-200) recommended the use of magnets for "expelling gross humors." Probably the first practical use of magnetism outside the medical area was the magnetic compass, which appeared in Europe in the twelfth century. Scholars are in disagreement about the origin of the compass, which may have been first used by the Chinese or by Norsemen. The magnetization of an iron needle by stroking it with a lodestone was first described in writing by Alexander Neckam (1157-1217), an English monk.

The first detailed experimental study of magnetism is to be found in a letter of Petrus Peregrinus (fl. c. 1269) to one Sygerum de Foucaucourt, a knight and neighbor. Little is known about the life of Peregrinus, whose original name was Pierre de Maricourt, Peregrinus being a title awarded to religious pilgrims and to individuals who served in the crusades. He is believed to have been a university graduate, and, at the time he wrote the letter in 1269, was apparently serving as a sort of military engineer in the army of Charles of Sicily as it lay siege to the Italian city of Lucera. English philosopher and scholar Roger Bacon (c.1214-1292) noted that Peregrinus had a reputation as a scholar prior to the siege, but the Peregrinus's earlier writings have been lost. Since Peregrinus is highly praised in a document Bacon wrote in 1267, but is not mentioned in his earlier writings, it is likely that Bacon met Peregrinus after 1260, possibly at the University of Paris. According to Bacon, Peregrinus was skilled in minerals, metalworking, and agriculture, and he had worked for three years on "burning glasses," lenses that would concentrate the Sun's rays to start a fire.

According to Peregrinus, the letter to de Foucaucourt was intended to be part of a longer treatise on the construction of physical instruments. The letter is divided into two parts, the first on the basic principles of magnetism and the second on the construction of magnetic devices. In the first part, Peregrinus begins with a defense of the experimental method, warning against untested theory and speculation. He then suggests that the best lodestones are those from northern Europe that are somewhat blue in color and that the strength of a lodestone is to be measured by the weight of iron that it can lift.

Peregrinus next describes his construction of a spherical magnet and demonstrated that it had two poles, which could be identified by placing iron needles on the surface and tracing the direction in which the needle pointed. He thus obtained, in effect, lines of longitude that intersected at two points, the two magnetic poles. He found that at these two points fragments of an iron needle can be made to stand on one end. He then describes several studies done with magnets contained in small bowls and allowed to float on water in a larger bowl. He notes that a magnet will orient itself so that its north pole points to the "north pole of the heavens," and establishes that opposites attract each other. He also describes the creation of new poles when a magnet is broken in two, interpreting the attraction between opposite poles as a natural affinity.

Peregrinus considered the possibility that a magnet's north pole might be attracted to a point on Earth but rejected this idea. Instead, he concluded that the magnet poles pointed towards the celestial poles, the points around which the stars in the sky appear to move. At the time, it was commonly thought that the pole star, Polaris, was located exactly at the celestial pole, and it was natural to assume that the magnet pole was attracted to this star. Peregrinus was aware that Polaris, as seen from Earth, actually moves in a small circle about the celestial pole and so had to assume that the magnet's motion was affected by all parts of the celestial sphere.

In the second part of his letter, Peregrinus describes the construction of two types of magnetic compass, both of which represented a significant advance beyond the existing instruments. In one, the magnet was enclosed inside a wooden case and floated on water in a circular vessel, the periphery of which was divided into 360 points. Sighting pins on the case allowed the user to determine the angular position of the Sun, Moon, and stars relative to magnetic north. In the second, the compass points were marked on the transparent lid of a jar, and the magnetized needle rested on a pivot inside the jar. A nonmagnetic bar attached to the magnetized needle indicated the east and west directions. At the end of the letter, Peregrinus describes a "perpetual motion machine" based on the behavior of magnets that he actually constructed. It did not work, but he attributes the failure to his own lack of skill in construction.

Impact

In the universities of thirteenth-century Europe, two routes to reliable knowledge were recognized: divine revelation through the scriptures and argumentation with due deference to ancient authority. The gathering of facts by observation and experimentation was at best suspect. It is possible that Peregrinus was reluctant to have his discoveries more widely known, lest he be accused of witchcraft. Bacon, for instance, had his own experimental studies interrupted by religious authorities and spent 14 years in prison for suspected heresy. He described Peregrinus as one of two perfect mathematicians and as "the only Latin writer to realize that experiment rather than argument is the basis of certainty in science." It is in this sense that Peregrinus is considered by some historians as the first true experimental scientist. More than three centuries would elapse before the so-called Scientific Revolution would establish the experimental method as the ultimate test of scientific truth.

The construction of a reliable ship's compass was a major factor in launching the great age of exploration by sea in the fifteenth and sixteenth centuries. Christopher Columbus (1451-1506) carried a compass like one described by Peregrinus on his journey to the New World and was quite interested in the mechanism by which it functioned. On his journey, Columbus was able to make important observations of the magnetic variation, that is the difference between compass north and geographical north. His crew, noting the deviation with respect to the pole star, became alarmed, but was calmed when Columbus explained that the pole star in fact moved with respect to true north.

The next major advance in the understanding of magnetism after Peregrinus occurred more than 300 years later, when the English physician William Gilbert (1544-1603) conducted the experiments described in his treatise *De Magnete*, published in 1600. Gilbert's debt to Peregrinus is obvious. Gilbert magnetized a large iron sphere, which he called a "terrella," and by studying the response of a compass needle as it was moved around the sphere, made a convincing case that Earth was itself a large magnet. By Gilbert's time, also, the magnetic variation was familiar to both sailors and scholars, as was the fact that a freely suspended magnetized needle would point or "dip" some degrees below the horizontal. Gilbert demonstrated that these phenomena too were consistent with a magnetized Earth. Gilbert's treatise concludes with a discussion of "magnetic rotation," a type of perpetual motion that he, like Peregrinus, believed would occur if a perfectly magnetized sphere were suspended so that it could move freely.

Although Peregrinus and even his predecessors were aware of the qualitative differences between magnetic attraction and static electricity, it was Gilbert who made the cleanest distinction between the two phenomena. For the following two centuries, magnetic and electric phenomena were considered unrelated. Then in 1800 the primitive battery invented by Italian physicist Alessandro Volta (1745-1827) made it possible to sustain a steady electric current. And a mere 20 years later, Danish physicist Hans Christian Ørsted (1777-1851) discovered that the current flowing through a wire could cause the deflection of a nearby compass needle. Ørsted's discovery triggered a period of intense investigation of electromagnetic phenomena that would last throughout the nineteenth century.

DONALD R. FRANCESCHETTI

Further Reading

Benjamin, Park. *A History of Electricity*. New York: Arno Press, 1975.

Kelly, Suzanne. "Peter Peregrinus." In *The Dictionary of Scientific Biography*, vol. 5. New York: Scribner's, 1970: 532-39.

Verschuur, Gerrit L. *Hidden Attraction: The Mystery and History of Magnetism*. New York: Oxford University Press, 1993.

Astronomical Tables: Applications and Improvements During the Middle Ages

Overview

Medieval astronomers were frequently called upon to resolve practical questions pertaining to social or religious matters. This was especially true in the Islamic world, where the motions of heavenly bodies were, and still are, closely tied to religious law. Astronomers also had to respond to the technical demands of astrologers who occupied an important place in Islamic society. Efforts throughout the Middle Ages to address these and related needs adequately led to improvements in existing astronomical tables and produced important theoretical developments that had applications far beyond the specific problems they were intended solve. The culmination of this work was the *Alfonsine Tables,* introduced in Paris around 1320.

Background

Ever since the time of the Babylonians, theoretical models have been developed for predicting the time of occurrence and location of celestial phenomena. Unfortunately, without calculators or computers, performing even the simplest calculations with these models was cumbersome and time-consuming. Astronomical tables were constructed to simplify the procedure.

Medieval astronomical tables were based almost exclusively on Ptolemy's (100?-170?) geocentric models. Ptolemy developed his geometrical models in the *Almagest.* Computing planetary positions based on his theory required knowing the numbers that specified the actual geometry of the models. Seven parameters were necessary, five of which were independent for each planet. Knowing these parameters and the geometry of the model, it was possible to find the celestial longitude of any planet.

In the *Almagest,* Ptolemy showed how to construct tables that made this procedure much more tractable. The initial starting position, or radix, for each celestial body was determined for some fixed time and displayed in tabular form. The mean motion of each body was then derived from the underlying theory and tabulated. In addition, tables of correction factors were provided. Also included were specialized tables for conjunctions of the Sun and Moon, eclipses, parallaxes, and other phenomena.

The procedure for using the tables was straightforward. To find the location of a planet at some time in the future, the mean motion tables were used in conjunction with the planet's radix to calculate how far it had moved from its initial position. However, this only indicated the approximate location of the planet. Since the motions of the Sun, Moon, and planets are not uniform, a series of corrections was necessary to fix the position more accurately. In the case of a planet, retrograde motion needed to be accounted for. The necessary adjustments for this could be made by simply locating the appropriate correction factor in the relevant table and applying it.

To further facilitate their use, Ptolemy abstracted the tables of the *Almagest* from their underlying theory. Several of the parameters and procedures were then modified in accordance with the results obtained in his *Planetary Hypotheses*. He then assembled these improved tables as the *Handy Tables* (c. 150). This served as the basis for Islamic astronomical tables, and the specific tabular form that Ptolemy devised persisted well into the seventeenth century.

Impact

Applied astronomy was very important in Islam for many reasons. Since the Islamic calendar is lunar, the beginning of the holy month of Ramadan and various religious festivals are regulated by the appearance of the lunar crescent. The times of the five daily prayers, required of all Muslims, are determined through the observation of celestial bodies. Other religious prohibitions and obligations are similarly tied to astronomical phenomena. Furthermore, the prayer wall in mosques (Muslim places of worship) is always aligned along the *qibla,* the direction of Islam's most sacred site—the Kaaba in Mecca. The Koran also encouraged Muslims to use the stars for guidance—a remark largely responsible for Islam's ceaseless parade of astrologers, who required astronomical details to ply their trade properly. Judicious decisions pertaining to agriculture, geography, navigation, and the like also relied on the results of the astronomers.

Before the founding of Islam in 622, the Bedouins of the Arabian peninsula had developed an extensive body of atheoretical knowledge about the motions of the Sun, Moon, and planets, as well as information on the seasons and fixed stars. In the early years of Islam, this folk astronomy was applied by legal scholars to address the religious practices just mentioned. As Muslim scientists became increasingly familiar with Hellenistic, Hindu, and Persian astronomical lore, more sophisticated methods were developed to deal with these secular and religious needs.

The works of Ptolemy were first translated into Arabic during the early ninth century, and Islamic astronomers quickly adopted the Ptolemaic geocentric view. Their own observations though—made at a different latitude and some 700 years after the construction of Ptolemy's *Handy Tables*—revealed significant deviations from Ptolemy's predictions. Thabit ibn Qurra (836?-901), al-Battani (858?-929), and others isolated and corrected erroneous parameters. New estimates were made of the speed of precessional movement, the obliquity of the ecliptic, the solar eccentricity, and the position of solar apogee. Based on these revised values, new tables were computed for various meridians and radices given for the date of the *hijrah* (the prophet Muhammad's emigration to Medina) in 622.

These tables were collected into handbooks known as *zijes.* Along with the tables for mean motions, radices, and correction factors, they included accurate calendric, prayer, and *qibla* tables. They also often included various trigonometric tables, indispensable for solving problems of spherical astronomy, as well as star catalogs. The introduction to a *zij* would usually discuss the underlying model and specific parameter values used to construct the tables along with canons for their use.

Many of the problems Islamic astronomers were called upon to solve required new mathematical methods. For instance, determining the *qibla* involved complex problems in spherical trigonometry. Al-Khwarizmi (800?-847?), in *Zij al-sindhind,* and al-Battani, in *Zij-i Djadid Sultani,* advanced astronomical theory by providing tables of sine functions to assist in solving such problems. Al-Battani's *Zij* also contained sophisticated tables of special trigonometric functions for solving problems involving spherical triangles. *De motu stellarum,* the Latin version of this work, printed in Nuremberg (1537), was important in the development of European astronomy.

The determination of the exact time prayer was to begin also involved spherical geometry. Ibn Yunus (940?-1009), in *al-Zij al-Hakimi,* made impressive strides in this direction. He compiled useful timekeeping tables that were widely imitated. They also helped establish the timekeeping institution of the *muwaqqit,* which was later to be associated with mosques and madrasas (Koranic schools). The timekeeping tables of Shams al-Din al-Khalili (1320?-1380?) represent the crowning achievement of medieval Islamic solutions to problems in spherical astronomy. Some of his tables for regulating prayer times were used well into the nineteenth century.

The *Toledan Tables* were compiled in the eleventh century under the direction of Cadi Ibn

Sa'id. These tables are usually attributed to al-Zarqali (1028-1087), who participated in the project. The tables, however, have no unified underlying astronomical theory. Disparate methods and incompatible parameter values were used to compute the various tables. For instance, the tables of differences of ascension and the tables of right ascension are calculated using different values for the obliquity of the ecliptic—the former using Ptolemy's value from the *Handy Tables,* the latter using al-Battani's value from *Zij-i Djadid Sultani.* Also, the equations of the Sun, Moon, and planets follow al-Khwarizmi's tables, while the theory of trepidation follows Thabit ibn Qurra's tables. Nevertheless, the *Toledan Tables* were immensely popular and adapted to many locations in Europe. They also influenced the production of almanacs, which were designed not to provide the means for calculating planetary positions, but rather to give those positions explicitly.

The *Toledan Tables* provided the model for King Alfonso X of Castile's (1221-1284) tables. Compiled during his reign, these tables were composed in Castellan Spanish. New observations were made to establish a consistent set of parameters and the tables recomputed on the Toledo meridian. Completed in about 1272, Alfonso's tables closely followed al-Zarqali's, but Alfonso's Spanish version exerted no influence on astronomy outside of Spain. Nasir al-Din al-Tusi's (1201-1274) *Zij-i ilkhani,* based on 12 years of observations at the Maragha observatory, also appeared in 1272. It, unlike Alfonso's work, was very influential, particularly in the East.

The Castellan Alfonsine tables were hardly circulated and only the canons for their use are extant, the actual tables having been lost. This stands in stark contrast to the hundreds of manuscript copies and thousands of printed versions of what is known as the Latin *Alfonsine Tables.* This work has long been incorrectly attributed to Alfonso. Evidence now strongly suggests that this is a completely independent work compiled by a group of Parisian astronomers at least as early as 1327.

A number of important innovations appear in these tables. They were formulated with a sliding sexagesimal system that allowed the mean motion tables to be used with different sets of radices. Thus, it was possible to make predictions using any calendar. In fact, the tables came with no fewer than 10 different sets of radices taken from various calendars, including the Islamic lunar calendar and the Christian Julian calendar. This flexibility goes a long way toward explaining the universal appeal these tables were to have. That which most clearly differentiated this work from its predecessors was its advanced precessional theory, which took into account sidereal motion. Consequently, the *Alfonsine Tables* generated planetary positions in tropical coordinates.

The *Alfonsine Tables* became the most influential handbook of practical astronomy in Europe. It was the basis for almost every almanac and ephemeris until superseded by the *Tabulae Prutenicae* (1551) of Erasmus Reinhold (1511-1553). The *Tabulae Prutenicae* was the first practical set of planetary tables based on Nicolaus Copernicus's (1473-1543) heliocentric theory. Reinhold cast his tables in essentially the same format as the *Alfonsine Tables,* so its users would have to make no commitment to Copernican heliocentrism.

STEPHEN D. NORTON

Further Reading

Books

Kennedy, Edward S. *Astronomy and Astrology in the Medieval World.* Aldershot, Great Britain: Variorum, 1998.

King, David A. *Astronomy in the Service of Islam.* Aldershot, Great Britain: Variorum, 1993.

Kunitzsch, Paul. *The Arabs and the Stars.* Northampton: Variorum Reprints, 1989.

Márquez-Villanueva, Francisco, and Carlos Alberto Vega, eds. *Alfonso X of Castile, The Learned King, 1221-1284.* Cambridge, MA: Department of Romance Languages and Literatures of Harvard University, 1990.

Samsó, Julio. *Islamic Astronomy and Medieval Spain.* Aldershot, Great Britain: Variorum, 1994.

Periodical Articles

Kennedy, Edward S. "A Survey of Islamic Astronomical Tables." *Transactions of the American Philosophical Society* 46, Part 2 (1956): 123-75.

King, David A. "On the Astronomical Tables of the Islamic Middle Ages." *Studia Copernica* 13 (1975): 37-56. Reprinted in *Islamic Mathematical Astronomy,* by David A. King. Aldershot, Great Britain: Variorum, 1993.

Poulle, Emmanuel. "The Alfonsine Tables and Alfonso X of Castile." *Journal for the History of Astronomy* 19 (1988): 97-113.

Toomer, G. "A Survey of the Toledan Tables." *Osiris* 15 (1968): 5-174.

Ptolemaic Astronomy, Islamic Planetary Theory, and Copernicus's Debt to the Maragha School

Overview

Ptolemy's (100?-170?) *Almagest* was first translated into Arabic during the early ninth century. Islamic astronomers initially accepted and worked within the Ptolemaic framework, isolating and correcting erroneous parameters. Objections were later raised concerning Ptolemy's failure to reconcile the mathematical models of the *Almagest* with the physical spheres they were intended to represent. A group of thirteenth-century Islamic astronomers, known as the Maragha school, revolutionized medieval theoretical astronomy by developing planetary models that resolved the Ptolemaic difficulties. The chief concerns and technical solutions of the Maragha school are evident in the later work of Nicolaus Copernicus (1473-1543).

Background

Plato (428?-347 B.C.) first challenged astronomers to explain the apparently irregular movements of celestial bodies in terms of uniform circular motions. Eudoxus (408?-355? B.C.) accepted this challenge "to save the phenomena" and developed a system of concentric spheres with Earth as their common center. Each planet, as well as the Sun and Moon, was attached to a single sphere. This, in turn, was part of a set of interconnected spheres, each of which rotated about its own axis at a different rate and orientation. The combined motions were then adjusted to approximate the observed movements of the body in question. Eudoxus employed 27 spheres: three each for the Sun and Moon, four each for the five planets, and one for the fixed stars.

Callipus (370?-300?) improved the Eudoxian system by adding spheres. Aristotle (384-322 B.C.) further modified it, but, unlike Eudoxus, he maintained that the spheres were material bodies. Accordingly, certain presuppositions of Aristotelian physics needed to be satisfied. This required 22 additional spheres. Unfortunately, these models failed to explain certain phenomena.

In the second century A.D. Ptolemy proposed a more satisfactory system in the *Almagest.* He endorsed Aristotelian physics and its conclusions regarding Earth being at rest and the center of a spherical universe. He also acknowledged that planetary motions could only be explained kinematically by uniform circular motion or combinations thereof. However, instead of concentric spheres, Ptolemy employed eccentric orbits, epicycles, and equants.

In its simplest form, a planet's motion might be represented by an eccentric orbit—a circle about Earth, known as the deferent, with Earth offset from the center. In addition, a planet could be made to travel on an epicycle—a smaller circular orbit whose center moves along the circumference of the deferent. These geometrical devices had been exploited to great effect by Hipparchus (170?-120? B.C.). In fact, Ptolemy adopted Hipparchus's solar model without alteration and refined his lunar and stellar models. Ptolemy's planetary models, though, were the product of his own genius.

To account for the anomalous behavior of the planets, Ptolemy introduced the equant. The equant lies along the diameter defined by Earth and the deferent center. It is the same distance as Earth from, but on the opposite side of, the deferent center. According to the *Almagest,* uniform circular motion was about the equant, not the deferent center.

In *Planetary Hypotheses* Ptolemy supported a physical interpretation of the system propounded in the *Almagest.* He accepted that the universe was composed of physically real, concentric spheres, but made no attempt to correlate the motions of these bodies with his mathematical models.

Impact

The *Almagest* was first translated into Arabic in the early ninth century. Islamic astronomers quickly adopted Ptolemy's geocentric view and directed their researches toward refining his mathematical models. Thabit ibn Qurra (836-901), al-Battani (858?-929), and others associated with al-Ma'mun's (786-833) House of Wisdom uncovered erroneous parameters and provided corrected values based on new observations. Attention then shifted to deeper methodological and theoretical considerations, especially the relationship between Ptolemy's models and their physical interpretation.

The reexamination of Ptolemaic astronomy reached its full maturity during the eleventh century. The most extensive and sophisticated attack on it was leveled by Ibn al-Haytham (965-1040?), also known as Alhazen, in *al-Shukuk ala Batlamyus*. Al-Haytham's objective was to harmonize the mathematical and physical aspects of astronomy. Accordingly, he rejected the use of mathematical models whose motions could not be realized by physical bodies. In particular, he objected to Ptolemy's use of the equant because it required uniform circular motion of a sphere about a point other than its center. Since this was physically impossible, the models of the *Almagest* could not be describing the actual motions of celestial bodies. Al-Haytham argued that they should therefore be abandoned in favor of more adequate models. The Persian Abu Ubayd al-Juzjani (?-1070?) went a step further by constructing his own non-Ptolemaic models.

Al-Juzjani's efforts were ultimately unsuccessful, but the search for alternate models continued during the twelfth century in the Islamic west. Astronomers of Andalusian Spain rejected geometric models that failed to satisfy Aristotle's axiom of uniform circular motion about a fixed point. According to Aristotelian physics, this fixed point was the center of the universe and coincided with Earth's center. Ptolemy's eccentrics, epicycles, and equants all violated the principle because they required the celestial spheres to revolve about fixed points other than Earth's center. Thus, the Ptolemaic approach was rejected as fundamentally flawed.

Sometimes referred to as the "Averröist critique," this line of criticism received its canonical formulation in the work of Ibn Rushd (1126-1198), or Averröes, who had been a student of Abu Bakr ibn Tufayl (1105?-1184). Ibn Tufayl claimed to have devised an arrangement that satisfied the philosophical presuppositions of Aristotelian physics without recourse to eccentrics or epicycles. Though nothing remains of this work, Ibn Tufayl's ideas inspired another of his students, al-Bitruji (1100?-1190?). Al-Bitruji proposed a system of concentric spheres reminiscent of that of Eudoxus. However, neither this nor any other configuration devised by the Andalusian school succeeded in making numerical predictions of planetary positions as accurate as Ptolemy's. The Andalusian approach was soon abandoned and the Ptolemaic system remained dominate.

The Golden Age of Islamic astronomy extended from the mid-thirteenth to the mid-fourteenth century. During this time, attempts were made to reform Ptolemaic astronomy by Nasir al-Din al-Tusi (1201-1274), Mu'ayyad al-Din al-Urdi (?-1266), Qutb al-Din al-Shirazi (1236-1311), Ibn al-Shatir (1305?-1375?), and others. Since the first three of these astronomers were associated with the Maragha observatory in northern Iran, and the last, Ibn al-Shatir, subsequently continued their work, they are today referred to as the Maragha school. The models they developed resolved the Ptolemaic problems regarding uniform circular motion.

Al-Tusi was the first to formulate a model of uniform circular motion that had the same predictive accuracy as Ptolemy's. He achieved this result with a device of his own invention—the "Tusi Couple." Consisting of two spheres rolling one within the other, the Tusi Couple was placed at the end of the equant vector—the line extending from the equant point to the center of an epicycle. When set in motion, the couple caused the length of the equant vector to vary, which caused the epicycle to trace out a path very close to that of the Ptolemaic deferent. Thus, the introduction of the deferent circle and non-uniform circular motion about it could be avoided. Furthermore, all component circular motions in al-Tusi's models were uniform about their own centers.

Al-Urdi, independent of and possibly prior to al-Tusi, developed a variant solution. His method proved mathematically equivalent to al-Tusi's, but required one less rotating sphere. This arrangement was later adopted by one of al-Tusi's disciples, al-Shirazi. The fourteenth-century Damascus astronomer Ibn al-Shatir continued the Maragha reforms. He combined the geometrical devices of al-Tusi and al-Urdi together with secondary epicycles to construct more accurate planetary models. His models retained the effect of the equant, but with the celestial bodies now rotating with uniform circular motion about Earth's center.

The motivation behind the Maragha school's tradition of model building has often been misunderstood. Though their models did restore uniform circular motion, they did not criticize Ptolemy because his models violated this principle. Their planetary models were designed to overcome the much more serious and fundamental problem of internal inconsistency.

Ptolemy accepted uniform circular motion as the only means of explaining the movements of celestial bodies. Nevertheless, he then introduced into his mathematical models the equant, which made such motions physically impossible. The latter point had been previously noted by al-Haytham. But, unlike al-Haytham, the Maragha school was not concerned with restoring uniform circular motion as such. Their goal was a mathematical methodology capable of accounting for the observable world in a manner consistent with whatever assumptions were made about the universe's physical nature. Thus, the Maragha school revolutionized medieval theoretical astronomy by rejecting Plato's limited goal of "saving the phenomena."

The question of Copernicus's debt to the Maragha school was raised in 1957 in an article by Victor Roberts. Roberts showed that the lunar model of Copernicus was essentially identical to that of Ibn al-Shatir's. Since then, ongoing research has established further similarities. In the *Commentariolus,* Copernicus presented planetary models that substituted the combination of two epicycles and a deferent for the Ptolemaic deferent-equant arrangement just as Ibn al-Shatir had done. A more striking resemblance appears in their models for Mercury. Here, Copernicus employs the Tusi Couple to vary Mercury's orbital radius exactly as in Ibn al-Shatir's model. Finally, Copernicus's solar model was mathematically equivalent to Ibn al-Shatir's, except that the positions of the Sun and Earth were reversed.

Copernicus and the Maragha school thus used the same mathematical devices and often applied them at precisely the same points. But there is a much deeper connection. They both stressed the need for internally consistent mathematical models that can be physically interpreted. Though no direct connection has yet been established, scholarly opinion suggests that these similarities cannot be attributed to mere coincidence. However, as significant as the Maragha influence might have been, it in no way diminishes the originality of Copernicus's heliocentric hypothesis.

STEPHEN D. NORTON

Further Reading

Books

Kennedy, Edward S. *Astronomy and Astrology in the Medieval World.* Aldershot, Great Britain: Variorum, 1998.

King, David A. *Islamic Mathematical Astronomy.* Aldershot, Great Britain: Variorum, 1998.

Saliba, George. *A History of Arabic Astronomy: Planetary Theories During the Golden Age of Islam.* New York: New York University Press, 1994.

Swerdlow, Noel, and Otto Neugebauer. *Mathematical Astronomy in Copernicus's De Revolutionibus.* New York: Springer, 1984.

Periodical Articles

De Bono, Mario. "Copernicus, Amico, Fracastoro and Tusi's Device: Observations on the Use and Transmission of a Model." *Journal for the History of Astronomy* 26 (May 1995): 133-54.

Gingerich, Owen. "Islamic Astronomy." *Scientific American* 254 (April 1986): 74-83. Reprinted in *The Great Copernicus Chase and Other Adventures in Astronomical History* by Owen Gingerich, Cambridge, MA: Sky Publishing and Cambridge University Press, 1992.

Hartner, Willy. "Ptolemy, Azarquiel, Ibn al-Shatir, and Copernicus on Mercury." *Archives internationales d'histoire des sciences* 24 (1974): 5-25.

Roberts, Victor. "The Solar and Lunar Theory of Ibn ash-Shatir: A Pre-Copernican Model." *Isis* 48 (1957): 428-32.

Aristotelian Physics, Impetus Theory, and the Mean Speed Theorem

Overview

Prior to the seventeenth century, many of the most fundamental problems of physics concerned difficulties associated with local motion—changes in place or position. Medieval attempts to explain how and why such changes took place were developed within an Aristotelian framework. These efforts progressively undermined and eventually led to the rejection of certain tenets of Aristotle's (384-322 B.C.) doctrine of motion. They also culminated in what many consider the single most important contribution of medieval scholars to physics—the mean speed theorem.

Background

Aristotle dichotomized the universe into a terrestrial or sublunar region, encompassing Earth and extending to the sphere of the Moon, and a celestial or supralunar region, extending from the sphere of the Moon to the fixed stars. All matter in the terrestrial region was thought to be composed of four elements—earth, water, air, and fire—while the celestial region was assumed to be filled with the fifth or divine element, ether. The ether was believed to be immune to all changes except local motion. Ordinary matter was subject not only to local motion but other types of change as well.

Aristotle distinguished between two types of local motion—natural and violent. Natural motions are those that a body exhibits when unimpeded. Violent motions occur when a body is displaced from its natural resting place. According to Aristotle, celestial bodies naturally move in circles or combinations thereof. Since it was thought that the ether in no way hindered this motion, it was concluded that these bodies do not exhibit violent motions. Bodies composed of ordinary matter behave quite differently and in a manner intimately tied to the structure of the sublunar world.

The terrestrial region of Aristotle's universe was composed of four concentric areas, each the natural place for one of the four elements. When displaced, it was believed that each element naturally moves rectilinearly toward its concentric ring (if unimpeded). The outermost ring was the natural place of fire, below that the ring of air, below that the ring of water, and below that the ring of earth. Aristotle attributed different degrees of heaviness or lightness to the basic elements to explain their tendency to seek their natural places. The element earth was deemed absolutely heavy. As such, it naturally moved downward toward Earth's center from the regions above. Similarly, the absolute lightness of fire caused it to rise from below to its natural place above the ring of air.

Explaining violent motion required a different mechanism. Aristotle believed the force responsible for motion had to be in constant physical contact with the moved body. When violent motion was initiated, the originating motive impulse was easily identified—the bow string for a shooting arrow, the hand for a thrown rock, etc. Aristotle maintained that the original mover not only sets the arrow or the stone in motion, but also activates the surrounding medium—in this case air. The air parts before the arrow or rock and circles back to maintain a continuous motive force behind the object. This force gradually diminishes due to the resistance of the medium. When it completely dissipates, the arrow or stone falls downward according to its natural motion.

Aristotle formulated specific rules to describe the consequences of this doctrine. He stated that the speed of a body in violent motion is directly proportional to the motive force and inversely proportional to resistance. The latter included the resistive power of the body in motion (a concept left undefined) and the resistance offered by the external medium. Accordingly, the speed of a body could be doubled by doubling the applied force or halving the resistance.

This doctrine of motion also led Aristotle to deny the existence of a vacuum. Since speed was proportional to the medium density, an indefinite rarefaction of the medium would produce a corresponding indefinite increase in speed. But if the medium were to vanish completely, then the speed of a body would be infinite. This was clearly absurd. Even more serious, motion in a void violated Aristotle's claim that violent motion necessarily occurred in a medium. Furthermore, Aristotle took it as axiomatic that bodies of different weights necessarily fall at different speeds—their speeds being directly proportional to their weight. However, he realized that without a material medium, lighter bodies would move just as swiftly as heavier bodies. To avoid these conclusions, Aristotle rejected the void and postulated a universe filled everywhere with matter.

Impact

John Philoponus (sixth century A.D.) strongly objected to Aristotle's theory of motion. He was especially critical of the role Aristotle assigned to the medium. He argued convincingly that it was not the agent or cause of violent motion. If Aristotle were correct, then it would be possible to move an arrow or rock by simply agitating the air behind them. This, however, was contrary to experience. He concluded that violent motion occurs by the mover transferring to the object of motion an incorporeal kinetic power—later known as impetus. By divorcing motion from the external medium, Philoponus held open the possibility that motion through a vacuum was possible, thus undermining one of the chief reasons for rejecting the void.

The idea of an impressed force was further developed by Islamic scholars who referred to it as *mail.* Ibn Sina (980-1037) propounded a version of *mail* theory that delineated three different types: psychic, natural, and violent. Only natural and violent *mail* are relevant to the present discussion. They were used by Ibn Sina to explain Aristotle's natural and violent motions. Ibn Sina argued that bodies in motion received *mail* directly in proportion to their weight. He further maintained that violent *mail* would endure in a body indefinitely in the absence of external resistance. A consequence of this was that motion in a void would continue without end since there would be nothing to stop it. Ibn Sina rejected the existence of a vacuum because such motion had never been observed.

Other anti-Aristotelian critiques were promulgated in the Islamic world. One of the most notable was advanced by Ibn Bajja (1095?-1138?). Possibly influenced by Philoponus, Ibn Bajja denied Aristotle's claim that speed was necessarily inversely proportional to the density of the medium. He pointed out that celestial bodies traveled with different speeds through the ether, which supposedly offered no resistance to motion. Thus, he concluded differences in speed were not the result of an external medium. In fact, Ibn Bajja argued, the only function of the medium was to retard motion.

Thomas Aquinas (1225-1274) elaborated upon Ibn Bajja's ideas. He attempted to demonstrate that motion in a void, or resistanceless media such as the ether, would not be instantaneous as Aristotle had feared. He argued that motion from one point to another required traversing the intervening points successively. This was only possible if motion were finite. Consequently, infinite speeds must be impossible in a vacuum. Aquinas, along with Roger Bacon (1214?-1294), rejected all attempts to explain violent motion by means of incorporeal forces imparted to moving bodies.

It was not until the fourteenth century that Philoponus's impressed-force theory gained general acceptance. Known as impetus theory, it became popular around 1320 and was taught at the University of Paris. Franciscus of Marchia (?-1344?) proposed one version in which a self-dissipating impetus was imparted not only to the body set in motion, but also to the surrounding medium. Thereafter, Jean Buridan (1295?-1358), who is perhaps responsible for the term "impetus," developed the theory so fully that he must be considered its main proponent.

Buridan conceived of impetus as a motive force transferred to bodies. He considered its magnitude directly proportional to body weight and velocity. Furthermore, he followed Ibn Sina in ascribing permanence to impetus, claiming it endured indefinitely unless diminished by external resistance. This clearly implied that, if all resistance were eliminated, a body in motion would continue in motion indefinitely at a constant speed. Buridan's formulation bears a striking resemblance to the concept of inertia and may have prepared the way for its development. Unfortunately, he denied the possibility of indefinite rectilinear motion, which is an essential feature of inertia. He did, however, propose that impetus might be the cause of the eternal circular motions of celestial bodies.

Buridan further applied impetus theory to the accelerated motion of free fall. Buridan identified a body's heaviness or *gravitas* as the cause of its natural uniform fall. A body's *gravitas* not only initiated the downward motion, it also augmented it through accumulated increments of impetus. Each increment generated a corresponding increase in velocity, which in turn generated a further increment and so on, resulting in continuously accelerated motion. Buridan's account is essentially Aristotelian since the cause of motion—impetus—is proportional to velocity and not acceleration as in Newtonian physics.

The most significant medieval results on local motion were achieved by members of Merton College, Oxford University. In 1328 Thomas Bradwardine (1290?-1349) developed a new law of motion that proved very influential. Richard Swineshead worked out the implications of Bradwardine's rule and together with William Heytesbury (1300?-1380), John Dumbleton, and other Mertonian scholars produced the correct definitions for uniform speed and uniformly accelerated motion. Applying these definitions, the Mertonians derived the single most important medieval contribution to physics—the mean speed theorem. The theorem relates uniformly accelerated motion to uniform velocity, making it possible to express the distance traveled by the former in terms of the distance traveled by the latter. Nicole Oresme (1320?-1382), in *On the Configurations of Qualities* (c. 1350), provided geometrical and arithmetical proofs of the theorem. These were widely disseminated throughout Europe in the fourteenth and fifteenth centuries and probably influenced Galileo (1564-1642), who made the mean speed theorem the foundation of his new science of motion.

STEPHEN D. NORTON

Further Reading

Grant, Edward. *The Foundations of Modern Science in the Middle Ages.* Cambridge: Cambridge University Press, 1996.

Grant, Edward. *Much Ado About Nothing: Theories of Space and Vacuum.* Cambridge: Cambridge University Press, 1981.

Lindberg, David C. *The Beginnings of Western Science.* Chicago: University of Chicago Press, 1992.

Murdoch, John E., and Edith D. Sylla. "The Science of Motion." In *Science in the Middle Ages,* edited by D. Lindberg. Chicago: University of Chicago Press, 1978.

Pedersen, Olaf. *Early Physics and Astronomy.* Cambridge: Cambridge University Press, 1993.

Theodoric of Freiberg and Kamal al-Din al-Farisi Independently Formulate the Correct Qualitative Description of the Rainbow

Overview

The fourteenth century witnessed many important contributions to physics, including the mean speed theorem, the graphical representation of functions, and a reformulation of impetus theory that prepared the way for the concept of inertia. As significant as these were, they were primarily the result of metaphysical speculations by scholastic philosophers. As such, they did little to advance experimental methodology. However, the fourteenth century also produced what many consider the greatest successes of experimental science during the Middle Ages—a correct qualitative description of the rainbow. Surprisingly, this was discovered almost simultaneously by Theodoric of Freiberg (c. 1250-c. 1310) in Europe and Kamal al-Din al-Farisi (c. 1260-c. 1320) in Persia.

Background

The only significant extant ancient theory of the rainbow available during the Middle Ages was that propounded by Aristotle (384-322 B.C.). According to Aristotle, the rainbow resulted from sunlight reflected from the surface of a cloud to the eye of an observer. Unlike reflection from smooth mirrors that produce images, he argued that the uneven surface of clouds could only reflect colors. Furthermore, the specific colors of the rainbow were produced by a mixture of light and darkness while their order in the bow depended on the ratio of the Sun-to-cloud to cloud-to-eye distances. Finally, the circularity of the rainbow was seen as part of the circumference of the base of a cone whose apex was the Sun and whose axis passed through the observer's eye and terminated at the center of the base.

Aristotle's account remained the most sophisticated mathematical treatment of the rainbow for almost eighteen centuries. However, his emphasis on reflection from the cloud as a whole proved a major stumbling block for later research. Ibn Sina (Avicenna, 980-1037) was one of the first to challenge this idea. He argued the cloud was not the locus of the rainbow; rather, it was the particles of moisture in front of the cloud that reflected light. Though his analysis suggested the possibility of a geometric analysis of reflection by a single raindrop, Ibn Sina failed to pursue this possibility.

Another Arabic scientist important to the story of the rainbow was Ibn al-Haytham (Alhazen, 965-c. 1040). In *Kitab al-Manazir* (Treasury of optics), he articulated a comprehensive scientific methodology of the logical connections between direct observations, hypotheses, and verification. This allowed the geometric analysis of physical phenomena to be translated into concrete experiments involving the manipulation of artificially created devices. Al-Haytham exploited this methodology to full effect in his optical investigations, which surpassed all previous research in the field. He conducted extensive experiments on refraction using a water-filled, spherical globe. Unfortunately, he failed to see the analogy between the glass globe and a raindrop.

Robert Grosseteste (c. 1175-1253) rejected the idea that the rainbow was formed by reflected light. He maintained that the rainbow was formed by the refraction of light rays through a cloud. The cloud acted like a lens to focus the rays on another cloud where they appeared as an image. He attributed the bow's colors to refractions through the successively denser layers of a convex cone of moisture. Though incorrect, Grosseteste's introduction of refraction into the theory of rainbows was a major advance.

Albertus Magnus (c. 1200-1280) was the first to suggest that refraction by individual drops played a role in the formation of the rainbow. He also noted that a transparent, hemispherical vessel filled with black ink projected a brightly colored semicircular arc when placed in sunlight. Albertus equated the degrees of opacity in this vessel with the different densities within Grosseteste's cone of moisture. In essence, he saw the vessel as a diminutive cloud instead of an individual raindrop.

The first quantitative contribution to rainbow studies was made by Robert Bacon (c. 1214-c. 1294). Using an astrolabe, Bacon showed that the maximum altitude of the bow is approximately 42° (the modern value is 44°). Notwithstanding this result, Bacon made no further efforts to geometrically analyze the rainbow. It seems he confused, as many before him had, the physics and physiology of colors. He believed that colors produced from crystals by refraction were objective since their location did not vary when observed from different locations, as was the case with the rainbow. Since the rainbow has no definite location, Bacon felt that it could not be the result of refraction and therefore must be a subjective phenomenon.

Impact

Al-Haytham's *al-Manazir* exerted a strong influence over optical studies in both the Arab world and Latin-speaking West. Nevertheless, those who correctly emphasized the role of the individual raindrop in the formation of the rainbow saw no way of implementing al-Haytham's methodology. They were unable to devise a satisfactory experimental design that would allow them to analyze the optical geometry of rainbow formation, that is, until Theodoric of Freiberg (Dietrich von Freiberg) provided the key insight.

In *De Iride* (On the rainbow, 1304-11) Theodoric argued that a globe of water could be treated as a magnified raindrop instead of a miniature cloud. He realized, of course, that a glass sphere only approximated a raindrop. Therefore, if he was going to be able to rely on his experiments, he needed to show that effects produced by any differences could be ignored. Most significantly, a raindrop does not have a glass envelope. Consequently, light passing through the glass sphere will be refracted four times instead of twice, as occurs in a raindrop. However, since water and air refract light by about the same amount, deviations will be small and thus can be safely ignored. Similarly, Theodoric was experimenting with a stationary globe whereas actual raindrops would presumably be falling. A suggestion of Albertus dealt with this potential problem. He argued that the drops would be falling so fast and replacing each other so quickly that it would be reasonable to replace them with a stationary curtain of transparent drops. Theodoric could now feel confident that his experimental results could be applied to the rainbow.

By placing a transparent sphere of water in a darkened room and directing a beam of sunlight on to it, Theodoric was able to carefully study the paths of light rays through a raindrop. His observations indicated that the primary rainbow was formed from light rays that had been refracted twice and reflected once. An initial refraction took place when the light ray entered the raindrop. There was then a reflection inside the drop followed by another refraction as it exited. This provided an understanding of the circular form of the rainbow. It also allowed Theodoric to explain the displacement of the rainbow when viewed from a different position—as one moves, a different set of raindrops is required to form the rainbow.

Theodoric also explained the production of the secondary rainbow. The light rays forming this bow undergo two refractions and two internal reflections. The geometry of the situation immediately explained the inversion and paleness of the colors in the secondary bow. The additional reflection reverses the order of the colors as well as weakening the light.

The aspect of Theodoric's research that most clearly foreshadowed modern scientific methodology was his attempt to explain how the colors of the rainbow were generated. He employed two pairs of contraries—obscure/clear and bounded/unbounded—to explain the origin of colors. Using these principles, he formulated various hypotheses that he tested through a series of experiments. Though his proffered explanation of colors was ultimately unsuccessful, his procedures exhibited some of the interplay between theory and experiment that has become the hallmark of modern science.

The Persian scientist al-Farisi produced a correct explanation of the rainbow independent of, and possibly prior to, Theodoric. He was encouraged by his teacher, Qutb al-Din al-Shirazi (1236-1311), to make a study of al-Haytham's optical works. Al-Farisi corrected al-Haytham on some points, rejected his theories in other places, and developed his ideas when possible. In particular, he surpassed al-Haytham's work on the rainbow by modifying his methodology.

Al-Haytham's methodology required that experiments be performed directly on the objects or phenomena of interest. This was impossible for the rainbow. Still, al-Farisi thought it might be possible, under the right conditions, to construct a suitable analog, subject it to direct observation, and apply the results to the phenomenon of interest—in this case the rainbow. This then is the great achievement of al-Farisi and Theodoric, that they extended the efficacy of experimentation beyond the direct manipulation of objects of interest.

Having accepted Ibn Sina's view that the locus of the rainbow is a myriad of water droplets, al-Farisi realized that a glass sphere filled with water could be used to study a raindrop. In accordance with his new methodology, he crafted experiments with this analog that led him to the correct qualitative description of the primary and secondary rainbow. Some scholars have claimed that al-Shirazi initially discovered the correct qualitative description of the rainbow while al-Farisi merely elaborated his teacher's ideas. However, Roshdi Rashed has argued convincingly against this view.

The results of al-Farisi's work remained obscure and exerted little influence on future rainbow research. Theodoric's work initially fared little better. Though his ideas were not actually lost, they had practically no impact on later-fourteenth- and even fifteenth-century optical theories about rainbows. In 1514 Jodocus Trutfetter published an account of Theodoric's theory, replete with diagrams. René Descartes (1596-1650) may have been familiar with this or some such similar treatment since many aspects of his own account of the rainbow closely resemble those of Theodoric. Regardless, Descartes applied the newly discovered law of refraction to transform Theodoric's qualitative theory into a comprehensive quantitative treatment. He was thus able to deduce the radii of both the primary and secondary bows as well as the ordering of their colors.

STEPHEN D. NORTON

Further Reading

Books

Boyer, Carl. *The Rainbow: From Myth to Mathematics*. New York: Thomas Yoseloff, 1959.

Crombie, A. C. "Experimental Method and Theodoric of Freiberg's Explanation of the Rainbow." In *Grosseteste and Experimental Science*. Oxford: Clarendon Press, 1953: 233-59.

Grant, Edward, ed. *A Source Book in Medieval Science*. Cambridge, MA: Harvard University Press, 1974.

Rashed, Roshdi. "Kamal al-Din al Abu'l Hasan Muhammad ibn al-Hasan al-Farisi." In C. C. Gillispie, ed., *Dictionary of Scientific Biography*, vol. VII. New York: Charles Scribner's Sons, 1973: 212-19.

Wallace, William. "Dietrich von Freiberg." In C. C. Gillispie, ed., *Dictionary of Scientific Biography*, vol. IV. New York: Charles Scribner's Sons, 1971: 92-95.

Wallace, William A. *The Scientific Methodology of Theodoric of Freiberg*. Fribourg, Switzerland: The University Press, 1959.

Periodical Articles

Sayili, A. M. "Al-Qarafi and His Explanation of the Rainbow." *Isis* 32 (1940): 14-26.

Advancements in Optics, 700-1449

Overview

The interest in the study of the laws and phenomena of optics was one of the most popular and potentially important scientific pursuits of the Middle Ages. As such, it was an influential avenue to the development of experimental scientific philosophy. Grounded in the optical thought of the Greeks in both European and Islamic intellectual traditions, the initial stages of the advance in optical study were the translation of ancient thought, Islamic commentary on the subject about the twelfth century, and European contributions built upon Islamic influences. The most original Islamic investigator was Ibn al-Haytham (in the Latin West, Alhazen), and his influence passed to both later Islamic and European scholars. By the thirteenth century, Greek theories of vision and concepts of light, the reflection and refraction of light, application of optics with burning mirrors and lenses, and most importantly, understanding the atmospheric phenomena of the rainbow and the halo became areas of research to European thinkers, especially the English Franciscans. The physical problem of the rainbow mechanism was the major area of practical research and was finally completely explained in Dietrich of Freiberg's theory of the early fourteenth century. Other strides were made in theory and application of accurate lenses and mirrors, as well.

Background

The study of optics, or the principles that determine the image-forming properties of reflecting surfaces (as water and mirrors) and transmitting media (water, glass, and its formation, as in lenses) in relation to light, is of ancient origin. In China, written evidence of such study (the *Mo Jing* or *Mohist Canon*) dates between 450 and 250 B.C., which parallels early Greek optical thought. In both East and West, theory focused on light sources, vision, shadows, and reflection. Vision theory rather than the study of light phenomena was the central theme of Greek optics (from the Greek *optika*, relating to the eye). Mathematician Euclid of Alexandria (about 300 B.C.) noted that light traveled in a straight line. Euclid may have been the origin of Greek belief and its dissemination that vision was a matter of rays being emitted by the eye (called "opseis") rather than light entering the eye to form images. Engineer Hero of Alexandria (first century A.D.) studied lines and angles of light reflection from mirrors as aspects of geometrical optics. And astronomer Claudius Ptolemy (fl. A.D. 139-161 studied theoretical and atmospheric refraction, noting the angle of refraction and incidence as unequal.

Greek optical theory, heavily influenced by Aristotle (384-322 B.C.), passed rather piecemeal with other Greek science to Europe through neo-Greek and later Roman thinkers, such as Seneca (A.D. 1-65) and Pliny the Elder (A.D. 23-79). The Latin encyclopedists, Cassiodorus (480-575) and Isidore of Seville (560-636), drew from these, providing a rather limited foundation of ancient science to the Latin Middle Ages. A much more thorough rediscovery and translating of Greek thought from Greek, Syriac, Persian, and Sanskrit sources passed into Arabic during the ninth century at Baghdad. These translations would serve to focus commentaries by Islamic thinkers, and these thinkers followed Aristotle in defining aspects of nature under the discipline of physics. The mathematical grounding of traditional optical study took on more of this physical character as investigation turned to experimental techniques.

Impact

Early Islamic thinking on the various subjects of optics included substantial thought from Yaqub al-Kindi (c. 801-870), hints of broadening of the physical subject matter of optics as in al-Farabi (d. 950), and discussion of applications, such as in Abu Sad ibn Sahl (c. 984). But Ibn al-Haytham (965-1039), Alexandrian mathematician and astronomer, was the first in the study of optics to emphasize critical deductive and inductive reasoning by way of experimentation. Among many works, he composed his comprehensive *Book on Optics* (c. 1027), which contained his treatment of vision and light. Ibn al-Haytham did not believe rays emanated from the eye but that light rays were received by it, and he correctly explained vision. In his theory of light, Ibn al-Haytham extended optics to the study of light propagation and the full definition of physical and geometrical optics with extensive experiments in reflection and refraction (he outlined the basic laws of refraction). Other treatises specialized in the geometry of spherical and parabolic mirrors and lenses and applications of these for causing heat and burning, the analysis of the light of the moon and stars, and the atmospheric optical phenomena of the rainbow and halo. From this initial presentation of a wider definition of optical subjects, the narrow meaning *optica* would be eventually replaced with the more suitable word, *perspectiva*.

The origin of the rainbow was a central theme of medieval optical study, particularly because Aristotle provided a detailed theory of it in his important four-part treatise on the physical earth *Meteorologica*. Ibn al-Haytham included experiments on the rainbow in his *Optics* but wrote a separate detailed work on atmospheric optics called *On the Rainbow and Halo*, providing further experimental depth. To Aristotle, the rainbow was a reflection phenomenon from clouds made up of uniform drops that acted as a continuum surface, something like a convex mirror. Ibn al-Haytham, who established the precedent of a laboratory for experimental optical research, decided from observing the reflection of light from plane and spherical mirrors that the rainbow was the result of reflection as from a spherical concave mirror. He simulated the rainbow colors by transmitting sunlight through glass spheres of water, which he thought simulated spherical concave mirrors and individual clouds (still acting as Aristotle's continuum, not individual drops). He did not consider his innovative spheres as representing cloud droplets, and he ignored the possibility of refraction taking part in the phenomenon, which were keys to later rainbow theory.

Ibn al-Haytham also dealt with other questions of atmospheric optics. He appears to be the first thinker to note the property of refraction of light in the atmosphere (*A Question Relating to Parallax*). He applied this bending of light as the reason for the distortions of celestial objects found by astronomical observers near the horizon. He seems to be the first medieval thinker to apply basic refractive theory in relation to the atmosphere as a means of determining a reasonable height of the atmosphere. Actually, his contemporary the Cordoban Moor al-Jayyani ibn Muadh (c. 989-1079) developed a more scientific method of studying atmospheric height by way of refraction. His treatise *On the Dawn* (also known as *On Twilight and the Rising of Clouds*) was mistakenly attributed to al-Haytham. Ibn

Muadh noted that sunlight seen before dawn and after sunset was a refractive phenomenon. He estimated the angle of depression of the sun at dawn and sunset to be a fairly accurate deduced angle of 18°. From this he determined the height of atmospheric moisture (believed responsible for twilight), which was considered proportional to atmospheric height.

Ibn al-Haytham's level of optical comprehensiveness was not to be seen again in the Islamic world. Smaller studies and application of optics appearing in astronomical and meteorological works, particularly with the continued interest in the rainbow, characterized subsequent Islamic scholars. His near contemporary, the Persian physician Ibn Sina (Avicenna, 980-1037), devoted over 20 volumes to physical science and independently dismissed Aristotle's cloud continuum and Ibn al-Haytham's spherical mirror analogy in explaining the rainbow mechanism. Anticipating an important clue in the water drops themselves, he theorized the rainbow was the result of reflection of light from the total amalgamation of water drops released by clouds as they dissolved into rain. This idea followed from his careful observation of sunlight diffracted by water drops formed while watering a garden.

It remained for two Islamic thinkers some two hundred years later to follow in al-Haytham's optical footsteps, both representing Islamic science in the philosophical view in the later medieval period. Astronomer Quib al-Din al-Shirazi (1236-1311) inspired his student Kamal al-Din Farisi (c. 1260-1320) to delve critically into al-Haytham's optical theories. Farisi eventually and comprehensively appraised all of al-Haytham's optical works in his book *Revision of Optics,* in which he openly noted al-Haytham's incorrect theories. Particularly, he researched the rainbow mechanism using al-Haytham and Ibn Sina's ideas as a base to arrive at most of the correct theory with lucid conceptual physics and logical geometry superior to al-Haytham. He realized al-Haytham's spheres of water were like cloud drops, not a mass of cloud, and that the rainbow was the result of two refractions of sunlight on and in the cloud drop with one reflection, though he could not satisfactorily explain the rainbow colors. The full solution slightly antedated him and came from late medieval Latin Europe.

The fragments of Greek science that had filtered to the Latin West, providing a sketchy foundation for optical study, were finally fleshed out with the vast effort of translating Greek sources and commentaries from Arabic into Latin in the twelfth and thirteenth centuries. The core of this work was carried out under Gerard of Cremona (1114-1187) and his group of translators at Toledo. Because Islamic translations from Greek thinkers were sometimes summaries rather than true full translations, Gerard and those succeeding him, concentrated on complete translations out of Greek of Aristotle, Plato, Archimedes, Euclid, and, among others, Ibn al-Haytham's great optical treatise. Compared to Islamic familiarity with Aristotle, the presentation of his works in Europe was something of a novelty, and it spurred a new level of European scientific scrutiny, one area being in optical research.

The most important twelfth- and thirteenth-century emphasis on optics was the work of English Oxford Franciscans and their associates at the University of Paris, developing seminal divergence from Aristotelian physics. Their initial guiding light was not a Franciscan, but Robert Grosseteste (c. 1168-1253), bishop of Lincoln and a chancellor of Oxford University. Grosseteste was an early proponent of Aristotle's cause-and-effect logic, and he emphasized the importance of observation and analysis. Grosseteste was well acquainted with the physical/geometrical optics of *perspectiva* passed from Islamic thinkers and he experimented with the magnifying glass. He was evidently the first European investigator to conclude from his own observations that refraction of light on cloud drops would explain the rainbow but provided no detailed theory.

Following close behind came famous disciples: fellow Franciscans Roger Bacon (c. 1214-1294) and John Peckham (after 1230-1292), German Albertus Magnus (c. 1193-1280), and Polish scholar Witelo (b. c. 1230). Bacon experimented with magnification using convex lenses and was the first to suggest lenses could correct poor eyesight. In his *Opus Maius* (c. 1267) he conjectured that the speed of light was finite and traveled as sound did and, though short on precise theory, he believed that reflection and refraction in "numberless drops of water" (cloud drops) generated the colors of the rainbow. Peckham, who would become archbishop of Canterbury, contributed to the advance of optics by his precise simplification of the abstractions in al-Haytham's *Book on Optics* in his *Perspectiva communis* (1277-79). This work carried along contemporary emphasis on the importance of both reflection and refraction in the generation of the rainbow.

The non-English associates of the Franciscans at Paris were seminal figures in evolving the medieval science corpus. Albertus Magnus (Albert the Great, a Dominican becoming acquainted with Grossteste's ideas at the University of Paris about 1240) followed the scheme of developing formal scholastic analysis of Aristotle with commentaries on his major physical science works. He devoted much space in analyzing the *Meteorologica*, particularly on the rainbow, where again theoretical definition was confined to stating there was reflection and refraction of "rays of light" in "descending raindrops." Witelo (or Vitelo) studied at Paris about 1253 and became engrossed in al-Haytham's *Book on Optics*, which inspired his own great optical work *Peri optikes* or *Perspectivae* (between 1270 and 1278), the standard western text for some 300 years on basic optical principles. In discussing the rainbow he stressed the reflection of light on clouds and refraction through cloud drops as a necessary medium acting as "spherical lenses." Witelo fairly defined the halo as refraction (actually the more diffuse phenomenon of diffraction) of light through the sun or moon. And he developed a method of machining parabolic mirrors from iron.

Though interest in the properties of lenses and mirrors continued through the period, the resolution of the rainbow was still the fundamental optical question. Kamal al-Din Farisi had come close to the complete answer, but the more comprehensive theory of Dietrich (or Theodoric) of Freiberg (c. 1250-1311), a Teutonic member of the Order of Preachers who had studied at Paris about 1297, preceded him. Unlike traditional medieval optical literature, Dietrich's treatise *On the rainbow* (about 1304) was not padded with commentary on other thinkers' ideas. He, like Farisi, followed al-Haytham's lead with glass spheres filled with water as an experimental base and realized that these acted as individual cloud drops. He correctly described and depicted the primary rainbow as two refractions on and one internal reflection in a cloud drop, and the secondary rainbow (the first appearance of such in detail) as requiring an additional internal reflection (thus the inversion of the rainbow colors). He further became the first thinker to explain correctly each of the colors of the rainbow phenomenon as requiring a definite angle of incidence of sunlight, enhancing the understanding of the nature of light. He also correctly deduced that the angle/color relationship explained the circular appearance of the rainbow.

Amid a great mass of optical commentaries, Dietrich's impressive empirical presentation was but weakly echoed in the commentary of the *Meteorologica* of Themon Judaeus (fl. 1370) of the Paris School. The Paris School was a group of important physical theorists focusing on the causal relations in physical motion, and they did not extend their investigation to optics. And unfortunately, through the sixteenth century, Dietrich was essentially forgotten amid the printing flurry of Aristotelian commentaries of past thinkers. But within the span from Ibn al-Haytham to Dietrich, optics had been molded into a true discipline of physical science.

WILLIAM J. MCPEAK

Further Reading

Books

Crombie, A. C. *Robert Grosseteste and the Origins of Experimental Science 1100-1700*. Oxford: Clarendon Press, 1953.

Grant, Edward. *Studies in Medieval Science and Natural Philosophy*. London: Variorum Reprints, 1981.

Lindberg, David C. *Theories of Vision from Al-Kindi to Kepler*. Chicago: Univeristy of Chicago Press, 1976.

Peckham, John. *Perspectiva communis*. Edited by David C. Lindberg. Madison: University of Wisconsin Press, 1970.

Sabra, A. I. *The Optics of Ibn Al-Haytham*, (Books I-III). London: The Warburg Institute, University of London, 1989.

Wickens, G. M. *Ibn Sina: Scientist and Philosopher*. Bristol: Burleigh Press, 1952.

Article

Smith, A. Mark. "The Latin Version of Ibn Muadh's Treatise *On Twilight and the Rising of Clouds*." *Arabic Sciences and Philosophy* 2 (1992): 83-132.

Alchemists Seek Gold and Everlasting Life

Overview

Alchemy was the precursor to the science of chemistry; it incorporated many elements of magic and religion. Originating in the Middle East, it first appeared in Europe after the Moorish conquest of Spain. A key focus of alchemy was the search for methods to transform one substance to another—especially turning other metals into gold. Alchemists believed gold to be the key to long life, and perhaps even immortality.

Alchemy continued to be practiced until about 1700, when it was gradually replaced by a less mystical and more practical approach to science.

Background

Thousands of years ago, early civilizations (in Egypt and China, for example) were already familiar with several chemical processes, including the fermentation of wine and the extraction of metals from ore. However, the knowledge of the craftsmen plying these processes was limited

SOMETHING FOR NOTHING: THE QUEST FOR PERPETUAL MOTION

In the thirteenth century French architect Villard de Honnecourt introduced a new device that he said would turn forever. It was a wheel with a clever arrangement of weights and folding arms; as it turned, the weights would travel downward at the outer edge of the wheel and return to the top on the interior. By doing this, Villard felt that more energy would be gained by the falling weights than would be expended by the rising ones, thus generating more energy than was used to turn the wheel. Since then, there have been many variations on this theme, and other attempts to make what are called "perpetual motion machines"—machines that will run forever. It's a tempting quest: if such a machine could be invented, energy problems would vanish. Unfortunately, perpetual motion is a false hope, shown impossible by the laws of thermodynamics. As we now know, it is not possible to build a machine that requires no energy to run, and, in fact, all machines must lose energy in some form. Even superconducting circuits do not carry electrical current for "free"—the energy required to lower the temperature of the wires and to keep them supercooled more than offsets the gains offered by superconductivity.

P. ANDREW KARAM

and strictly practical. They did not understand actions and reactions of the substances during any of these processes.

The philosophers of these civilizations began giving thought to the nature of matter itself. In China, a book entitled the *Shu Ching*, written more than 2,000 years ago, described everything as being composed of earth, fire, water, metal, and wood. Similarly, the ancient Greeks held that the four *elements* of matter (which should of course not be confused with the chemical elements of modern science) were earth, fire, water, and air.

At the same time, the Egyptian city of Alexandria was Greek-speaking and an important center of learning. Alchemy began to develop there, as scholars attempted to explain some of the ancient Egyptian capabilities of which most knowledge had been lost. One of the early alchemists in Alexandria, known as Mary the Jewess, was particularly interested in laboratory equipment. Among her inventions was the double boiler, which is still known in French-speaking countries as the *bain-marie* or Mary's bath. Merchants in Alexandria also pursued alchemy; they were interested in making base metals look like gold by combining them with reactive mercury and yellow sulfur.

When the Arabs conquered Egypt in 642, alchemy spread within Islamic culture. The man considered to be the father of Islamic alchemy was Abu Musa Jabir ibn Hayyan of Baghdad (721-776). He valued experimentation and synthesized several new compounds. Arab alchemists continued working with mercury and sulfur, believing that they could be combined in various proportions to produce different metals. Alchemy was brought to Europe in the eighth-century Moorish conquest of Spain. However, it didn't become widely known there until the 1100s, when Spanish scholars translated Arabic works into Latin. Many common chemical terms, such as alcohol and alkali, as well as the word alchemy itself, derive from the Arabic.

Impact

Alchemy was based on the belief that all materials were made from the "elements" earth, air, fire, and water, and that these were affected by exposure to heat, cold, moisture, or dryness. Alchemists thought if they could manipulate the conditions properly, they could change the balance of the elements in the substance, thus causing a transmutation. They avidly sought what they called the philosopher's stone, a magical substance that would make the transmutation easier.

Gold was key to the philosophy of alchemy not only because of its rarity and corresponding value, but also because it didn't decay, rust or tarnish. Thus the idea of changing base metals into gold became associated with the concepts of incorruptibility and regeneration. The Chinese believed that eating from golden dishes led to

An early illustration showing alchemists at work. *(Bettmann/Corbis. Reproduced with permission.)*

long life. In twelfth-century Christian Europe, the association of alchemy with immortality and resurrection took on a particular religious significance. Transmutation into gold was symbolic of God's grace, and the transformation symbolized the effect of religious conversion on the soul.

However, the association of alchemy with religion was neither new nor exclusively Christian. Mary the Jewess as well as the Islamic alchemists had referred to their knowledge as being bestowed upon them by God. "Spiritual alchemy" has survived in the teachings of the

Rosicrucians. This fraternity came to America in 1694 and exists today as a nonsectarian philosophical order.

Astrology was an important part of the alchemists' studies because the positions of the planets and stars were believed to influence the outcome of their work. Each celestial body was thought to correspond to a particular metal. The Sun was associated with gold, the Moon with silver, Mars with iron, Venus with copper, Jupiter with tin, Saturn with lead, and Mercury with the metal mercury, also called quicksilver.

Alchemists invented the process of distillation, which became one of their most important techniques. It was seen as separating the essence or "spirit" of a substance from its inert components. "Above, the heavenly things," wrote the Alexandrian alchemist Zosimus, "below, the earthly." Using distillation, Jabir had produced *aqua regia* (a mixture of hydrochloric and nitric acids), the first known liquid in which gold would dissolve. Aqua regia was viewed as the alkahest, or universal solvent, the next best thing to the philosopher's stone. The reagents developed by alchemists were much stronger acids than the vinegar and juices previously available to them, allowing a much wider range of experimentation.

The goals of alchemy included finding an elixir of life that would cure disease and fend off death. It existed in an age when the actual causes of disease were unknown and medical options were few. Not surprisingly, then, there were many charlatans who assumed the guise of alchemists in order to profit from the hope or desperation of others. But alchemy was also pursued by many of the philosophers and scholars of the day.

Roger Bacon (1214-1294), an English monk and Oxford don, was one of these scholars. He is regarded as one of the greatest alchemists of his time. Like Jabir, who had written "He who makes no experiments will attain nothing," Bacon distinguished between knowledge deduced by logic and that obtained by experiment. His carefully documented and interpreted laboratory work laid the foundation for the modern scientific method. He sought the transmutation of base metals to gold, but did not regard the failure of this effort as the failure of his experiments. Like a man digging for treasure in his garden, he reasoned, the digging itself was making the field more fertile.

Bacon's scientific investigations, particularly his emphasis on experimentation and reason rather than revealed knowledge, did not make him popular with Church authorities, and in 1284 he was jailed by Pope Urban IV. He was released a decade later, when he was 80 years old. Soon afterward he died, possibly due to an explosion in his laboratory.

Despite the theological symbolism associated with alchemy, many alchemists had similar problems with religious authorities. The Spanish alchemist Ramon Llull (in English, Raymond Lully; c. 1235-1315) was charged with heresy. As if to prove his detractors wrong, he went to North Africa in an attempt to convert the Moors. This served only to antagonize Islamic authorities as well, and he was stoned to death in Tunis in 1315. Another Spanish alchemist, Arnau de Villanova (c. 1235-1312), fell afoul of the Inquisition and was charged with heresy and witchcraft. Pope John XXII issued a papal bull denouncing alchemy in 1317. This did little to affect its popularity, however, and it continued to be pursued throughout the Renaissance.

By the 1500s, scientists were beginning to turn away from the more mystical elements of alchemy and apply the practical knowledge it had bequeathed them to the study of medicines and disease. These iatrochemists (from the Greek word for physician, *iatros*) sought to learn about the chemical effects of medicines on the human body.

The Swiss physician Phillipus Paracelsus (1493-1541) pursued iatrochemisty, still believing in the fundamental nature of earth, air, fire, and water. He also accepted the alchemists' principles of mercury, sulfur, and salt, which was valued for its action as a preservative. A mercury-based medication for syphilis, similar to the one he devised was still being used at the beginning of the twentieth century. Another of his medications that was employed for centuries was the painkiller laudanum, distilled from the poppy plant.

Chemists and physicians formulated medicines from the alchemists' "principles" for many years. Unfortunately, since mercury is poisonous, they often did more harm than good. In fact, heavy metal poisoning had always been common among alchemists themselves, leading to impaired judgment and sometimes insanity.

The age of alchemy as a science is generally considered to have ended with the publication of the book *The Sceptical Chymist* by the Irish scientist Robert Boyle (1627-1691). Boyle showed by experiment that earth, air, fire, and water were not, in fact, fundamental elements. Thereafter, while alchemy continued to exist, it became associated primarily with magic and the occult. Mainstream chemists were often embarrassed by it, while continuing to make use of the substances and techniques it had contributed.

Traces of the alchemists remain in our language and literature. For example, the use of the word "spirits" for alcohol reflects its origin in the refining process of distillation. Alkaline earth metals, a categorization seen in the modern periodic table, get their name from the old "element," earth. The archetypal wizards of legends and fairy tales were modeled after alchemists.

SHERRI CHASIN CALVO

Further Reading

Books

Brock, William H. *The Norton History of Chemistry*. New York: W.W. Norton & Co., 1992.

Knight, David. *Ideas in Chemistry: A History of the Science*. New Brunswick, N.J.: Rutgers University Press, 1992.

Time-Life Books. *Secrets of the Alchemists*. New York: Time-Life Books, 1990.

Article

Karpenko, Victor. "Alchemy as *donum dei*." *HYLE—An International Journal for the Philosophy of Chemistry* 4, no. 1 (1998): 63-80.

The Alchemy of Mineral Acids

Overview

Geber was the pseudonym of a fourteenth-century alchemist whose books were highly influential during the Middle Ages. He is credited with the discovery of sulfuric acid, whose preparation he described along with that of other strong acids. These acids were capable of dissolving many metals—a property that helped to spur interest in alchemy throughout Europe. Today, these compounds are among the most important industrial chemicals.

Background

Little is known of the alchemist Geber, although he is believed to have been Spanish and to have published his work during the early part of the 1300s. Geber was not his real name (he is often known as the False Geber or Psuedo-Geber), but one he took from Jabir ibn Hayyan (c. 721-815), an influential Arabic alchemist. (Geber is the Latin form of the Arabic name Jabir.) Before the invention of movable type in the fifteenth century, books were produced largely by hand and in only small numbers. As a result, medieval alchemists such as Geber sometimes attributed their work to earlier, well-known scholars in order to increase the likelihood of their books being published.

Geber is known to have produced at least four books: *The Sum of Perfection*, *The Book of Furnaces*, *The Investigation of Perfection*, and *The Invention of Verity*. (The word perfection in the titles refers to the perfection of metals—the attempt by alchemists to convert base metals such as lead into gold.) Geber's work is unusual in that, unlike the authors of other works on alchemy of the time, he does not merely present the ideas of previous scholars. Instead, he apparently had considerable firsthand experience in working with chemicals and performing his own experiments. In his books, he describes basic chemical procedures such as filtration and distillation (a process in which a liquid is converted to a gas and then condensed back to a liquid). Geber gives instructions for preparing certain compounds, and his writing attempts to teach these methods to others. Compared to the writing of earlier medieval alchemists, his work is fairly clear and less cloaked in mystery. At least partly as a result, his books were widely read among the alchemists of the next two centuries.

Geber's most important contribution to science, however, was his work with acids. Acids are an important class of compounds. According to one definition, an acid is any substance that increases the concentration of hydrogen ions when it is dissolved in water. Certain acids are capable of dissolving a variety of metals. Common organic (carbon-containing) acids, such as citric acid and acetic acid, had long been known to scientists by the Middle Ages. Citric acid was obtained from citrus fruits such as lemons; acetic acid was obtained from vinegar. Organic acids, however, are weak acids. They are very limited in their ability to dissolve metals.

Mineral acids (those that do not contain carbon) are much stronger. They include sulfuric, nitric, and hydrochloric acids, all of which were described by Geber. In fact, he is credited with the discovery of sulfuric acid. To produce this chemical, he described heating what were called vitriols (certain sulfur-containing compounds) to a high temperature. This heating resulted in a chemical reaction that produced sulfur and an oxygen-containing gas as its byproducts. When the gas was cooled, it would condense with water vapor, forming a liquid that could dissolve metals. This liquid, which Geber called oil of vitriol, consisted of sulfuric acid (H_2SO_4) dissolved in water.

Geber described a similar process, using a chemical called niter (today known as potassium nitrate) along with vitriols, to produce nitric acid, which was found to dissolve silver. Geber also explained how mixing sal ammoniac (ammonium chloride) with nitric acid produced an acidic liquid *aqua regia*. Aqua regia consisted of a mixture of nitric and hydrochloric acids that could even dissolve gold.

Impact

The discovery of mineral acids, along with the translation of earlier Arabic and Greek alchemical texts into Latin, marked the beginning of European alchemy during the Middle Ages. Some consider the discovery of sulfuric acid to be the greatest achievement in chemistry during the medieval period.

One of the central goals of alchemy that was promoted by Geber was the transformation of base metals into gold. (A base metal was considered to be any metal that was not gold.) In order to do so, it was believed that base metals would first need to be "unmade", or dissolved, as one step in their transformation. Therefore, part of the alchemists' quest was to find a so-called universal solvent, a substance that was capable of dissolving any other substance. The strong mineral acids provided them with a solution to this problem. Because aqua regia could dissolve gold along with all other known metals, they considered it to be a close approximation of the universal solvent. Since alchemists could now "unmake" gold, it seemed within the realm of possibility of that they would soon be able to reverse this process and create gold from other substances.

However, alchemists incorrectly assumed that when a metal was dissolved in acid, it was broken down into its elements. Geber, for example, held to the belief that all metals are composed of the elements sulfur and mercury, an idea that had been proposed by his role model Jabir. Because only the proportions of the two substances were thought to differ from one metal to another, it was believed that one metal could be transformed into another simply by changing these proportions. What the alchemists didn't realize was that most of the metals with which they worked—gold, silver, iron, tin, and lead—were actually elements themselves. As a result, they cannot be broken down into simpler substances. (Dissolving metals in acid causes them to lose electrons, changing neutral metal atoms into ions—atoms bearing an electrical charge. However, this process does not change the identity of the element itself. Metallic lead, for instance, dissolves as lead ions, not as the ions of some other element.)

The discovery of mineral acids helped to bring respect to the study of chemicals and their properties. It also allowed alchemists to perform experiments that were previously impossible. However, the experiments involving these acids continued to focus primarily on potential gold-making techniques. As a result, mineral acids were produced only in small quantities by individual alchemists usually working in isolation. (One practical application was soon found for nitric acid. When a silver-coated gold coin was placed in a nitric acid solution, the silver would appear to be "transformed" to gold as the silver dissolved. This trick was used by swindlers to convince wealthy patrons to invest their money in gold-making schemes. Once the funds were handed over, however, the swindlers would never be heard from again.)

It would be several centuries before this class of chemicals had more than a minor impact on society at large. Sometime in the mid-1700s, it was discovered that sulfuric acid was useful in dyeing wool with indigo. Normally, wool does not absorb the rich blue color of indigo very easily. However, when the wool was first treated with sulfuric acid, the dye would bind to the cloth quite well. Because indigo blue (the same color still used to dye many blue jeans today) proved to be such a popular color, great quantities of sulfuric acid were soon in demand. Cheaper and quicker methods of manufacturing the acid were soon developed. As the price of sulfuric acid fell, chemists found more and more uses for it.

Meanwhile, in England, hat-makers had discovered that a solution of mercury and nitric acid could be used in the manufacture of popular felt hats. And in the Netherlands, it was found that when tin was dissolved in aqua regia and then added to a purple dye, a brilliant scarlet hue was produced that could be used to color silk.

It wasn't until the late 1700s that chemists first began to investigate the chemical nature of acids. They began to ask how and why acids are capable of behaving as they do instead of simply investigating what they can do. At that time, Antoine Lavoisier proposed that acidity was caused by the presence of oxygen in a compound. His proposal turned out to be incorrect, but it turned attention toward further study of this class of compounds. It was not until the twentieth century that a full understanding of the chemical nature of acids was finally achieved. Today, Geber's oil of vitriol is the most important industrial chemical worldwide. Its chief use is in the manufacture of fertilizers, but it is also used in the production of drugs, explosives, detergents, and automobile batteries. Nitric acid is used to make many of the same products.

STACEY R. MURRAY

Further Reading

Asimov, Isaac. *Asimov's Biographical Encyclopedia of Science and Technology*, 2nd rev. ed. New York: Doubleday & Company, Inc., 1982.

Cobb, Cathy and Harold Goldwhite. *Creations of Fire: Chemistry's Lively History from Alchemy to the Atomic Age*. New York: Plenum Press, 1995.

Farber, Eduard. *The Evolution of Chemistry: A History of its Ideas, Methods, and Materials*, 2nd ed. New York: The Ronald Press Company, 1959.

Partington, J. R. *A Short History of Chemistry*, 3rd ed., revised and enlarged. New York: Dover Publications, Inc., 1989.

Stillman, John Maxson. *The Story of Early Chemistry*. New York: D. Appleton and Company, 1924.

The Earth and Physical Sciences of Shen Kua

Overview

Shen Kua was a Chinese scientist, mathematician, and soldier in the eleventh century. He found success in many endeavors, but is perhaps best known for his work in the earth sciences, where he made many valuable contributions. Among these were the invention of the magnetic compass, remarkably accurate speculations about the origins of fossils, and essays covering many aspects of other geologic features. Much of his work was not known outside China, and would not be repeated anywhere else for over 700 years.

Background

Chinese civilization dates back thousands of years and is the source of the oldest recorded history on Earth. For much of human history, Chinese government, agriculture, science, and technology were equal or superior to that of the West. In many ways, whether a civilization makes scientific and technological advances depends on whether it has a central government, urban civilization, and productive agriculture, because in order to discover and invent, a society needs scientists and inventors. This requires agriculture that is sufficiently productive so a significant portion of the population is not tied to the land, a relatively efficient method of distributing food to the cities, and some way of encouraging and supporting scientists and inventors.

In China, the development of relatively efficient agriculture and a large population made it possible to form a coherent government centuries before this occurred in Europe, and Chinese society was relatively stable for long periods of time. Because of this, China not only developed the world's first real bureaucracy, but also some of the world's first scientists.

The T'ang dynasty, which had been marked by a stifling bureaucracy that discouraged innovation, began to collapse about 750. This process lasted over 200 years, and resulted in a China that was a hodgepodge of small kingdoms. In the middle of the tenth century, the Sung dynasty began to consolidate power and reestablished a central Chinese government. Also at this time, China's center of political power shifted from the northern cradle of Chinese civilization to the more prosperous, and more liberal, south. Shen Kua was born during this time of re-invigoration.

Like many urban Chinese at that time, Shen Kua was primarily educated at home by his mother. He entered the government as a civil servant and quickly proved himself by offering wise advice on many difficult issues. Among the problems he tackled successfully were draining swamps to reclaim land for agriculture, using silt dredged from rivers as fertile farming soil, and helping with military preparations. As he rose in the Chinese civil service, he also became interested in the sciences, including astronomy and mathematics. Unfortunately, he also became entangled in factional politics, and this led to his dismissal from government service.

Shen Kua spent the last years of his life at his estate, which he named "Brush Talks from the Dream Brook," writing a collection of documents that summarized much of what he had learned during his life. These essays covered an astonishing array of topics, many of them drawn from his personal experience and researches. Shen Kua wrote about earth sciences, astronomy, economics, government, mathematics, astronomy, and other areas. In all his essays, Shen Kua showed an impressive grasp of the principles of a wide variety of fields and made many suppositions that have since proved to be relatively accurate.

In particular, Shen Kua was first to suggest that fossils represented the remains of dead animals, and that the rocks in which they were found had originally been beneath water. He was also first to suggest that, by depositing silt for long periods of time, rivers and streams could form land and rocks that could gradually extend into the sea, causing the continent to become larger. And, noting bamboo fossils in an area that was dry and barren, he suggested that the climate must have changed at some time in the past because bamboo only grew in humid places. Shen Kua also was first to suggest that coal might be a better fuel than charcoal, given that China's forests were being cut down to burn for charcoal at a rapid rate. He also noticed that the soot from burning petroleum could be used to make ink, and some of the petroleum "mines" he initiated still produce petroleum for this purpose (and others). All these observations would not be rediscovered in Europe for 800 or 900 years. Perhaps Shen Kua's most important contribution was the observation that magnetized needles can be used to find direction.

In the area of astronomy, Shen Kua insisted on maintaining daily records of a variety of astronomical phenomena. From these observations, solar and lunar eclipses, planetary motions, sunspot activity, and even supernovae, were recorded faithfully long after his death. He also showed a gift for visualizing and describing the motions of astronomical bodies, and was first to attempt to describe these motions using trigonometry.

Impact

Although we now know that Shen Kua wrote reasonably accurate descriptions of a variety of geological processes, his works did not have a significant impact on eleventh-century China. In addition, by the time his works were first read by Europeans, most had been rediscovered by European scientists and were no longer "new." However, Shen Kua is generally recognized as one of the greatest Chinese scientists of all time.

In his geologic theorizing, he suggested that ancient climates, such as those that once supported bamboo groves in a now-barren desert, must have changed to explain the fossils he saw. He also suggested that, for marine fossils to be found far inland and at relatively high elevations, the sea must have retreated, or the rocks must have been lifted up out of the water. He further noted that, because of the amount of silt carried to sea by China's rivers, the land must be extending into the ocean a little at a time. All these observations are still held to be reasonable descriptions of physical processes that have taken place on Earth since its formation. Although he could not have known some of the mechanisms by which these processes took place, he had a remarkably keen understanding of the geological processes that had shaped the China he lived in.

Perhaps the most important contributions made by Shen Kua were the daily astronomical observations he insisted upon. These observations have been used recently to help understand long-term changes in solar activity. They constitute an almost unbroken record of sunspots stretching back nearly a millennium, and seem to be highly accurate and reliable. At least two supernovae have been very accurately dated using Chinese records of strange "new" stars that appeared in the heavens, and other Chinese observations have provided data that support theories and calculations regarding some orbital and rotational properties of Earth and the Moon.

Unlike European scientists many centuries later, Shen Kua did not explain his observations by referring to catastrophic events. Where European scientists insisted on a biblical interpretation of natural phenomena (e.g., suggesting that Noah's Flood explained sedimentary rocks and fossils), the Chinese seem to have had no such tradition. This does not mean that Chinese scientists were "better" than Europeans or vice versa, but it helps to demonstrate that science takes place within a culture and not in isolation. European scientists fit their theorizing into a biblical framework because the Bible was so profoundly important to European culture, and scientists were a part of that culture. In contrast, the Chinese creation story did not involve a huge flood (although many non-European creation stories did), and the Chinese religion emphasized worship of a larger number of gods and their ancestors. These religious and cultural differences, combined with a longer cultural history, helped to shape the way Shen Kua and his fellow scientists observed their world.

Shen Kua's scientific work provided some necessary information to the Chinese, even though much of it was either ignored or neglected for centuries. Today, he is remembered primarily for being a keen observer of nature and a brilliant thinker about the meaning of his observations. His works were part of the beginnings of a Chinese tradition in science that has a history longer than any other in the world. They also provide an excellent example of how scientists from separate cultures, in this case Chinese and European, can observe the same phenomena (for example, sedimentary rocks and fossils), yet come away with completely different explanations and theories because of their different cultural traditions.

P. ANDREW KARAM

Further Reading

Needham, Joseph. *Science and Civilization in China*. Cambridge: Cambridge University Press, 1954.

Temple, Robert. *The Genius of China: 3000 Years of Science, Discovery, and Invention*. New York: Simon & Schuster, 1986.

Ordering Knowledge in the Medieval World

Overview

During the Middle Ages in Europe, the perception and organization of knowledge underwent a significant transformation that helped make possible the ideas and accomplishments of the Scientific Revolution during the Renaissance and early modern period. As the works of Aristotle (384-322 B.C.) and other ancient philosophers were brought together with the teachings of Christianity and the mathematic and scientific contributions of the Islamic world, a diverse array of knowledge found its way into medieval university curricula and encyclopedias for the learned. The ideas and worldview presented in these institutions and volumes offer insight into how medieval scholars helped turn the ancient world into the modern one.

Background

Science, or natural philosophy as it was better known for many centuries, flourished during the heyday of Greek civilization. The Greek conquests lead by Alexander the Great (356-323 B.C.) in the fourth century B.C. brought about a mixing of Greek and Roman culture that ultimately resulted in a decline in interest in scientific and mathematical pursuits. While Greek scholars had pursued the study of a wide range of philosophical and scientific subjects with a high degree of sophistication and technical detail, Roman readers and students had more practical and basic interests. As Rome's Latin culture gradually displaced the Greek scholarly tradition, scientific ideas and methods were distilled into encyclopedias or commentaries intended for the leisure reading of Roman gentlemen. From one of these encyclopedias, written by Marcus Terentius Varro (116-27 B.C.) in the first century B.C., came the model of organizing knowledge into nine liberal arts: grammar, rhetoric, logic (these came to be known as the "trivium"), arithmetic, geometry, astronomy, music (these were known as the "quadrivium"), medicine, and architecture. The last two were dropped by later writers and scholars, leaving the seven liberal arts to form the framework of knowledge and education in the Roman Empire.

The years of the disintegration and collapse of the Roman Empire saw a decline in education. Schools and the spirit of education continued to exist in some regions of the empire, but in others all connections to the knowledge and traditions of the classical period were lost. As Christianity and its European institutions began to strengthen in the fourth century A.D., monasteries became new centers for learning. Monasteries ran schools for initiates (and sometimes for more affluent local children) and maintained libraries and scriptoria (rooms where manuscripts could be copied). The education offered in these schools was predominantly spiritual, relying mainly upon biblical texts, commentaries, and devotional writings. Classical learning was not entirely abandoned, however; encyclopedias written by seventh-century Christian scholars survive and reveal the use of both Christian and Greek sources to describe the natural and man-made worlds.

Christianity provided not only the institutions that perpetuated intellectual life during the Middle Ages, it was also the ideological underpinning of all scholarly work. From the fourth century onward, Christian writers and scholars were guided by Saint Augustine's (354-430) principle that the study of nature should serve as a "handmaiden" to religious devotion. By viewing, appreciating, and understanding nature, one could better appreciate God's works. While this subordinate position for science may seem oppressive to modern readers, it actually provided some justification and stimulus for the study of nature and eventually aided in the advance of scientific ideas and methods.

Monastic schools provided a limited context for the study of academic subjects during the early Middle Ages. During the reign of Charlemagne (742-814), however, social reforms brought new schools to monasteries and cathedrals throughout the Carolingian empire (which by then encompassed most of Western Europe). Within these schools the study of the seven liberal arts was revived, and for the first time scientific and mathematical subjects were enriched by some contact with texts produced in the Islam world. While Western Europe had ceased to advance the scientific subjects mastered by the Greeks—geometry, optics, astronomy, and other branches of mathematics—Islamic scholars preserved, studied, and improved upon these works during the ninth through thirteenth centuries. During the Western revival of interest in Greek scholarship in around the tenth century, and for the next 400 years, European scholars relied upon Islamic texts.

Charlemagne's schools were the first step in the establishment of a robust scholarly tradition in Europe. Thanks to political stability and increasing affluence due to technological and social innovations, the population of Europe grew rapidly beginning around 1000. Urban centers, which had withered in the early Middle Ages, once again held large portions of the population and provided a fertile ground for the development of culture and intellectual life. The schools that formed in these cities served a diverse group of students, whose needs were more practical than spiritual. The subjects of the "quadrivium" gained in importance, and along with law and medicine formed the core of the new curricula. Classical texts took a prominent place alongside Christian sources. As society became more complicated—for example, through the rise of commerce—attempts to rationalize and systematize affairs of all kinds prevailed. This had an influence on the organization of knowledge inside as well as outside the schools, and led to an intellectual ferment that, in a few hundred years, moved natural philosophy from the periphery to the center of not only the school curricula, but the world itself.

Impact

By the thirteenth century, major universities were flourishing in Paris, Oxford, and Bologna, with student populations ranging from 500 to 2,500 young men. These were typically organized into four faculties: an undergraduate faculty of liberal arts and three advanced faculties for the study of law, medicine, and theology. The liberal arts curriculum had expanded from the classical pattern to include moral philosophy, metaphysics, and natural philosophy. The technical methods of arithmetic and geometry were touched upon in the basic curriculum, as were those of astronomy (emphasizing astrological principles and the important methods of calendar establishment and timekeeping). Of more central importance, however, was Aristotelian natural philosophy. Aristotle's works on cosmology, physics, meteorology, and natural history were mandatory reading for all students. Because these subjects, then at the heart of the university curriculum, were learned by all through study of Aristotle's works and commentaries upon them, these ideas were universally understood and moved easily from one university to another, creating a cosmopolitan intellectual worldview that helped to unify scholars throughout Europe.

Universities subsequently became the seats of learning in medieval Europe. Scientific study, based on the texts being rediscovered and translated from the Greek and Islamic traditions, shared a place in the curriculum with Christian theology and Aristotelian logic. But while the subjects themselves remained organized into separate areas of study, methodological approaches began to be applied across disciplines. The critical evaluation of statements and ideas characteristic of Aristotelian logic was used to consider claims about nature, and even theological doctrine. This led to tensions between the Church and the universities, and several times during the thirteenth century the pope issued bans on the study of Aristotle. But Aristotle's works proved irresistible to scholars, and much effort was devoted to reconciling Aristotle's claims with those of the Bible. The "handmaiden" ideal motivated scholars who tried to bring Aristotle's scientific claims in line with Christian theology as they worked to show that this vast body of knowledge about the world could be put to use to serve man and the Church.

From the thirteenth century to the fifteenth century the outward organization of knowledge changed little. University curricula remained fixed, and tensions between Aristotelian philosophy and Christian theology persisted, with the latter maintaining supremacy in the work of nearly every scholar. However, within the separate areas of study, such as physics and astronomy, critical analysis and investigation of particular claims made in ancient texts led gradually to new ways of thinking about the natural world. Careful curiosity about the details of the physics of motion, for example, prompted fourteenth-century investigators to frame questions and attempt solutions that would later inspire Galileo (1564-1642). Astronomical observations and calculations made in the late Middle Ages made possible the groundbreaking work of Nicolaus Copernicus (1473-1543) and Johannes Kepler (1571-1630).

By name and definition, the "scientific revolution" of the fifteenth through seventeenth centuries shows a strong break with earlier traditions. But it is undoubtedly true that without the organization of knowledge in the medieval universities—the high esteem given to critical judgment by Aristotelian philosophy, the importance of using such criticism to reconcile Aristotelian ideas with Christian doctrine, and the freedom to consider particular scientific claims without regard to contentious theological issues—the tools to make the revolution would not have existed.

LOREN BUTLER FEFFER

Further Reading

Crombie, A.C. *Medieval and Early Modern Science.* Garden City, NJ: Doubleday, 1959.

Dijksterhuis, E.J. *The Mechanization of the World Picture.* Oxford: Clarendon Press, 1961.

Grant, Edward. *Physical Science in the Middle Ages.* New York: Wiley, 1971.

Kelley, Donald, ed. *History and the Disciplines: The Reclassification of Knowledge in Early-Modern Europe.* Rochester, NY: University of Rochester Press, 1997.

Lindberg, David. *The Beginnings of Western Science.* Chicago: University of Chicago Press, 1992.

Lindberg, David, ed. *Science in the Middle Ages.* Chicago: University of Chicago Press, 1978.

Singer, Charles. *A Short History of Anatomy and Physiology from the Greeks to Harvey.* New York: Dover, 1957.

Siraisi, Nancy G. *Medieval and Early Renaissance Medicine.* Chicago: University of Chicago Press, 1990.

The Rise of Medieval Universities

Overview

The European university is a particular organization that emerged out of the conditions of medieval society. Students and teachers in Europe applied the medieval trend of guild organization to protect themselves from local laws, high prices, and prejudices. Wider needs within medieval society for people with skills and learning boosted student numbers, and universities grew to meet the demand.

Background

The collapse of the Roman Empire in the fourth century created a period of anarchy and economic crisis across Europe. The intellectual climate changed drastically, and large numbers of books and papers were lost or destroyed. The overall need for learned men fell in parallel with the decline of trade, economics, and local administration. Greek and Roman learning was preserved in Eastern Europe in the Byzantine Empire, and over time Islamic scholars absorbed and spread the ancient texts throughout the Middle East. In Western Europe the few surviving texts were scattered in monastery libraries. However, the early medieval monks were more interested in theological and philosophical texts than pagan mathematics or science, so few copies were made of such works. Over the centuries many surviving ancient texts decayed into dust, or were destroyed in wars and other disasters.

Latin was the language of the monks and the surviving texts were rewritten in abbreviated medieval style of Latin, often based on poor translations from Greek. Over time the curriculum of medieval learning became set, based on large compendiums of simplified Greek knowledge compiled by encyclopedists such as Boethius (480-524). Medieval learning was based on the seven liberal arts. The quadrivium (four) were mathematically based, comprising arithmetic, music, geometry, and astronomy, but these were much less popular than the linguistic trivium (three) of grammar, rhetoric, and logic, which led to further study in theology, philosophy, medicine, and law. The main demand for higher education was within the church, and the majority of students were clergy, as were their teachers.

In the eleventh century new contact with the East, in the form of the Crusades, helped to recover lost ancient knowledge. While the Crusades were mainly destructive and religious-driven wars, there were some positive outcomes for European society. Western scholars came to realize that Islamic intellectuals had a storehouse of ancient learning wider than their own. The Arabic scholars had added new material to the classics, either on their own, or by absorbing the intellectual traditions of nearby cultures such as Hindus and Babylonians. There was also contact with the Muslim world in Spain, the southern half of which was an Islamic state. Many European scholars traveled to Spain to learn Arabic and other so-called oriental languages.

European economics and politics slowly began to develop, and the growth in trade and government administration saw an increased need for literate and numerate scholars. The survival of ancient texts in Western monasteries had made them the focal points of medieval learning. The cathedral schools, especially those in capital cities or at pivotal trade routes, began to grow with the slow rise of trade and economic stability. These became centers for copying the new texts recovered from the East. While originally intended for religious study, various reforms made these schools accept secular students as well. As student numbers climbed, these centers of learning gradually evolved into universities.

Impact

The word university originates from the term *universitas*, which originally meant any collection of professionals in a guild or organization. The motivations behind these corporations were to provide their members with protection from rival groups, and enable price regulation and monopolies. Over time the term became narrowed to mean strictly a society of academics.

There is some debate among scholars about which particular place can be called the first university. The medical school at Salerno, in southern Italy, is often cited as the first university, or at least one of the first universities. Salerno was well known as a health resort from the ninth century. It was also a meeting place of Greek, Latin, Arabic, and Jewish learning, being a port situated on important trade routes. It became a *universitas* sometime in the twelfth century, and obtained formal recognition in 1231, but remained solely a medical school and did not influence the style and organization of later universities.

The *universitas* that was to inspire the majority of other institutions in southern Europe was Bologna. The Italian town had a law school of great renown, which attracted students from all over Europe, often from wealthy backgrounds. Like many medieval towns, Bologna discriminated against foreign residents. They were taxed at higher rates, charged more for lodging and food, had harsh laws imposed upon them, and were liable for military service. Near the end of the twelfth century the foreign law students at Bologna formed a union to provide protection from these local customs and laws. The students had to fight for their rights, and it took a three-year strike before their absence caused the authorities to give in to their demands. Students, it was discovered, were a vital part of the local economy, and so they could demand better treatment, or take their money elsewhere. To keep the students at Bologna they were granted cheap rent, food, and taxes, as well as exception from military service and the right to set teaching fees.

In Paris, at around the same time, the teachers of that city formed themselves into a corporation, a *universitas magistorum*. Students in Paris tended to be French, but their teachers were often foreign, and so organized themselves for protection and mutual benefit. Students were allowed to join the guild as junior members and, if they passed their examinations, could slowly advance up the corporate hierarchy. Paris was the model that later northern European universities followed.

Universities began to spread across Europe. Often disputes within a university led to migrations of teachers and students and the formation of new universities. Migrations from Bologna led to the founding of Padua (1222). Further moves from Padua led to the creation of a university at

Vercelli (1228). Some historians claim that up to half the universities of medieval Europe originated from such disputes. Universities also sprung up seemingly on their own, although usually following the organizational principles of either Bologna or Paris. By 1500, there were 62 recognized universities in Europe.

The fortunes of universities were closely tied to the towns they existed within, or near. Many famous schools, such as Oxford and Cambridge, were founded at busy commercial centers. There was often conflict between the town authorities and the academic guilds. Many riots occurred in the early history of universities, referred to as "town versus gown battles." One of the questions at stake was who had legal authority over academics. Over time it became accepted that scholars could not be arrested or tortured by town authorities, except for murder. In effect, universities became independent entities with their own code of conduct and discipline.

In the early universities, lectures were usually held in the master's room, or a hired hall, as these universities owned no buildings of their own. Classes consisted of a master reading aloud and commenting on an established text, while the students copied down the lecture word for word. This gave the students both the original text and a learned commentary on the work. Lecturers who spoke too softly, or too quickly, were often shouted at by their students, and in some cases attacked. As the lecturers relied on the fees paid by their students, teachers could be boycotted, and driven by economic necessity to alter their teaching or leave. The use of Latin as the academic language meant that academics could study and teach in any European country. University students and teachers were very mobile, often traveling to several institutions in their careers, and helped create a European wide sense of learning.

Universities taught the seven liberal arts and at least some of the advanced topics of theology, law, medicine, and philosophy. Many universities began to include practical courses in response to public demand. Courses in the art of letter writing trained the clerks, money-counters, and administrators of the flourishing economy.

However, the era of growth did not last, as the fourteenth century was beset with famines, disease, and war. The conflict that came to be called The Hundred Years' War disrupted trade, and the plague known as the Black Death killed approximately a third of Europe's population. The universities continued as well as they could, although many were forced to suspend classes for extended periods. These disruptions had wider social implications, for while the twelfth century had been a time of expanding intellectual horizons, particularly with the influx of Arabic and ancient knowledge, the university curriculum now became fixed and rigidly taught.

By the sixteenth century many critics regarded the universities as places of backward, unimportant studies. University academics were accused of following their ancient sources too closely, while ignoring the dramatic changes in European religion, politics, economics, and wider discoveries of the world. Yet the universities survived and even flourished, for social changes had once again increased the demand for educated men to fill positions in commerce and administration, and the universities held a monopoly on higher learning. Universities continue to evolve today, and yet still retain some of their earliest characteristics, as formed in the medieval period.

DAVID TULLOCH

Further Reading

Cobban, A. B. *The Medieval Universities: Their Development and Organization*. New York: Routledge, 1975.

Murray, Alexander. *Reason and Society in the Middle Ages*. Oxford: Clarendon Press, 1978.

Rudy, Willis. *The Universities of Europe, 1100-1914: A History*. London and Toronto: Associated University Presses, Inc., 1984.

Schachner, Nathan. *The Mediaeval Universities*. New York: 1938.

Radcliff-Umstead, Douglas, ed. *The University World, A Synoptic View of Higher Education in the Middle Age and Renaissance*. Pittsburgh: 1973.

Biographical Sketches

St. Albertus Magnus

c. 1193-1280

German Natural Scientist, Philosopher, and Theologian

Although surpassed in philosophy and theology by his humble student, Thomas Aquinas, and perhaps also in natural science by his jealous rival, Roger Bacon, his contemporaries generally recognized Albert von Bollstädt as the greatest mind of the thirteenth century, expert in all known branches of knowledge. Even during his lifetime he was called "Albert the Great" in English, "Albert der Grosse" in German, and "Albertus Magnus" in Latin. Known as the "Universal Doctor" by virtue of the broad scope of his learning, he was canonized a saint in the Roman Catholic Church in 1931 and proclaimed the patron saint of natural scientists in 1941.

Born in Lauingen, Swabia, Germany, the oldest son of the Count of Bollstädt, probably in 1193, but perhaps in 1200 or 1206, he is sometimes called Albert von Lauingen because of his birthplace. After early schooling worthy of a young nobleman, his family sent him to the University of Padua, Italy. He joined the Dominican Order in 1223 in Padua, attracted by the lessons of the master general of that order, Jordan of Saxony. He may have continued his studies in Bologna, Italy, before returning to Germany to teach in the monasteries of Hildesheim, Freiburg im Breisgau, Regensburg, Strassburg, and, starting in 1243, Cologne.

One of his students in Cologne was Thomas Aquinas, who arrived probably in 1244. The two became lifelong friends. When Albert was called to the Convent of St. Jacques at the University of Paris in 1245, Thomas followed him in order to remain his student. About this time he wrote a commentary on Peter Lombard's *Sentences*, the major theological textbook of the twelfth century. He received his doctorate of theology in Paris in 1245, taught there for three years, then returned to Cologne in 1248. Again Thomas followed him.

Albert intended to spend the rest of his life quietly as a scholar and teacher in Cologne, but he suffered a few interruptions. He was appointed provincial general of his order in 1254, and served until he was able to resign in 1257. Pope Alexander IV named him Bishop of Regensburg in 1259. Albert was installed to that office in 1260, but resigned as soon as Alexander died in 1261. From 1263 to 1264, Pope Urban IV commanded him to travel throughout Germany to promote a new Crusade that never came to pass. In 1274, he obeyed the summons of Pope Gregory X to the Second Council of Lyon, France. Finally, in 1277, sick and elderly, he forced himself to go to Paris to protest vigorously against Bishop Étienne Tempier's attempt to condemn some of the propositions of Thomas Aquinas.

Albert became aware of the works of Aristotle in Paris in 1245. At that time, Aristotelian philosophy and science were just beginning to become known in the West, and not without controversy. Aristotle's ideas were transmitted from the Arabic and Greek worlds to the Latin West in large part through the commentaries of the twelfth-century Muslim philosopher and physician Averroës (Ibn Rushd; 1126-1198), whose interpretations, as well as Aristotle's own pre-Christian paganism, were often at odds with Roman Catholic doctrine. The teachings of Aristotle on natural science were banned in Paris between 1210 and 1234. Albert approached Aristotle critically, refuted both the Averroist views and the pagan elements, and thus made Aristotle more palatable to Christian thinkers. He eventually wrote commentaries on all of Aristotle's known works. He and Thomas Aquinas were the two founders of Christian Aristotelianism in medieval scholastic thought.

Albert's most useful contribution to science was as a botanist. His observations, descriptions, and classifications of many kinds of plants were studied as recently as the nineteenth century. He discovered the flow of sap in trees, practiced grafting, and correlated levels of light and heat with rates of plant growth. Many historians consider his botanical treatise *De vegetabilibus et plantis* to be his major work.

Among Albert's writings are commentaries on the Psalms, the prophets, and the Gospels; *Summa theologiae,* which prefigured that of Thomas Aquinas; seven books each on logic, the biological sciences, the physical sciences, and psychology; one book each on ethics, politics, metaphysics, and cosmology; a commentary on Pseudo-Dionysius the Areopagite; and several volumes of theological treatises and sermons.

ERIC V.D. LUFT

Page from Albertus Magnus's *De Natura Rerum.* (Gianni Dagli Orti/Corbis. Reproduced with permission.)

al-Battani

c. 858-929

Arab Astronomer and Mathematician

Al-Battani is considered the greatest astronomer of the medieval Islamic world. He is best known for his astronomical handbook *Kitab al-Zij,* which introduced new trigonometric methods for performing astronomical computations. He devised improved instruments and made accurate observations that allowed him to give corrected values for several astronomical constants, including the obliquity of the ecliptic and time of the equinoxes.

Known to the West as Albategnius, al-Battani was born in or near Harran in northwestern Mesopotamia (modern Turkey) around 858. His father was the noted instrument-maker Jabir ibn Sinan al-Harrani. No information exists regarding al-Battani's formal education. However, it seems reasonable to assume that his facility in devising improved instruments, which included a new type of armillary sphere, was nurtured through technical training provided by his father. His ancestors were Sabians, a religious sect adhering to a mixture of Christian and Islamic doctrines, but al-Battani was fully committed to Islam. Most of his research was conduct at al-Raqqa on the Euphrates River, where his family had moved when he was a youth. In 929 he journeyed to Baghdad to petition the caliph regarding some injustice to the people of al-Raqqa. He died at Qasr al-Jiss on the return trip.

Of the many works al-Battani wrote on astronomy and mathematics, his best known, *Kitab al-Zij*, contains most of his important findings. The *Zij* consists of a star catalog and trigonometric tables, as well as solar, lunar, and planetary tables together with canons for their use. The treatise was written from a Ptolemaic perspective. However, al-Battani turned a critical eye toward Ptolemy's (second century A.D.) practical results, and designed improved instruments to collect more accurate data. These observations revealed errors in Ptolemy's *Almagest* and allowed al-Battani to provide corrected values for many of the main parameters of planetary motion.

Al-Battani's careful observations showed that the solar apogee, the point at which the Sun is smallest in apparent size and farthest from Earth, had shifted from the position indicated by the *Almagest*. Contradicting Ptolemy, this implied that the solar apogee was slowly moving. Though al-Battani did not explicitly state this, he has often been credited with the discovery. Thabit ibn Qurra (836-901) had earlier found the solar apogee to be moving at a rate of 1° every 60 years, but when he found the rate of precession (the conical gyration of a spinning body's axis of rotation) to be the same, he concluded they were identical. Al-Battani's value for the solar apogee was no better than Ibn Qurra's. Thus, he was in no better position to distinguish between the Sun's motion and precession. In fact, the first to correctly and unambiguously state the proper motion of the solar apogee was al-Zarqali (1029?-1087?). Al-Battani also provided better values for the obliquity of the ecliptic and the precession constant.

Among his many other accomplishments, al-Battani rectified the Moon's mean motion in longitude, provided better estimates of the apparent solar and lunar diameters and their annual variation, and demonstrated the possibility of annular solar eclipses. He also developed a sophisticated method for determining the magnitude of lunar eclipses. His redetermination of the time of equinox allowed him to make an improved estimation of the tropical year, which he calculated as 365 days, 5 hours, 46 minutes, and 24 seconds—short by only 2 minutes and 22 seconds. Al-Battani also improved astronomical computations through his introduction of the trigonometric sine function and formulae for the solution of problems involving spherical triangles.

The *Zij* proved influential in the development of European astronomy. It was originally translated into Latin during the twelfth century by Robert of Chester (fl. c. 1141-1150). However, the only extant Latin version was made about the same time by Plato of Tivoli. This version was first printed in Nuremberg (1537) under the title *De motu stellarum* (On Stellar Motion).

STEPHEN D. NORTON

Jean Buridan
1300-1358
French Physicist and Philosopher

More than three centuries before Sir Isaac Newton (1642-1727), Jean Buridan anticipated Newton's first law of motion when he stated that an object in motion will remain moving. This put him at odds with the prevailing idea, passed down from Aristotle (384-322 B.C.), who had maintained that a moving object requires continual application of force to keep it in motion.

Born at Béthune, in the French district of Artois, Buridan studied philosophy at the University of Paris under William of Ockham (c. 1290-1349). He went on to an appointment as professor at his alma mater, then took a position as university rector from 1328 to 1340. In 1345, Buridan became an ambassador from the university to the papal court, at that time located in Avignon, France.

Buridan wrote widely, and his works include *Compendium logicae*, *Summa de dialectica*, *Consequentie*, and a series of commentaries to works of Aristotle. In his commentary on the latter's *Physics*, he challenged Aristotle's assertion that the air around an object in motion is what keeps it moving. Buridan was not the first to take issue with the Aristotelian view of motion: the Byzantine Johannes Philoponus (c. 490-570) had done so eight centuries before.

But Buridan took the point a great deal further, producing an amazingly accurate hypothesis regarding impetus: that one object imparts to another a certain amount of power, in proportion to its velocity and mass, that causes the second object to move a certain distance. He was also correct in stating that the weight of an object may increase or decrease its speed, depending on other circumstances; and that air resistance slows an object in motion.

In the realm of philosophy, Buridan was primarily concerned with the same issues of epistemology as his teacher, Ockham, though the two reputedly came to a parting of the ways on certain issues. Buridan's most important philosophical argument sprang from a discussion of Aristotle's *De caelo* (On the heavens), and borrows the image of a dog used by the Greek philosopher to make a point in that book. Buridan's illustration also makes use of a dog, though for some reason his argument has come to be known as "Buridan's ass." In any case, the illustration concerns the question of choices: if a dog is forced to choose between two equal amounts of food, which he desires equally and about which he possesses equal knowledge, he must either starve to death or make a completely random choice.

Actually, there is some dispute as to whether Buridan himself made the "Buridan's ass" argument, with some scholars asserting that it was created by opponents to make his ideas appear absurd. It has also been suggested that the question raised by the "Buridan's ass" illustration provides a framework for the scientific investigation of statistics.

JUDSON KNIGHT

Campanus of Novara

c. 1240-1296

Italian Astronomer and Mathematician

Campanus of Novara, also known as Johannes Campanus, is best known for his edition of Euclid's (fourth century B.C.) *Elements,* which remained the standard text for teaching Euclidean geometry throughout the late Middle Ages. His *Theorica planetarium* was also very influential, helping to popularize the planetary equatorium, an instrument for astronomical measurement, in Europe.

Campanus, who referred to himself as Campanus Novariensis, a Latinized version of his Italian name, was born in Novara, Lombardy, as his name suggests. Little is known of his early years, and his birth date can only be approximated. He was a close friend of Panteléon, patriarch of Jerusalem. When Panteléon was elected pope in 1261, becoming Urban IV, Campanus was made one of his chaplains. Urban also bestowed upon Campanus the rectorship of the Church of Savines (1263) and a canonicate of Toledo's cathedral (1264). Campanus also served as chaplain (1263-64) to the cardinal deacon of Saint Adrian, Ottobono Fieschi, and as parson of Felmersham in Bedfordshire, England. He later received a canonicate of Paris and served as papal chaplain to both Nicholas IV (1288-92) and Boniface VII (1294-1303). His final years were spent at the Augustinian Friars' convent at Viterbo, where he died in 1296.

Sometime between 1255 and 1259 Campanus completed his version of Euclid's *Elements.* It remained widely popular as a text for teaching Euclidean geometry until superseded in the sixteenth century by a translation made directly from the Greek. His edition was reissued at least 13 times and used for the first printed version, which appeared in Venice (1482). However, Campanus's version of the *Elements* was not an original translation, as he did not possess the necessary skill to make a direct translation from either Arabic or Greek. His version was a reworking of one or more translations. In particular, he seems to have relied heavily on an earlier translation by Adelard of Bath (1090?-1150?).

Campanus also produced a number of astronomical works, the most important being *Theorica planetarium.* Composed sometime during the papacy of Urban IV (1261-64), *Theorica* reviewed Ptolemy's (second century A.D.) ideas on the structure and dimensions of the universe. In the *Almagest,* Ptolemy provided absolute distances for the Sun and Moon, but only relative distances for the five known planets. In *Planetary Hypotheses,* Ptolemy argued that the farthest distance a planet traveled from Earth was the closest distance of approach for the next planet in order above it. Thus, knowing the absolute distance of the lowest body—the Moon—one could calculate the dimensions of the universe. Campanus borrowed Ptolemy's data, made the necessary calculations, and reported the results in *Theorica.* He also calculated the total area, in square miles, of the sphere of the fixed stars.

Theorica also contains the first mention in Latin astronomy of the planetary equatorium—a device for finding the positions of celestial bodies at any given time. Campanus provided instructions for the construction and use of this device. It is unlikely that he developed the idea for this instrument on his own since descriptions of more complex versions of the equatorium were written almost 200 years before *Theorica* appeared. Furthermore, the extreme difficulty in manufacturing a device of Campanus's design suggests that not even a prototype was produced. Nevertheless, Campanus's description was the primary stimulus for the subsequent development of the instrument.

Theorica was a highly technical work and primarily of interest to only professional astronomers. Campanus did, however, compose a popular version, titled *Sphere* (*Tractatus de spera*), that required no specialized knowledge to understand. *Computus maior,* another well known work by Campanus, was a calendric treatise that addressed the difficulties in computing the date of Easter. *Sphere* and *Computus* were reprinted many times in the two centuries following Campanus's death.

STEPHEN D. NORTON

Abu Nasr Muhammad ibn Muhammad ibn Tarkhan ibn Awzalagh al-Farabi

c. 870-950

Turkish Philosopher, Mathematician, and Musician

Al-Farabi provided the first comprehensive Arabic classification of the sciences. One of Islam's greatest philosophers, he was widely known as "the Second Master" (Aristotle being the first). His defense of rationality and attempt to reconcile the philosophies of Plato (c. 427-347 B.C.) and Aristotle (384-322 B.C.) were highly influential in the Islamic world. An accomplished musician, he was also one of the foremost music theorists of the Middle Ages.

Known to the West as Alfarabius, al-Farabi's date of birth is unknown, but it is generally believed to have been around 870. It seems he was born in the village of Wasij in the Farab district of Trasoxiana (modern Turkestan). His father was a military officer in the service of the Samanid emirs. Little is known of al-Farabi's childhood. However, it is likely he studied music at Bukhara before going to Marv (or possibly Harran) to study with the Nestorian Christian Yuhanna ibn Haylan. His studies under Ibn Haylan were continued at Baghdad, where he also studied logic with Abu Bishr Matta ibn Yunus (d. 940) and Arabic grammar with Abu Bakr ibn al-Sarraj (875?-928?). Sometime during his thirties, al-Farabi traveled to Constantinople, where he studied Greek philosophic and scientific works. He eventually returned to Baghdad to lecture and write. Political turmoil forced al-Farabi to leave Baghdad in 942. He remained in Egypt for many years before joining the Hamdanid prince Sayf al-Dawlah in Damascus. He died there a year later in 950.

While not denying the revealed truths of religion, al-Farabi was convinced philosophical truths were the supreme form of wisdom available to man. He felt such truths manifested themselves most clearly in the natural and mathematical sciences of the Greeks. Consequently, he championed the use of reason and sought to establish the preeminence of natural philosophy within the framework of Islamic society.

To this end, al-Farabi elaborated on the hierarchy of knowledge, developing a classification of the sciences. In *Kitab al-ibsa al 'Ulum* (*Catalog of Sciences*), he described the nature and scope of the various sciences. He placed philosophy at their head, which he claimed guaranteed the truth of knowledge by virtue of apodictic reasoning. Since al-Farabi was the first great Muslim commentator on Aristotle, it comes as no surprise that his main divisions of the sciences were Aristotelian. Al-Farabi's systematic classification of the sciences was adopted with only minor changes by Ibn al-Haytham (c. 965-1039), al-Ghazali (1058-1111), and Ibn Rushd (1126-1198).

Kitab al-ibsa al-'Ulum provided the medieval West with its first sophisticated system for classifying the sciences. It was first translated into Latin by Dominic Gundissalinus between 1130 and 1150. A later translation, *De ortu scientiarum,* by Gerard of Cremona (c. 1114-1187) was more widely circulated.

In "Attainment of Happiness," al-Farabi discussed the order in which the sciences should be studied. He had himself studied many branches of science, writing an alchemical treatise and commentaries on Euclid's (330?-260? B.C.) *Elements* and Ptolemy's (second century A.D.) *Almagest.* However, most of his specialized writings on natural science were polemical.

Al-Farabi devoted special attention to the mathematical science of music. His masterwork, *Kitab al-Musiqa al-kabir* (Book of music), was certainly the greatest Islamic discourse of its kind written to that time, and possibly the greatest medieval Arabic treatise on music. In it he discussed musical intervals and their combinations, as well as examined matters of rhythm.

Al-Farabi, following in the footsteps of al-Kindi (801?-866?), defended reason in the interpretation of the Koran. Accordingly, he applied Greek philosophy to Islamic moral and political problems. In *Al-Madina al-fadila* (The ideal city), he showed the relationship between Plato's ideal community and Islam's divine law and argued happiness could be attained through politics.

STEPHEN D. NORTON

Robert Grosseteste

c. 1175-1253

English Physicist and Philosopher

Though he is perhaps best known as the teacher of Roger Bacon (1213-1292), Robert Grosseteste distinguished himself as a scientist in his own right. He wrote on astronomy, discussing comets and advancing a theory of tides. He also presented his own theories concerning light and sound. He described light as the basic substance of the universe and postulated, with considerable accuracy, that sound was a vibrating motion passing through the air.

Grosseteste was born the son of poor parents, and at an early age was forced to earn his own living, at times resorting to begging. The mayor of his hometown, Lincoln, England, eventually recognized his intellectual abilities, and arranged to have him enrolled in school. There he distinguished himself so much that he went on to an academic career that took him successively to the universities of Oxford, Cambridge, and Paris.

In about 1215, Grosseteste was appointed chancellor at Oxford, a position in which he served until 1221. He became the first rector of the Franciscan monks at that institution in 1229 or 1230, then in 1235 received consecration as Bishop of Lincoln. Throughout his remaining career, he distinguished himself for his opposition to abuses of power by both King Henry III and Pope Innocent IV, whom he openly described as "the Antichrist."

Bacon, who studied under Grosseteste in the 1230s, described his teacher as one of the most learned men of his day. Grosseteste excelled in his studies of Greek and Hebrew, and wrote several works on Aristotle (384-322 B.C.).

Starting in about 1215, Grosseteste engaged in a number of scientific studies. Within a decade, he had conducted experiments in optics, using mirrors and lenses, in an attempt to explain the qualities of light in general, and of the rainbow in particular. In 1230, he published *De generatione sonorum,* his treatise on sound. In it, he advanced his theory of vibrations, which would be corroborated by later studies in the modern era.

As with many men of science during the Middle Ages, Grosseteste's curiosity earned him disapprobation as a suspected magician—a reputation no doubt compounded by his outspoken opposition to church and secular leaders of the day. It was reputed that he had published a study entitled *Magick,* and that he built a head of brass that could be used for discerning answers to questions and foretelling the future. Not only the scholar Gerbert (a.k.a. Pope Sylvester II, 945-1003) but even Grosseteste's student Bacon, repeated this bizarre tale.

JUDSON KNIGHT

Abu 'Ali al-Hasan ibn al-Hasan ibn al-Haytham

965-c. 1040

Persian Astronomer, Mathematician, and Physicist

Al-Haytham was the greatest Arab scientist of the Middle Ages. His *Kitab al-Manazir* was the most important and influential work on optics between the time of Ptolemy (second century A.D.) and Johannes Kepler (1571-1630). Al-Haytham also made significant contributions to astronomy, mathematics, and medicine.

Al-Haytham, known to the Latin West as Alhazen, was born in 965 in Basra (in modern Iraq). At an early age he became perplexed by the conflicting claims of competing religious sects. Frustrated by his failure to resolve these differences, he concluded that truth was only attainable through rational inquiry into empirical matters. However, he was unable to devote himself entirely to science. Hoping to attract the attention of the Fatimid caliph al-Hakam (996-1021) and to secure a more favorable position for himself, he claimed to have devised a means for regulating the flow of the Nile. Impressed, the dangerously unbalanced al-Hakim retained his services. However, it soon became clear that the project was hopeless. Al-Haytham admitted failure but, fearing for his life, feigned madness. Confined to his house, he maintained the ruse until the death of al-Hakim. In later years, he earned a living copying scientific and mathematical manuscripts.

Al-Haytham's theoretical and experimental investigations in optics surpassed all previous research in the field. In working out his radically new theories of light and vision, he articulated a comprehensive scientific methodology of the logical connections between observations, hypotheses, and verification. A distinctive feature of this methodology is its endorsement of experimentation through the manipulation of artificially created devices.

Kitab al-Manazir (*Optica thesaurus*) is al-Haytham's greatest work. At the time of its writing, theories of light were intimately connected with theories of vision based on the notion that vision required direct contact between the visual organ and objects of vision. Different accounts of how this contact occurred were promulgated and developed into opposing schools of thought in ancient Greece. Adherents of the intromission theory of vision believed objects emitted thin films or images of themselves through the intervening space to the eye. Exponents of the extromission theory believed the eye emitted an invisible fire that "touched" objects of vision to reveal their colors and shape.

In the *Kitab,* al-Haytham argued against the extromission view on the grounds that a material effluence flowing from the eye could not possibly fill the heavens fast enough to make vision possible. An insightful objection to traditional intromission theories had previously been provided by al-Kindi (801?-866?). Al-Kindi showed that if each point on a surface of an object radiates light in all directions, then each point on the eye will be stimulated by more than one point in the visual field, resulting in total confusion.

Al-Haytham's solution was a modified intromission theory. He formulated a comprehensive ray theory of light that allowed him to establish a one-to-one correspondence between each point on the eye's surface and points in the visual field. Specifically, he showed that only one ray was perpendicular to any given point on the eye's surface. All other rays would be refracted and thus weakened to point where they failed to stimulate the eye's visual power.

Al-Haytham also worked out the geometry of image formation in spherical and parabolic mirrors, observed that the angles of incidence and refraction do not remain constant, and conducted the first experiments on dispersion of light into its constituent colors. His reputation in mathematics derives from his sophisticated geometric analysis of mirrors, especially the "Alhazen problem," which involves finding the point of reflection for any two points opposite a mirror's surface.

In astronomy, al-Haytham noted that the Ptolemaic system failed to preserve uniform circular motion about Earth's center. Furthermore, he rejected the use of mathematical models whose motions could not be realized by physical bodies.

Al-Haytham correctly explained the increase in the apparent size of the Sun and Moon near the horizon in terms of atmospheric refraction. He also considered the atmosphere's density, developing a relationship between it and height. This allowed him to make an estimate of the atmosphere's height.

STEPHEN D. NORTON

Abu 'Ali al-Husayn ibn Abdallah ibn Sina
980-1037

Persian Physician, Philosopher, Mathematician, and Astronomer

Ibn Sina, known in the West as Avicenna, was the medieval Islamic world's most important philosopher-scientist. His unique codification of traditional learning into an Aristotelian framework exerted a strong influence on scholasticism in the West. His *al-Qanun fi al-Tibb* (Canon of medicine) remained the standard medical text until the seventeenth century, while his *Kitab al-Shifa* (Book of healing) synthesized logic, physics, mathematics, and metaphysics.

Ibn Sina was born in 980 in Afshana, near Bukhara in central Asia (now Uzbekistan). A child prodigy, he was able to recite the Koran by age 10, and by 16 had sufficiently mastered contemporary medical knowledge so that he was able to practice medicine. He was later a jurist at Korkanj, an administrator at Rayy, and both physician and vizier to the Prince Shams al-Dawlah of Hamadan (1015-22). After Shams al-Dawlah died, Ibn Sina was imprisoned. He was released when Ala al-Dawla of Isfahan temporarily occupied the city. Ibn Sina spent his remaining years as Ala al-Dawla's physician. He died of a mysterious illness in 1037 at Hamadan while on campaign with his patron.

Ibn Sina began *al-Qanun fi'l-Tibb* while at Rayy. Its five books provide a systematic overview and synthesis of all medical knowledge, as supplemented by Ibn Sina's own research and ideas on scientific methodology. He formulated the rules of agreement, difference, and concomitant variation as well as rules for isolating causes and analyzing quantitative effects. The quality and completeness of the *Qanun* was such that it superseded all other Muslim medical treatises. Translated by Gerard of Cremona (1114-1187) during the twelfth century, the *Qanun* acquired almost undisputed authority during the Middle Ages.

Ibn Sina's *Kitab al-Shifa* (1021-23) was a composite work of philosophy and science. In this volume he synthesized Aristotelian and Neoplatonic thought with Muslim theology. The *Kitab* also contains many original scientific ideas. He proposed a corpuscular theory of light, which implied a finite speed for light. He distinguished between different forms of heat and mechanical energy and contributed to the development of the concepts of force, infinity, and the vacuum. He also investigated the relationship between time and motion, concluding they must be interrelated since time can have no meaning in a world devoid of motion.

Kitib al-ma'adin, appended to a medieval translation of Aristotle's *Meteorologica,* appears to be an alternate version of parts of the *Kitab.* It contains Ibn Sina's views on the formation of stones and mountains and is the main source for medieval mineralogy and geology.

Based on his observation that certain springs could petrify objects, he postulated the existence of a mineralizing or petrifying virtue within the Earth. He believed mountains were formed through sedimentation and accretion of petrified clays as the seas retreated from land, and valleys were created by the erosive action of water. He also hinted at a cyclical process of mountain formation and decay. Ibn Sina's division of minerals into salts, sulfurs, metals, and stones was retained until the end of the eighteenth century.

Ibn Sina was well versed in alchemy. He subscribed to Jabir ibn Hayyan's (721?-815?) theory that all substances were derived from sulfur (idealized principle of combustibility) and mercury (idealized principle of metallic properties). However, he actively sought to separate medicine from alchemy's less tenable claims; and contrary to his contemporaries, he did not believe it possible to transmute one substance into another.

In the realm of astronomy, Ibn Sina made observations, invented a device similar to a vernier scale to improve the precision of his instruments, and edited the *Almagest,* adding figures to illustrate parallax and developing Ptolemy's (second century A.D.) geometrical methods. He also made original contributions to the mathematical analysis of music.

STEPHEN D. NORTON

Jabir ibn Hayyan (Geber)

c. 721-c. 815

Arab Alchemist and Physician

Jabir ibn Hayyan, often known as Geber, is sometimes confused with a fourteenth-century Spanish mystic who also called himself Geber. In fact the latter deliberately took on the name of his distinguished predecessor, and thus is typically known as "the false Geber." As for the true Jabir, he is widely credited as the Father of Chemistry, the first alchemist to take his studies beyond superstition and into the realm of pure science. Among his many practical discoveries were arsenic, sulphur, and mercury.

Born Abu Musa Jabir ibn Hayyan, Jabir practiced alchemy and medicine professionally in the town of Kufa, now in Iraq, beginning around 776. Little else is known of his biography, except the fact that at one point he worked under the patronage of a vizier from the Barmakid dynasty, and that he was in Kufa when he died. Far more is known concerning Jabir's work, including the fact that he produced some 100 writings, of which 22 were on the subjects of chemistry and alchemy.

By recognizing that it was necessary to use specific amounts of a given substance to produce a particular reaction, and by emphasizing experimentation, observation, and reproducible methods, Jabir laid the groundwork for chemistry as a science. He identified three basic types of chemical substances: spirits such as arsenic and ammonium chloride, which vaporize with heat; metals; and compounds that may be crushed into powder. Today this distinction between volatile substances, metals, and nonmetals remains in use. In addition to his discoveries of specific chemicals—which include nitric, hydrochloric, citric, and tartaric acids—Jabir contributed to knowledge concerning distillation, sublimation, crystallization, calcination, and evaporation.

He also discovered the fact that heating a metal adds to its weight, and his experiments with the darkening effects of light on silver nitrate provided a forerunner for the idea of photographic negative images. Among the instruments developed by Jabir was the alembic, a vessel used in distillation. He also was the first to use manganese dioxide for making glass, and developed aqua regia as a means of dissolving gold. In addition, he created a number of practical applications for chemistry in areas such as steelmaking, rust-prevention, gold lettering, cloth-dyeing, leather-tanning, and waterproofing.

Finally, evidence of Jabir's lasting impact can be discerned from the many Arabic terms—most notably, *alkali*—that made their way into European languages through his writings. Of Jabir's many books, among the most influential in the West have been *Kitab al-Kimya,* translated in 1144 by Robert of Chester as "The Book of the Composition of Alchemy," and *Kitab al'-Sab'een,* which Gerard of Cremona (1114-1187) translated. Numerous other translations appeared in Latin, and during later centuries works of the false Geber were mixed in with those of the true Jabir.

JUDSON KNIGHT

Jordanus Nemorarius

d. c. 1260

European Mathematician

Jordanus Nemorarius was the first mathematician to correctly formulate the law of the inclined plane. His writings on geometry were important for explorers who relied on the astrolabe for navigation. Furthermore, Jordanus used letters in place of numbers in his books on mathematics, and was able to articulate general algebraic theorems in this manner. However, his system of algebraic notation was only a distant antecedent to the algebra that is used today. Thus, while Jordanus's work was influential, his system of notation was unused by later mathematicians.

The inspired, but rudimentary, forms expressed in Jordanus's writings are significant because of later developments they anticipated. For instance, since Jordanus substituted letters for numbers, he may be considered to have anticipated algebra as it was subsequently perfected by figures such as René Descartes (1596-1650) in the seventeenth century. However, Jordanus's work did not combine these letters with operational symbols.

Likewise, in a manner characteristic of medieval mathematicians and scientists, Jordanus maintained strong connections between algebra and geometry. To this extent, algebraic equations and formulas were derived from the interpretation of geometric formulas or phenomena. For example, Jordanus's significance as the first to correctly formulate the law of the inclined plane hints at the connections between geometry and algebra in his work. Furthermore, his use of letters to denote the magnitudes of stars in works on mathematical astronomy, such as *Tractatus de Sphaera* and *Planisphaerium,* suggest that his observation of the natural world influenced his speculations into mathematics.

However, for medieval scientists and explorers, these connections were of essential importance. Jordanus's work with the inclined plane was particularly significant for exploration. The astrolabe was the most important astronomical instrument of the era and was used to solve complex problems of spherical trigonometry. Jordanus's work with stereographic projections and his articulation of the relations that exist between circles and inclined planes aided explorers and mathematicians in refining and understanding the measurements taken by the astrolabe.

Though Jordanus's texts were of vital importance for medieval science, almost nothing is known of his life. Along with Leonardo Fibonacci (1170?-1240?), Jordanus is regarded as one of the dominant scientific figures of the medieval era. However, his identity itself remains a mystery.

In Latin, the phrase de Nemore (from which Nemorarius is derived) means "of the forest" or may even be used to refer to a forester. However, the same phrase could also refer to someone who comes from a specific area. In the thirteenth century there were many towns and villages named Nemoris or Nemus. Such place-names could be found all across Europe in countries such as Italy, Spain, Belgium, Germany, and France. Evidence based on manuscripts of Jordanus's work suggests to some investigators that he was Italian. Others have argued that Jordanus was German, and identify Jordanus de Nemore with Jordanus de Saxony, an important Dominican friar who was active in the early thirteenth century.

The main problem with this association results from the amount of work produced by both Jordanus de Nemore and Jordanus de Saxony. It seems impossible for one man to have produced the work of two fundamentally different men. Likewise, people did not live very long in the medieval era. Jordanus de Saxony, for example, died while in his fifties. Such facts make the connection between these two men less likely.

Futhermore, the life of Jordanus de Saxony is relatively well documented and does not support the view that he was Jordanus de Nemore. Jordanus de Saxony was German, studied theology in Paris, and became a successful leader of the Dominican order. His training suggests an aptitude in theology and the liberal arts, but does not indicate that he could have exhibited the mastery of mathematics and science found in the works attributed to Jordanus de Nemore.

DEAN SWINFORD

Abu Yusuf Ya Qub ibn Ishaq al-Sabbah al-Kindi (Alkindus)

c. 801-c. 866

Arab Philosopher, Astronomer, Mathematician, Physician, and Geographer

Al-Kindi, known in the West as Alkindus, was one of the Islamic world's most important and prolific scholars. The scope of his work was encyclopedic, encompassing alchemy, astronomy, philosophy, mathematics, medicine, geography, music, pharmacology, and more. Al-Kindi's ideas were highly influential in the Latin West during the Middle Ages due to a number of translations by Gerard of Cremona (1114?-1187).

Al-Kindi was born around 801 to a noble family of the Kinda tribe of Yemen. He was the first important Islamic philosopher of Arabic (Bedouin) origin and is commonly referred to as the first Arab philosopher. He was educated at the important intellectual center of Kufa (in modern Iraq) and later at Baghdad, where he attracted the attention of Caliph al-Ma'mun (786-833). Al-Ma'mun made him a member of his scientific academy, called the House of Wisdom (Bayt al-Hilkmah). After the Caliph's death, his successor, al-Mu'tasim, selected al-Kindi to tutor his son. Al-Kindi fell out of favor with al-Mu'tasim's successors and spent the last years of his life in relative isolation. He died sometime during the reign of al-M'utamid.

As a member of the House of Wisdom, al-Kindi played an important role in the preservation of Hellenistic science. He redacted many translations of Greek treatises, and his summaries and commentaries established the Neoplatonic interpretation of Aristotle (384-322 B.C.) that dominated Arabic Aristotelianism until the time of Ibn Rushd, or Averroës (1126-1198). Much of al-Kindi's effort was devoted to restoring incomplete works and gaps in the body of inherited scientific literature. Additionally, he sought to augment, extend, and correct this knowledge when necessary.

De aspectibus was al-Kindi's primary optical treatise. Working within the Greek optical tradition, he leveled a devastating critique against the intromission theory (of forms), which asserted that vision is the result of objects emitting thin films or images of themselves through the intervening space to the eye. (The rival position was the extromission theory, which argued that the eye emits an invisible fire that "touches" objects of vision to reveal their colors and shape.) He demonstrated that the intromission theory was incompatible with the laws of perspective. If the theory were true, then a circle viewed on edge would be perceived in all its circularity; but experience shows that it is perceived as a line. Al-Kindi also argued that objects do not emit radiation from their surfaces as a whole. Rather, each point on a surface radiates light in all directions. Though implicit in early theories, al-Kindi was the first to state this principle explicitly.

Al-Kindi wrote two treatises on mineralogy, *Risala fi anwa al-jawahir al-thaminah wa ghayriha* (Treatise on various types of precious stones and other kinds of stones) and *Risalah fi anwa al-hijarah wa'l-jawahir* (Treatise on various types of stones and jewels). These works were the first of their kind in Arabic. He also produced the first Arabic book on metallurgy, *Risalah fi anwa al-suyuf al-hadid* (Treatise on various kinds of steel swords). In the study of acoustics, he developed a means for determining pitch and showed that the different notes combining to produce harmony have a specific pitch. Furthermore, he argued that sound is caused by air waves striking the eardrum. His musical treatises were among the best such works in Arabic.

An accomplished physician, al-Kindi naturally delved into pharmacology. According to Galen (130-c. 200), each of the four qualities of medicine (heat, cold, dryness, and humidity) could assume, at different times, one of four degrees of intensity. Al-Kindi attempted to quantify these intensities. His research on the links between dosage variations and their qualitative effects led him to create a systematic quantitative system for determining the composition of medical preparations.

STEPHEN D. NORTON

Ramon Llull

c. 1232-1315

Spanish Alchemist and Scholar

Ramon Llull, also known as Raymond Lully, was a quintessential medieval figure: passionate in faith and love, eager to tilt at windmills, a believer in alchemy and its attendant mysticism. Yet from the landscape of Llull's mind, shadowed as it was by superstition and extra-scientific lore, emerged the conceptual prototype for the most modern of all machines. Nearly seven centuries before Alan Mathison Turing (1912-1954) proposed his "Turing machine," helping to usher in the computer age, Llull suggested the idea of a machine that could generate objective truths.

Born on the island of Majorca, Llull was the son of a Spanish knight who had received an estate from John I of Aragon. The teenaged Llull was given the title "Seneschal of the Isles," but he soon fell into disrepute for his licentious behavior. Among the many women he romanced was Eleonora de Castello, who was married, and when she was stricken with cancer he came to believe that this was a judgment from God. He therefore set aside his old ways and took holy orders, though it appears he did not become a priest; rather, he remained a layman, with a wife and children.

As ardent in the church as he had been in the pursuit of Majorca's ladies, Llull soon conceived a plan of going to North Africa as a missionary to convert the Muslims. He left Genoa for Tunis in 1291. This was also the year when the last crusader stronghold at Acre in Palestine fell, marking an end to the numbered crusades in the Holy Land. Crusading fervor had been dying for a long time, however, and this perhaps explains why the pope had refused to support Llull's mission, despite entreaties from the Spanish would-be missionary.

Undaunted, Llull went on his own to Tunis, but was forced to leave, and spent time in Paris, Naples, and Pisa, preaching to raise support for his next missionary trip. In Algiers in 1308 he encountered more success, winning many converts, but this so outraged the city's Muslim majority that they ran him out of town. He fled to Tunis, where he was recognized from his previous visit and promptly thrown in prison. He spent some time in prison, continuing to preach the gospel, before a group of Genoese merchants obtained his release. Upon his return to Europe, Llull again went to the pope for help, and was again refused. (By then the papacy was far more concerned with intra-European affairs than with converting the Muslims: in 1309, the papal seat had been moved from Rome to Avignon in France, sparking a conflict that would nearly tear the Catholic Church apart in the centuries that followed.)

Once again failing to obtain papal support, Llull rested for a time in Majorca before journeying to Tunis. He had only begun to preach there when an angry mob attacked and beat him, leaving him for dead on a beach. Again a group of Genoese sailors rescued him, and put him on board a ship bound for Majorca. His wounds failed to heal, however, and he died within sight of his home on June 30, 1315.

In his lifetime, Llull gained a legendary reputation as an alchemist. It was reputed, for instance, that he had created a large sum of gold for the king of England. He also left behind numerous works on alchemy and other shadowy scientific or pseudo-scientific pursuits, among them *Alchimia magic naturalis, De aquis super accurtationes, De conservatione vitoe,* and *Ars magna*.

The last of these contained a set of discussions significant to modern-day computer science. As a devoted cabalist, one who studied the Jewish scriptures with the idea that the writings hid a deeper meaning encoded in the letters of the text itself, Llull was fascinated with developing a mechanism for discerning hidden knowledge. In the future it might be possible, he suggested, to construct a machine that would generate ideas and then prove or disprove them. The *Ars magna* contains an illustration of a wheel designed to be rotated as a means of generating and testing new concepts, and Llull even created a rotating set of concentric rings in an attempt to make his truth-generating machine a physical reality.

He developed this astonishingly forward-looking concept in service to a purpose that fueled much of his career. If it were possible to mechanize the process of objective truth-seeking, Llull believed, then all observers would be forced to accept the conclusions generated by the machine—conclusions which, he was confident, would prove the superiority of the Christian God over Allah. Among Llull's many admirers in subsequent centuries was the mathematician Gottfried Wilhelm von Leibniz (1646-1716).

JUDSON KNIGHT

Abu al-'Abbas Abdallah al-Ma'mun ibn al-Rashid

786-833

Persian Caliph

Al-Ma'mun was the seventh Abbasid caliph and a great patron of the sciences in the Islamic world. He established an influential scientific academy in Baghdad where Arab scholars made important contributions to mathematics, astronomy, and others fields. Under Al-Ma'mun's sponsorship, the translation of Greek and Hellenistic scientific texts into Arabic reached its peak.

Al-Ma'mun was born in Baghdad in 786 to the celebrated Caliph Harun ar-Rashid and an Iranian concubine. Al-Ma'mun's younger half-brother, al-Amin, was born to one of ar-Rashid's Arabic wives. Ar-Rashid selected al-Amin as his successor to the caliphate in Baghdad; but Al-Ma'mun was to have suzerainty over the eastern provinces, wielding his power from Merv in Khorasan (now Turkmenistan). However, at the death of ar-Rashid in 809, Al-Ma'mun rejected al-Amin's right of succession and a merciless civil war ensued. Al-Ma'mun besieged his brother in Baghdad in April 812. The city fell in September of the following year, and al-Amin was killed.

At the time al-Ma'mun assumed control of the Abbasid empire, Islam was deeply divided over who was Mohammed's rightful heir. Sunnites accepted the Abbasid caliphs, while Shiites championed the descendants of Ali, Mohammed's cousin and brother-in-law. Al-Ma'mun sought to reconcile these groups by designating 'Ali ar-Rida, a descendant of Ali, as his heir. This failed to mollify Shiite extremists and outraged Sunnite partisans. Rebellion soon erupted in Baghdad. Al-Ma'mun quickly quelled the uprising, but 'Ali ar-Rida mysteriously died in August 818, most likely poisoned by al-Ma'mun.

After restoring order, al-Ma'mun abandoned his policy of reconciliation. Instead, he sought to promote acceptance of a more flexible caliphate by imposing Mu'tazili doctrine on his subjects. Supporters of Mu'tazilism emphasized rational methods of inquiry and borrowed freely from Greek and Hellenistic philosophers. In keeping with the spirit of this movement, al-Ma'mun established a scientific academy at Baghdad called the House of Wisdom (Bayt al-Hilkmah).

The academy's library was the most ambitious institution of its kind since the foundation of the great library at Alexandria. Al-Ma'mun stocked it with every available book and even sent an expedition to Byzantium to acquire additional manuscripts. He encouraged the translation of philosophical and scientific works from Greek and Syriac into Arabic.

Hunayn ibn Ishaq al-Ibadi (808-873) was the director of the House of Wisdom. He primarily translated medical texts, especially those of Hippocrates (460?-370? B.C.) and Galen (130-200?), while his son translated Euclid's (330?-260? B.C.) *Elements* and Ptolemy's (second century A.D.) *Almagest.* Thabit ibn Qurra (836-901) was the principal translator of mathematical texts.

The academy's observatory attracted Islam's greatest astronomers. Al-Khwarizmi (780?-850?) worked there and produced an important set of astronomical tables as well as his celebrated work on algebra. Al-Farghani (800?-870?) also worked there. His *Elements of Astronomy* did much to spread the more elementary parts of Ptolemy's work.

A major impetus for the academy's astronomical work came from religious observances. Various problems in mathematical astronomy, specifically spherical geometry, were addressed in attempting to accurately determine the *qible,* the location or direction of Mecca, which was necessary for Muslim prayer. Al-Ma'mun also commissioned his astronomers to determine the Earth's size. After measuring a degree of the meridian in the plain of Palmyra and another location, it was determined that the Earth's diameter was 6,500 mi (10,461 km), approximately 18% too small. Al-Ma'mun's astronomers also redeterimined the value for the inclination of the ecliptic. A large map of the world was also produced for al-Ma'mun using a crude cylindrical projection.

Al-Ma'mun died at Tarsus in 833 while campaigning against the Byzantines. Though his Mu'tazili reform served to undermine the authority of the Abbasid caliphs, the translations, commentaries, and original research produced by his House of Wisdom contributed greatly to the development of science.

STEPHEN D. NORTON

Nicholas of Cusa

1400-1464

German Philosopher, Mathematician, Astronomer, and Futurist

Nicholas Cryfts (or Krebs), known as Nicholas of Cusa (the German city Cues or Kues, his birthplace), was one of the first great polymathic (learned in many areas) minds of the early Renaissance, and one of the first "Renaissance men" with all the spirit implied in the title. Cleric, statesman, philosopher of seminal humanism, mathematician, astronomer, and futurist, Nicholas's early schooling included the universities of Heidelberg, Padua, and Bologna (1416-1423). Although he studied law for a short time, he turned to the clerical life with a move to theology at the University of Cologne (1425). Then followed diplomatic and administrative duties on behalf of the church, which led to his being honored with the designation cardinal priest (1449) by Pope Nicholas V, although he declined at first.

Nicholas of Cusa had begun formulating a theory of knowledge based on his premise of the incompleteness of humanity's knowledge of the universe. In regard to this, and having both the nominalist's and mystic's distrust of realism, he rejected the entrenched late medieval scholasticism with its dependence on Aristotelian method as a proper baseline. Instead he defined three stages to knowledge: phantasy (fantasy, meaning of the senses), reason (abstraction and discursive knowledge), and intellect (what he called mystical or ultimate knowledge found in the only perfect reality, that is, from God).

By 1439, Nicholas completed the first of his twelve philosophical/scientific treatises in which he underlined the limits of human understanding of science (defined generally as knowledge) and also the means to surpass those limits. This was the famous *De docta ignorantia* (On learned ignorance) wherein the knowledge of intellect can reconcile the differences in the states of finite and infinite, called "the coincidence of opposites." He used the squaring of the circle as an example: a square circumscribed or bounded by a circle will, if the number of sides are increased toward infinity, approach the shape of the circle. In other words, a line and a segment of a circle almost become the same (coincidence of opposites)—but not quite. The latter provided a philosophical and metaphysical aspect to the example as well. Through intellect, humanity can approach perfect wisdom but can never achieve it. The obvious mathematical parallel to the example provides seminal concepts in the study of infinity and infinitesimal theory.

This treatise also contained his early ideas on the cosmos, particularly his intuitive idea that logic was better served with the earth revolving around the sun and not being the center of the universe. This idea was probably based on Paris School physical theorist Nicole Oresme (c. 1320-1382). He also decided that the stars were other suns (although, he also defined the earth as a star) and that other habitable worlds like earth orbited them. He anticipated the question of the infinity of the universe, calling it unbounded but perhaps not spatially infinite. Another of his treatises (1436) foretold the need for calendar reform that came to pass with the Gregorian calendar reform of 1582. About 1444 he began more serious instrumental astronomical study. He improved on the earlier medieval Alphonsine Tables with a more practical method of finding positions of the sun, moon, and planets. A small scientific treatise *Idiota de staticis experimentis* (Static experiments) is important in physical statics and meteorology with regard to Nicholas's observations and experiments dealing with objects absorbing moisture and gaining weight and using that as a measure of atmospheric humidity.

Nicholas's intellectual pursuits were tempered by his duties as a papal legate. In 1460 he ran afoul of the unscrupulous Duke Sigmund of Austria and the Tyrol with regard to church authority. He was imprisoned by the Duke and ill-treated to the point that he never fully recovered. On his way to Pope Pius II in 1464, he died at Todi in Italy, but he left behind the gifts of a hospital (at Cues), his extensive library, and his scholarly achievements, all of which are still in use to this day.

WILLIAM J. MCPEAK

William Ockham
c. 1285- c. 1349
English Philosopher, Theologian, and Political Writer

William Ockham, known as William of Ockham (or Occam), had a significant effect on the decline of medieval Scholasticism, the separation of church and state, and the eventual rise of scientific thinking. Although his writing and teaching led to his excommunication, the importance of his thought was later recognized and he is regarded as a major philosopher of the Church. He is best known for his use of the Law of Economy, known as Ockham's Razor.

Ockham was trained in logic and, at an early age, became a member of the Franciscan Order. He studied at Merton College of Oxford University and taught there from 1309-1319. His ideas were controversial, and he left without his master's degree. He taught in Paris from 1320-23 and published a number of writings opposing the Church's involvement in secular political activities, thereby angering the Papal Court, which was in residence in Avignon, France, at that time. He was also denounced by the former Chancellor of Oxford for heretical teachings. In 1324, he was summoned to Avignon by Pope John XXII for investigation of heresy and remained under house arrest there until he escaped to Germany and the court of Holy Roman Emperor Louis IV in 1328. He continued to support Louis's opposition to the Pope's involvement in the secular affairs of the empire, even accusing the Pope of heresy, and was excommunicated. He died in Germany, apparently of the plague that was sweeping Europe.

Ockham advocated a major reform in medieval philosophy. He opposed much of the Aristotelian system of the Scholastics on the basis that it was excessively deterministic and did not allow enough freedom to either God or humans. Although he had great personal faith in God and supported the Church's rule in all matters of the spirit, he also had faith in the human use of logic. In opposition to the prevailing philosophical and religious thought of the day—principally the Scholastic schools of Thomas Aquinas and John Scotus—Ockham showed by the use of logic that many basic religious beliefs, even the existence of God, could not be proved by philosophical reasoning. He contended that these religious beliefs could be based only on divine revelation (in which he firmly believed). He also held that the human mind could never be certain of many, if not most, philosophical propositions, but must rely on logical arguments of the probability of their truth. He is thus called a *skeptic* and a *probabilist*. He is also called a *nominalist* or *terminist* because he denied that the way we know something (i.e. the forms of knowledge) coincide exactly with the actual things (i.e the forms of being).

Ockham not only sought to reform Scholasticism through his philosophical ideas, he also believed that Scholasticism had become much too complex with its convoluted subtle arguments and needed to be simplified. He used as a basic principle the law of economy or parsimony: *Pluralitas non est ponenda sine necessitate*, or, it is better, in explaining something, to use as few assumptions as possible. This law had been stated earlier by Durand de Saint-Pourcain, but because of Ockham's extensive use of it as a basic principle underlying all phenomena, it is now known as Ockham's Razor. Since Ockham's day, this rule has become one of the fundamental principles of science. Whenever there are several competing explanations or theories, the simplest one, the one that requires the fewest assumptions, is selected as the correct choice.

In addition to his Razor, Ockham's skepticism, his separation of religious and secular authority, and his probabilistic and nominalistic philosophy have significantly influenced the eventual development of modern science.

J. WILLIAM MONCRIEF

Petrus (Peter) Peregrinus de Maricourt

fl. 1269

French Physicist and Engineer

Petrus Peregrinus de Maricourt conducted the first systematic experiments on magnetism and invented improved nautical compasses. He was one of the few medieval scientists to have conducted experimental inquires.

Little is known of Peregrinus's life. His real name was Petrus de Maharncuria (Pierre de Maricourt), indicating he was from Méharicourt in Picardy, France. It appears he was of noble birth. The only event of his life that can be dated with certainty is the completion of *Epistola de magnete* (Letter on the magnet), which describes his experiments with magnets. He signed this document, "Completed in camp, at the siege of Lucera, in the year of our Lord 1269." Based on this it has been suggested that Peregrinus was an engineer in the army of Charles of Anjou, King of Sicily, who at the time was directing the siege of Lucera in southern Italy. Since the Church officially declared the assault a crusade, and any who participated in a crusade could be awarded the honorific "Peregrinus" (the Pilgrim), it is likely that he earned his cognomen as a result of services rendered during the siege.

Peregrinus's *Epistola* is the earliest extant experimental treatise on magnetism. Though the north-south orientation of magnets had been known and used since at least the eleventh century in China and twelfth century in Europe, Peregrinus's treatise provides the earliest extant account of magnetic polarity. He also described various methods for determining the poles of magnets. This in turn allowed him to outline the laws of magnetic attraction and repulsion.

One method for determining the poles that Peregrinus described requires placing an iron needle on a spherically shaped magnet. After the needle aligns itself, one draws a line along its length, dividing the sphere in half. Repeating this process with the needle at different positions, one finds that the lines converge at two points opposite each other on the sphere. Since the network of lines resembles Earth's lines of longitude, which pass through the North and South Poles, the points of convergence on the magnet are called the "poles" of the magnet. Peregrinus appears to have been the first to apply the term *polus* to these points. Furthermore, he stated that when a magnet is floated in a vessel of water its north pole aligns with the north celestial pole while its south pole aligns with the south celestial pole.

Peregrinus described the mutual attraction of opposite poles when brought together and the repulsion of like poles. He further noted that when a magnet is broken in two the parts function as separate magnets with their own north and south poles. When the pieces are rejoined, the resulting magnet behaves like the original. He also observed that strong magnets can reverse the polarity of weaker ones.

The *Epistola* was intended as part of a larger work on instruments. Accordingly, the second part deals with the construction of devices exploiting magnetic effects. Along with an ill-conceived perpetual motion machine, Peregrinus described two improved compasses. His wet compass involved enclosing an oval magnet in a wooden case and floating it on water in a circular container. Markings were placed along the container rim, and a rule with perpendicular sights was attached to the cover. This allowed mariners to determine the direction of their ships as well as the azimuth of the Sun, Moon, and stars. His dry compass consisted of a magnetized needle on a pivot within a circular jar. As with the wet compass, the transparent cover for this instrument was marked and provided with a perpendicular rule for sighting. These improvements greatly increased the utility of the compass for navigation.

STEPHEN D. NORTON

Michael Constantine Psellus

1018-c. 1078

Byzantine Scholar and Statesman

Michael Constantine Psellus is known both for his role in Byzantine politics and for his wide-ranging scholarship, which influenced the later Italian Renaissance. Not only did he write the *Chronographia,* a history of the years 976-1078, but he commented on most known areas of science and mathematics, and is regarded as the last great Greek scholar of astronomy.

Psellus served in a number of key positions in the court at Constantinople, beginning as imperial secretary to Michael V in 1041. Upon the accession of Constantine IX in 1042, he became secretary of state, a position he maintained until the end of Constantine's reign in 1054. He also served as professor of philosophy at Constantinople for nine years beginning in 1045.

The year 1054 was an important one both in Byzantine history and in Psellus's life. It was then that the breach between the Greek Orthodox and Roman Catholic churches became permanent, and Psellus—who favored the split with Rome—was deeply affected by this upheaval. He left the university, adopting the name Michael in addition to his given name of Constantine Psellus.

Perhaps he intended to embark on a life of spiritual contemplation, but by 1055 Psellus was back in the midst of the Byzantine political maelstrom, having been recalled to service by the empress Theodora. He served as her prime minister, and later took up that role during the reign of Michael VII Ducas (1071-1078), who had been his student during his days at the university. As prime minister, he urged Michael to avoid any attempt at rapprochement with the Roman church.

Both before and during his years as a statesman, Psellus wrote widely. He reorganized the university system with a curriculum that placed a greater emphasis on classical antiquity, in particular Homeric myth and the philosophy of Plato (427-347 B.C.). He also wrote on philosophy, theology, grammar, and law.

Psellus also wrote commentaries on the natural sciences, medicine, and mathematics. His writings on the latter have provided scholars with valuable information concerning his ancient countrymen Pythagoras (c. 580-c. 500 B.C.) and Diophantus of Alexandria (c. 200-c. 284). He is also remembered for his work on astronomy, in which his role was primarily one of a recorder and historian rather than a theoretician.

Perhaps Psellus's greatest contribution to thought was his emphasis on Platonic idealism as opposed to the more firmly scientific worldview of Aristotle (384-322 B.C.). In the long run, this emphasis may have had a negative impact on scientific thinking, but in the short run at least he helped to refresh the pools of inquiry, which, in turn, influenced Italian thinkers of the Renaissance. These thinkers resurrected Platonic ideas after a long period of dormancy.

In 1078, a shift in dynastic politics, with the rise of the Macedonian line, spelled an end to Psellus's days in the Byzantine court. He died soon afterward.

JUDSON KNIGHT

Robert of Chester

fl. c. 1141-c. 1150

English Mathematician and Astronomer

Robert of Chester was one of the foremost medieval translators of Arabic scientific works into Latin. Through his translations he introduced Arabic algebra and alchemy to Western Europe. However, he is best known for making the first Latin translation of the *Koran,* the primary religious text of the Islamic religion.

Robert was born in the early twelfth century. As was common for scholars at the time, he was known to his contemporaries by many names, including Robert Retinensis, Robertus Ketenensis, de Ketene, Ostiensis, Astensis, Cataneus, and Robert Cestrensis. Chester apparently refers to where he was educated, as it is believed that he was originally from the town of Ketton in Rutland. Little else is known of his life. In 1136 he went to Barcelona to study with Plato of Tivoli (Plato Tibertinus), and by 1141 he was living near the Ebro studying alchemy and astrology with his close friend Hermann the Dalmatian (Hermannus Secundus). Robert was later Archdeacon of Pampelona in northern Spain (1143). He returned to England at least twice, once in 1147 and again in 1150. The date of Robert's death is not known.

In 1141 Peter the Venerable, Abbot of Cluny, found Robert and Herman engaged in their researches and convinced them to undertake the study of Islamic religion and law and to translate the Koran. Peter intended to use the fruits of their work in his diatribes against the "infidel" Muslims. Though Robert considered this a digression from his astronomical and mathematical researches, he completed the translation of the *Koran* by himself in 1143.

On February 11, 1144, Robert completed his translation of Jabir ibn Hayyan's (721?-815?) alchemical text *Kitab al-Kimya.* Retitled *Liber de Compositione Alchemiae* (The book of the composition of alchemy), it was the first Latin translation of an Arabic alchemical work to appear in Western Europe. The treatise recounts the story of the Umayyad prince Khalid ibn Yazid (died c. 704) who, according to legend, was the first Muslim to take a serious interest in alchemy. He supposedly studied with the Christian alchemist Morienus, who was a disciple of the well-known seventh-century Alexandrian alchemist Stephanos.

Robert introduced Western Europe to algebra with his Latin translation of Muhammad ibn Musa al-Khwarizmi's (800?-847?) *al-Jabr wa'l-Muqabalah.* Al-Khwarizmi's *Algebra* was the primary influence on the development of European algebra, determining its rhetorical form and much of its specialized vocabulary. The modern term *algorithm* is derived from the translation's first line, which begins *"Dicit Algoritmi."* The translation also contains the first use of *sinus*—from which the term "sine" is derived—in its modern trigonometric sense. Robert's translation appeared in 1145 (Gerard of Cremona [1114-1187] produced another translation in 1150). Because Abraham Bar Hiyya ha-Nasi's (fl. before 1136) *Hibbur ha-meshihah we-ha-tishboret*—the earliest exposition of algebra composed in Europe (written in Hebrew)—also appeared in 1145, this is generally considered the birth year of European algebra.

Robert also translated a number of astronomical works, including Ptolemy's (second century A.D.) *De compositione astrolabii* (On the composition of the astrolabe, 1147), al-Kindi's (801?-873?) *De judiciis astrorum* and al-Zarqali's (d. 1100) *Canones.* The latter work was one of several sets of astronomical tables Robert translated and updated. On a return trip to London in 1147, Robert made observations and readjusted these tables to the meridian of that city for the year 1150. He had previously made similar adjustments for Toledo based on the tables of Rabbi ben Ezra. Extracts from his various tables appear in *De diuersitate annorum ex Roberto Cestrensi super tabulas toletnas.* This volume also indicates that Robert had constructed a sexagesimal multiplication table containing all the products from 1 × 1 to 60 × 60.

STEPHEN D. NORTON

St. Thomas Aquinas

c. 1225-1274

Italian Philosopher

The writings of Thomas Aquinas represent the pinnacle of medieval Scholasticism, a school of thought that attempted to bring together Christian faith, classical learning, and knowledge of the world. In his life's work, the *Summa theologica,* Thomas Aquinas addressed new ideas that seemed to threaten the stability of Christian faith, among them the rising attitude of scientific inquiry among European scholars. The *Summa* attempted to delineate the realms of reason and faith, to affect an understanding between them, and to place God as the "Prime Mover" governing all realms. Thomas's ideas, initially scorned by church authorities, soon became received doctrine, and had a profound influence on European thinking for many centuries.

Born in the Italian town of Aquino—hence the name "Aquinas"—Thomas was the youngest son of a Norman count. At the age of five, he was placed in the Benedictine monastery at Monte Cassino, but an armed conflict between the papacy and the Holy Roman Empire created dangers there, and thus by the time he was 14 he was forced to move to Naples. There he enrolled in the school that would become that city's university.

In 1244, Thomas joined the Dominican monastic rule against the protests of his family, and in 1245 began studying with the Dominicans at the University of Paris. There he came under the influence of Albertus Magnus (c. 1200-1280). By this time Thomas had been given the nickname "the dumb ox" because he was tall, large, and slow in movement. But his famous teacher, impressed by the young man's abilities, made the remarkable prediction that "the bellow of this ox will be heard around the world."

Thomas went on to study with Albert at the latter's home in Cologne from 1248 to 1252, then returned to Paris to earn his degree in theology. In 1256, he became a theology instructor at the university. During this period, he produced *Summa contra gentiles*, a guidebook for Dominican missionaries in Spain and North Africa who encountered the conflicting ideas presented by Muslims, Jews, and Christians outside the Catholic faith.

In 1261, Thomas moved to Rome to serve as a lecturer at the papal court. While there, he began writing the *Summa,* an undertaking that would consume the remainder of his life. Part one, completed during this time, concerned the existence and attributes of God. He then returned to Paris, just in time to become involved in a brewing controversy concerning the influence of Averroës (Ibn Rushd, 1126-1198) among European thinkers.

Initially Thomas challenged the Arab philosopher's position that reason and faith could coexist. In time, however, he adopted a similar viewpoint, maintaining that reason can aid the believer in discovering certain truths about God, and thus, reason can be used to prove the existence of God as creator of the world. This unorthodox idea, combined with his growing interest in the works of Aristotle (384-322 B.C.), exposed him to censure among church circles during this period.

By 1271 or 1272, Thomas had completed the second portion of his *Summa,* concerning questions of happiness, sin, law, and grace. Having returned to Naples in 1272 to set up a Dominican study house attached to the university, Thomas went to work on the third part of the *Summa,* this one concerning the identity of Christ and the meaning of his work. On December 6, 1273, his own work suddenly stopped, and he explained to others that everything he had done seemed meaningless. Whether he suffered a physical breakdown, experienced a spiritual insight, or simply ran out of ideas is not known.

The completed *Summa*—the greatest of the several books Aquinas produced—ran to about 2 million words, or the equivalent of about 8,000 double-spaced, typewritten sheets. Among the areas touching on science was his overall discussion of God as physical creator, or "Prime Mover," an idea derived from his studies of Aristotle. On a specific level, Thomas made the erroneous observation in the *Summa* that metals are formed by rays from the Sun, Moon, and planets, each of which governs a particular metal.

His health failing, Thomas set out in 1274 to attend a church council in France. He was struck on the head by a branch falling from a tree over the road, and may have suffered a concussion. He stopped at a castle belonging to his niece to recover, and soon afterward was taken to a monastery, where he died on March 7, 1274.

At the time of his death, Thomas's work was under question by the church, which took issue with his attempts to reconcile reason and faith. Four decades later, however, proceedings were under way to canonize him. In 1323 he was declared a saint, and in 1567 was named a Doctor of the Church, or one of the leading church fathers. Though his views would come to seem hopelessly static to the restless minds of the Renaissance, in his time Thomas was a forward-looking thinker who attempted to incorporate what were then far from orthodox ideas. Thus in many ways he opened the door for greater scientific inquiry in the last centuries of the Middle Ages.

JUDSON KNIGHT

Ulugh Beg

1394-1449

Tartar Astronomer and Mathematician

Ulugh Beg made Samarkand one of the leading cultural and intellectual centers of the world. In that city he established a *madrasa* (Islamic institution of higher learning) that emphasized astronomical studies. He also constructed an observatory that became the leading center for astronomical research in the fifteenth century. Working with his assistants, he produced the important *Zij-i Djadid Sultani* astronomical tables and star catalog.

The grandson of the Tartar conqueror Tamerlane, Ulugh Beg was born in Sultaniyya (in modern Iran), on March 22, 1394. His birth name, Muhammad Taragay, was immediately superseded by the cognomen Ulugh Beg, meaning "Great Prince." In 1409 Ulugh Beg's father, Shah Rukh, appointed him governor of Maverannakr (present-day southeastern Uzbekistan), the chief city of which was Samarkand. Though actively engaged in the economic, political, and military affairs of his territory, Ulugh Beg was more interested in scientific pursuits.

Among his many construction projects was a magnificent two-story *madrasa* in Samarkand. Another was built in Bukhara. Completed around 1420, these institutions still stand and fulfill their original function. Ulugh Beg interviewed and selected the scientists who were to teach at these schools. Over 90 scholars were on the faculty at Samarkand, where it is said Ulugh Beg himself lectured. He also determined the curriculum, whose most important subject was astronomy. To support research in this area, he decided to build an observatory.

Construction on the observatory began in 1424. Its design closely followed that of the Maragha observatory (built in 1259) in Tabriz. Built on a circular foundation more than 262.5 ft (80 m) in diameter, the Samarkand observatory's three-stories rose to a height of 108 ft (33 m). A trench 8 ft (2.5 m) wide and 36 ft (11 m) deep at its lowest point was excavated to accommodate the huge meridian arc that was to be the observatory's main instrument. When completed, the arc extended from the trench's deepest point to just below the observatory roof. Its radius was 131 ft (40 m), making it the largest astronomical instrument of the fifteenth century.

The 262.5-ft (40-m) meridian arc has been called a quadrant by some scholars, but it was most likely a Fakhri sextant. Designed primarily for solar observations, it was used for various lunar and planetary measurements as well. The observatory contained other instruments, including an armillary sphere, triquetrum, astrolabe, quadrant, and a large parallectic ruler. Observations made at Samarkand were used to establish the inclination of the ecliptic, the point of the vernal equinox, the precession of the equinoxes, and other basic astronomical constants. The Muslim astronomer Ali-Kudsi was the observatory's director.

Ulugh Beg's most important work was the *Zij-i Djadid Sultani*. It contains a theoretical section and the results of observations he and his assistants made. Calendric, planetary, and trigonometric tables are included as well as a new star catalog containing 1,018 stars. The positions of many of these stars were determined from observations made at Samarkand, while others were taken from the catalog of al-Sufi, who had apparently taken them from Ptolemy (second century A.D.). Ulugh Beg's catalog was only the second original star catalog since Hipparchus (170?-120? B.C.). The completion date for the *Zij-i Djadid Sultani* is traditionally given as 1437, but Ulugh Beg continued working on it until 1449.

After a brief reign as the ruler of Turkestan (1447-49), Ulugh Beg was assassinated on October 27, 1449, by order of his son, Abdul al-Latif. Ulugh Beg was buried as a martyr in the Gur Emir mausoleum of Tamerlane. Unfortunately, the Samarkand observatory fell to ruins by the end of the fifteenth century and was forgotten. It was only rediscovered in 1908 by V.L. Vyatkin.

STEPHEN D. NORTON

Abu Ishaq Ibrahim ibn Yahya al-Zarqali

1028-1087

Spanish Arab Astronomer

Abu Ishaq Ibrahim ibn Yahya al-Zarqali (latinized as Arzachel) was a Spanish Arab, born in 1028, who became known as the most prominent astronomer of his time. His *Toledan Tables* were responsible for invigorating the science of astronomy because it made possible the computation of planetary positions at any time based on observations. He edited these tables from his own observations in Toledo, Spain, as well as the tables of other Muslim and Jewish astronomers.

He also detailed the use of astronomical instruments. He developed a flat astrolabe, called a Safihah, which is used to find the altitude and position of stars; his description of this device was translated into Hebrew and Latin. He was the first to show clearly that the motion of the solar apogee, or aphelion, which is when the sun is furthest from earth, amounts to 12.0 seconds per year. The actual value is now known to be 11.8 seconds per year.

Al-Zarqali corrected the work of the second century Greek astronomer Ptolemy who had calculated the length of the Mediterranean Sea at 62°; al-Zarqali calculated 42° which is nearly correct.

Al-Zarqali's work was quoted a number of times in Nicolaus Copernicus's *De Revolutionibus Orbium Celestium* (On the revolution of the celestial orbs) in which he forwarded his thesis that the sun, and not the earth, was the center of our solar system. Al-Zarqali died in 1087.

MICHAEL T. YANCEY

Biographical Mentions

Abu Ma'shar
787-886

Arab astrologer, also known as Albumasar or Albumazar, who influenced Western thinking on cosmology during the Middle Ages. In his *Introductorium in astronomiam* and *De magnis coniunctionibus,* he maintained that the world was created when the seven planets were in conjunction with Aries, and predicted that it would end when the same phenomenon occurred in Pisces. Translated into Latin and later vernacular tongues, his works were widely circulated in Europe and he became the inspiration for literary depictions of astrologers by several minor authors of the early modern era.

Abu Nasr al-Farabi
c. 870-c. 950

Persian Turkistani follower of al-Kindi (c. 801-c. 866) who was a polymathic scholar (learned in many disciplines) and philosopher leaning toward Neoplatonism and interested in synthesizing that with Aristotelianism. While many of his works are lost, 117 books on various subjects survive. Al-Farabi was known to have composed many commentaries on Plato and Aristotle. One that is known for its useful fundamentals and classification of scientific study was the *Kitab al-Ihsa al-Ulum*. This book contained an early commentary on Aristotle's *Meteorologica*. Al-Farabi also contributed to the study of logic by distinguishing its two parts of idea and proof.

Abu al-Wafa
940-998

Persian mathematician and astronomer, also known as Abul Wefa, who developed the first wall quadrant, used for observing the movement of heavenly bodies. In a work written in 960, he discussed the oscillation or inequality in the motion of the Moon, a phenomenon also called variation. Contrary to the claims of some later historians, however, he did not discover this phenomenon. Abu al-Wafa, who spent much of his career working in a Baghdad observatory, is also known for his translations of ancient Greek texts and for his writings on applied mathematics.

Albert of Saxony
c. 1316-1390

German philosopher and mathematician who wrote on mechanics and geology. Around 1350, Albert presented a theory of impetus that distinguished between uniform and irregular motion. In the realm of geology, he maintained that a process of uplift on land compensates for the force of erosion exerted on it by the oceans. Presenting an early version of isostasy, he held that rock formations tend to find their natural level by sinking or rising according to their densities.

Leon Battista Alberti
1404-1472

Italian architect and writer who, in addition to his work in mathematics, wrote on geology. In about 1450 Alberti discussed the erosive effect of the atmosphere, and in so doing served as a forerunner for the modern science of hydrogeology. Though not a professional mathematician or scientist, he also laid the groundwork for the scientific study of perspective in his *Trattato della pittura* (1436).

Alfonso X of Castile
1221-1284

Spanish monarch who fostered learning in astronomy and other disciplines. Alfonso, nicknamed Sabio (the Learned), assumed the thrones of Castile, the leading state in Christian Spain, and León in 1252. He encouraged the preparation of a revised set of planetary tables that became the basis for virtually all astronomical observations in Western Europe during the next two centuries. In 1483 the Alfonsine Tables, as they were called, were printed for the first time. Alfonso was elected Holy Roman emperor in 1257, but the election was not approved by Pope Alexander IV. As a leader, he promulgated a new code of laws, which formed the basis for Spanish jurisprudence, and captured the cities of Cartagena and Cádiz from the Moors.

Alfred of Sareshel or Shareshill
fl. 1215

English medieval scholar (also known as Alfredus Anglicus) who was one of the earlier medieval European translators of Arabic commentaries on Aristotle's scientific works into Latin. These were fundamental to initial European commentaries that followed Latin translations of Aristotle's original works. Alfred provided a Latin translation and expansion of Ibn Sina's (Avicenna; 980-1037) commentary on Aristotle's fourth book of his *Meteorologica* and went on to provide the first European commentary (though of a summary nature) on the *Meteorologica* (about 1215).

Dante Alighieri
1265-1321

Italian poet who in his *Divine Comedy* (1308-21) popularized ideas on geometry and astronomy. Perhaps the most significant literary work of the later Middle Ages, the *Divine Comedy* is an allegory representing the poet's journey through Hell, Purgatory, and Paradise, with the Roman poet Virgil serving as guide through the first two locales, and the last guided by Beatrice Portinari, a woman of Dante's native Florence toward whom he cherished an ideal love. The book is a compendium of medieval thought, featuring figures and ideas from the preceding three millennia of Western civilization. In choosing to compose it in Italian rather than Latin, Dante established the foundations of Italian literature and, indeed, all modern Western vernacular literatures. Like everything else about the work, the scientific underpinnings are exceedingly complex, drawing on the cosmological ideas of Ptolemy and others. In the realm of geology, Dante used notions developed by Ristoro d'Arezzo, who wrote that mountains are created by the stars' upward pull on the Earth.

Jacob Anatoli
c. 1194-1258

French Jewish philosopher and physician who translated several Arabic works into Hebrew. In addition to numerous writings by Ibn Rushd (a.k.a. Averroës; 1126-1198), Anatoli translated the *Jawami,* or *The Elements of Astronomy,* by al-Farghani (fl. c. 860). His Hebrew translation would later serve as the basis for the translation of the *Jawami* into Latin.

Andalo di Negro
fl. 1320s

Italian astronomer and astrologer, much admired by the writer Giovanni Boccaccio (1313-1375), whose work occupied a position somewhere between science and pseudo-science. In 1323 Andalo wrote *Canones super almanach profati,* in which he discussed the astronomical tables put forth by Jacob ben Tibbon (1236-1312) in the *Luhot.* But he also promoted "astrological medicine," the belief that a physician could determine the nature and future course of an illness by studying planetary positions. He did, however, admit that the horoscope of the patient would be of little help because it would be virtually impossible to determine the exact hour of birth.

Arnau de Villanova
c. 1235-1311

A physician who studied at Montpellier and went on to become a medical master (lecturer) there, Arnau believed that observation and experience were more reliable than the untested words of ancient writers. He helped define the medical curriculum at Montpellier, translated a number of important Arabic medical texts, and wrote on medicine and astronomy. His medical services were in demand from royalty and popes. His unusual religious views often got him into trouble, but he was tolerated because of his practical medical skills.

Cecco d'Ascoli
c. 1269-1327

Italian astrologer and mathematician who taught both subjects at the University of Bologna (1322-24). Born Francesco degli Stabili, Cecco taught astrology at a number of institutions around Italy before moving to Bologna. He presented a defense of astrology after Dante Alighieri (1265-1321) attacked it in the *Divine Comedy*, and accused Dante of heresy; ironically, Cecco himself was burned at the stake as a heretic. His most important writing was an allegorical and encyclopedic poem entitled *L'acerba.*

Bartholomeus Anglicus
fl. c. 1220-c. 1250

Also called Bartholomew the Englishman, he was a professor of theology at the University of Paris who joined the Franciscan order (about 1225) and began composing the first comprehensive encyclopedia of sciences of the time. Medieval scholars defined science as knowledge, so topics included theology and philosophy along with medicine, zoology, botany, geography, mineralogy, and other earth sciences. Bartholomew's presentation of these latter reflect the influence of Aristotle's *Meteorologica*, and he followed a few earlier thinkers in defining wind as air in motion, not as Aristotle's earthly exhalation. The work was a great success, and was copied over and over and printed at least fourteen times before 1500.

Bede
c. 673-735

English historian, educator, and cleric whose *Historia Ecclesiastica Gentis Angolorum* (Ecclesiastical history of the English people), completed in 731, is the principle source for information about Anglo-Saxon life and religion. While a master of Greek, Latin, mathematics, astronomy, music, and biblical commentary, Bede was a humble monk and a dedicated teacher. His abilities and guidance turned the monastery at Jarrow in Northumbria into a major center of learning in the early medieval world, sending its teachers out across Britain and Europe to establish more schools. Bede's work and influence were so important that Alcuin of York gave Bede the title "venerable" and King Alfred ordered the translation of the *Historia* from Latin into Anglo-Saxon.

Bernard of Verdun
fl. late 1400s

French scholar who, along with Giles of Rome, helped establish the basis for scientific method. The two men, working independent of one another, both suggested that disputes concerning the competing cosmological systems of Aristotle and Ptolemy should be settled empirically, by observing evidence. Bernard, also known as Bernardus de Virduno, published an early account of the torquentum, an astronomical measuring instrument, in the *Tractus super totum astrologium.*

Bhaskara
1114-1185

Indian astronomer, mathematician, and physicist whose *Karanakutuhala* or *Calculation of Astronomical Wonders* was one of the most significant texts in medieval Hindu astronomy. In about 1150 Bhaskara—sometimes referred to as Bhaskara II to distinguish him from an ancient mathematician of the same name—made one of the first descriptions of a machine for perpetual motion.

al-Biruni
973-1048

Arab mathematician and astronomer who collaborated with Abul Wefa in determining the longitudinal difference between Baghdad and Kath (in modern Uzbekistan). A student of both Hindu and Arab astronomy, al-Biruni's writings provide an overview of advances made by both civilizations. He also wrote an astronomical encyclopedia and in other works discussed weights and measures, specific gravity, and gemology.

al-Bitruji
d. 1204

Spanish Arab scholar, also known as Alpetragius, whose writings on astronomy contradicted the ideas of Ptolemy and resurrected those of Aristotle. In 1217 Scottish scholar and mathematician Michael Scot translated al-Bitruji's work into Latin, an effort that furthered growing awareness among European scholars of the faults in Ptolemaic cosmology.

Giovanni Boccaccio
1313-1375

Italian writer, best known for the *Decameron* (1353), who discussed geology in his *Filocolo* (c. 1340). In the latter text, he wrote on the origin of fossils and maintained that the sea had once covered the Earth. The willingness of Boccaccio, an educated man for his time, to accept ancient myths is instructive regarding the medieval mind: at one time he reported that the remains of Polyphemus, the Cyclops described in Homer's *Odyssey,* had been located in a cave in Sicily. According to Boccaccio, the giant was 300 ft (91.5 m) tall.

Chu Hsi
fl. c. 1175

Chinese scholar who stated in his *Chu Tsi Shu Chieh Yao* that fossils were once living organisms. The work also contains discussion of cosmology, and Chu maintained that the universe had been created by "violent friction," which in turn caused the Earth, Sun, and other bodies to remain in motion. He accurately stated that "Should Heaven stop only for one instant, Earth must fall down."

Giacomo Dondi
1318-1389

An Italian astronomer, physician, and clockmaker, Dondi had studied medicine and went on to become physician to the Emperor Charles IV. He was professor of astronomy at the University of Padua, also teaching astrology, philosophy and logic. Dondi wrote about diet during times of plague, hot springs, and salt extraction. He is best remembered for designing and building an astronomical clock over sixteen years, the written description of it being the second oldest surviving account of a mechanical clock.

John Duns Scotus
c. 1266-1308

Scottish philosopher and theologian who helped establish the framework for the scientific method by distinguishing between causal laws and empirical generalizations. The author of numerous works, including commentaries on Aristotle, Duns Scotus outlined a philosophical theory that came to be known as Scotism, in opposition to the Thomism of Thomas Aquinas and his adherents. Among the notions put forth by Duns Scotus was the idea that matter, as matter, has some verifiable existence and reality all its own. The stubbornness and obstructionism of his followers, in the face of the changes associated with the Renaissance, led to the coining of the term "dunce" to describe someone unable or unwilling to recognize an obvious fact.

John Scotus Erigena
c. 810-c. 877

Irish theologian and philosopher who in his *De divisione naturae* (862-66) put forth the theory that Mercury, Venus, Mars, and Jupiter all orbit the Sun—an extraordinarily daring notion in his time. Erigena, who taught at the court of French king Charles II (the Bald) near Laon, also wrote a major work, *De predestinatione* (851), on questions of predestination, salvation, and free will. *De divisione,* the most important of his writings, marked an early attempt to reconcile non-religious views on Earth's origins (in this case, the Neoplatonist theory of emanationism) with Christian creationism. His work gained great influence, but in 1225 Pope Honorius II condemned it due to its tone of pantheism, or the idea that God and the natural laws of the universe constitute a single entity.

False Geber
1300s

Spanish alchemist who first prepared sulfuric acid, the most widely used inorganic chemical in the modern world. This, and other acids obtained from minerals, are much stronger than the organically-based acids known to the Greeks and Arabs, thus making possible many new chemical reactions. As was common practice at time, "false Geber" attributed his work to an earlier figure of repute—Jabir ibn Hayyan (latinized as Geber)—to gain respect. Consequently, little is known about the life of this important medieval alchemist, not even his real name.

Abu'l-Abbas Ahmad ibn Muhammad ibn Kathir al-Farghani, also known as Muhammad ibn Kathir, Ahmad ibn Muhammad ibn Kathir, and Alfraganus
c. 820-after 861

Due to some confusion over his name, a few historians suggest the lives of a father and son have been mingled. An astronomer-astrologist he also supervised the building of a canal (the Great Nilometer) at Cairo. Legend has it that he made the canal too high at one end preventing water flow. He wrote about sundials, astronomy, the astrolabe, and commentaries on Ptolemy, but most of his writings are now lost. His surviving astronomical work, *Elements*, was translated into Latin twice, and was very popular.

Kamal al-Din Farisi
c. 1260-c. 1320

Islamic mathematician and optical thinker who made important contributions to number theory and the optics of the rainbow. A pupil of the astronomer and mathematician Qutb al-Din al-Shirazi (1236-1311), he is best known for his nearly correct explanation of the optical mechanism of the rainbow, which followed from his exhaustive critical study of ibn al-Haytham's (965-1039) great *Optics* and writing his own *Revision of Optics.* He correctly noted the fundamental idea that the appearance of a rainbow was due to two refractions of light on a cloud droplet but he was not sure of the number of reflections and was incorrect in his explanation for the colors of the rainbow. Al-Farisi also made important contributions to number theory, introducing ideas concerning factorization and combinatorial methods.

Feng Tao
881-954

Chinese government minister generally credited with initiating the first printing of the Confucian Classics—equivalent to the first printing of the Bible in Europe—in 932. The result was that the foundational works of Confucius became relatively accessible, at least to the upper echelons of society, and that scholarship and knowledge spread throughout China. Feng's lasting reputation was hurt, however, by the fact that during the Five Dynasties and Ten Kingdoms period, which began with the fall of the T'ang Dynasty in 907, he served no fewer than five different imperial houses. Given the Confucian emphasis on loyalty, his opportunism made him something of a disgrace in the eyes of later scholars.

Giovanni da Fontana
fl. 1410-1449

Venetian engineer who wrote several works, including an encyclopedia of philosophy and science based on Aristotle's organization, but more importantly a work on the use of instrumentation for earth sciences. The *Metrologum de pisce cane et volacre*, loosely translated as "Methods of measuring plane surfaces, depths of water, and altitudes of air by mechanical and pyrotechnical animals aided by clocks," was an ingenious work on more accurate types of clocks used to time the ascent of rockets to determine atmospheric height and the descent and rise of submerged floats to determine water body depths. He also had futuristic ideas about flying with wings and walking on the bottom of the sea.

Franco of Liège
fl. 1000s

Flemish mathematician best known for his attempt to solve the problem of the quadrature of the circle (constructing a square with the same area as a given circle). Though the "squaring of a circle" is now known to be impossible, many had vainly attempted to solve the problem. In his *De quadratura circuli* (c. 1050), Franco recounted his efforts, which involved cutting-up a parchment circle and pasting the pieces together to form a square.

Gerard of Brussels
fl. c. 1225

This mathematician and physicist wrote *Liber de motu* between 1187-1260. This work describes various geometrical figures (lines, areas and solids) in rotation, the basis of kinematics. Gerard uses Archimedian principles, yet there is some doubt that he had read any Archimedes. A number of later writers on kinematics were influenced by his ideas, including Thomas Bradwardine and Nicole Oresme.

Hermann von Reichenau
1013-1054

German scholar who wrote a number of works on mathematics and astronomy. Among Hermann's writings were *De utilitatibus astrolabii* (On the uses of the astrolabe) and *De mensura astrolabii* (On measurement with the astrolabe), as well as several mathematical texts. His most famous book was a history of the world from the time of Christ up to the year 1054.

Hermannus Contractus
1013-1054

German scholar who wrote treatises on the astrolabe, the abacus, and the number game rithmomachia. Hermannus, son of Count Wolverad of Swabia, had from birth suffered from "contracted" or deformed limbs—hence his surname. He studied in the monastic school at Reichenau, where he later became abbot. As a teacher, Hermannus had a large student following.

Hulagu Khan
c. 1217-1265

Mongol ruler and founder of the Il-khanid dynasty who, while playing a major role in the destruction of medieval Iranian and Iraqi civilization, fostered learning through his assistance to al-Tusi and others. Grandson of Genghis Khan, Hulagu was sent westward by his brother Mangu, Genghis's successor as Great Khan. Hulagu broke the power of the Assassin sect (1256), executed the last of the Abbasid caliphs and destroyed Baghdad (1258), then, at the hands of the Mamluks in Nazareth, became the first Mongol leader to suffer a serious defeat (1260). Afterward he established the capital of his new dynasty at Maragheh, now in Azerbaijan, where he sponsored al-Tusi's creation of an outstanding observatory.

Ibn Bajja
c. 1095-c. 1138

Arab philosopher, also called Avempace, who defended Johannes Philoponus's critique of Aristotle's ideas concerning motion. Aristotle had asserted that a physical body will remain in motion only as long as force is applied, whereas Johannes maintained that a body will keep moving in the absence of friction or opposition. Ibn Bajja was among a growing number of medieval scholars, including Ibn Sina (Avicenna), who upheld Johannes against Aristotle. Johannes's view was later substantiated by Isaac Newton's laws of motion.

Ibn Muadh al-Jayyani
989-after c. 1079

Spanish Moor mathematician and astronomer who made important contributions to geometrical mathematics. Born in Córdoba, Spain, Ibn Muadh was evidently in Cairo from 1012 to 1017 and moved to Jaén in Spain where he wrote his most important works. *On ratio* was a defense of the fifth book of Euclid's *Elements* in which he introduced the concept of "number" as one of the five mathematical magnitudes essential to defining ratio. His *The Book of Unknown Arcs of Sphere* was evidently the first treatise on spherical trigonometry. In his *On the Dawn or On Twilight and the Rising of Clouds* he determined by refraction that the sun was 18° below the horizon at dawn and end of twilight, providing a means of estimating the height of the atmosphere.

Ibn Yunus
950-1009

Arab astronomer, astrologer, and mathematician whose *al-Zij al- Hakimi al-kabir* contains remarkably accurate astronomical calculations. In addition to descriptions of 40 planetary conjunctions and 30 lunar eclipses, the book contains tables for the Muslim, Coptic, Syrian, and Persian calendars. Ibn Yunus, who spent much of his career working for caliphs of the Egyptian Fatimid dynasty, was still in good health when he predicted that he would die in seven days. He completed all unfinished business, locked himself in his house, and recited the *Koran* until he died—on the day he had predicted he would.

John of Rupescinna
d. 1362

French alchemist who in his *Liber lucis* (c. 1350) wrote about the therapeutic properties of the quintessence of wine, or alcohol. The work also contained an explanation of how to build an alchemical furnace. Also known as John of Roquetaillade, John wrote a number of works and was imprisoned several times by popes due to his denunciation of clerical abuses.

Abd al-Rahman al-Khazini
fl. 1100s

Arab astronomer who compiled a set of astronomic tables for the years 1115-16, called the *Sanjaric Tables* (1115). He also wrote on the specific weights of objects and studied the principles of balance established by Archimedes.

Konrad von Megenburg
c. 1309-1374

German scholar who in his *Buch der Natur* (1349-50) stated that earthquakes are the result of winds accumulating in the earth's interior and forcing their way outward. Later printed, the book went into several editions and constitutes an excellent example of the organic concept of nature—an inheritance from Aristotle—that prevailed in much late medieval Western thought.

Kuo Shou-ching (Guo Shoujing)
1231-1316

Chinese astronomer, mathematician, and hydraulic engineer best known for preparing the Shoushi calendar, which is equivalent to the modern Gregorian calendar. Kublai Khan entrusted Kuo Shou-ching with the regulation of China's irrigation and waterways in 1271. He was then instructed to reform the calendar, which he completed by 1280. His efforts involved designing an observatory and sophisticated astronomical instruments that he used to observe the Sun, the 28 *xiu* (lunar lodges), and many unnamed stars.

Levi Ben Gershom
1288-1344

French Jewish astronomer and mathematician who in his *Sefer takunah* (1340) attempted to replace the cosmology of Ptolemy with a new, rationalized Jewish cosmology. One of the few Jewish scholars to enjoy a successful career in Christian Europe, Levi observed a lunar eclipse in 1335 and a solar eclipse in 1337. He also developed the Jacob's staff, a mechanism for measuring the angular distance between heavenly bodies, and used the camera obscura in making astronomical observations.

Ramon Llull, also known as Raymond Lully
1232-1316

This Spanish polymath and philosopher was converted by his wife from a courtly way of life to a religious one. He established colleges for the study of Arabic, Hebrew, and other languages, to oppose "the infidel." He wrote at least 292 works in various languages, including mystical works, poetry, allegorical novels, and works on theology, philosophy, arithmetic, geometry, astronomy, astrology, grammar, rhetoric, logic, law, and medicine. His attempts to unify all knowledge into a single system influenced a number of later thinkers.

Manegold of Lautenbach
c. 1030-1103

Alsatian philosopher who argued that Christian scholars should turn away from studies in physics and should not attempt to understand nature using reason alone. Manegold was almost a textbook case of the worldview that modern people associate with medieval times. In the Investiture Controversy, a church-state struggle that pitted Pope Gregory VII against Holy Roman emperor Henry IV, Manegold proved so biased in his allegiance to the pope and his disdain for the emperor—who he compared to an incompetent swineherd—that even the *Catholic Encyclopedia* refers to him as "the rude, fanatical Manegold of Lautenbach." Some modern political scholars, however, see in his defiance toward the emperor an early form of civil disobedience.

Abu Ja'far al-Mansur
709-775

Arab ruler, second caliph of the Abbasid dynasty, who fostered learning in astronomy and other disciplines. It was al-Mansur who began construction of the Abbasid capital at Baghdad, which replaced the old Umayyad center in Damascus as the new seat of power in the Islamic world. Once established in Baghdad, al-Mansur became an ardent patron of scholarship, encouraging translations of Greek scientific and mathematical classics into Arabic. In 773 he commissioned the translation of the sixth-century Hindu astronomical treatise *Siddhanta*.

Abu al-Hasan Ali al-Masudi
c. 956

A geographer and historian who traveled widely for most of his life, to see the wonders of the world and to gain direct experience through observation. He wrote at least 37 works on history, geography, jurisprudence, theology, genealogy, government and administration, but only two survive intact. He considered experience better than the authority of ancient writers, and challenged traditionalism. He saw geography as the basis of history, saying that geography helped determine the life, structure and character of a region.

William Merle
fl. 1300s

English scholar, also known as Morley or Merlee, who kept the first regular, systematic records of weather and was one of the first to attempt a scientific forecast. The Babylonians had recorded wind directions in c. 900 B.C., and in the sixth century B.C. the Greeks kept records on rainfall. Merle, however, was the first individual figure notably associated with meteorology. He maintained his weather records at Oxford, England, from 1337-44.

David al-Mukammas
fl. 900s

Arab Jewish philosopher who adapted Greek and Arab concepts, and was the first Jewish thinker to apply the ideas of Aristotle. His principal work is *Ishrun maqalat,* or the "Twenty Chapters," in which he maintained that science provides knowledge of reality and consists of two elements, theoretical and practical knowledge. At one point Mukammas converted to Christianity, but he later returned to Judaism, and in his writings he argued against both the Christian and Muslim belief systems.

Alexander Neckam
1157-1217

English theologian and poet, also known as Nequam, who produced the earliest European account of the compass. He mentioned the magnetic needle in *De utensilibus* (On instruments, c. 1187) and noted its use as an aid to navigation. In *De naturis rerum* (On the nature of things, c. 1190) he presented miscellaneous Greek and Islamic scientific facts that at the time were unknown in Western Europe. Neckam taught theology at Oxford and from 1213 was abbot at Cirencester in Gloucestershire.

Peter John Olivi
1248-1298

French scholar and philosopher who was an early proponent of the theory of impetus. For centuries Aristotle's claim that a moving object requires continual application of force to keep it in motion had gone unchallenged. According to this explanation, wind and air themselves act as a form of propellant. The sixth-century Greek scholar Johannes Philoponus had posed the first notable challenge to this idea, but Olivi—soon followed by his countryman Jean Buridan—appears to have been the first scholar in Western Europe to do so.

Nicole d'Oresme
c. 1325-1382

French mathematician who developed an early type of coordinate geometry and discussed Earth's rotation. In *Questiones super libros Aristotelis de anima* (c. 1370), Oresme addressed Aristotle's ideas concerning the earth's motion. Aristotle had claimed that the planet is stationary, but Oresme—before he went back on his first position, the correct one—initially stated that a daily rotation would be both feasible and explainable in physical terms. *Questiones* also examines the speed and reflective qualities of light. In about 1375 Oresme wrote a work on ballistics.

John Peckham
d. 1292

English scholar who wrote on practical applications of optics, as well as on other scientific subjects. A member of the Franciscan order, Peckham taught at Oxford and later Rome before becoming archbishop of Canterbury in 1279. In his scientific writing, most notably his *Perspectiva communis*, he showed the influence of Ibn al-Haytham (Alhazen; 965-1039) and Witelo (c. 1230-1275). During the fourteenth century, the writings of Peckham and Witelo gained a much wider audience than those of Roger Bacon (1213-1292) with regard to optics. It would only be in the mid-sixteenth century that scholars took a new interest in Bacon's work on the subject, which, as it turned out, offered much more promise with regard to development of optical instruments.

Petrus Bonus
fl. 1300s

Italian alchemist and physician whose *Pretiosa margarita novella* (c. 1330) is an important work on alchemy. In the text Petrus, also known as Bonus Lombardo or Buono Lombardo of Ferrara, described the prevailing alchemical theories of the day. Some two centuries later, in 1546, the alchemist Janus Lacinius edited the book and published it as *The New Pearl of Great Price,* which has remained in print almost ever since.

Georg von Peuerbach
1423-1461

Austrian astronomer and mathematician who studied comets and eclipses, and whose student was the noted German astronomer Regiomontanus (1436-1476). In 1456 he observed what came to be known as Halley's comet, and a year later he and Regiomontanus observed a lunar eclipse from a site near Vienna. Peuerbach developed several astronomical instruments, as well as a large star globe. At the time of his death, Peuerbach was at work on a study of Ptolemy (second century A.D.), *Epitome in Almagestum ptolemaei,* which Regiomontanus later completed.

Ar-Razi, also known as
Abu Bahr Muhammad ibn Zakariyya
c. 854-c. 925 or 935

Persian physician and alchemist who was a director of hospitals and who wrote books on smallpox, measles, and comprehensive manuals collecting ancient Greek medical knowledge. He was also a philosopher and religious critic. These writings are now lost, but his religious criticisms made him the enemy of many. He was perceived to have radical views that praised science over religion, Plato over Aristotle, and that criticized Galen. His reputation declined under repeated attacks until his medical works were translated into Latin and proved popular in Europe.

Richard Anglicus
fl. 1200s

English alchemist who in his *Correctum alchymiae* (c. 1250) divided minerals into two classes according to their physical origins. He may have been the same person as Richard of Wendover, who served as bishop of Rochester from 1235-51 and died in 1252 or 1256.

Ristoro d'Arezzo
c. 1220-1282

Italian scholar who put forth ideas of varying accuracy concerning geology. Ristoro maintained that mountains are formed because the stars pull earth upward, a notion later reflected in the *Divine Comedy* of Dante Alighieri. On the other hand, Ristoro was quite correct in stating that the Earth's interior is in a molten state. Near the end of his life, he found jumbles of rock and gravel in the Alps—a geological disturbance possibly created during the end of the last Ice Age—that he cited as evidence of the biblical flood.

Johannes de Sacrobosco
1195-1256

English astronomer and mathematician, also known as John of Holywood or John of Halifax, whose *Tractatus de sphaera* (1220) helped popularize the ideas of Ptolemy (fl. 100s). In 1472 the book became Europe's first printed work on astronomy, and despite a number of errors, it continued to be widely used until the seventeenth century. Sacrobosco called for calendar reform in *De anni ratione* (1232), and discussed the use of the quadrant by astronomers in *Tractatus de quadrante.*

Michael Scot
before 1200-c. 1235

Details of the life of this translator and astrologer are unknown. The "Scot" in his name may mean he was either Scottish or Irish. He described himself as "astrologer to the emperor." Around 1217, he traveled to Spain and learned Arabic and possibly Hebrew. Michael translated the works of al-Bitruji, some previously "lost" writings of Aristotle, and various other books from Arabic into Latin. His original writings were on astrological and general lore, often including magical elements.

Ala al-Din Abu'l-Hasan ali ibn Ibrahim
ibn al-Shatir
c. 1305-c. 1375

Arab astronomer who sought to restore uniform circular motion to planetary theory by replacing Ptolemy's eccentric deferent and equant with secondary epicycles. This eliminated a major defect of Ptolemaic lunar theory by reducing the variation of the Moon's distance. Though developed within a geocentric framework, Ibn al-Shatir's models are mathematically equivalent to those later developed by Nicolaus Copernicus; however, no direct influence has been established. Ibn al-Shatir also constructed various instruments, including sundials, astrolabes, and quadrants.

Shen Kua
1031-1095

Chinese polymath and astronomer who studied medicine, but became renown for his engineering ability. He became a court engineer, and eventually rose to be head of the powerful Finance Commission. However, due to factional politics he was impeached, and his political career faded. Little of his writing survives, but we know he wrote on many topics including music, astronomy, medicine, optics, magnetism, geography, painting, and tea. A keen mathematician, he applied mathematical principles to many areas, including imperial policy and military tactics.

Qutb al-Din al-Shirazi
1236-1311

Persian mathematician, astronomer, and physician who devised a geometrical model for planetary longitudes that involved a minimum of rotating vectors. Al-Shirazi wrote on geometry, medicine, philosophy, theology, and optics. Some scholars credit him with providing the first correct qualitative explanation of the rainbow, though others attribute this to his pupil, al-Farisi.

Su-sung
fl. 1000s

Chinese astronomer and court official who built an extremely sophisticated astronomical clock. A diplomatic envoy from the Sung dynasty court to that of a "barbarian" emperor occupying northern China, Su-sung was embarrassed in 1077 to discover that his hosts' calendar was more accurate than that of his own emperor. He subsequently requested and received permission from the Sung emperor to build what he called a "heavenly clockwork." His astronomical clock, powered by a water wheel, was completed in around 1090. It stood the equivalent of five stories in height and included an armillary sphere and several carved figures used to indicate the time. With the change of emperors following the death of his patron, however, the clock was allowed to fall into disrepair.

Abd al-Rahman al-Sufi
903-986

Arab astronomer who revised the catalogue of fixed stars established by Ptolemy and prepared an accurate map of the sky that became a standard work in the West for several centuries. Known in Europe as Azophi, al-Sufi wrote about a southern group of stars, today known as the Nebecula Major or the Greater Magellanic Cloud, based on reports from Arab sailors in the Malay Archipelago. Al-Sufi had an enormous influence on the Arab astronomical studies of his time. A small mountainous ring on the Moon is named after him.

Themon Judaeus
fl. 1370

One of the four University of Paris scholars central to the so-called Paris school of physical thought. Evidently from Munster in Westphalia, he was the student of Albert of Saxony (c. 1316-1390), who in turn was a student of the accepted founder of the group, Jean Buridan (c. 1300-c. 1358). Themon followed them with commentaries on Aristotle's physics and the celestial realm. His commentary on the *Meteorologica* was more substantial than theirs and disagreed with Aristotle that rather than a terrestrial phenomenon, the Milky Way was a heavier portion of celestial space and therefore celestial. He also preserved the basics of Theodoric (or Dietrich) of Freiberg's correct theory of the rainbow.

Theodoric of Freiberg
c. 1250-c. 1310

German natural philosopher who studied at the University of Paris, gaining a Master of Theology. His major work, *De iride et radialibus impressionibus* (On the rainbow and radiant impressions), written around 1305, was unusual in its use of experimentation to replicate a natural phenomena. Dietrich established the source of rainbows as light interacting with raindrops, and his work was not surpassed until the time of Descartes and Newton. He also wrote works on theology, philosophy, optics, and astronomy.

Paolo Toscanelli dal Pozzo
1397-1482

Italian physician, mapmaker, and astrologer who suggested to Christopher Columbus (1451-1506) that he could reach Asia by sailing westward. In 1433 Toscanelli observed a comet and accurately depicted its position in relation to the stars. Chinese and Polish astronomers also reported seeing the same comet.

Simon Tunsted
d. 1369

English scholar of Minorite order who advanced in study to become a doctor of theology and wrote on music theory and earth science. Born at Norwich, he became master of Minorites at Oxford (1351). He wrote a significant Latin commentary on Aristotle's *Meteorologica* containing new nominalistic aspects in a format that was incorrectly attributed to the well-known Scottish nominalist Duns Scotus (c. 1266-1308).

Nasir al-Din al-Tusi
1201-1274

Arab astronomer and mathematician who established an observatory at Maragheh (now in Azerbaijan) under the patronage of the Mongol conqueror Hulagu Khan. Al-Tusi designed a number of instruments for the observatory, which became a center of scientific learning in Muslim Central Asia. There al-Tusi prepared the *Ilkhanic*

Tables, which contained 12 years worth of astronomical observations. He also developed new hypotheses on the movement of planets, ideas some scholars consider the most important contributions to the understanding of cosmology between the time of Ptolemy (second century A.D.) and that of Nicolaus Copernicus (1473-1543).

Wilhelm von Hirsau
d. 1091

German cleric and scholar who in about 1090 wrote a treatise on astronomy and built what would become a famous model of the planetary system. Abbot of Hirsau from 1069, Wilhelm supported the vigorous reforms of Pope Gregory VII and instituted reforms at his own monastery, where he promoted liberal studies. His works include *Constitutiones Hirsaugienses, Dialogi de musica,* and *De astronomia.*

William of Saint Cloud
fl. c. 1285

French astronomer about whom little is known except his surviving astronomical writings. We know he observed a conjunction of Saturn and Jupiter in 1285 and was writing around 1290. He wrote a calendar dedicated to Queen Marie of Brabant, which contradicted the church astronomical calculations of the time, and an *Almanach* that gave planetary positions accurate to within a minute of a degree. His use of observation over authority was novel at the time, and inspired later observational astronomers.

Witelo, or Vitello
c.1230-after 1275

A natural philosopher and optical scientist, Witelo probably studied arts at the University of Paris, and later canon law at Padua. During an Easter holiday (probably sometime between 1262 and 1265), he wrote *Tractatus de primaria causa penitentie et de natura demonum*, an astronomical work that plots the movement of the sun, the lunar cycle, and the variation of day and night. He later wrote a number of other works in natural philosophy. Only one, *Perspectiva*, survives, a huge tome on optics, rich with geometrical theory.

Yahya ben al-Bitriq
c. 754-775

Islamic scholar and translator who was known for his translations of ancient Greek and Roman writings into Arabic. Although many of his translations were in regard to medicine under the caliphate of al-Munsur, he also translated other scientific works. In particular, he made the first translation of Aristotle's *Meteorologica* into Arabic, although it was not an exact translation, and he sometimes reduced it to summarization and paraphrase. Yet this translation would be that used by Gerard of Cremona (1114-1187) to provide the first Latin text of the work to the West, which would be the basis for European commentaries until late in the thirteenth century.

Bibliography of Primary Sources

Albertus Magnus. *De Rebus Metallicus et Mineralibus* (1260). In this work Albertus compiled a list of some 100 minerals; the work retained its authoritative status for several centuries.

Albertus Magnus. *De vegetabilibus et plantis.* Albert's major work, his observations, descriptions, and classifications of many kinds of plants were studied as recently as the nineteenth century.

Alfonsine Tables. Latin name for a work formerly attributed to Alfonso, but now believed to be the work of Parisian astronomers from the 1300s. Because of several innovations in the work, it became possible to make predictions using any calendar—the Islamic lunar or the Christian Julian or others. It became the most influential handbook of practical astronomy in Europe until about 1551.

Aryabhata. *Aryabhatiya* (c. 409). A collection of verses that describe the entirety of Indian mathematics and astronomy, covering spherical trigonometry, arithmetic, algebra and plane trigonometry. In this work, Aryabhata calculated π to four decimal places.

Bacon, Roger. *Opus Maius* (c. 1267). In this work Bacon conjectured that the speed of light was finite and traveled as sound did and, though short on precise theory, he believed that reflection and refraction in "numberless drops of water" (cloud drops) generated the colors of the rainbow.

Battani, al-. *Zij-i Djadid Sultani.* Advanced astronomical theory by providing tables of sine functions to assist in solving complex mathematical and astronomical problems. Also contained sophisticated tables of special trigonometric functions for solving problems involving spherical triangles. *De motu stellarum,* the Latin version of this work, printed in 1537, was important in the development of European astronomy.

Boyle, Robert. *The Sceptical Chymist* (1661). The age of alchemy as a science is generally considered to have ended with the publication of this book. Boyle showed by experiment that earth, air, fire, and water were not, in fact, fundamental elements. Thereafter, alchemy became associated primarily with magic and the occult.

Brahmagupta. *Brahma-sphuta-siddhanta* (The opening of the universe) (c. 628). An important work on mathematics and astronomy. The beginning chapters discuss pure mathematics, while the majority of the text discusses astronomical topics such as solar and lunar eclipses and the motions and positions of the planets.

Campanus of Novara. *Theorica planetarum* (c. 1260). The first European description of a planetarium. The book also discussed the longitude of the planets, and included determinations of the time for each planet's retrograde motion, instructions for using astronomical tables compiled earlier by Arab scientists, and calculations of planets' sizes and distance from Earth.

Chu Hsi. *Chu Tsi Shu Chieh Yao* (c. 1175). In this work, the author suggested that fossils were once living organisms.

Dietrich (or Theodoric) of Freiburg. *De Iride* (On the rainbow) (c. 1304). Dietrich described correctly the physics of a primary rainbow and gave the first detailed description of the secondary rainbow. Furthermore, he became the first thinker to explain correctly that each color of the rainbow phenomenon required a definite angle of incidence of sunlight (which enhanced the understanding of the nature of light) and correctly deduced that the angle/color relationship explained the circular appearance of the rainbow.

Gerard of Cremona. Between about 1140 and 1187, he translated many works into Latin including Ptolemy's *Almagest*, Euclid's *Elements*, al-Khwarizmi's *Algebra*, Aristotle's *Physics*, *On the Heavens*, and *On Generation and Corruption*, and Avicenna's (Ibn Sina) *Canon of Medicine*. Because Gerard had such a good command of the languages and a clear understanding of the subject matter, he was able to produce translations that were true to the original meaning and nuances of the Arabic works.

Gilbert, William. *De Magnete* (1600). In this groundbreaking work, Gilbert, English scientist and physician to Queen Elizabeth I, was the first to describe Earth as a giant magnet. He demonstrated the expected behavior of the compass needle at various points on Earth's surface. This book allowed sailors to understand how the magnetic compass they had used for hundreds of years actually worked.

Grosseteste, Robert *De generatione sonorum* (The generation of sound) (1230). In this, his first treatise on sound, Grosseteste advanced his theory of vibrations, which would be corroborated by later studies in the modern era.

Ibn al-Haytham (Alhazen). *Kitab al-Manazir* (Book on optics) (c. 1027). This book contained his treatment of vision and light, and his belief that rays did not emanate from the eye, but that light rays were received by the eye. He correctly explained vision in this work.

Ibn al-Haytham (Alhazen). *On the Rainbow and Halo*. A detailed work on atmospheric optics, this work provided further experimental depth to his *Book on Optics*.

Ibn al-Haytham (Alhazen). *A Question Relating to Parallax*. This work dealt with other questions of atmospheric optics. He appears to be the first thinker to note the property of refraction of light in the atmosphere. He applied this bending of light as the reason that astronomical observers see distortions of celestial objects near the horizon.

Ibn al-Haytham (Alhazen). *Al-Shukuk ala Batlamyus* (c. 965-c. 1040). An extensive and sophisticated attack on Ptolemaic astronomy. Al-Haytham's objective was to harmonize the mathematical and physical aspects of astronomy. Accordingly, he rejected the use of mathematical models whose motions could not be realized by physical bodies. In particular, he objected to Ptolemy's use of the equant because it required uniform circular motion of a sphere about a point other than its center. Al-Haytham argued that the models of the *Almagest* should be abandoned in favor of more adequate models.

Ibn Sina (Avicenna). *Kitab al-Shifa* (Book of Healing) (1021-23). A composite work of philosophy and science. In this volume he synthesized Aristotelian and Neoplatonic thought with Muslim theology. The *Kitab* also contains many original scientific ideas. He proposed a corpuscular theory of light, which implied a finite speed for light. He distinguished between different forms of heat and mechanical energy and contributed to the development of the concepts of force, infinity, and the vacuum. He also investigated the relationship between time and motion, concluding they must be interrelated since time can have no meaning in a world devoid of motion.

Jabir ibn Hayyan. *Kitab al-Kimya*. Translated in 1144 by Robert of Chester as "The Book of the Composition of Alchemy," this was the most influential of Jabir's books in the West.

Jabir ibn Hayyan. *The Sum of Perfection*, *The Book of Furnaces*, *The Investigation of Perfection*, and *The Invention of Verity*. Jabir's work shows that he apparently had considerable firsthand experience in working with chemicals and performing his own experiments. In his books, he describes basic chemical procedures, gives instructions for preparing certain compounds, and attempts to teach these methods to others. His books were widely read among the alchemists of the fourteenth and fifteenth centuries.

Jayyani, ibn Muadh al-. *On the Dawn*, or *On Twilight and the Rising of Clouds*. Ibn Muadh developed a more scientific method of studying atmospheric height by way of refraction. He noted that sunlight seen before dawn and after sunset was a refractive phenomenon and estimated the angle of depression of the sun at dawn and sunset. From this he determined the height of atmospheric moisture (believed responsible for twilight), which was considered proportional to atmospheric height.

John Peckham. *Perspectiva communis* (1277-79). Peckham contributed to the advance of optics by his precise simplification of the abstractions in al-Haytham's *Book on Optics* in his work. It carried along contemporary emphasis on the importance of both reflection and refraction in the generation of the rainbow.

Kamal al-Din Farisi. *Revision of Optics*. Farisi appraised all of al-Haytham's optical works in this book, and openly noted al-Haytham's incorrect theories. Particularly, he researched the rainbow mechanism using al-Haytham and Ibn Sina's ideas as a base to arrive at most of the correct theory in a manner superior to that of al-Haytham.

Kindi, al-. *De aspectibus*. Al-Kindi's primary optical treatise. Working within the Greek optical tradition, he leveled a devastating critique against the intromission theory (of forms), which asserted that vision is the result of objects emitting thin films or images of themselves through the intervening space to the eye. He demonstrated that the intromission theory was incompatible with the laws of perspective. Al-Kindi also argued that each point on an object's surface radiates light in all directions. Though implied in early theories, al-Kindi was the first to state this principle explicitly.

Kindi, al-. *Risala fi anwa al-jawahir al-thaminah wa ghayriha* (Treatise on various types of precious stones and other kinds of stones) and *Risalah fi anwa al-hijarah wa'l-jawahir* (Treatise on various types of stones and jewels). Two treatises on mineralogy, these works were the first of their kind in Arabic. He also produced the first Arabic book on metallurgy, *Risalah fi anwa al-suyuf al-hadid* (Treatise on various kinds of steel swords). In the study of acoustics, he developed a means for determining pitch and showed that the different notes combining to produce harmony have a specific pitch. Furthermore, he argued that sound is caused by airwaves striking the eardrum. His musical treatises were among the best such works in Arabic.

Llull, Raymon. (Raymond Lully). *Ars magna* (The great art). Prefiguring modern computer science by centuries, Llull suggested that it might someday be possible to construct a machine that would generate ideas and then prove or disprove them. *Ars Magna* contains an illustration of a wheel designed to be rotated as a means of generating and testing new concepts. If it were possible to mechanize the process of objective truth-seeking, Llull believed, then all observers would be forced to accept such conclusions.

Michael Constantine Psellus. *Chronographia.* A history of the years 976-1078 by a Byzantine statesman and the last of the great Greek astronomers.

Mukammas, David al-. *Ishrun maqalat* (Twenty chapters) (c. 900). Here Mukammas maintained that science provides knowledge of reality and consists of two elements, theoretical and practical knowledge.

Nicholas of Cusa. *De docta ignorantia* (On learned ignorance) (1439). The first of Nicholas's 12 philosophical/scientific treatises in which he underlined the limits of human understanding of science. This treatise also contained his early ideas on the cosmos, particularly his intuitive idea that logic was better served with the earth revolving around the sun and not being the center of the universe.

Nicholas of Cusa. *Idiota de staticis experimentis* (Static experiments). A small but important tract on physical statics and meteorology, outlining Nicholas's observations and experiments on moisture absorption and subsequent weight gain as a measure of atmospheric humidity.

Peuerbach, Georg von. *Epitome in Almagestum Ptolemaei.* Peuerbach's monumental work on Ptolemy, completed by his student Regiomontanus after Peuerbach's death.

Peuerbach, Georg von. *Tabulae ecclipsium.* Peuerbach's tabulation of eclipse calculations.

Peuerbach, Georg von. *Theoricae nova planetarum.* This discussion of the epicycle theory of the planets (first presented by Ptolemy) included Peuerbach's assertion that the planets revolve in transparent but solid spheres.

Ptolemy. *Almagest.* In this book, Ptolemy described his theory that the sun, planets, and stars revolve around Earth (Ptolemaic system). He used epicycles (movement of the planets about smaller circles imposed on greater circles) that allowed practical adjustments of planetary motion to account for retrograde motion and varying planetary brightness.

Reinhold, Erasmus. *Tabulae Prutenicae* (1551). The first practical set of planetary tables based on Nicolaus Copernicus's heliocentric theory. Reinhold cast his tables in essentially the same format as the *Alfonsine Tables,* so its users would have to make no commitment to Copernican heliocentrism.

Richard of Wallingford. *Quadripartitum* and *Tractus de sectore.* The first texts on trigonometry written in Latin.

Richard of Wallingford. *Tractus Albionis.* Describes the astronomical instrument Richard called the Albion, an equatorium used for calculating planetary positions according to the system of epicycles established by Ptolemy.

Surya Siddhanta (c. 400). Astronomical text containing the first known tabulation of the sine function.

Thomas Aquinas. *Commentary on Aristotle's Physics.* With this book, Aquinas achieved a synthesis of Greek pagan religion with Christian theology and so challenged traditional reasoning that asserted that understanding came only from belief.

Thomas Aquinas. *Commentary on the Book of Causes.* Aquinas attempted to harmonize the work of Aristotle with medieval theological doctrine with his discussion of first causes.

Thomas Aquinas. *Summa theologica* (1266-1273). An attempt to delineate the realms of reason and faith, to effect an understanding between them, and to place God as the "Prime Mover" governing all realms. Thomas's ideas, initially scorned by church authorities, soon became received doctrine, and had a profound influence on European thinking for many centuries.

Tusi, al-. *Treatise on the Quadrilateral.* Probably the first mathematical work in history to treat trigonometry as a discipline in its own right, rather than as a mere application of astronomy.

Tusi, al-. *Zij-i ilkhani* (Ilkhanic tables). Written in Persian and translated into Arabic later, this book contained observations on the planets, recorded over 12 years.

Ulugh Beg. *Zij-i Djadid Sultani.* (Traditionally, 1437 is given as the completion date, but Ulugh Beg continued to work on it until 1449.) This work contains a theoretical section and the results of observations he and his assistants made. Calendric, planetary, and trigonometric tables are included as well as a new star catalog containing 1,018 stars. This catalog was only the second original star catalog since Hipparchus (c.170-c.120 B.C.).

Unknown. *Book of Causes.* Some scholars maintain that it originated in ninth or tenth century Baghdad, others that it was written in twelfth-century Spain. The book asserted that each cause resulted in a certain order of effects, and argued against miracles, saying that God could not do anything contrary to the order he had already established.

Witelo (or Vitelo). *Peri optikes* or *Perspectivae communis* (between 1270 and 1278). A great optical work and the standard Western text on basic optical principles for some 300 years. He wrote on the mechanics of the rainbow, defined the halo as refraction (or diffraction) of light through the sun or moon, and developed a method of machining parabolic mirrors from iron.

Zarqali, Abu Ishaq Ibrahim ibn Yahya al-. *Toledan Tables.* These tables made the computation of planetary positions possible at any time based on observations and were responsible for invigorating the science of astronomy. Al-Zarqali edited these tables from his own observations in Toledo, Spain, as well as the tables of other Muslim and Jewish astronomers.

Technology and Invention

Chronology

c. 700 Arab advances in sailing and navigation: development of the lateen or triangular sail, designed for situations in which wind power is limited; and the "karmal," a hand-sized rectangular board with a knotted cord attached to its center, for finding a ship's latitude.

c. 800-c. 1000 A number of developments in European transportation: introduction of the harness or horse collar from Asia (c. 800); horseshoes made of iron (c. 900); and the whiffletree, a horizontal bar for joining the harnesses of two animals and equalizing their pull (c. 1000).

919 A Chinese scientist discovers gunpowder, and the first bombs are created in the century that follows.

984 Chinese engineer Ch'iao Wei-Yo, working on the Grand Canal that links the Yangtze and Yellow rivers, builds the first water lock for raising and lowering boats.

c. 1000 The magnetic compass is developed in China.

1034-1221 The Chinese develop movable-type printing, first using type made of baked clay and later of wood; multicolor-printing appears in 1107, but Europe lags far behind, only adopting block printing in about 1290.

c. 1150 The first windmills appear in Europe.

c. 1240 Chinese scientists develop the first rockets, which they use in an unsuccessful effort to stop the Mongol invaders.

1272 The first spinning wheel is used in Bologna, Italy.

1288 The first known guns are made in China; firearms are first mentioned in Western accounts 25 years later, in 1313.

c. 1290 The longbow comes into use in England, and with its long range (240 yards or 219 meters), it will change the face of battle even before use of firearms becomes widespread.

1315 The first mechanical clocks begin to appear in Europe.

1418 Xylography, the art of engraving on wood and printing from those engravings, first appears in Europe on the eve of Gutenberg's great leap forward in printing technology.

Overview: Technology and Invention 700-1449

The medieval era often is considered a time of modest achievements, literally a middle period between the substantial intellectual achievements of Greece and Rome and the Renaissance. Yet this sweeping generalization was not true for technology, which underwent great changes during these years. The decline of the Roman Empire, the emergence of the feudal system, the growing frequency of international trade, and, European exposure to the Middle East through the Crusades changed the way people lived. The Middle Ages were a period of substantial achievement in agriculture, new power sources, mechanization, military weapons, transportation, and construction. These developments so transformed technology that they produced a medieval industrial revolution.

Warfare

The introduction of the stirrup, previously known only in Asia, transformed warfare and may have helped restructure society in eighth-century Western Europe. It allowed a warrior to use a lance and other arms more forcefully against an opponent while keeping his seat. This new mounted shock combat was so successful that it spread throughout Western Europe, where an armored cavalry became the military standard. Some scholars argue that the expense of outfitting a warrior with horse, armor, and associated weapons drove communities to band together to support this new military machine. In exchange for their support, these knights provided defense and security for their communities and the feudal system, so characteristic of the Middle Ages, evolved. Others contend that feudalism had so many variations in so many places that it's impossible to tie its development to one particular development.

Whatever its societal effect, mounted shock combat remained the prevailing mode of battle until the thirteenth century, when the Welsh longbow finally allowed archers to disable armored warriors. Military tactics changed yet again, capitalizing on the longbow's advantages and developing strategies to defend against it. By the fourteenth and fifteenth centuries warfare was transformed again when the introduction of gunpowder from the East led to the development of the cannon and the matchlock gun. Although firearms did not become reliable weapons until the Renaissance, their appearance in the late medieval period established a new link between the military might of a state and its quest for empire. Government interest in a strong military led to support for metallurgy, weapons development, and standing armies. This began an era of gunpowder empires in which military might led to territorial acquisition.

Agriculture

Several significant innovations in agricultural techniques changed the way food was produced during the Middle Ages. The development of the wheeled moldboard plow and coulter allowed farmers to cultivate the heavy, wet soils of Northern Europe. The new plow needed at least four (and as many as eight) oxen to pull it. Its size and weight also made turning at the end of a furrow difficult. To make plowing easier, fields changed from their traditional square shape to long narrow strips. With the invention of the padded horse collar a hundred or so years later, farmers were able to use horses rather than oxen to pull the plow. Because horses were faster and could work longer than oxen, the amount of land that could be farmed—and the amount of food it produced—increased. With the further adoption of horseshoes, horses became both beasts of burden and a means of transportation for the agrarian community.

In addition to the new plow and the wider use of horses, farmers began to adopt the three-field crop-rotation system in which one-third of the land was sown in spring crops [peas and beans], one-third in fall crops [wheat, oats] and one-third left as pasture or fallow land. Each parcel of land was used in a cycle called the triennial system. This more efficient use of the land increased food production and renewed the soil, giving people more, better, and varied food as well as vegetable protein, which is important for a healthy diet.

These important developments increased food supplies, supported a larger population, and encouraged the growth of commercial centers and towns. The resulting increase in wealth through increased trade and commerce helped move Western Europe from an agrarian society to a more urban-based culture in the latter part of the medieval era.

Power from Water and Wind

In Roman times, slaves and animals were the chief sources of power. The Christian Church, however, frowned on slavery so the practice diminished in Western Europe (although many forms of serfdom were hardly better than slavery). To compensate, natural power sources such as water- and windmills replaced animate power. These mills, which dotted the landscape, were used to grind grain, saw timber, full (shrink, thicken, and press) cloth, drive bellows, and power forge hammers. The mills' widespread distribution and use made them the prime power sources of Western Europe, each providing the equivalent of about 20 horsepower.

A Mechanical Age

The extensive use of water- and windmills was made possible by advancements in mechanical technology, especially the use of gears, cranks, cams, camshafts, flywheels, and connecting rods, which transformed the motion of a water mill wheel or windmill sail into a variety of useful tasks such as lifting, pounding, or sawing. The mechanical acumen crasftsmen gained from building and maintaining these mills also played a key role in the development of spring- and weight-driven clocks. These intricate mechanical devices became common in the fourteenth century both as time keepers—usually having only an hour hand on the clock face—and as models of the universe. Many towns competed with each other to build the largest, most intricate clock with various moving figures and sounds. Several extant working models in various European cities continue to dazzle visitors as marvels of medieval mechanical ingenuity.

Ocean-going Travel

Three innovations in ship design created new transportation technology that made ocean travel much more feasible. The first was the adoption of the triangular or lateen sail that had first been used in the East. When paired with the traditional square sails on Western ships, the lateen sail allowed boats to tack or sail into the wind. Ships no longer had to wait in port for favorable winds before they could leave.

The second innovation, the stern-post rudder, gave sailors more control of their ships. Located in the middle of the boat's stern and riding below the water, the rudder was less subject to the motion of the waves and was easier to operate than the dual oars it replaced. Eventually larger rudders allowed larger ships to transport more cargo. With the incorporation of the compass, the third innovation in ocean travel, ships could sail on cloudy days in a variety of weather conditions beyond coastal waters. As ship design advanced, lighter, stronger craft with skeleton frames began to explore and navigate the world's oceans. This, in turn, further stimulated economic and commercial activity in Europe.

Stone Structures

A defining hallmark of medieval technology is the skill with which its craftsmen designed and built in stone. This empirical technology, developed over decades of trial and error, is a testament to the ingenuity and talent of masons, journeymen, and other workers. Their knowledge diffused throughout Western Europe as they moved from building site to building site, sharing their experiences with other tradesmen.

The extent of this empirical talent is evident in the sturdy stone structures built during the Middle Ages. Town walls and the castles within them served feudal communities well as citadels of safety in the centuries before gunpowder transformed warfare. The medieval landscape became synonymous with these structures.

The most beautiful of medieval stone structures, Gothic cathedrals, with their soaring stone arches, ribbed vaulting, and elegant flying buttresses defined the epitome of church architecture in the era. Built to maximize natural interior light, their interplay of engineering, religion, and art was created by master masons who sought to build churches of dramatic height to awe and inspire parishioners and pilgrims in a time when religion dominated much of society and culture. These cathedrals dominated the skyline of many medieval cities and towns, and their durability is a testament to the high technological achievements of the Middle Ages.

Conclusion

The medieval period was a high point in the history of technology, a time when technology affected the lives of Western Europeans as fully as the Industrial Revolution would centuries later. The Crusades and increased ocean travel had introduced Europe to the technology of the Far and Middle East during the latter part of the era. When these new methods met local needs and talent, the two combined to produce lasting and impressive results. With their myriad achievements in warfare, agriculture, power sources, mechanics, and water transportation, medieval craftsmen created a technological revolution that is often overlooked. Fortunately, many examples of those achievements still exist. They remind us that technology can advance without much theoretical grounding and that skilled, dedicated, and motivated people play a key role in technological progress.

H. J. EISENMAN

The Invention of Block Printing and Early Forms of Movable Type

Overview

While written language is unquestionably one of the most important of all human achievements, the ability to reproduce written materials quickly and efficiently ranks not far behind. Only when written works could be duplicated in quantities and speeds exceeding those achievable through laborious handwritten copies did writing become a medium for the widespread dissemination of knowledge—the more copies of material available, the more people who have access to them, the more likely the spread of literacy. The challenge, particularly in civilizations with large, complex systems of writing, was to develop a method for quickly and efficiently arranging those symbols, using the arrangement to create printed material, then re-arranging the symbols for further use. The Chinese, beginning about A.D. 700, introduced innovations to carved seals, which in turn led to block printing, whereby an entire page of text is carved on a block. By A.D. 1041 block printing had given way to the earliest known system of movable type, a full four centuries before Johannes Gutenberg (1398?-1468) invented the printing press in Western Europe.

Background

As with other early civilizations, notably the Mesopotamians, the Chinese had long used carved materials to stamp official documents and correspondence. Carved seals or stamps made their appearance in China as early as 300 B.C. (and are in still in use today.) Unlike the Mesopotamians, who pressed their stamps and seals into damp clay, the Chinese, by at least A.D. 650, had begun using ink, a medium obviously more portable than clay and which, when combined with paper, one of China's greatest early inventions, provided the foundation for the emergence of printing.

Paper itself had been invented in China by Ts'ai Lun (50?-118?), a member of the ruling order, who developed papers from materials that included hemp, tree bark, and scraps of cloth. The lightness of paper, as well as its adaptability for multiple purposes, helped it become common almost immediately throughout China, replacing the bulky, heavy bamboo strips and expensive silks on which written words had previously been recorded. By the middle of the eighth century, papermaking had found its way to Arab nations, and from there proceeded over the next five centuries into the West.

Of particular consequence to the development of printing was paper's ability to accept inked imprints, reproducing the symbols carved on stamps and seals. Among the most important uses of those seals was their application for mass-producing hundreds of copies of religious and philosophical comments and insights. While restricted to a few dozen words at most, these copies were produced rapidly and in large volume—something that would have been impossible to accomplish if copied by hand.

From relatively simple seals and stamps containing a few words and phrases, Chinese printers next made the leap to carving more extensive bodies of text on larger blocks of wood. By 868 block printing led to the creation of the first known book to be produced by printing—the *Diamond Sutra,* a 16-ft (4.9-m) long scroll containing Buddhist teachings. (Block printing had become important in Japan about a century earlier, and remains an important art form there to this day.)

Within a century block printing had advanced to the point where major editions of vital Chinese historical, religious, philosophical, and literary works were undertaken. The most notable of these was accomplished over two decades (932-53) when Feng Tao (881-954), a government minister, oversaw the block-printed duplication of a huge set of books containing the teachings of Confucius. The edition consisted of 130 volumes and is credited by some scholars with helping to dramatically increase the spread of literacy in China. Certainly it helped spur the publishing of more books—it has been suggested that over the next eight centuries the Chinese produced more printed books than all the rest of the world's cultures combined.

Like paper, block printing spread throughout the world, first to Japan and Korea, then onward to the West. One product of block printing that was popular in China—playing cards—also found early favor in the West, appearing in Europe as early as the 1370s.

The success of block printing did not halt attempts to further improve the printing process. Block printing, while vastly faster and more efficient than hand-copying, remained time-consuming: each page had to be carved separately, requiring many blocks for documents and books of even moderate length.

By 1041 an ambitious attempt was made to overcome the limitations imposed by the fixed, permanent nature of carved printing blocks. Ironically, the experiment rested on a return to the original printing medium, clay, but with a vital difference. Pi Sheng, an eleventh-century Chinese alchemist, began to experiment with shaping symbols in a mixture of clay and glue, then baking the formed symbols until they were hardened.

Once the shapes were baked they were arranged in the desired order on an iron tray covered with resin and other materials. The tray was warmed and the heat of the iron softened the resin until the baked-clay symbols settled into place. When the tray was cooled, the resin hardened once more and the symbols were fixed in place. After the tray was used to print the desired number of copies, it could be re-warmed, loosening the type and freeing it for rearrangement and reuse. Movable type had thus been invented.

Block printing continued to be the dominant printing technology, however. In part, this was a consequence of Chinese written language, which contained tens of thousands of characters, or logograms. Indeed, around 1313 a magistrate named Wang Chen commissioned the carving of more than 60,000 characters from individual blocks of wood to facilitate the printing of an enormous history of Chinese technology. To further increase printing efficiencies, Wang Chen also developed a system of cases designed to arrange and house type blocks for convenient access during the preparation of printing trays. As with Pi Sheng's innovations, though, Wang Chen's approach to reusable type found only limited acceptance in China.

Movable type fared better in Korea, where in 1403 the nation's king, Htai Tjong, ordered more than 100,000 symbols and characters to be cast in bronze. Two other enormous castings were made in Korea before Gutenberg independently discovered this process in the West.

Impact

Block printing played an important role both in the advance of literacy in China and the overall course of literacy throughout the world. So long as printed materials were rare and accessible to only a few, those materials were not only restricted to upper, educated classes, but they carried a symbolic weight—a sort of magical quality—often exceeding the value of the information they bore. By enabling rapid mass duplications the inventors of printing stripped printed material of some of its magic, but delivered to those materials a far greater gift—that of dispersal, of access, and content. Printed matter reached more people than any previous written medium, and the people, in turn, were able to focus on the content of the printed pages rather than their mystical or royal qualities. Print was the great lever, lifting all who could read. More pragmatically, printing on paper enabled not only the preservation of books, but also paper money, playing cards and other forms of entertainment, the transmission of immediate news, the codification and dissemination of laws, and the sharing of knowledge among cultures. It can be argued that printing, from its origins in clay and carved wooden blocks, is the most crucial of all technologies.

KEITH FERRELL

Further Reading

Carter, Thomas Francis. *The Invention of Printing in China and Its Spread Westward.* New York: Ronald Press, 1955.

Gernet, Jacques. *A History of Chinese Civilization.* Cambridge: Cambridge University Press, 1996.

Needham, Joseph. *Science and Civilization In China: Mathematics and the Sciences of the Heavens and the Earth,* Vol. 3. Cambridge: Cambridge University Press, 1959.

The Evolution of Timekeeping: Water Clocks in China and Mechanical Clocks in Europe

Overview

Early in history, humans sought methods to tell time. A concept rather than a physical entity, time eluded accurate measurement for many centuries. One of the first successful timekeeping devices was the water clock, which was perfected in China in the eighth century. It wasn't until nearly seven centuries later that mechanical clocks began to make their appearance. Mechanical clocks not only made timekeeping much more precise, which was important for scientific purposes, but also introduced it to the masses when centrally located clock towers equipped with bells loudly struck the hour.

Background

One solar day spans one rotation of the earth on its axis. This natural unit of time is still the basic unit of timekeeping. For a variety of reasons, however, humans from past to present have desired smaller increments for determining the time. Thousands of years ago, humans began to separate the day into sections. At first, they assigned such broad categories as late morning or early afternoon, or identified the time of day by its association to mealtimes. By 2100 B.C., Egyptians had begun dividing the day and night each into 12 parts. Derived from the Greek word *hora*, an hour denoted the interval between the rising of specific stars at night. Since the period from dawn to dusk and from dusk to day were not identical—changing from season to season and even day to day—the length of an hour changed accordingly. As the days become longer or shorter, the time covered by these so-called temporal hours varied. For example, 12 daytime temporal hours in the summer might cover 14 hours of daylight, whereas the following 12 nighttime temporal hours would be crowded into the remaining 10-hour period.

Many societies used ancient sundials to measure time intervals. Originally employed to identify the changing of seasons, they were further developed to measure increments within a day. Sundials rely on the sun to cast a shadow onto a marked platform. As the sun moves across the sky, the shadow advances across the platform and denotes the temporal hour.

Chinese inventors developed the first method for measuring time consistently and without reliance on sunlight, day length, or star movement. Since about 3000 B.C., the Chinese used water clocks to gauge the passage of time. Water clocks are also known as *clepsydrae*, the Greek word for "water thief." A simple water clock is an apparatus that slowly drips or runs water from a small hole in one vessel into another that is stationed below it. By marking the water level in the lower vessel after a day had passed and then dividing it into equal portions, the clockmaker could use the device to tell time fairly accurately. Tests have indicated that early water clocks were correct to within 15 minutes each day.

Water clocks in China continued to progress into more sophisticated and accurate devices. Their development took a leap forward in the eighth century during the K'ai-Yuan reign when a Buddhist monk named I-Hsing (I-Xing) along with Liang Lin-Tsan, an engineer and member of the crown prince's bodyguard, began work on a clock escapement to control the speed and regularity of the clock's movements. The clock, a bronze model of the celestial sphere (a representation of how the stars appear from Earth), used drops of water to move the driving-wheel mechanism, and keep track of hours, days, and years. The clock also connected to a bell and drum, to provide a sound alert every 15 minutes.

Another notable clock in Chinese history was the astronomical clock of Chang Ssu-Hsün in 1092. Built into a 33-foot (10-m)-tall tower, the clock used water to power a complicated escapement mechanism that was similar in appearance to a Ferris wheel with water buckets in the place of seats. A water tank dripped water into one bucket at a time. As the bucket filled, it became heavy enough to trip a lever and rotate the wheel. When the wheel rotated, the next bucket moved under the water tank for filling. Chang also included in his clock design an armillary sphere, which consisted of rings to mimic planet orbits. In addition, the clock mechanism triggered 12 jacks, or puppets, to appear in sequence to ring bells and hit a drum to announce the time.

The next major advancement in timekeeping came with the development of mechanical clocks, probably in the late thirteenth century. These clocks depended on neither the sun nor water to keep time. Some used pendulums, while other smaller clocks relied on repeated winding to run. English records indicate that a mechanical clock was operating in a Bedfordshire church in 1283. Similar reports refer to five other mechanical clocks in English churches before 1300. Within the next 50 years, the mechanical clocks became common throughout Europe.

Impact

While temporal hours and early timekeeping methods were sufficient for many societal uses, humans continued their quest for better modes of telling time. Early astronomers and mathematicians in particular needed accurate time increments that remained static from day to day and season to season. Without precise measurements they could not determine speed, which was crucial for navigational and astronomical observations and applications.

The advent of the water clock did much to change the way humans viewed time. Now the time of day did not depend on whether the sun was sufficiently able to penetrate the clouds and cast a shadow onto a sundial or whether the night sky was dark enough to view the stars' positions. An hour could now represent a constant length of time, and could be further divided into smaller fragments. When I-Hsing and Liang invented the escapement, they greatly refined clock performance. Chang then took I-Hsing and Liang's contribution to the next level by making an even more intricate escapement, which was named the "heavenly scale." Water clocks continued to be popular in China and many other countries well into the fourteenth century. (Currently, The Children's Museum of Indianapolis boasts the largest water clock in North America with a 26.5-foot [8-m]-tall device.)

Despite improvements to the mechanism over the centuries, water clocks never attained perfection. They repeatedly needed resetting to the correct time, as well as near-constant maintenance. Winter was particularly trying. During these colder months, if the water was not replaced with mercury or some other liquid with a lower freezing point, the water would turn to ice and the clocks would stop.

In Europe, the development of clocks took a different turn. Instead of looking to water as a power source, Europeans took another path. According to *History of the Hour: Clocks and Modern Temporal Orders,* "The principle of the Chinese escapement is pivoting balance levers that stabilized a stop-and-go motion. The principle of the European escapement, which employs the centrifugal force of an oscillating inert mass, does not resemble it in any way whatsoever."

These weight-driven mechanical clocks injected time into European society. Clock towers sprung up in cities and loudly rang the hour for all the residents to hear. The earliest tower clocks were rather inaccurate—they lost or gained up to two hours each day—and had only one hand to denote the general time of day. For years, clockmakers struggled to regulate the mechanism's oscillation without much success. This problem did not deter the public from demanding better timekeeping devices. By 1500, clockmakers found a way to make the mechanism small enough so that the wealthy could purchase models for their homes. These clocks, many of which were used as alarm clocks, kept time by springs that were wound about once a day. The clocks kept time fairly well, although hours went by more and more slowly as the spring unwound.

With timekeeping becoming commonplace, societal dependence grew. Meetings, church services, and other appointments could now be scheduled at certain hours, instead of general times of day. Scientists could now begin to make much more accurate time measurements, physicians could do simple diagnostic tests as determining pulse rate, and navigators could use time to determine their position at sea. As time became more important, people began demanding more accurate clocks. Despite persistent attempts to perfect mechanical clocks, it wasn't until 1656 when Dutch mathematician Christiaan Huygens (1629-1695) used pendulums as a timekeeping mechanism, that clocks were able to tick off minutes accurately. Huygens's original design was correct to within a minute a day. By 1670 William Clement of London had refined the pendulum clock to keep time to within a second each day.

These improvements set the stage for later advancements that by 1761 had generated John Harrison's (1693-1776) marine chronometer accurate to 0.2 seconds per day, and by 1889 Siegmund Riefler's pendulum clock was true to 0.01 seconds a day. High-performance quartz-crystal clocks appeared in the 1930s, followed by the atomic clocks of more recent years.

LESLIE A. MERTZ

Further Reading

Books

Dohrn-van Rossum, G. *History of the Hour: Clocks and Modern Temporal Orders.*Translated by T. Dunlap. Chicago: The University of Chicago Press, 1996.

Maran, S. ed. *The Astronomy and Astrophysics Encyclopedia.* New York: Van Nostrand Reinhold, 1992.

Needham, J., Wang Ling, and D. de Solla Price. *Heavenly Clock: The Great Astronomical Clocks of Medieval China.* Published in association with the Antiquarian Horological Society. Cambridge: Cambridge University Press, 1960.

Other

National Institute of Standards and Technology. "A Revolution in Timekeeping." http://physics.nist.gov/ GenInt/Time/revol.html.

National Institute of Standards and Technology. "Earliest Clocks." http://physics.nist.gov/GenInt/Time/early. html.

The Technology of the Medieval Islamic World

Overview

For centuries during the European Middle Ages, non-European cultures continued to make progress in science and technology, only marginally affected by Europe's troubles and relative stagnation. Among these cultures, the Islamic world stands out for having made particularly important contributions, in part because of the location of Islamic nations between the learning centers of the East and Europe. During these centuries, Islamic scholars not only retained the best of ancient and classical European discoveries, but they also augmented these with many discoveries of their own and some that were imported from India and China. Among the areas in which Islamic scientists made significant contributions are metallurgy, glassmaking, architecture, chemistry, military engineering, and what is now known as civil engineering. Some of these developments were transferred to Europe through trade, others during the Moorish occupation of the Iberian peninsula, and still more during the Crusades. The introduction of Islamic science and technology into Europe through these various routes of encounter was an important factor that helped bring Europe into the Renaissance.

Background

The Roman Empire collapsed in the fifth century A.D., destroyed by a combination of barbarian armies and cultural decay. Without the steadying force of the Roman government, most of Europe fell into relative anarchy and lawlessness, a state that was relieved by the imposition of religious discipline by the Catholic Church. During the following centuries, Europe was in a state of consolidation; recovering from the barbarian invasions, the loss of a central administrative government, and a series of plagues, famine, and natural disasters.

The ensuing period became known as the Dark Ages, a term that is no longer considered appropriate because learning did not actually stop during these centuries. Monks, scholastics, and others managed to not only keep alive the knowledge of the past, but also made some progress on their own. Unfortunately, education was a rare commodity in those years, and scientific and technical progress in medieval Europe was dramatically slowed. However, this was not the case in all parts of the world.

In particular, the Arab (and later, Islamic) world benefited from the knowledge they had obtained from Greek scholars, as well as from their proximity to the scholarship taking place in the East. Through their contact with the Greeks and Egyptians at the library of Alexandria, the Arabs learned of Greek writings and scientific observations dating back many centuries. At the same time, because their borders touched upon the Persian and Hindu lands, Arab scholars were privy to the intellectual traditions of these cultures and, through them, they even learned of many Chinese advances. The Arabs did not, however, simply copy and translate texts from other cultures; rather, they used these texts either as the starting point of their own work, or they used them to augment their researches. In either case, Arab scientists and engineers not only adapted the best that other cultures had to offer, but they also developed their own science and engineering, some of which lasts to the present day.

One point that must be made is the difference between "Arabic" and "Islamic." Specifically, "Arabic" pertains to Arabs, who are a distinct ethnic group, whereas "Islamic" refers to any followers of the Muslim religion. It is analogous to the distinction between being an American and being a Christian. Not all Arabs were Muslims and not all Muslims were Arabic.

However, with the rise of Islam in the seventh century, these two terms became closely identified because Islam was born among the Arabs and, within a few decades, virtually all Arabs were Muslims. Over the next few centuries, Islam spread through the Indian subcontinent to the east and to the Atlantic Ocean in the west, bringing it into contact with all the great Old World cultures of that time. Because of this, Islamic scientists were in a rare position of being exposed to the best learning of the world, at exactly the time that Europe was in relative chaos.

In the centuries that followed, a number of technological advances arose in Islamic nations. For example, to support increasing urban populations, there required an increased dependence on agriculture. In the arid climate of the Middle East, this necessitated new methods for obtaining, storing, and distributing water to irrigate crops as well as to support the water needs of the city dwellers. Increased urbanization and the advent of a major new religion also required improvements in building techniques to construct homes and temples, while the increased traffic going to and from the cities called for new and better roads and bridges. And, as Islam became more powerful politically, it strove to expand its influence in whatever manner possible, including war. This, in turn, led to a number of innovations in military engineering, as well as advances in metallurgy and metalworking. Finally, but not least, other advances served to make life somewhat more comfortable (primarily for the wealthy or the powerful)—glassmaking, icemaking, and early air conditioning (nonmechanical, of course).

These advances, and others, made their way into Islamic life over the course of several centuries. Some became central to daily life and remained that way for centuries, and many of these innovations were carried to new lands by Islamic conquerors and traders. Most of them, in one form or another, were adopted directly or in modified form by Islam's trading partners, neighbors, and subjects, and they had a very significant impact on the societies of the time, and for subsequent centuries.

Impact

The effects of many of these inventions were felt in the cities as well as among the farmers. Nations on the receiving end of the armies of Islam obviously became acquainted with the military technology developed by Islamic military engineers, and their trading partners benefited from some of the other improvements. Finally, the areas that fell under the political or military domination of Islam had the opportunity to incorporate some of this technology into their societies, often benefiting in the process. The remainder of this essay will examine each of these segments of medieval society and how they were affected by the technological advances that developed in this time.

Agriculture and Rural Life

Some of the most important inventions involved water, as water is essential to life and to agriculture. In some societies living in arid conditions, more time is devoted to finding and storing water than virtually any other activity because, without water, no society can survive. The ability to either raise water from beneath the ground, to store rainwater, and to transfer water from storage locations to the fields was of vital importance.

Water wells and primitive ways of retrieving water from them have existed since before recorded history. The most basic way to retrieve water is by simply lowering a bucket into the well and pulling it up again. But this is very labor-intensive and is not sufficient, in and of itself, for any purpose more intensive than providing water for a family and a few livestock. On the other hand, mechanical methods of raising water can greatly improve the efficiency of the process and, in conjunction with some other methods, can provide enough water to irrigate fields.

Some of these water retrieval methods involved better ways of raising water from beneath the ground. Mechanical pumps, mechanical lifts to raise buckets, geared wheels turned by animals, and wheels with buckets attached were among the mechanical methods invented for this purpose. Civil engineering was used, too. In some cases, a horizontal tunnel was dug underground that intersected a downward-sloping water table at one end. This then could carry water for great distances to feed an irrigation canal, while secondary wells could be dug to intersect the tunnel, providing water at intervals along its length. In other areas, rivers or seasonal riverbeds were dammed, impounding water that could then be used. And, in most places, cisterns were constructed to catch and hold rainwater for houses, settlements, or farming. Finally, many of these also required the construction of canals to carry the water from where it was collected to where it was needed. These canals had to be constructed to minimize losses of water due to evaporation or soaking into the ground, both presenting difficult obstacles.

All of these developments helped facilitate water collection, storage, use, or some combination of these three activities. By doing this, they made agriculture more efficient while simultaneously making the farmers' lives a bit easier. This did not necessarily give farmers more free time; it simply let them farm more land or devote more time to managing their crops and livestock better. The net result may have been a slightly easier life for the farmers, but more importantly, farmers could raise more food, helping the cities to grow.

Roads also had an impact on farming life because they made transporting goods easier and more reliable. This, in turn, meant that farmers spent less time actually transporting and selling their crops, again giving them more time to tend to their farms. In addition, better roads made the cities more accessible to the farmers, increasing the ability of farmers to purchase goods they may not otherwise have had access to.

Finally, the many improvements in agricultural practices and technology also helped farmers to work more efficiently. These included the use of manure as a fertilizer, the development of new devices to help plow the land, to sort grain, and to process foods. This last item is more important than it sounds because it includes devices such as presses for making olive and other vegetable oils, mills for grinding grains into flour, making butter and yogurt, and so forth. Primarily, these devices helped make food preparation simpler, but they also helped make the transportation and storage of many food items more reliable. For example, olive oil will remain fresh and useable longer than olives themselves, and the oil is easier to transport from place to place. Similarly, flour is more convenient than whole grains. In other cases, yogurt is often easier to digest than whole milk, especially for the majority of people who cannot digest lactose. As a whole, then, it seems fair to say that improvements in food technology helped farmers by making it easier for them to transport their goods for trade or sale in the cities and, at the same time, these improvements also helped make food storage and preparation easier and more efficient for farmers themselves.

Cities and Urban Life

In a sense, improvements in farming were a necessary prerequisite for urban life. Almost by definition, cities are not places where each family can provide its own food, so, to exist at all, cities are almost totally dependent on farmers. For this reason, the impact of advances in agricultural or food technology on city dwellers can be overlooked and other inventions discussed instead.

Perhaps the most noticeable and important advances that affected urban life were those in architecture and building materials. New or improved building materials helped make better housing, improved public buildings, more impressive palaces, and elaborate places of worship. These helped to make the cities better places to live and provided a better home for the equivalent of governmental agencies.

Although wood was used for construction when it was available, the forests of Lebanon had largely vanished by the tenth century, and much of the Middle East is bereft of forest. This forced the use of clay, fired brick, unfired (or raw) brick, and stone for construction through most of the Islamic world. Although much heavier than wood, these materials did not provide significant added strength and were often difficult to work with. They had, however, the advantage of holding the day's heat in the cold desert night, and conversely, remaining cool during the hottest part of the day. In addition, the use of brick in particular was possible in virtually all parts of the Islamic world because all that was needed for brick-making was clay, straw, and a kiln. In addition to the materials mentioned above, some buildings used metals in their construction.

These materials were used in novel ways by Islamic architects during the medieval period. For example, the Dome of the Rock, a temple in Jerusalem, is constructed of masonry with a dome made of wood. The wood is covered with lead, which is then covered with brass for a construction that is sturdy and remarkably resistant to the effects of weather. Other temples and mosques used similar combinations of materials and employed architectural innovations such as domes and arches to help support the building's weight, for aesthetic reasons, or both. Other architectural features developed by Islamic artisans (among others) included the use of interior courtyards, the use of "stacked" arches, and the use of arched bridges to carry traffic across rivers, streams, and gullies.

Military Advances

Islamic inventiveness was also evident on the battlefield during the medieval period, playing a role during Islam's expansion and in its self-defense during the Crusades. In particular, Islamic military engineers developed new weapons, stronger defenses, and more reliable weapons, while Islamic metallurgists helped improve the metals that went into so many of their weapons. Unfortunately, a detailed description of all such Islamic military innovations is not possible here, but a brief description of the more important inventions and their overall impact will be attempted.

Two of the most important inventions in the history of warfare (and among the most important in human history) are the invention of gunpowder and firearms. It is fairly certain that both were invented in China, but they reached the Islamic world long before Europe. Islamic military engineers lost little time in adopting gunpowder and turning out incendiary devices, artillery, handheld guns, grenades, and more. All of these gave the Islamic armies a decided advantage in battle and during sieges, and all were used to great effect on multiple occasions against the invading crusaders, as this technology had not yet reached Europe.

Another invention was the trebuchet, a catapult in which the payload was projected with a sling instead of a rigid cup or bowl. Trebuchets were powerful weapons that were used to fling both solid rock and incendiary or explosive devices against invading armies or into besieged fortresses. Like firearms, trebuchets were used to great effect in battle.

Islamic artisans also worked mightily to improve on more traditional weapons, such as the sword, bow and arrow, crossbow, and lance. They devised a bow made of a combination of materials that could pierce chain mail at a distance of a quarter mile, and their lances and arrows were often tipped with heads made of Damascus steel, renowned for its superior properties. Muslim swords, also made of Damascus steel, were so cherished that many were given names and passed down from father to son, and crossbows were often used as long-range weapons to defend against besieging armies. All of these improvements helped give Muslim armies a significant advantage over those equipped with lesser weapons.

All in all, the weapons wielded by Islamic soldiers were almost never inferior to those of their opponents, and were usually superior in some manner or another. This gave Islamic armies a decided advantage in nearly all of their battles, and was a deciding factor in many. One result of this military superiority was facilitating the rapid spread of Islam throughout the Middle East, across northern Africa, and eastwards to Indonesia in a relatively short period of time. Another was that the attacking crusaders were frequently unsuccessful in their attempts to expel Muslims from the Holy Lands, and the Spanish took several centuries to expel the Moors from Spain, in part because of Muslim superiority in military technology.

Domestic Comforts and Luxuries

Other inventions had little impact on the necessities of daily life, but had a decided impact on the quality of life, especially for the wealthy. Among these innovations were glassmaking, a primitive sort of air conditioning, and icemaking.

Glass, a technology that originated in either the Middle East or Egypt several centuries before the time of Christ, remained a near-monopoly of Syria until the twelfth century A.D. In the Middle East, glass was used for making drinking vessels, ornaments, bottles, and similar items, mostly for those who could afford them. In the twelfth or thirteenth century, Venice purchased glassmaking technology from the Muslims, establishing a European monopoly that lasted until the seventeenth century. Thus, glassmaking benefited not only the Islamic world, but was also a factor that helped Venice become a major economic power in its day.

Icemaking emerged as another benefit for the wealthy. In the days before refrigeration, it was still possible to make ice, relying on some basic principles of physics and the cool desert nights. Once made, ice was sold to rulers, merchants, and others who could afford it and used for cooling drinks, chilling food, and so forth. Similar principles could be used to help cool rooms, making use, for example, of the fact that evaporating water will lower the temperature of air passing over it. This basic physical principle, along with architecture that utilized the thermal inertia of building materials and cooling shadows, was used in a number of different guises to help cool rooms during the heat of the desert days.

Beyond the Islamic World

Finally, nations outside of the Islamic world could not help but be affected by the technological advances developed by the Muslims. Whether these other nations were trading partners, wartime adversaries, defeated enemies, or occupied territories, they were exposed to and could draw upon lessons learned from their Islamic neighbors, whether friend or foe. One result of this was the relatively rapid spread of Islamic technology across most of Europe and the Middle East.

Trading partners, of course, not only gained valuable information and technology, but provided it, too. Situated as it was between the advanced civilizations of India and China to the east, and the recovering civilizations of Europe and Egypt to the west, Muslims not only reaped the benefits of both worlds, but combined good ideas from both and, in turn, transferred such ideas to East and West. This led, for example, directly to the transfer of glassmaking to the Venetians, as well as to improving European steelmaking, Indian architecture, and civil engineering projects in many parts of the world. Both Islam and its trading partners benefited from this exchange of ideas and technology, allowing both sides to advance more rapidly than either would have done on its own.

Similarly, Islamic weapons and fortifications were frequently copied by their foes, and were sometimes put into use against their designers. In addition, Islamic military technology seems to have been assimilated in a short period of time by all European kingdoms, keeping any single kingdom from gaining a decisive and irrevocable advantage over its neighboring foes.

Finally, Islamic technology was adopted by the peoples conquered by Islamic armies. Islamic water-raising devices continue to be used in parts of Spain, for example, and the terminology for agricultural techniques in that country reveal Arabic roots. Many of the buildings and other civil engineering projects constructed by the Muslims still exist and, in some cases, are still in use, and buildings inspired by Islamic designs and built using Islamic techniques are more common yet. In short, Islamic technology has left a long-lasting mark on many societies, and in many cases, this was an improvement over earlier conditions.

In summary, then, it may be said that Islamic scholars and engineers made a wide array of technological improvements available to their own society and others that they contacted. These changes not only helped spread Islam rapidly throughout the known world, but also helped to change the life of city dwellers and farmers alike. Because of these innovations, virtually every nation from what is now China to the Atlantic Ocean was affected and, in many ways, improved.

P. ANDREW KARAM

Further Reading

Al-Hassan, Ahmad, and Donald Hill. *Islamic Technology: An Illustrated History.* Cambridge: Cambridge University Press, 1986.

The Spread of Papermaking Technology into Europe

Overview

While technology has significantly altered the equipment used to manufacture paper, the basic operations remain the same. Papermaking involves a five-step process where suspended cellulose fibers are filtered onto a screen to form a sheet of fiber. The sheet is pressed, dehydrated, and then modified based on the intended use. After its invention, probably sometime before 100 B.C., knowledge of the papermaking processes slowly spread throughout the world. As more and more cultures discovered it, paper quickly became the medium of choice for recording written information.

Prior to the invention of paper, various kinds of materials were used as a means of recording written information. Different cultures used clay, wood, bark, leaves, stone, metal, papyrus, parchment, vellum, and cloth as recording mediums at one time or another. The Sumerians first developed early cuneiform writing in the form of pictographs on clay tablets before 4000 B.C. Other cultures from around the world adopted tree bark for record-keeping use in one way or another. Extensive use of bark for written records has been found in Pacific Rim cultures, Indonesia, America, and the Himalayas. Large tree leaves were used to record information in India and Asia, while rice pith paper was traditionally used in China.

The invention of papyrus has played an important role in history. The word paper itself was originally derived from the Greek and Latin words for papyrus. The oldest documented papyrus rolls that have writing on them are over 5,000 years old. Papyrus helped to shape and strengthen Egyptian society. It was an important symbol in Egyptian architecture and religion, and the availability of papyrus sheets for recording information was an important asset to Egyptian rulers. Without papyrus, the course of Mediterranean history and literature would have been vastly different. The same is true for the effect that paper had on European history.

Background

While A.D. 105 is often sited as the year papermaking was invented, historical records indicate that paper was used at least 200 years prior to that and archaeological evidence places papermaking perhaps even 200 years before that (300 B.C.). Early Chinese paper appears to have been made by from a suspension of hemp, mulberry, and bamboo. Eventually other fibers and dye were added to the paper to improve quality and longevity. The first reported use of paper for toilet purposes comes from sixth century China where it was made from rice straw.

From China, papermaking technology slowly spread throughout Asia. Records indicate that Korea produced paper in the sixth century. The pulp used to make paper was prepared by pounding fibers of hemp, rattan, mulberry, bamboo, rice straw, and even seaweed. According to tradition, a Korean monk introduced Japan to papermaking by presenting it to the emperor in 610. The Japanese first used paper only for official records and documentation, but with the rise of Buddhism the demand for paper grew rapidly.

Taught by Chinese papermakers, Tibetans began to make their own paper as a replacement for their traditional writing materials, a large palm leaf. Even though they adopted paper, Tibetan books traditionally still reflect the long, narrow format of the original palm-leaf books. Papermaking spread to other cultures as well. The process was introduced to central Asia, Persia, and India through the trade routes. The first recorded use of paper in central Asia dates from 751. Skilled Chinese papermakers were captured in battle in Turkestan, and forced to make paper during their imprisonment.

The arrival of paper in Samarkand is significant because paper could now be distributed as a commodity through the Arab world. The first paper manufacturing facility in Baghdad was established in 793. Waterpower was used in this factory to pound the pulp, which led to the widespread use of paper throughout the Arab culture by the beginning of the eleventh century. Paper was used not only for books, but as wrapping material and napkins as well. As the Islamic culture spread throughout Europe, papermaking techniques followed.

Paper first penetrated Europe as a commodity in the tenth century from the Islamic world. Certainly as Muslim armies ventured further into European countries such as Spain, the technology was brought with them. In addition, trade routes involving Italian seaports that had active commercial relations with the Arab world and overland routes from Spain to France were vehicles that spread papermaking techniques throughout Europe. Papermaking centers were originally established in Italy by 1275 and later came to other nations. However, paper came slowly to the rest of Europe and it was not until the fifteenth century that it came into widespread use.

Western Europeans were initially suspicious of paper. The Christian world more than likely thought of it as a manifestation of Muslim culture and rejected it. In fact, in 1221 Holy Roman Emperor Frederick II declared all official documents written on paper to be invalid. This decree also helped to protect the interests of powerful European landowners who monopolized the markets for parchment and vellum. It would not be until the introduction of the printing press in the fifteenth century that Western Europeans would fully embrace paper.

While paper was not fully accepted by European society, it was almost immediately recognized that the water mill would be quite useful in processing of pulp to make paper. By the time papermaking technology reached Europe, there were over 6,000 water mills in England alone and many others spread throughout Europe. It is believed that many of these mills had multiple purposes, and were able to perform a variety of tasks, one of which may have been papermaking.

The early papermaking process involved a number of different steps. The material chosen by European papermakers was most often rags made from cotton or linen fiber. These were cleaned and then heated in an alkali solution. The rags were then washed and pounded to pulp by the mill. Most paper mills at that time used water to power them. The flow of water would spin a wheel and the energy from this would be transferred to a hammer that would pulverize the rags. Once a pulp had been obtained, bleach would be added to whiten the suspension. The papermaker would then dip a paper mold into the suspension and lift it out horizontally to trap the fibers against the screen mold. After the sheet had formed, it was couched (removed) from the mold and readied for pressing. A stack of paper sheets would be placed on a large wooden press, and all of the mill workers would tighten the press. Typically, a stack of sheets could be pressed into two-thirds of its original height. After pressing, the sheets could be hung out to dry in the upper reaches of the mill away from the dusty conditions near the bottom. The final processing steps involved dipping the paper in gelatin to make it less absorbent and then finishing each sheet by hand rubbing it with a smooth stone. Eventually, technology advanced to such a level that even the final finishing could be accomplished by water-driven smoothers.

Impact

The invention of paper had a significant impact on Asian and Middle Eastern societies. Both China and Japan used paper extensively to the benefit of their society. Paper was used for both official and religious documents as well as in art. In many early cultures, only nobility was allowed to own paper. Paper was also used for sanitary purposes as toilet paper and napkins. Surprisingly, paper initially had little impact on Western Europe when it was first introduced.

There were quite a few factors that impeded the acceptance and common usage of paper. First, paper was seen as an invention of hated Islamic society, and many felt that by accepting paper they were also accepting that society. In fact, as previously mentioned, documentation of laws and events on paper was declared illegal in early thirteenth century by the church. Another factor that also played a role was the influence of wealthy landowners. They raised sheep and cattle for the purposes of making the writing medium commonly used at that time, parchment and vellum. Thus, they stood to lose a large amount of money if paper was readily accepted. Paper was not accepted until the influence of the Islamic culture began to wane in Europe and the utility of paper was made readily apparent with the invention of the printing press.

Once accepted, paper quickly became the preferred recording medium. It could be mass-produced for a fraction of the cost of other materials, so it was available to everyone. With the advent of the printing press, written words and their power were obtainable by all segments of society.

With paper becoming so readily available, a greater emphasis was put on education so that more people could read and write. Ideas could be recorded and disseminated much more easily than they had in the past. Thus, society on the whole had an increase in the level of education. Literature, as well as religious and scientific works was available to everyone so that their lives could be enriched. Communication between people was also made much easier, now that everyone had access to paper. So while it took some time for paper to be utilized, once it had been adopted it changed society in significant and positive ways.

JAMES J. HOFFMANN

Further Reading

Hunter, D. *Papermaking: The History and Technique of an Ancient Craft.*Dover, N.H.: Dover Publications, 1978.

Hunter, D.*Paper-Making through Eighteen Centuries.* New York: Reprint Services Corporation, 1930.

Limousin, O.*The Story of Paper: What Is Paper Made Of?* New York: Young Discovery Library, 1989.

Improvements in Iron Processing and the Development of the Blast Furnace

Overview

The development of more efficient means for producing and working with iron, and the production of high-grade iron, is one the key advances in human history. Because iron is harder than bronze—the metal most commonly used before iron—it is better able to hold sharp edges, making it a superb metal for weapons and tools. Armies bearing iron weapons held a significant advantage over forces equipped with bronze weapons. This superiority gave a name to an epoch—the Iron Age—which began in approximately 1000 B.C.

Equally important, but not so keenly appreciated as stronger weapons and tools, were the technological advances required, in turn, to produce better iron. Because pure iron does not occur naturally on earth, the skills required for its extraction from raw ore included advances in chemistry, heat and temperature control, toolmaking, and mining. While forms of iron had been in use for centuries, it was the development of the blast furnace that enabled both the production of higher quality iron and the use of that iron for various implements and purposes. Particularly crucial was the fact that blast furnaces not only purified iron more completely, they also produced molten iron, which could be poured into molds, resulting in cast tools, implements, and weapons—a dramatic leap over previous methods of hammering iron into shape.

Background

Iron is, after aluminum, the second most common metal in the earth's crust, and the fourth most common of the elements (after oxygen, silicon, and aluminum.) Unfortunately for early extractors of iron, the metal does not exist independently—natural iron is always found in combination with other materials such as carbon. A purer iron, as much as 90% pure, is found in meteorites, and it is meteoritic iron that proved most useful to early civilizations. Obviously, though, meteoritic iron would always be in short supply. The use of meteoritic iron has been traced back to 3000 B.C., with extractions of iron from mined ore occurring in Persia around 2000 B.C.

While early smelting processes (the use of heat to extract metals from ores) proved useful for some metals, the processes were not so effective for the extraction of iron. Iron was too tightly bound to the other materials in the ore. The iron tended to separate into a mass called a bloom. The bloom was then removed, heated separately, and while still hot was hammered into desired shapes; the force of the hammer blows served not only to shape the iron, but also to strengthen it. Because the iron had to be worked by hand implements, the result was called "wrought" iron. Wrought iron proved superior to bronze, although too brittle and impure to serve well in most uses. The process was also inefficient, tending to produce blooms of only a few kilograms. An advance was needed in smelting techniques.

The advance arrived on several fronts. Part of the difficulty in producing a purer form of iron involved the high temperatures required to separate iron from other substances. The melting point of iron is 2,800°F (1,538°C), a temperature difficult to achieve in ancient furnaces, which were often little more than bowls carved out of hillsides. Bellows were used to pump air into some bowl furnaces, increasing the heat of the charcoal fires, but the fires still burned too cool. Even early stone-built shaft furnaces—essentially chimneys that funneled air more forcefully over the fire—failed to achieve temperatures high enough to remove all of the impurities, or slag, from the iron ore. Bellows were common in Roman shaft furnaces.

While actual blast furnaces—in which high-pressure air is forced into to the shaft, vastly increasing the temperature of the fire—did not appear in Europe until the fifteenth century, blast furnaces are recorded in China as early as 300 B.C. Though the iron that emerged from these blast furnaces remained brittle as a result of impurities, the iron was nonetheless molten: it could be poured into molds shaped to produce the desired object when the iron cooled. Blast furnaces also produced larger quantities of iron more quickly than traditional bloomeries.

By A.D. 1100 major blast furnace operations had been established in the Chinese prefecture (state) of Chizou, with smaller blast furnaces in operation throughout China. Many, if not most, of the Chinese furnaces were built into hillsides, simplifying the challenge of creating a tall stack for the furnace. The largest of the Chinese furnaces of that period provided living quarters for more than 1,000 workers, whose tasks were divided among mining and working the furnaces. Smelting sites were constructed farther and farther from the mines themselves, so as to be closer to the forests whose dwindling supplies of wood were used to create charcoal.

Deforestation was one consequence of the large quantities of charcoal required for blast furnace operation. The depletion of readily available wood may have spurred the Chinese to use coal (coke) as early as A.D. 400, although widespread use of coal did not occur until roughly 1100. Poetry of that time celebrates the use of coal to produce molten iron, the quality of the weapons cast from the iron, and, tellingly, the relief felt by the forests as charcoal gave way to coal for fueling the furnaces.

In addition to weapons, the Chinese cast coins, tools, and implements, some of which (or accounts of which) made their way to the West. Whether or not Chinese blast furnace technologies were themselves copied by Western ironworkers is not known, but the first European blast furnaces made their appearance in the mid-1300s, and reached England about a half century later. Europe did not shift from charcoal to coal/coke firing until the seventeenth century, however, and not completely for a century after that.

Progress toward the modern European blast furnace came in the form of several innovations to the shaft furnace. Larger and larger blooms were able to be produced, some weighing over 220.5 lbs (100 kg). High bloomery furnaces raised the shaft to more than 9.8 ft (3 m) and produced even larger blooms.

By 1205 at least one blast furnace was in operation in Germany. Constructed primarily of loam, the furnace, and other similar ones built over the next two centuries, had more in common with bloomeries than with subsequent blast furnaces. Among the important innovations of the early German furnaces, though, was the use of water-powered bellows to drive greater volumes of air at higher pressure.

By the mid-fourteenth century European blast furnaces were beginning to take shape. Taller shafts and mechanically driven bellows forced larger amounts of air into the furnace with a resultant increase in furnace temperature. Larger volumes of charcoal were also permitted by the larger furnace size, further raising the temperature. The higher temperature produced higher carbon molten iron.

The liquefied iron was guided into runoff channels constructed at right angles to the furnace. Because the arrangement of the channels around the furnace resembled piglets suckling at a sow, the product came to be known as pig iron. The fact that the iron was cast in molds gave it another name—cast iron.

Cast iron proved a valuable military innovation, as cannon barrels cast of iron could be larger and more durable. For other applications, cast iron was reheated, reducing the amount of carbon it contained; the heated iron was then worked with hand tools, combining the product of the blast furnace with traditional methods of working wrought iron. Since the iron was more pure, however, the products were less brittle.

While blast furnace technology was still relatively new to Europe in 1450, its importance could not be denied, nor the demand for its products stopped. The true age of iron was begun and, with its beginning the iron foundation of the Industrial Revolution, still three centuries in the future, was laid.

Impact

Iron changed the world. Better iron, the product of blast furnaces, changed the world more dramatically. Iron—and steel—would become the most important manufacturing materials of the modern world. Improvements in iron meant improvements in weapons, often to devastating effect. Iron plows offered an agricultural advance equaled only by the earliest cultivation of crops. Airtight iron vessels would permit experiments with pressure and steam that led to the steam engine. Demand for iron products led to a voracious demand for charcoal, which would result in the deforestation of much of Europe and England, forcing the development of coal mining, which in turn led to deeper and deeper mines. Those deep mines would, in the centuries ahead, require powerful engines to remove ground water. And those engines—steam engines—were themselves made possible only by advances in metalworking, advances whose origins may lie in bronze, but whose greatest accomplishments flowed from molten iron.

KEITH FERRELL

Further Reading

Brock, William H. *The Norton History of Chemistry.* New York: W.W. Norton, 1993.

Mumford, Lewis. *Technics And Civilization.* New York: Harcourt, Brace, 1934.

Sass, Stephen L. *The Substance of Civilization: Materials and Human History from the Stone Age to the Age of Silicon.* New York: Arcade, 1998.

The Invention of Gunpowder and Its Introduction Into Europe

Overview

Black powder, now known as gunpowder, was the chief tool of war until the modern discovery of explosives such as nitrocellulose and nitroglycerin. While gunpowder is still used in mining and fireworks, it is a much less valuable commodity now than it was hundreds of years ago. Gunpowder is a mixture of potassium nitrate (saltpeter), carbon (charcoal), and sulfur. When combined in the proper amounts, the gray powder will burn rapidly or explode with enough force to hurl a projectile, if confined in a partially closed container, when touched with an open flame or hot metal.

While it may never be actually known for certain who invented the first explosive, there is ample evidence that it originated in China during the tenth century. The Chinese are believed to have initially used black powder in their religious ceremonies. It had been a common practice to bang bamboo together to make a crackling noise in an attempt to drive away demons. Black powder was used to intensify this sound by sprinkling it on a fire. There is also support of the idea that they used black powder in fireworks and signals. The reported mixture by weight of 75% saltpeter, 15% charcoal, and 10% sulfur has changed little over time. As time wore on, the Chinese likely used black powder for military purposes. It served as a propellant for small pebbles and rockets. By the mid-thirteenth century, there are historical records that indicate the Chinese built primitive cannons where they shot stone projectiles from bamboo tubes. However, further development in the uses of black powder stagnated.

Some historians contend that the Arabs invented black powder. While the historical records are sketchy prior to the twelfth century, it has been documented that they developed the first working gun prior to 1300. It was a bamboo tube reinforced with iron, which used a charge of black powder to fire an arrow. This kind of sophistication points to the idea that the Arabs had probably been working with black powder for quite some time prior to 1300.

How the knowledge of gunpowder came about in Western Europe remains a mystery. There are two primary schools of thought on this issue. First, some historians claim that the Europeans could have gained their knowledge of the technique from a variety of sources such as the Chinese, Arabs, or possibly the Mongols. The second idea is that Europeans could have independently discovered gunpowder, without the knowledge trickling in from other lands. Not only is the knowledge source in dispute, but the date black powder became widely known is still in question.

A strong case can be made that black powder was known to at least one European as early 1242. It has been argued that the formula for black powder was contained in a manuscript written by Roger Bacon (1220-1292) at that time. The explicit instructions for making black powder and saltpeter is said to be contained in a cryptic Latin anagram that is difficult to decipher because he may have desired to keep the method secret. It is possible that since Bacon was a scholar, he could have picked up the formula in his reading of Arabic writings. However, it is evident that black powder had arrived in Western Europe by the late thirteenth century.

While there is some evidence that other cultures recognized the military potential of gunpowder, none developed weapons to make use of it as fast as the Europeans. Some scholars attribute the invention of firearms to an early fourteenth century German monk named Berthold Schwarz. Firearms are frequently mentioned in fourteenth century manuscripts from many countries and there is a record of the shipment of guns and powder from Ghent to England in 1314. The first official word of a cannon appears soon after that in Florentine document from 1326. By the middle of the century, the use of the cannon was widespread and its design similar to that of modern day.

Early firearms began to evolve in distinct directions. Cannons got larger and personal firearms got smaller. However, despite their size differences, both operated on similar principles. A small touchhole was drilled into the barrel and filled with the fine powder. Ignition of the charge could be attained by touching a slow burning cord to the powder or by providing a spark.

Background

Originally, the manufacture of black powder was accomplished by hand. It required uniform blending and mixing; otherwise the powder would not ignite. The three ingredients were ground together with a mortar and pestle. The following step involved the use of hand held wooden stamps to pulverize the material. This stamping process was gradually replaced by a mechanized one using water mills. In 1435, the first powder mill driven by waterpower was erected near Nuremberg, Germany.

Black powder is relatively insensitive to shock and friction and must be ignited by flame or heat. In the early days devices such as torches, glowing tinder, and heated iron rods were used to ignite the powder and, in most cases, a fuse was used. It consisted of a trail of the powder that led to the main charge in order to give the person lighting the fuse time to get to a safe place.

Because the burning of black powder is a surface phenomenon, the degree of granulation determines the burn rate. A fine granulation burns faster than a coarse one. While a fine granulation produces a fast burning rate that is effective, it tends to create excessive pressures in the gun barrel. Thus, in order to be a safe propellant in firearms, black powder must be made coarse. Europeans in the fifteenth and sixteenth centuries noted this and began manufacturing gunpowder in large grains of uniform size. They controlled the burning speed by varying the size of the granules.

The earliest gunpowder was made by grinding the ingredients separately and mixing them together dry. This was known as serpentine powder. The behavior of serpentine powder was highly variable, depending on a number of factors that were difficult to predict and control. Serpentine powder was also affected by vibration, which caused the serpentine powder to separate into layers according to relative density. The sulfur settled to the bottom and the charcoal rose to the top. Remixing at the battle site was necessary to maintain the proper proportions for ignition, an inconvenient and hazardous procedure that produced clouds of noxious and potentially explosive dust. Attempts were made to improve serpentine powder.

Corned powder was the next step in the refinement of gunpowder. Shortly after 1400, the ingredients of gunpowder were combined in water and ground together as a slurry. This proved to be a significant improvement in several respects. Wet incorporation was more complete and uniform than dry mixing so that the process "froze" the components permanently into a stable grain matrix so that separation was no longer a problem, and wet slurry could be ground in large quantities by water-driven mills with little danger of explosion. The use of waterpower also sharply reduced cost and allowed large quantities to be produced quickly. After grinding, the slurry was dried and then broken into grains. The grain size varied from coarse to extremely fine. Powder too fine to be used was reincorporated into the slurry for reprocessing. Corned powder burned more uniformly and rapidly than serpentine, so it was a stronger powder that rendered many older guns dangerous.

Impact

Few inventions have had an impact on human affairs as dramatic and decisive as that of gunpowder. The development of a means of harnessing the energy released by a chemical reaction in order to drive a projectile against a target marked a watershed in the harnessing of energy to human needs. Before gunpowder, weapons were designed around the limits of their users' muscular strength; after gunpowder, they were designed more in response to tactical demand.

Technologically, gunpowder bridged the gap between the medieval and modern eras. It served as a backdrop for many technological developments. As an example, the steam engine owes its roots to the precise technique of cannon boring. Gunpowder bridged the gap between the old and the new intellectually as well. The development of weapons to harness the power in gunpowder was the first significant success in rationally and systematically exploiting an energy source whose power could not be perceived directly with the ordinary senses. As such, early gunpowder technology was an important precursor of modern science.

Gunpowder permanently revolutionized European life. It hastened the decline of feudalism by changing the emphasis of battle from the cavalry to that of siege and field artillery. Gunpowder threatened the rule of the church with a competing secular power and feelings of nationalism. Immense firepower made Europe the leader in colonization. The course of human history would be vastly different had there been no gunpowder.

JAMES J. HOFFMANN

Further Reading

Ball, B. S. *Weapons and Warfare in Renaissance Europe: Gunpowder, Technology, and Tactics.* Baltimore: Johns Hopkins University Press, 1997.

McNeill, W. H. *Age of Gunpowder Empires, 1450-1800.* New York: American Historical Association, 1990.

Partington, J. R. *A History of Greek Fire and Gunpowder.* Baltimore: Johns Hopkins University Press, 1998.

The Bow in Medieval Warfare

Overview

The basic design of the bow most likely appeared around 3000 B.C. The fundamental technological principle of increasing the accuracy and velocity of a projectile far beyond that which is possible using only the force of the human arm remains unchanged. The simple design of a wooden stave, curved and pulled taught with a tension string, was used as a primary weapon in European warfare until the advent of reliable personal firearms in the eighteenth century. In other areas of the world, the bow enjoyed an even longer tenure as a favored weapon.

The new and redesigned bows of the Middle Ages—the short, long-, and crossbows—and the new class of archers who wielded them were both immortalized in chivalric literature and vilified in public discourse. Technologically simple modifications to the weapon to increase its tactical advantage on the battlefield raised compelling questions of the ethics and possible drawbacks of military technology. In Europe, the quest to create better weapons to pierce increasingly strong armor and successfully besiege fortress-like castles made the bow perhaps the most modified piece of technology of the Middle Ages.

Background

The medieval short bow was, with the exception of variation in the materials used for its construction, the unaltered descendent of its classical predecessor. The weapon was effective at shorter ranges, within 100 yards (91 m) in capable hands. Hit directly, an unarmored or lightly armored soldier would sustain grievous wounds. The short bow helped in some of the key battles of the early medieval period, from the repulsion of Viking raids to the Battle of Hastings in 1066. As the monarchies of Western Europe grew in power, and increased their military pursuits—most especially during the Crusades—demand for more powerful weaponry was met by increasingly resistant defenses.

The crossbow first appeared in Europe in tenth century Italy, however, the technological idea was most likely of foreign origin. The crossbow was constructed by turning a bow horizontal on a fixed stock and adding a projectile guide and a release trigger. Despite the advantages of small metal bolts used as the projectile instead of traditional wooden arrows, initial models of the crossbow were difficult to draw and set, resulting in slow firing times. Despite the slow turnover of shots, the crossbow proved instantly devastating on the battlefield. The crossbows used in the early Crusades had a range of 300 yards (274 m), could pierce metal armor, and even kill a horse under its rider.

Improving upon the shortbow's lack of range and power and the crossbow's tedious loading and slow firing time, the longbow emerged in Europe in the thirteenth century. The weapon did not appear with regular frequency on the battlefield until the fourteenth century and until then was limited to more localized use, especially in England where it quickly became a favorite weapon. The design of the longbow dated back to antiquity, with similar weapons described in Greek narratives. Though used in the same manner as the short bow, the medieval longbow sometimes spanned over 6 inches (15 cm) and required upwards of 100 pounds (45 kg)of tension to draw back the string. More taut bows with greater firing power could be produced by replacing the normally used vegetable fibers (usually hemp or linen) with animal sinew. Thus, the use of the weapon required skill and brawn. The longbow was a breakthrough in medieval weaponry. It could send an arrow over 300 yards (247 m) when fired by a skilled archer. The bow could be drawn and aimed so as to change the firing angle and velocity with relative ease given its size, and it was possible for an archer to fire a dozen or more per minute.

Simply adding more length and tension to the bow—beyond that of longbow—to increase its power was possible, but not practical. The mechanized crossbow of the fourteenth century achieved both an optimum of range and force without increasing exertion on the archer by incorporating a firing lever into the design. The string was drawn with a crank device that permitted more tension with considerably less effort than previous crossbows. Though the firing mechanism was no faster than older models, the mechanized bow could be fired from a variety of angles and even while the archer was reclining—thus adding an element of stealth to an already dreaded weapon.

Impact

To gauge all of the ramifications of the increased use of the bow and the introduction of long- and crossbows in the Middle Ages is nearly impossible. There are several examples of military exploits and territorial acquisitions that would not have been achieved with only swordsmen. The basic principles of bow technology were also applied to large siege weaponry such as slingshots and the counterbalanced, stone-hurling trebuchets that aided in penetrating and destroying stone fortifications.

Despite the technical advancements of archery and military technology during the Middle Ages, the short bow remained a staple of armaments. Relatively light, more readily fired, and with moderately sized projectiles, the short bow was an enormous tactical advantage on any foe in an age when hand-to-hand combat with metal weapons was the predominant mode of warfare.

The longbow was the mass-deployment weapon of choice in late medieval warfare—even in the earliest days of mortars and cannons. The longbow was at greater distances too inaccurate to be used as an individual weapon. Groups of longbowmen firing together could create, as a French nobleman from the fourteenth century described, "[a] rain of arrows to shower down upon the opponent." The distance that could separate a unit of longbowmen from the front of battle was the critical factor in its success. Its purpose was to thin out, or at least weaken, the ranks of advancing men so that there would be significantly fewer people for swordsmen, cavalry, and foot soldiers to face in hand-to-hand combat. Designed to be a medium-range bow with substantial killing power, the high firing arc of the projectile greatly reduced its penetrating power at long range. Conversely, that same property facilitated the use of the longbow as a siege weapon because it could clear modestly tall fortifications.

The deployment of the longbow in battle required a corps of physically strong and skilled archers. The demand for these archers created a new class of soldier. A longbowman, because he was removed from the front of battle, did not need to have armor, a horse, or other weapons to go into battle. These battle trappings were expensive and often excluded from the group of swordsmen and cavalry all but the nobles and knights. The yeoman archers needed little other than their weapon and skill, thus permitting men of a wider segment of society to go on campaign. Some of the most famous and decisive victories of the longbow era, such as the English defeat of the French at Agincourt, were facilitated by units of yeoman archers.

Both the short and longbows were instrumental in the largest military operation of the Middle Ages, the Crusades. However, they were met with the force, accuracy, and agility of sometimes superior constructions of bows. The distinctive design of the Byzantine and Arab composite bows (imagine drawing an S with a backward below it) gave them the ability to not only fire projectiles with greater force than the similarly sized short bow, but unlike the longbow, it could be used on horseback. A greater tactical advantage was that the new composite bow could be made easier to draw and fire by adjusting the curvature at the ends of the bow. Thus, a special corps of skilled and physically superior archers was not needed to effectively use the weapon.

The medieval crossbow was unique not only for its mechanical operation, but its controversial nature. Perhaps no instrument of war drew greater debate about its ethical use, nor did any weapon so stigmatize the person who wielded it. The crossbow was denounced so vehemently that in 1139, its use was declared an offense worthy of excommunication from the Church. The deployment of crossbowmen was often the last resort of a feudal lord or king in military campaigns, and the crossbowmen themselves were considered suspect and morally corrupt. To assure that there was an ample supply of crossbowmen when the demand arose, a ruler had to offer attractive professional perks for those willing to compromise their reputations and take up the arm. Medieval crossbowmen often made double the pay of other men-at-arms, and sometimes even received grants of land for their tenured service. Regardless of the persistence of the stigma attached to it, the crossbow remained the most feared, used, and effective weapon. Similar discourse about the deadliness, ethics, and chivalric honor of a particular weapon and its carrier did not arise again until the introduction of firearms centuries later.

ADRIENNE WILMOTH LERNER

Further Reading

Lindberg, David C.*The Beginnings of Western Science.* Chicago: University of Chicago Press, 1992.

The Invention of Guns

Overview

The invention of guns followed the development of the explosive black powder in China. The first guns were simple tubes from which to shoot explosive charges, but gradually they were made easier to load, aim, and fire. Guns revolutionized warfare and effectively ended the age of the armored knight and the castle stronghold. They have had a profound effect on human history.

Background

Gunpowder is an explosive mixture of 15% charcoal, 10% sulfur, and 75% potassium nitrate, or saltpeter. It was already being used in ninth-century China for making fireworks. Chinese books from as early as 1044 include recipes describing the necessary proportions of the three ingredients. The first primitive guns were probably bamboo tubes, fragile and ineffective firearms used in futile attempts to stop the Mongol invaders.

Europeans obtained gunpowder in the thirteenth century. A formula for making it was discovered in writings dating from the year 1249 and attributed to Roger Bacon (1220-1292). However, most scholars now believe that it was not developed independently in Europe, but was brought there from China, perhaps by Arabs or Mongols. The Europeans, having gotten hold of gunpowder, seemed more determined than the Chinese to develop it into ever more efficient weapons.

Impact

The first European firearms were cannons, and they helped to bring about the end of feudalism. Defenders behind castle walls previously could withstand long sieges from the outside, as long as they had sufficient food and water. Every nobleman had his own stronghold and was little bothered by any central authority. Beginning with the siege of Metz in 1324, castle walls started to crumble under bombardment by cannonballs. Monarchies were able to vanquish troublesome barons and enforce their power.

Cannons were very effective weapons in a siege, but soldiers soon wanted guns they could carry. At first, simple "hand gonnes" were used side-by-side with traditional weapons such as crossbows, pikes, and lances. The development of small arms quickly changed how military battles were fought. The metal armor that had protected knights from spears and lances was not sufficient protection against guns. Armor began to be perceived as not being worth its weight and impracticality, and its use gradually declined. Full metal suits eventually disappeared, to be replaced by helmets and breastplates.

So it was that the explorers who sailed out during the Age of Discovery wore little armor, but carried guns on their ships. With firearms it became easier to colonize lands that were already inhabited, like Africa and the Americas. The indigenous people, even though they were far more numerous than the invaders, could be overcome with these frightening weapons they had never seen before.

All firearms operate by the same basic process. A firing mechanism causes the gunpowder to detonate, and the explosion shoots a projectile out through a long tubular barrel. Today, some guns use compressed air or a spring instead of an explosive to launch the projectile, but gun technology was greatly refined over the centuries.

The first guns were essentially portable cannons: simple tubes, or barrels, generally made of iron or brass. These were loaded from the front with gunpowder and a lead ball. Near the sealed rear end, or breech, was a small hole for igniting the charge. Lighting the gunpowder by hand through the tiny opening was tricky in the heat of battle, and these guns had a range of only about 35 yards (32 m) even if the powder did ignite.

In the 1400s, the first mechanical firing mechanism, the matchlock, was developed. Its movable S-shaped arm, or serpentine, held a lit, slow-burning wick. Movement of the arm caused the wick to be lowered into a small pan of priming powder. The resultant "flash in the pan" ignited the main charge through a hole in the breech. The lock protected the working elements, and the mechanism freed the user's hand for aiming. However, the wick was easily extinguished by wind or rain, and its glow could give away the gunner's position at night. Still, because they were relatively inexpensive and easy to manufacture, matchlocks were in common use long after more advanced firing mechanisms were developed. They were the guns with which the Europeans colonized the New World. Introduced in India and Japan by the Portuguese before 1500, they were still being used there into the nineteenth century.

The first gun designed with a buttstock allowing it to be fired from the shoulder was the harquebus, a type of matchlock gun invented in Spain in the mid-1400s. The musket was developed about a century later as a larger version of the same basic firearm. Like its predecessors, it was loaded from the front; breech-loading guns were not invented until the nineteenth century. Early guns were all quite limited in terms of their range, accuracy, and reloading rate, but they did intimidate the enemy, and by about 1525 guns had already been the decisive factor in several battles between Spain and France.

Further innovations were to follow. The flintlock used a sparking mechanism rather than a lit wick to ignite the gunpowder. These guns, the first to have a trigger, were easier to aim and could be used while on horseback. However, they were still unreliable in the rain. This problem was not solved until 1807, when a Scottish clergyman and hunter named Alexander Forsythe invented the percussion firing mechanism. It depended upon a mercury fulminate compound that would explode upon impact. The percussion firing mechanism led to the development of cartridges that combined the mercury fulminate, gunpowder, and projectile in a single package and standard caliber gun barrels to accommodate them.

To increase the accuracy and range of the weapons, rifling was developed. Rifling involves incising spiral grooves in the inside of the barrel to cause the bullet to spin as it is ejected. As anyone who has tried to throw a football knows, spinning it on its axis helps it fly farther and straighter. Medieval archers had observed this phenomenon with their arrows as well, and rifling was tried out as early as the fifteenth century. However, large-scale manufacture of the rifle was not feasible until the nineteenth century.

It is almost impossible to overstate the impact of guns on human civilization, even beyond their effect on the methods of waging war. Although humans have always had a tendency towards violence, guns made it easy to kill from a distance. They are often used legally for hunting, for defense of home and family, or for recreational target shooting. Tragically, they are even more often associated with criminal activities. History-changing political assassinations, which in earlier times required close access to the intended target, now require only a suitable vantage point. Societies worldwide continue to struggle with the Pandora's box opened by the invention of guns.

SHERRI CHASIN CALVO

Further Reading

Boothroyd, Geoffrey. *Guns through the Ages.* New York: Sterling Publishing Co., 1962..

Diamond, Jared. *Guns, Germs and Steel: The Fates of Human Societies.* New York: Norton, 1997.

Needham, Joseph. *Science and Civilization in China.* Cambridge: Cambridge University Press, 1962.

Pope, Dudley. *Guns.* London: Spring Books, 1969.

Temple, Robert. *The Genius of China: 3,000 Years of Science, Discovery and Invention.* New York: Simon and Schuster, 1986.

Williams, Suzanne. *Made in China: Ideas and Inventions from Ancient China.* Berkeley, CA: Pacific View Press, 1997.

The Chinese Invention of Gunpowder, Explosives, and Artillery and Their Impact on European Warfare

Overview

The development of feudalism in Europe was accompanied by the introduction of the heavily armored, horse-mounted knight and the fortified castle. While Eastern technology helped pave the way for these developments, it also helped to ensure their eventual obsolescence. Gunpowder was a Chinese invention that revolutionized warfare. The Chinese used explosives on a wide scale beginning in the tenth and eleventh centuries. The cannons, flamethrowers, and grenades that they used in battle were quickly adopted by European forces for battles on land and at sea. However, Europeans refined the applications of gunpowder and improved the devices that used gunpowder, producing weapons that dramatically transformed the nature of warfare.

Background

The formula for gunpowder is deceptively simple. It is formed from the combination of three materials: saltpeter, sulfur, and charcoal. Saltpeter, or potassium nitrate (KNO_3) is regarded as the most important ingredient, and constitutes up to 75% of the recipe. In order to make gunpowder, one has only to mix these compounds together as powdered solids. Once ignited, this mixture produces temperatures that range from 2,100° to 2,700° Celsius. Ignition produces a volume of gases resulting in 274 to 360 cubic cm of gas per gram of powder. These gases produce explosions and propel bullets and cannonballs.

It was not until the Song Dynasty, which began in A.D. 960, that this recipe was documented in China. Gunpowder and firearms as we know them were developed during this period. Historians regard the Song Dynasty as a highly prosperous and relatively peaceful time, and assert that the Song Dynasty marks the entrance of China into the modern era. A materialistic culture arose during the Song Dynasty, and the circulation of money became widespread. The civil service examination for Chinese officials also became extremely important during this era and is indicative of a level of intellectual consensus that had not previously existed in China.

While written records that mention sulfur and nitrates, two of the ingredients of gunpowder, appeared as early as the Han Dynasty (948-951), it was not until the Sung Dynasty that records mention the creation and use of military explosives. Earlier documentation concerning these materials refers to their usefulness in elixirs made by alchemists. Such uses for chemical compounds in earlier eras reveals an intellectual reliance on superstition. The development of gunpowder in the Sung Dynasty, on the other hand, hints at a modern interest in the practical applications of intellectual speculation.

Impact

One of the earliest known uses of rocketry in Chinese warfare dates to the fall of the Ch'in dynasty during the thirteenth century. The great Mongol leader Khan Ogodei had gained power and was intent on eliminating the Chin and their fierce resistance to his armies. In 1232 the Mongol army held the Ch'in capital of Pien, also known as K'ai-feng, under siege. While the city did eventually fall to the Mongols, its inhabitants were able to defend themselves effectively. Indeed, this was one of the first battles in recorded military history in which firearms were used by both sides. At this stage of development, gunpowder was used primarily in ceramic grenades that were hurled by catapults. Used by the defenders of Pien, the grenades proved deadly to the Mongol warriors and their horses. The defenders of Pien used catapults because, at that point, Chinese cannons, like the early cannons implemented by the Europeans, had only a limited effectiveness.

The Chinese defenders of Pien used another weapon—the flamethrower—that, unlike early ceramic grenades, was used primarily by the Chinese and was not widely borrowed by European armies. Medieval Chinese artisans are credited with the invention of a flamethrower, which was referred to as the fire lance. In order to form a fire lance, Chinese inventors pasted together nearly 20 layers of strong yellow paper and shaped these into a pipe over 24 in (60 cm) in length. They then filled this pipe with iron filings, porcelain fragments, and gunpowder, and fastened the pipe to a lance. Soldiers who handled these flamethrowers carried with them onto the battlefield a small iron box containing glowing embers. In battle, the soldiers used these embers to ignite the fire lances. These weapons produced flames over 9.84 ft (3 m) long. Also, the porcelain shards and iron filings that were packed into the tube shot out in a deadly cloud of shrapnel.

Such a weapon clearly anticipated the European-developed handgun, in that it was portable and could be operated by a single soldier. However, it was not the weapon that most directly influenced European military technology. Instead, in Europe, the cannon became the most common device to rely on gunpowder. By the middle of the fourteenth century, cannons were a common sight on European battlefields. But the earliest cannons were not especially effective. For the most part, European cannons were based on the Chinese design, and were not particularly accurate or powerful.

It was not until the fifteenth century that cannons capable of seriously damaging fortified castle and city walls were developed. By this time, siege artillery that used gunpowder replaced siege engines, such as the catapult, which was characteristic of medieval warfare. The new cannons were smaller and much easier to transport than large siege engines. Many siege engines, such as the French trebuchet, were too heavy to move. These had to be constructed onsite by the attacking army. The early cannon could be transported to the battlefield via carriage. However, these early cannons were not particularly maneuverable during battle. As a result, they did not seriously impact the style of warfare initially. While cannons slowly replaced siege engines, they did not change how battles were fought. Because they were locked in place once they were set up, these early cannons were also easy targets for heavy artillery. Furthermore, because of their weight and design, these cannons could not be adjusted very easily once they had been placed on the battlefield. Attached to primitive carriages, such cannons were nearly impossible to aim.

At the same time that cannons began to appear, the portable handgun was developed by European armies. The advancements that allowed the handgun to dominate warfare were, for the most part, European in origin. Gunpowder and early cannons were imported from China, but the Chinese did not develop or refine their firepower for several centuries. Indeed, by the sixteenth century the Chinese bought the majority of their firearms from the Portuguese.

Early handguns were little more than miniature cannons mounted on sticks. They were practically insignificant in battle when compared to the crossbow or the longbow, both of which had a powerful impact on diminishing the superiority of heavily armored cavalry units. It was not until the middle of the fifteenth century that the development of the matchlock improved the viability of the handgun. The matchlock was little more than a trigger-operated hook that allowed the operator to aim the gun with both hands. While this type of gun, the harquebus, greatly improved the rate and accuracy of fire, the longbow and crossbow were still superior weapons.

However, while the initial effects of firearm usage were limited, the same cannot be said for medieval navies. The cannon profoundly influenced naval warfare and led to the domination of the sailing ship over the galley. Prior to the cannon, the sailing ship suffered major disadvantages in close-quarter combat. Difficult to maneuver and dependent on the wind, early sailing vessels were no match for galleys on calm steady waters. Once cannons were fitted to sailing ships, however, the smaller, more maneuverable galleys were unable to approach.

The large cannon did not drastically alter land combat, but it revolutionized warfare at sea. The large guns could cause serious structural damage to ships. They operated like the rams that were fitted onto galleys, but allowed the attacking ship to maintain some degree of distance. Cannons damaged the rigging and attacked the buoyancy of ships. Large cannonballs destroyed ships in the same way that they toppled castle walls. Furthermore, cannonballs that pierced wooden ships showered the interior of the struck ship with dangerous splinters.

Interestingly, smaller cannons and antipersonnel guns had little impact on naval warfare tactics. Armies on land quickly implemented and improved these technologies, but demonstrated ambivalence to the heavy cannon. The reverse was true for navies, for which the large sea cannon was the most important weapon of the time.

Often, ships were outfitted with more than 30 cannons on each side. These cannons fired balls that weighed 10 to 20 lbs (4.5 to 9 kg). The bow-to-bow fighting style of galleys in naval warfare was replaced by fighting ships aligned broadside to broadside and exchanging volleys of cannon fire. The heavy artillery on sailing vessels allowed sailors to crush the highly maneuverable, but fragile galleys that had dominated the Mediterranean for so long.

This new style of warfare determined more than the dominant type of sailing vessel. The heavy reliance on the naval cannon also abolished the need for infantry combat between soldiers and sailors on opposing ships. Prior to the development of the naval cannon, ships carried large numbers of armed soldiers who attempted to overwhelm the fighting force of the ships they attacked.

The defeat of the Spanish armada by the English navy in 1588 demonstrates the conclusive transition to artillery combat at sea. The Spanish armada had many more soldiers and ships than the English fleet. However, while the Spanish still relied on traditional boarding practices, the English were dependent on cannon fire. The English used their superior firepower to whittle away the Spanish forces. Likewise, the maneuverability of the English ships prevented the Spanish from engaging in close-range shock tactics.

The Chinese invention of gunpowder resulted in numerous weapons and applications that transformed battle. While it took a long time for armies to fully realize the potential offered by gunpowder, the new weapons made possible by its invention and availability eventually determined the victors of many important conflicts.

DEAN SWINFORD

Further Reading

Hall, Bert S. *Weapons and Warfare in Renaissance Europe: Gunpowder, Technology, and Tactics.* Baltimore: Johns Hopkins University Press, 1997.

Jones, Archer. *The Art of War in the Western World.* Chicago: University of Illinois Press, 1987.

Li, Dun J. *The Ageless Chinese: A History.* New York: Scribner, 1978.

McNeill, William H. *The Pursuit of Power: Technology, Armed Force, and Society Since A.D. 1000.* Chicago: University of Chicago Press, 1982.

van Creveld, Martin. *Technology and War: From 2000 B.C. to the Present.* New York: Free Press, 1989.

The Evolution of Medieval Body Armor

Overview

Body armor is protective clothing that has the ability to repel weapons used in combat against the wearer. Armor is designed to ward off attacks from both sharp and blunt trauma. While the body armor used today is primarily designed to give protection from projectiles moving at high velocities (i.e. bulletproof vests), armor has traditionally fallen into one of three categories: soft, mail, or rigid armor. Soft armor was usually comprised of layers of leather, fabric, quilting or felt, and often, a combination of these materials was used. Mail armor was usually made from iron or steel rings that were interwoven. This gave the wearer some flexibility without sacrificing a great deal of protection. Rigid armor was often made of metal, wood, or any other highly resilient material that offered maximum protection to the wearer. This category also includes the familiar plate armor worn by knights in the latter part of the Middle Ages. The protective plates were made of metal and eventually covered the entire body. The plates were often riveted together and connected internally by leather straps so that they offered maximum protection, yet still allowed the wearer to have freedom of movement.

It is assumed that the use of body armor predates historical references. It is known that as early as the eleventh century B.C., Chinese warriors wore armor made of rhinoceros skin. Most likely, prior to that time, other primitive combatants wore protective helmets and clothing made of leather, animal skins and other similar protective materials. Thus, it is more than likely that the first type of armor used by man was the soft type, but actual verification predates written history. Padded garments were the predecessors of more intricate and protective armor. As technology advanced, so did weaponry and defenses. The makers of weapons and armor were in a constant struggle to keep up with the advances in each field; an advance in armor was offset by an advance in weaponry. This struggle dictated the evolution of armor from a soft outer garment to the knight's full metal suit.

The armor of knights changed over the course of time. Initially, knights wore a helmet of quilted fabric covered with leather that may have been covered with mail. The 1300s witnessed the use of a stronger helmet with more protection for the skull and face. Plate armor was added to protect the vital organs and helped to cover areas left vulnerable by a lack of mail. Eventually, elongated pieces of plate were used to protect many of the joints and this evolved into the metal plates that are often associated with knights by the 1400s.

Background

While early body armor was primarily soft, metal protection was introduced with the advent of metal weapons. For obvious reasons, the use of metal to protect the wearer from a blow or piercing objects would be much more effective than soft material. The first metal armor consisted of overlapping bronze plates sewn onto a protective garment. This offered real protection against attacks. This type of armor eventually evolved into chain mail as stronger metals were discovered. Yet, the soft armor was not totally abandoned. Padded undergarments continued to be worn to help absorb some of the shock from blows to the body.

Chain mail consists of interwoven rings of metal that provide a flexible, yet strong defense from slashing weapons. The fabrication of mail was extremely costly and labor-intensive. Mail served to be the main armor of Europeans until the fourteenth century. Even after that, mail was worn to help protect the inevitable gaps that came from wearing plate armor.

The *byrnie* was one of the earliest forms of armor that used mail. It was originally a sleeveless suit of armor that covered the upper half of the body. The byrnie consisted of a soft backing overlaid with hardened metal rings. Another version of this was a garment that is now referred to as chain mail. It consisted of interlaced metal rings that proved to be much lighter than the byrnie, but less protective. Padded garments known as *gambesons* were often worn under the mail to make up for the lack of protection from crushing blows that was associated with chain mail. When this was extended to the head, it was known as a coif. The coif covered the head and neck, but left an open area for the face. This provided the head with a significant degree of protection.

The ensuing years saw the development of better coverage of the body with mail and more sophisticated helmets. The art of fabricating mail became so advanced that even gloves of mail could be made with fine mesh. Around the same time, plate armor was quickly overtaking mail as the preferred choice for protective garments.

The earliest plate armor appeared in the early 1200s in the form of thin plates worn beneath the gambeson. Later, external plate armor began to appear as covering for the joints. Around 1250, the first breastplate made of plate armor, called the cuirass, appeared in Europe. Plate protection then appeared on various parts of the body for the next 100 years until the entire body was protected by plate armor. However, soft armor and mail were still used under the metal plates.

Gothic-style plate armor appeared in the early fifteenth century and represented the pinnacle of personal armor protection. These armor suits gave full body protection covering its wearer literally from head to toe, with only a slit for the eyes and small holes in the helmet for breathing. The fabrication of these suits was so advanced that individual finger joints were made in gloves. The area of the shoulder perhaps demonstrated the greatest sophistication. The knight could easily wield a weapon with full range of movement, yet still have his body completely covered with armor.

As time went on, the armor became much thicker and heavier, in order to protect the knights against the latest advances in warfare. Interestingly, larger breeds of horses also appeared at this time in order to support the heavier armor. In addition, armor and mail were consistently used to protect the horses as well. The increased protection made the wearer much slower, required more energy for movement and made it difficult to stay cool. The full helmet obscured vision, hampered breathing and made it impossible to communicate during battle. Body armor was so intricate that it needed to be made by skilled craftsmen using expensive materials. A complete set of body armor was a huge cost to the wearer.

Eventually, the modernization of weapons made body armor obsolete. The thickness of the armor plate needed to stop an arrow gave way to the impractical thickness needed to stop projectiles shot from firearms. By the sixteenth century, body armor was no longer used and served primarily for ceremonial and decorative functions.

Impact

The advances in body armor had a significant impact on medieval society. In fact, it could be argued that armor helped to shape society and encourage feudalism. Knights played a major role in this social development and perhaps their greatest symbol was their armor. The armor was, in many respects, what allowed knights to be knights. A knight's position in society was determined somewhat by his ability to protect that position and for this, his armor served a knight well. Armor allowed knights to exert power over the lower class, both through the physical protective nature of the armor, as well as through the psychological advantage that they gained by wearing it. Armor was an integral part of being a knight and even served to help define behavior, through a code of ethics commonly referred to as chivalry. For example, by the late twelfth century, it was considered a dishonor for a knight to attack another knight without allowing the opponent to don his armor.

Armor also helped accelerate technological advances in weaponry. There was a constant battle between the development of offensive and defensive weapons in attempts to keep one type ahead of the other. Each advance in offensive arms was soon met by a change in the armor to help combat it. In turn, the offensive weapon was further advanced. This seesaw struggle pushed the development of arms and defenses to the limit, leaving many technological advances in its wake.

While it will never be known for certain, it is likely that medieval society would not have developed as it did without the advances in body armor. Armor had tremendous impact on knights and medieval European society as a whole by providing both a protective function and serving as a symbol for all to see. Armor helped to define behavior, create and enforce a social system, and encourage militaristic technology. In these ways, armor helped to shape the Middle Ages.

JAMES J. HOFFMANN

Further Reading

Ashdown, C.H. *European Arms and Armor*. New York: Barnes and Noble Books, 1995.

Nicolle, D. *Arms and Armour of the Crusading Era, 1050-1350: Western Europe and the Crusader States*. London: Greenhill Books/Lionel Leventhal, 1999.

Oakeshott, R.E. *The Archaeology of Weapon: Arms and Armour from Prehistory to the Age of Chivalry*. Dover, New Hampshire: Dover Publications, 1996.

The Development of Canal Locks

Overview

Canals have been used since ancient times to carry water where it is needed or allow transportation where natural waterways do not go. The canal lock was developed in China, and first used in Europe during the Middle Ages. Locks enable ships to go from one water level to another, thus making many more transportation routes possible.

Background

Canals are artificial inland waterways. They are built for water supply, sewage removal, crop irrigation, drainage, and transportation. The first known canals were dug in the Middle East thousands of years ago. King Sennacherib, who ruled Assyria in 704-681 B.C., had a 50-mile (80-km) long stone canal built to supply the city of Nineveh with fresh water. Several additional canals were built elsewhere in the Mesopotamian region, as well as in Egypt and Phoenicia. In about 510 B.C. the Persian king Darius I (550-486 B.C.) even attempted to connect the Nile with the Red Sea.

The ancient Romans embarked on major public works projects throughout their empire, and these included canals. Roman canals were used primarily for military transportation, although some were built to provide drainage. With the decline of Rome, the European canals fell into disrepair. In the late Middle Ages, however, as trade expanded, many were reclaimed and the system was enlarged, until it carried as much as 85% of Europe's commercial traffic. This was the period in which the extensive canal network of Venice was built.

The Chinese began building canals about 2,300 years ago to provide easier access to their major rivers and the ability to transport grain from the rich river valleys to the cities of the north. The Grand Canal, which was begun in 540 B.C. and opened in sections over the next 1,800 years, remains the longest artificial waterway in the world as well as the oldest still in use. It travels approximately 1,000 miles (1,600 km) between the Chang (Yangtze) River and Beijing (Peking). The Grand Canal is a *summit canal*; that is, it follows the contour of the landscape. The changing terrain required boats to go up and down hills, which they cannot normally do. The Chinese solved this problem with water traps called locks.

Impact

Locks are used to move ships from one water level to another, for example, in a canal between two lakes that are at different heights above sea level. The earliest locks, called flash locks, were simple movable barriers across the canal or river. The barriers, or lock-gates, slowed the water flow and allowed water to build up behind the gate. When the gate was opened, a boat going downstream would be carried smoothly over any downhill drops or shallow areas by the surge of water. However, a boat heading upstream would have to be hauled through the gate against the flow. Flash locks were used in China by the first century B.C. in canals near Nanyang. In A.D. 983, the provincial transportation commissioner Chaio Wei-Yo built the first recorded *chamber* or *pound lock* on the Grand Canal. A pound lock consists of a section of the canal that is enclosed with watertight gates at both ends so that the water level inside can be raised or lowered. A ship that needs to move to the lower water level would enter through the upper gate. When the gate closes behind the ship, water is allowed to drain into the canal below the lock by opening ports or sluices in the lower gate. When the water within the lock is at the same level as the water below the lock, the lower gate is opened and the ship goes on its way.

To go from a lower level to a higher one, the process is reversed. The ship enters through the gate at the lower end of the lock, and sluices in the upper gate are opened to allow water in from the upper level to raise the water level in the lock. When the two levels are equal, the upper gate is opened and the ship moves on.

The pound lock provided a way to get a ship from a lower water level to a higher level without dragging it. Since most of the water in a pound lock is held between the gates at any given time, it was much quicker to use than the flash lock, and required much less water flow to operate. Despite this improvement, water supply in canal operation was always a consideration. Side pools intermediate between the two water levels were sometimes used to store water when the lock was being emptied, and serve as a reservoir from which to fill it again. The Grand Canal had side pools as well as large holding tanks to ensure that water would be available to operate the locks.

In the eleventh and twelfth centuries, single locks began to be used in the Low Countries (today's Belgium, Netherlands, and Luxembourg) and in Italy. The first pound lock in Europe was built by the Dutch in 1373, at the junction of the Utrecht Canal with the River Lek at Vreeswijk. It used guillotine gates, which move up and down to open and close. Around 1500, miter gates began to appear in Italy. These open like double-leaf doors and close to form a V-shape, held shut by the pressure of the water pushing against them. The invention of miter gates is often attributed to Leonardo da Vinci because the earliest drawings of them are found in his notebooks. Miter gates can withstand the forces of water pressure better than guillotine gates, allowing locks to be made wider than they had been before.

Canal locks must often be built in a series so that a large water level difference can be handled gradually. For example, the modern Welland Canal must raise ships 325 feet (99 m) between Lakes Ontario and Erie. It does this using a series of eight locks. The pound lock enabled most of the world's canals to be built, because without it changes in elevation would have made canals impractical.

Canals have had a major impact on trade and politics by providing significant transportation short cuts. In eighteenth century England, canals improved transportation between the factories in the north, their raw materials, and their markets, thereby helping to foster the Industrial Revolution. This in turn provided the improved technological capabilities that allowed massive canal projects to be undertaken in the early 1800s.

Many canals were built in the interior of the United States to link the Great Lakes and the Mississippi River. Canal-building in the United States tapered off when railroads and eventually cars and trucks provided alternate means of transportation.

The vast country of Russia has also benefited greatly by building canals. Canals have made transportation easier across Russia's large expanse. By providing access to the White, Baltic, Black, Caspian, and Azov Seas from Moscow, canals have made the Russian capital into a major port despite its inland location.

Internationally, two canals have been particularly important. The Suez Canal in Egypt, dating from 1869, connects the Mediterranean and Red Seas, and was the belated realization of the project envisioned by Darius I. It provides European ships with a passage to India that avoids a 6,000-mile (9,660-km) trip around Africa. The path of the Suez Canal through Egypt's lakes and marshes allowed engineers to avoid using locks, which despite their advantages do cause delays. Other modern alternatives to locks include engineering the terrain itself or installing hydraulic lifts.

The Panama Canal, with its 12 locks, opened in 1914. It links the Atlantic and Pacific Oceans across the narrow Isthmus of Panama. Ships can travel between the East and West Coasts of North America through the Panama Canal without circumnavigating the entire continent of South America, a difference of almost 8,000 miles (12,900 km).

SHERRI CHASIN CALVO

Further Reading

National Geographic Society. *Builders of the Ancient World.* Washington, D.C.: National Geographic Society, 1986.

Needham, Joseph. *Science and Civilisation in China.* Cambridge: Cambridge University Press, 1962.

Temple, Robert. *The Genius of China: 3,000 Years of Science, Discovery and Invention.* New York: Simon and Schuster, 1986.

Williams, Suzanne. *Made in China: Ideas and Inventions from Ancient China.* Berkeley, Calif.: Pacific View Press, 1997.

The Spinning Wheel: The Beginning of the Medieval Textile Industry

Overview

The spinning wheel revolutionized the production of yarn, which increased productivity and led to the establishment of a thriving medieval textile industry. In turn, this helped set in motion forces that would create a perfect environment for the beginning of the Renaissance. Finally, the spinning wheel would help elevate the economic and social standing of medieval women.

Background

Textiles have played an important part in human history. Originally our ancestors used animal skins for protection against the elements. Over time this close connection to animals allowed early humans to use a number of fibers from goats, sheep, wolves, and rabbits to create the world's first textiles. The same eventually held true for fibers obtained from plants, such as flax, cotton, and hemp.

Archeologists believe that the history of textile manufacturing extends back almost two million years. Research has also established that wool was probably the original fiber used to develop the first textiles. This is due to the fact that sheep easily adapt to a vast number of environmental conditions. Woolen artifacts dating from the fourth millennium B.C. have been discovered in archeological sites from modern day Iraq to the plains of Central Asia. Historians believe that the early pastoral people of Central Asia were the first to domesticate sheep for this purpose.

The earliest manufacturing of linen cloth dates back to about 6500 B.C., on the Anatolian peninsula. The ancient Egyptians were also successfully cultivating flax and manufacturing high quality linen by the middle of the fifth millenium B.C.. Linen was not only used for clothing, but it also played an important role in the Egyptian religion. Researchers have discovered mummies wrapped in as much as 2,953 ft (900 m) of fine linen.

The use of linen migrated from Egypt to the great classical civilizations of the Mediterranean Basin. Both the Greeks and the Romans used linen in their clothing. Along with the expansion of the Roman Empire, the use of both linen and wool spread into Western Europe and would play an important role in both the late classical and the medieval economies.

Cotton entered the world of textiles from South Asia. It had been cultivated for millennia in the rich fertile soil around the Indus River Valley. It became a staple of textile manufacturing in the Mediterranean Basin, introduced by the returning veterans of Alexander the Great's (356-323 B.C.) army.

Attempts to cultivate cotton were made in Malta, Egypt, and the lands adjacent to the Persian Gulf. Each of these attempts achieved some success, but they were never able to attain the quality of Indian cotton. By the early medieval period, historians found evidence of cotton cultivation around the Black Sea and as far east as China, but none of these varieties could match the quality of the South Asian species.

China's major impact on the textile industry revolved around the production of silk. Archeologists have been unable to pinpoint the exact beginning of the use of silk, but evidence suggests that it was widely used by the middle of the second millennium B.C. Widespread Chinese trade in silk began about 1700 B.C. under the control of the Shang Dynasty. It was during this period that a highly organized silk industry developed. The politically powerful Shang bureaucracy established and controlled the entire process of this new industry. In the second century, with the founding of the Han Dynasty in China an extensive Eurasian trade complex was established between the Chinese and Roman Empires. The trading network, known as the Silk Road, was in full operation by 100 B.C., with silk products found as far north and west as Scandinavia.

Every one of these cultures used the labor of women for the task of spinning. Spinning is the process of twisting fibers together into a continuous thread. This skill was identified so closely with women that the implements used in the process became symbols of the feminine gender. Both the terms "spinster" and "distaff" have been widely used in the West for centuries, the former designating an unmarried woman, and the latter describing the maternal side of the family. Greek and Roman women spun yarn from a hand distaff. This was a slow process and severely limited the production of cloth. The invention and implementation of the spinning wheel in the later Middle Ages increased the amount of yarn available for the production of textiles. This would have an important social and economic impact on Europe.

Impact

The success of the spinning wheel created a textile revolution in Europe. So important were textiles to the economy that Europe experienced the formation of textile guilds. These organizations regulated both the quality and price of this valuable product. They afforded their members significant political, social, and economic power.

Trade fairs that specialized in textiles became the center of medieval economic life. The vast majority of new trade routes were created to connect these great "cloth fairs." The cities that grew up around these textile fairs would be the centers of change as Europe moved from a traditional agricultural economy to a new one based on commerce. This new economic reality was based upon textiles and trade and created the necessity for a new model of exchange based upon a vibrant set of economic variables.

The movement of large quantities of valuable goods created the need for an extensive long-range banking system whereby merchants from different countries could safely exchange and store the significant amounts of money needed to buy, ship, and sell on such a large scale. Over time, certain families established themselves as Europe's first class of international bankers. The most prominent of these families were the Medici and the Fuggers. These banking institutions established the price structure that formed the basis of the new commercial economy. Additional opportunities for wealth were created because of this new system. Great financial success could be achieved by concentrating solely on the movement of textiles. A revitalized shipping industry created a class of middlemen whose wealth was based upon the movement of textiles along the great water routes of Western Europe.

A new class of professionals emerged that were necessary for the day-to-day conduct of business within the textile industry. Great quantities of money in the form of profits, wages, and investments necessitated accurate bookkeeping. Experienced, talented men who kept accurate account of the movement of this money were of vital importance to this new economy. Lawyers also had an important role to play. New legal concepts centered on long-term contracts and international trade required the development of new statutes to regulate and protect the individuals taking part in this trade. The search for legal precedents concerning long distance trade was one of the important factors leading to the onset of the Renaissance. Lawyers on the Italian peninsula searched archives looking for ancient Roman legal manuscripts. The Roman Empire had an extensive trade network and developed legal codes that allowed the empire to function successfully. These legal scholars came into contact with the writings of Roman lawyers, scholars, philosophers, scientists and their commentaries on the intellectual leaders of Greece and Rome. This was the beginning of early modern Europe's love affair with the classical world, which today is known as the Renaissance. In addition, the financial backing for the research that created the European humanistic worldview came from wealthy banking and textile families such as the Medici.

The invention of the spinning wheel also helped change the lives of medieval women. Three distinct cultural forces influenced the medieval world: the classical world, Christianity, and Germanic culture. The classical world relegated women to second-class status. Both the Greeks and the Romans believed women were basically flawed and incomplete. Christian theology, which minimized both the pleasures of this world and the sanctity of women, helped perpetuate the view of women as second-class citizens. The celibate life of the Christian clergy emphasized poverty, chastity, and obedience. Medieval theologians depicted women as physically, mentally, and morally weaker than men. This misogynistic worldview sanctioned coercive treatment of women, including wife beating.

The most prominent early Christian, Saint Paul, described marriage as "a debased state." Saint Paul's famous statement, "It's better to marry than to burn," lowered marriage to a religious state whose sole purpose was to sanction sexual relations. Paul encouraged widowers not to remarry, but to dedicate their remaining time on earth to the life and welfare of the Church.

These negative practices toward women were challenged by ancient Germanic custom. The same German tribes that brought down the Roman Empire were among the most egalitarian societies in the world. Their laws allowed women to inherit both money and property. The spinning wheel gave women both economic and social power. Since women had traditionally dominated the craft of spinning, they naturally adapted very easily to this new technology. As their productivity and power increased, women were able to demand and receive important concessions within medieval society. The greatest success was that women were granted the freedom to form their own craft guilds. This allowed them to control both the quality and price of the product. In time, unmarried women could own their own shops and become economically self-sufficient. If they were widowed, women had the right to pass the business on to their daughters. This was the first step in the centuries-long march toward equality. Hundreds of years later, twentieth-century author Virginia Woolf would write, "a woman must have money and a room of her own if she is going to write." The craft of spinning and the guilds they produced were an important advancement in that direction. The spinning wheel also helped establish a textile industry that would be the first to successfully use the technology of the Industrial Revolution.

RICHARD D. FITZGERALD

Further Reading

Duby, George. *Rural Economy and Country Life in the Medieval West.* Philadelphia: University of Pennsylvania Press, 1998.

Gies, Francis, and Joseph Gies. *Women in the Middle Ages.* New York: Crowell 1978.

White, Lynn T. *Medieval Technology and Social Change.* Oxford: Clarendon Press, 1962.

The Magnetic Compass

Overview

The magnetic compass was an important advance in navigation because it allowed mariners to determine their direction even if clouds obscured their usual astronomical cues such as the North Star. It uses a magnetic needle that can turn freely so that it always points to the north pole of the Earth's magnetic field. Knowing where north is allows the other directions to be determined as well. The compass was invented by the Chinese, and was widely used for navigation beginning in about the thirteenth century.

Background

The phenomenon of magnetism was known to the ancient Greeks, but the magnetic compass was invented by the Chinese. The thirteenth century explorer Marco Polo (1254-1324) is said to have brought a compass with him when he returned to Venice after his twenty years of service in the court of Kublai Khan (1215-1294). He may indeed have carried home such a souvenir. However, the knowledge that a piece of the naturally magnetic iron ore magnetite (Fe_3O_4), called a lodestone, would align itself from north to south if allowed to move freely, seems to have arisen at least a century before Marco Polo in Europe and the Arab world.

Scholars continue to debate whether this discovery was independent, or whether the new technology was spread westward from China through trade or other contact between civilizations. Some speculate that the Chinese may have used lodestones for navigation in voyages to the east coast of India in about 100 B.C. Chinese references to a "south pointer" are found in texts as early as the first century A.D. The south pointer was a spoon carved from lodestone, which was allowed to rotate on a smooth brass plate until its handle pointed south.

Magnets align themselves along the north-south axis because the Earth itself is a huge magnet. The poles of the Earth's magnetic field roughly correspond to the rotational axis of the globe. This means that the north magnetic pole is in the approximate direction of the north geographic pole, or true north. A light magnet that can move freely will align itself in the north-south direction. However, a heavy bar magnet lying on a tabletop will not move because gravity and friction counteract the magnetic force.

Before the magnetic compass, sailors navigated by the position of the stars. They knew, for example, that the North Star, Polaris, remained in a fixed northerly position in the sky while the other stars seemed to move around it. But this method was completely useless when the stars were obscured by clouds or fog. As a result, early mariners preferred to stay near the coast as they traveled from place to place. Even the daring Vikings, the first Europeans known to have reached the New World, did so in the northern latitudes where the open water distance was shorter, and used North Atlantic islands like Iceland and Greenland as stepping-stones.

Impact

The first magnetic compasses were needles or other bits of iron that were magnetized by being rubbed on a piece of lodestone, and then attached to straw or cork so that they would float in a bowl of water. Sometimes the needle was simply hung by a thread. Left free to spin, the splinter of iron would always align itself in a north-south direction. At first this phenomenon was associated with sorcerers more than with sailors. In China, the compass was valued for its contribution to feng shui, the protection of a site from harmful influences by adjusting the orientation of buildings and furnishings. In medieval Europe, lodestones were believed to have magical powers. As a result, the Church initially denounced the magnetic compass as an instrument of Satan.

The earliest records of compasses being used for navigation date from around the eleventh or twelfth century. The Chinese military text *Wu Ching Tsung Yao*, written about 1044 by Tseng Kung-Liang, describes a technique used by soldiers lost at night or in bad weather. They took a thin sliver of iron, heated and then rapidly cooled it to magnetize it, and floated it in a bowl of water. The earliest known mention of the compass in a European text was by the Englishman Alexander Neckam (1157-1217) in his 1180 textbook *De Utensilibus* (On instruments). By the mid-1200s, compasses were being used by the Vikings and Arab merchants. The ability to navigate no matter what the weather gave sailors the courage to venture farther from the sight of land.

Navigational aids were of particular importance to the English, whose territorial aspirations exceeded the bounds of their small island nation, and who grew to rely heavily upon their navy. English mariners and inventors developed several refinements to make the compass more useful. At first, compasses were only supplemental instruments, resorted to when neither the Sun nor the North Star could be seen. The devices could be frustrating, because the bobbing cork introduced uncertainty and made the compass hard to read, particularly aboard a moving vessel.

In the late 1200s, the magnetic needle was attached to a pivot standing on the bottom of the compass bowl, restricting it to circular motion. Soon mariners began mounting a compass card on the pivot directly beneath the needle, marked off with 32 points of direction. The points of direction consist of the four cardinal points (north, south, east, and west) and the intercardinal points between them. For example, the intercardinal points between north and east are called north by east, north-northeast, northeast by north, northeast, northeast by east, east-northeast, and east by north. Modern compasses generally include fewer intercardinal points (omitting "north by east," and so on) and are instead marked off with the 360 degrees of a circle.

In many compasses the bowl was still filled with water or oil even though the pivot actually held up the needle, since the fluid helped to dampen the effects of extraneous motion. However, fluid-filled compasses tended to leak and were difficult to repair, so dry-card compasses were also built. To keep the compass level despite the motion of the ship, the compass was hung on gimbals, or rings mounted on its side. This independent suspension allowed it to swing freely, rather than being placed upon a fixed surface that would tilt with the ship.

The technical improvements to the magnetic compass made it easier to read and less subject to being tilted and shaken, but did not eliminate its basic source of inaccuracy. The Earth's north magnetic pole does not correspond precisely to its north geographic pole. The resulting inaccuracy in the compass measurement is called the variation. The extent of the variation depends on where the compass is on the Earth's surface, and changes over time. Today, the north magnetic pole is about 800 miles (1290 km) south of the geographic pole. In the fourteenth century, the magnetic and geographic poles were about 1600 miles (2575 km) apart. Metal objects in the vicinity of the compass also affect the direction in which it points. The effect of these local influences is called deviation.

Despite its drawbacks, the magnetic compass became an essential navigational tool. Eventually navigators began to understand that the compass did not quite point to true north. At about the same time the use of the compass was spreading, the Moors who had conquered Spain introduced Europe to the astrolabe, an instrument for measuring the positions of stars. The English friar Nicholas of Lynn, on a mapping assignment for King Edward III (1312-1377) in the mid-1300s, discovered that the two instruments disagreed on the direction of north by about 15 degrees. Later a meridional compass was designed with its compass card adjusted so that the needle aligned with the north directional point at a particular spot off the coast of Cornwall.

The compass was key to the long-distance voyages undertaken by Europeans beginning in the fifteenth century. Prince Henry of Portugal (1394-1460), called Henry the Navigator, established an observatory and navigation school and encouraged the idea of ambitious voyages to far-off lands. The Portuguese mariner Gonzalo Cabral reached the Azores in 1427. Unbeknownst to the explorer and his crew, they had traveled about a third of the way to the New World.

In 1519, another Portuguese explorer, Ferdinand Magellan (c. 1480-1521), led a Spanish-financed expedition that sailed around the world for the first time. Magellan himself did not live to complete the three-year voyage, having been killed in a battle in the Philippines. However, his expedition, given the technology of his time, is still considered by scholars to be among the greatest navigational triumphs in history, and it was the magnetic compass that helped make it possible.

In 1600, William Gilbert (1544-1603), English scientist and physician to Queen Elizabeth I, in his book *De Magnete*, was the first to describe the Earth as a giant magnet. In terms of this model, he went on to demonstrate the expected behavior of the compass needle at various points on the Earth's surface. After hundreds of years of using the magnetic compass, sailors could finally understand why it worked.

The development of iron and steel ships in the late nineteenth century made magnetic compasses less useful in navigation. A metal hull affected the local magnetic field and reduced the accuracy of the compass. Large modern ships and aircraft use compasses mounted in pedestals called binnacles, which contain magnets and steel pieces to counteract the effects of the metal hull. They also depend on gyrocompasses, spinning devices which, once they are set to point north, continue to do so regardless of the magnetic field. However, ordinary magnetic compasses are still widely used in boating, hiking, surveying, and other activities.

SHERRI CHASIN CALVO

Further Reading

Needham, Joseph. *Science and Civilisation in China*. Cambridge: Cambridge University Press, 1962.

Temple, Robert. *The Genius of China: 3,000 Years of Science, Discovery and Invention*. New York: Simon and Schuster, 1986.

Williams, Suzanne. *Made in China: Ideas and Inventions from Ancient China*. Berkeley, Calif.: Pacific View Press, 1997.

Development of the Lateen Sail

Overview

The lateen sail, developed during the first millennium, was introduced to medieval Europe where it revolutionized marine travel. Combined with the less-versatile square sail, the lateen sail was crucial to the development of navigable ships powered only by the wind. Their use in "fore-and-aft" rigged ships helped to launch an era of seagoing commerce, exploration, and warfare that continued through the end of the Age of Sail.

Background

The first boats were crude, powered by a combination of water currents and human muscle power. They could travel downstream with ease, could traverse calm bodies of water with slightly more difficulty, and could travel against the current with great difficulty, if at all. Simple rafts could carry relatively large amounts of cargo, but only at the expense of maneuverability, while canoes (and canoe-like vessels) were maneuverable, but not very commodious. However, these simple craft were the only options for seaborne travel over many centuries, if not millennia.

The next advance was marrying a square sail to these craft. By doing this, a vessel could take advantage of the wind to help push it along its way, sparing the crew the effort of rowing. Unfortunately, a sail could only be used when the wind was blowing approximately in the desired direction of travel; the rest of the time it was more hindrance than help. So the utility of these first square sails, which could only take the wind from one direction, was limited due to the vagaries of the wind.

This does not mean that early craft were crude or clumsy. The ancient Greeks, Phoenicians, Romans, and other great civilizations conducted commerce, travel, and war with oared galleys that were both durable and nimble.

However, these vessels were still dependent to some extent on muscle power and had only a limited ability to sail in any direction other than directly downwind.

The next breakthrough in sea travel came as early as the second century, with the invention of the lateen sail. Lateen sails were developed by the Arabs, then adopted in the eastern Mediterranean. Because they were used in the Mediterranean, northern sailors gave them the name "lateen" from "Latin." A lateen sail is a triangular piece of cloth. One side (two corners) is attached to a crossbar (called a yard) near the top of a mast, the third corner is fastened near the deck, and the mast or the crossbar pivots with some degree of freedom. The yard is often tilted at an angle to the ship's mast, so that the sail will often be in the form of a triangle with the point at the top of the mast. The lateen sail is called a "fore-and-aft" sail because it is usually rigged such that the fabric of the sail itself runs along the length of the ship, or from the front end of the vessel towards the back (in nautical terms, from forward to aft, or fore and aft). This setup allows the sail to take the wind from directly astern (behind the ship) or from either side, and eventually was modified so that a ship could even sail into the wind. The lateen sail was soon combined with the older square sail, providing a good combination of steering and motive power. All in all, the lateen sail and its derivatives revolutionized going to sea.

Impact

The lateen sail had a tremendous impact on sailing, and on many areas affected by sailing, such as commerce, travel, exploration, and the military. Developments in these areas helped make some nations, including England and Spain, great powers.

The most obvious impact of the lateen sail was on the navigability and seaworthiness of deep-water ships. The older square sails (like those used by Viking ships) would only let a ship be blown before the wind because they were fixed to a yard rigidly mounted to a mast that was fixed to the deck. So long as the wind did not vary much in direction, the ship could raise the sail, but in variable winds or when navigating around complex shores, sailors could not count on the wind blowing from an appropriate direction.

In contrast, the lateen sail was mounted so that it ran along the length of the ship. This meant that even wind blowing from the side could be used to propel the ship forward. In fact, later ships were designed so they could sail somewhat into the wind using their rudder and rigging. While the lateen sail was not solely responsible for such maneuvering, it was certainly helpful.

While the added maneuverability of ships at sea was helpful, the impact of this maneuverability was significant in many other areas. No longer was a ship wholly at the mercy of the winds, and no longer was its range solely dependent on the ability of the crew to row. Instead, a ship's captain could shape his course from port to port. In addition, a ship could carry more stores and cargo because the ability of a crew to propel huge weights by sheer muscle power was no longer important. These two developments alone made seaborne commerce much more reliable and profitable, and helped to build the fortunes of maritime nations.

These same factors also affected naval warfare. The added power of the wind let ships carry more stores so they could stay at sea for longer periods of time. In addition, ships could be made larger, and heavy cannons could be mounted on them, turning the ship into a formidable weapon of war. These developments culminated in the huge fleets built by the Spanish, Dutch, French, and English at various times through the nineteenth century, and led also to profound revolutions in naval warfare.

Another area in which the benefits of the lateen sail were evident was in the ability of nations to begin the maritime exploration of the world. Although not much exploration took place until near the end of the fifteenth century, sailors in the few centuries leading up to that time learned how best to exploit the benefits of the new sailing rigs. In general, the same capabilities that made lateen-rigged vessels suitable for warfare and commerce also suited exploration. As before, the ability to navigate more predictably and reliably, combined with the greater cargo-carrying capacity of wind-powered ships, gave captains the confidence to strike out into the unknown and the ability to do so for longer periods of time. Until the fifteenth century, most sea travel took place either within sight of land or along very well-known and well-traveled sea lanes. This began to change with the introduction of more seaworthy vessels. For example, the ships Columbus sailed to the New World and those of Magellan had lateen sails.

Carrying people is little different from carrying any other sort of cargo. The same characteristics noted above also made it possible for ships to become troop transports, and they made ships capable of transporting large numbers of colonists or colonial administrators to distant lands. Both these developments were other ways of projecting a seafaring nation's power to the far reaches of the world, and helped to further solidify the dominant positions of the major powers for several centuries.

All the benefits of the lateen sail had a significant impact on the political power structure of Europe and the Mediterranean in the Middle Ages, an impact that increased with time. Probably the first use of lateen-rigged sailing vessels to project political power on a large scale was during the Crusades, when many of the European armies sailed on such ships to "liberate" the Holy Land. In large part, the military and commercial success of Venice was due to the use of lateen-rigged ships. These vessels carried Venetian goods throughout the Mediterranean and brought foreign trade goods back to Venice for sale to the rest of Europe. Venice was one of Europe's first commercial empires, and used some of this wealth to become one of the Mediterranean's most powerful states of that era.

Another marine innovation, probably introduced sometime in the twelfth century by pirates, was the stern rudder. By moving the ship's rudder from the ship's starboard side (the right-hand side as you face forward, also the side of the steering board) to the stern (the rear of the ship), the rudder could be fastened directly to the sternpost, making it stronger and less likely to be damaged in storm or battle. This not only made steering easier, but also allowed ships to increase in size by at least threefold. The larger ships had greater carrying weight for trade goods, which made voyages even more profitable, and for weaponry.

In later centuries, other nations used sea power as the basis for great national status. Portugal and Spain were first, sailing lateen-rigged ships on voyages of exploration and later of conquest. These same ships returned great amounts of riches to their mother countries, including trade goods from the Orient and riches plundered from the New World to Spain. Eventually, the English and Dutch became great powers based almost solely on their proficiency at sea. Both nations established colonial empires. The Dutch also built commercial power, and the English, military sea power. By this time, other innovations had increased the efficiency and seaworthiness of sailing ships further still.

The lateen sail was crucial for the development of ships that were maneuverable and reliable under sail power alone. These improvements made it possible for ships to increase in size, giving them the ability to carry cargo more profitably and more reliably. They also made ships more important as weapons of war. All these factors gave seafaring nations an advantage over their landlocked neighbors, and helped to shape the balance of power in the Mediterranean and in Northern Europe for centuries.

P. ANDREW KARAM

Further Reading

King, Don. *A Sea of Words: A Lexicon and Companion for Patrick O'Brien's Seafaring Tales*. Owl Books, 1997.

Thubron, Colon. *The Venetians*. Time-Life Books, 1980.

Incan Roads in South America

Overview

At the time of the Spanish conquest in 1532, the Inca civilization was one of the most advanced in the New World. One of their achievements was a marvelous system of roads that linked their empire together into a coherent whole. Because of these roads, the Inca were able to move supplies, messengers, and troops anywhere in their empire quickly and efficiently. In many ways, these roads helped to hold the Inca Empire together.

Background

Civilization in the Andes is nearly 3,000 years old, beginning with the Chav'n culture of the Peruvian highlands which flourished around 1000 B.C. The Chav'n passed into history, as did several subsequent civilizations over the next 2,000 years. Around A.D. 600, the Tiahuanacan civilization arose near the shores of Lake Titicaca, in what is now Bolivia.

The Tiahuanacans were masters of the political and religious control of their population. A highly organized society, the Tiahuanacans also built impressive structures of stone to guard their civilization from the depredations of neighboring tribes. In some ways, these structures are the most impressive ever built in the New World, and the Inca built many of their own structures atop foundations constructed by the Tiahuanacans.

As Tiahuanacan civilization began to wane, the Inca (at that time only a minor tribe in Peru) were winning battles with the Chimu culture for control of Chimu territories, which stretched over 600 mi (965 km) of Peruvian mountains and coast. After assimilating the Chimu, the Inca turned their attention to the Tiahuanaca, and by somewhere around 1200 had conquered them as well. Although the Tiahuanaca lacked the Inca's drive to expand and conquer, they knew how to organize their population and harness their people for public-works projects. The Inca put this administrative expertise, along with the Tiahuanacan talent for engineering, to good use.

Over the next few centuries, the Inca continued to expand and assimilate new cultures. They eventually controlled a narrow strip of land that stretched nearly 2,000 mi (3,220 km) along the Andes mountains, from the Pacific Ocean to the eastern jungle. They instituted a system of political, religious, and administrative controls not unlike those used by the Romans to maintain their empire. Also like the Romans, the Inca maintained a system of roads that helped link their far-flung empire together into a coherent whole. Because the Inca never invented wheeled carts, their roads were never more than footpaths, but they were impressive technological accomplishments nonetheless.

The Incan road system linked together a geographically large and culturally diverse collection of cultures into a coherent political empire. To do this, the roads solved some interesting engineering problems, addressed some "human" needs, and served to highlight some of the most important administrative innovations of the Incan Empire.

The Incan roads were constructed only well enough that a man or a llama could walk easily, but this is not to say that the Inca merely built trails along the mountains. The Inca built nearly 15,000 mi (24,140 km) of roads. Nearly one-quarter of this total was made up of two primary roads that ran the length of the empire, one along the crest of the Andes and the other along the coast. The rest of the road system consisted of either roads linking these two thoroughfares or "side" roads to various villages or other sites of importance.

For the most part, the Incan roads followed the terrain without many stone bridges or tunnels. In a few places, the Inca filled in small gorges with rock, but valleys and gorges were usually traversed by great swinging bridges built of vines or rope securely fastened to rock at either end. Although these bridges would sway greatly in the wind or beneath the weight of a person, they were secure, anchored by ropes that might be as thick as a person. Bridges of this sort are still used in the Andes today.

Another Incan innovation solved some of the problems of messengers or troops traveling long distances. These people needed places to stop for food, shelter, or rest as they traveled, and the Andes are not a hospitable mountain range for foot travel. The solution was to divide the road into sections that could reasonably be traversed by a person in a day's travel by foot and by llama. This distance varied depending on the terrain and other factors, and these variations were taken into account when establishing the length of a section. At frequent stages along the road, about a day's travel apart, the Inca built stone structures to hold food and provide shelter for travelers. In this way, messengers, soldiers, and travelers were assured of a place to rest and a meal to eat at the end of a hard day's hike. Many of these way stations, complete with warehouses and barracks, still exist. At greater intervals along the roads were fortifications and garrisons of soldiers for the protection of the empire. In addition to their role in fending off external attack, these soldiers could also protect travelers and enforce unpopular laws.

It is not certain whether the Inca actually built all the roads in their road system or if they "inherited" some of them from tribes and civilizations they conquered. However, there is no doubt that the Inca assembled these roads into a single network and subsequently maintained them.

There is also no doubt that the Incan administrative systems were responsible for much of the road system's construction and its continued upkeep. The Inca maintained records of population, productivity, crops, lands, and other important parameters. Villages were divided into working groups of 100 men. Each working group was given assignments, and each village was responsible for maintaining the road nearest them. This maintenance involved not only keeping the road smooth and repaired, but also maintaining the integrity of the way stations and stocking them with food as necessary. This concept of requiring communities to maintain the roads near them was a characteristic Incan innovation, as was the manner in which it was done. Of all the New World civilizations, the Inca alone organized and harnessed their entire population for work on behalf of the state.

Impact

The Incan road system bound together the Inca Empire. Like the Romans, the Inca grew by conquest and assimilation. After a century or so, their empire included any number of distinct cultures, each with its own history, traditions, and customs. In the absence of frequent contact with the central government, it would have been very easy for peripheral tribes to simply go their own way, gradually whittling the empire down to a much smaller size.

The roads gave the highest ruler (also called the Inca) the ability to maintain frequent contact with all his subjects. Some of this contact was direct, for the Inca constantly traveled throughout his empire. In addition, these roads made it possible for the leaders of recently conquered tribes to travel to the capital city at Cuzco, where they saw at first-hand the strength of the Inca Empire and were indoctrinated into Inca religion and culture. These visits not only helped make them more willing subjects, but helped impress upon them the futility of resistance to assimilation.

The roads also made it possible for those in the capital to spread news and proclamations throughout the empire and to receive news and statistical information from those outside the capital. This helped to make each tribe of the far-flung empire feel an integral part of the Incan Empire. In addition, this constant flow of information helped the Inca and his advisors to administer his empire.

Finally, these roads ran very near the borders of the empire, so troops could be easily mobilized to combat uprisings, protect the frontier, or subdue new tribes to add to the empire.

Like the Romans, the Inca realized early that good roads served political, military, and economic ends. Because of this, they designed an extensive system of roads that reached into virtually every corner of their empire, linking even the most remote areas to the capital and making the Inca Empire one of the New World's most advanced civilizations.

P. ANDREW KARAM

Further Reading

Crow, John. *The Epic of Latin America*. Berkeley: University of California Press, 1992.

Feeding an Expanding World

Overview

The heavy plows needed to productively till the heavy soil of northern Europe's fertile valleys were not widely adopted until the Middle Ages, when the breeding of large draft horses began. More efficient plowing and the advent of three-field crop rotation resulted in a tremendous increase in agricultural productivity by the twelfth century. With larger farms starting to cluster in the valleys, village life flourished and larger populations could be sustained. Communal government and economics, manifested in the system of feudalism, became more important in the lives of the people.

Background

In the early Middle Ages, the *aratrum*, a simple wooden scratch-plow, was generally used to till fields in Europe. This type of plow worked well enough for the light soils of the Mediterranean, but was not well suited for the moist, heavy soil of northern Europe. If the soil was not plowed deeply enough, seed would blow away and crops would fail to thrive.

A better type of plow had existed since ancient times, and had been described by the Roman writer Pliny the Elder (A.D. 23-79), but was not in general use. The three-piece *plovum* was made of iron, and usually was equipped with wheels. It had a vertical sod-cutter, or *coulter*, a horizonal cutting blade, or *plowshare*, and a tilted *moldboard* for turning over the soil. This efficient plow, which cut deep furrows and dug up more weeds, had two major problems. First, iron was expensive, although improvements in metallurgy around the turn of the millenium had resulted in iron tools that were not only less expensive but stronger as well. Still, the iron plow was too heavy to pull with the draft animals available to most farmers. A team of as many as eight oxen, or castrated bulls, could haul it, but oxen were very slow and costly to buy and feed.

Since horses were primarily used for transportation and warfare, they had been bred mostly for speed. The swift breeds that were valued were neither the largest nor the strongest. They were not well suited for use with the heavy three-piece iron plow. But knights wearing increasingly heavy armor began requiring stronger horses, and size and strength became more important in horse breeding. Finally farmers began breeding an offshoot of the Carolingian charger for agricultural work. These horses were three or four times faster than oxen, stronger, and more versatile. Meanwhile, other advances had made it possible to use them more effectively.

THE WHEELBARROW

Sometime in the eleventh or twelfth century, Europeans were surprised and delighted to be introduced to a device with a pair of handles, a wheel, and a place to carry things resting on the wheel. This was, of course, something found in nearly every gardener's tool shed, every construction site, and every farm throughout the world today: the wheelbarrow. Its introduction was revolutionary—using a wheelbarrow, a single man could carry as much as any two men without this device. The benefits to farmers, the men building the day's castles and cathedrals, and ordinary tradesmean were phenomenal.

What was new to Europe, however, was a Chinese tradition going back over a millennium. The first known wheelbarrow was introduced in China in the first century B.C. Although only a few basic wheelbarrow designs and uses were ever used in the West, in its birthplace the wheelbarrow saw a profusion of forms and uses that have still not been matched elsewhere in the world. In particular, the Chinese use of wheelbarrows in warfare was ingenious: they used wheelbarrow convoys to transport supplies to the front lines, and used wheelbarrow-mounted shields to great advantage in warfare. Even today, many of the wheelbarrow designs in common use in China have not been adopted in the West.

P. ANDREW KARAM

In about the tenth century, the non-strangulating *horse collar* was developed. The horse collar distributed a heavy load around the animal's chest and shoulders, allowing it to pull much more weight than it could with a traditional harness. At about the same time, the nailed iron horseshoe was adopted. Horseshoes enhanced weight distribution and traction while protecting the horse's hooves. The cultivation of oats became more important, since oats made up the bulk of the horses' diet.

Impact

The increase in plowing capacity brought about by the introduction of the draft horse and the rediscovery of the three-piece iron plow made it possible for farmers to institute the most important agricultural advance of the Middle Ages, three-field crop rotation.

Crop rotation is a system in which the crops grown in particular fields are alternated from year to year with periods when a field is uncultivated, or fallow. A single crop tends to pull the same nutrients out of the soil until it becomes depleted of those nutrients and is no longer fertile. By growing plants that require different nutrients, or allowing a field to lie fallow, it is possible to reduce the depletion of the soil. Fallow fields served as pasture for animals, and their manure was plowed into the soil to act as a fertilizer.

In the early Middle Ages, farmers generally employed the old Roman system of two-field crop rotation. Three-field crop rotation, in which each field lies fallow in successive seasons, increased productivity by 50%, since two-thirds of the land was under cultivation every year rather than only half of it. Correspondingly, more plowing was required, which was not practical with the scratch-plow. With faster draft animals and the iron plow, the peasant family could farm more fields. They could plant a wider variety of crops, allowing them to spread their work out over the spring and autumn planting seasons.

The iron plow also allowed farming to move down into northern Europe's valleys, where the land was more fertile but the soil was too heavy for the scratch-plow to be an effective tool. The scratch-plow had encouraged farming in the uplands, where the lighter, less fertile soil had to be cross-plowed in square fields, gone over twice at right angles. Now many of these fields were abandoned, and the valleys were plowed up in long open strips. This resulted in the typical ridge-and-furrow pattern that is still familiar today. By the twelfth century, the three-piece plow and open strip farming had been adopted across most of northern Europe.

Since they could work more efficiently with the three-piece iron plow, farmers could take the time to clear more land for fields. They cleared thousands of square miles of forests and drained large tracts of Low Country (now Netherlands,

Belgium, and Luxembourg) marshes to create more farmland. The population increased, and became more concentrated. Sometimes an area grew overpopulated, and even more land had to be cleared to provide sustenance. Political and economic institutions grew up to regulate the division and working of the land and the distribution of the crops that were produced. The result was the manorial economy of the late Middle Ages, the full maturity of the feudal system.

Under the feudal system, stewardship of land flowed down from the king under a complex hierarchy of loyalty and obligation. The lord of the local manor was himself a *vassal*, or subordinate, of a higher-ranking *seigneur*. The petty lord generally had charge of a few square miles of estate called a *fief*, and peasants worked it communally under his authority. The amount of work each peasant owed the lord depended on his individual status.

Each peasant household was also permitted to work a plot of land as its own. These plots usually consisted of groups of strips, not necessarily adjacent, scattered around the open fields. The distribution helped to reduce individual households' risks from freak hailstorms, differences in soil quality, or other local conditions that might affect a particular section of land.

Manorial life was essentially communal. Often the peasants had worked together to clear new fields, or *assarts*, and they were used to working together on the lord's fields. The high costs of owning draft animals often encouraged shared efforts on the "private" fields as well.

Although village elders also played an important role, the lord controlled most aspects of everyday life. The village was part of his domain, and its church often paid him a portion of the peasants' tithes. His heavily fortified manor house or castle was the linchpin of the community's defense. He ran the mill in which grain was ground into flour and the ovens in which it was baked into bread.

The agricultural advances of the Middle Ages did not banish the threat of starvation. Bad weather, plant and animal diseases, and other problems still frequently resulted in famines. The Great Famine of 1315-1322 brought disaster across much of northern Europe. Still, the overall effect of the new farming technologies was a more plentiful and dependable food supply. In some areas, agriculture was productive enough to free some of the villagers from subsistence farming. Their efforts could be channeled into other pursuits, especially crafts. Products such as textiles and metal implements were manufactured for trade as well as for local use. In this way, medieval innovations in farming set the stage for the growth of industry.

SHERRI CHASIN CALVO

Further Reading

Hallam, H., ed. *Agrarian History of England and Wales*. Cambridge: Cambridge University Press, 1988.

Jordan, William Chester. *The Great Famine: Northern Europe in the Early Fourteenth Century*. Princeton: Princeton University Press, 1996.

Latouche, Robert. *The Birth of Western Economy*. E.M. Wilkinson, transl. New York: Barnes and Noble, 1961.

White, Lynn Jr. *Medieval Technology and Social Change*. Oxford: Oxford University Press, 1962.

The Development of Windmills

Overview

Windmills provided medieval society with a reliable source of energy that helped initiate a thirteenth-century Industrial Revolution. This device also helped create a mechanical view of reality that would dominate the West for centuries and eventually lead to the onset of the Scientific Revolution.

Background

From man's earliest days, energy has been a fundamental necessity for human existence. The adoption of fire by our ancestors enabled them to survive in northern sections of the Eurasian landmass during the last Ice Age. Fire was humankind's first inanimate source of energy. The heat and light that it generated allowed the first *Homo sapiens* to increase their protein intake by providing the opportunity to cook meat. It also kept them warm during the winter months, which extended the life expectancy of both infants and the aged. Finally, this new source of energy added to human safety by increasing the visual capacity of humans at night.

The transfer of collective knowledge, by which the solution to one problem is used to facilitate the answer to another, is an important human characteristic. Over time, early humans discovered that they could increase the strength of their wooden implements by using burning embers to harden the tips of their tools and weapons. Stronger digging sticks, in turn, increased the success rate of those who foraged for roots and tubers, and the same held true for weapons used in hunting and warfare. These tools extended human productivity by increasing the successful application of human energy.

The next energy revolution was part of a larger Agricultural Revolution that precipitated the emergence of civilizations. Humans were able to harness the energy of both rivers and animals to create an agricultural surplus for the first time in history. This was the result of the successful interaction among humans, animals, and the natural environment. The use of draft animals to pull heavy plows increased the harvest of domesticated plants. This new source of energy allowed the first farmers to bring large amounts of fertile land under cultivation. Over time, stronger hybrid plants were developed, and in a very basic way these early farmers helped direct the evolutionary process of the first agricultural species.

The application of energy also had a negative impact on the environment. Historians and scientists alike have concluded that humans were so successful at manipulating the environment that soil depletion became a major problem. This accelerated the process of desertification.

In time, technology would also increase the effectiveness of energy resources. The invention of boats and barges harnessed the energy of rivers for transportation. This had an important impact on the growth of human society. River energy allowed for greater transportation, communication, and trade. Along with material goods, ideas about religion, government, and information concerning new technology were made available to a greater number of people. Social scientists refer to this process as "cultural diffusion" and "technology transfer." From the very beginning of civilization trade and ideas moved in a very systematic way from society to society, and increases in energy accelerated that movement.

As the ancient Mediterranean world entered the classical period, there were additional improvements in the use of energy. Long range intracontinental trade became a basic feature of this time. In ancient Greece, the Athenian navy rose to military dominance based upon the extension of human energy. The successful development of a naval vessel called the "trireme" allowed Athenians to dominate the important waterways of the classical world. The trireme was both fast and maneuverable. Its energy source consisted of 170 oarsmen, who increased the speed of the ship to such an extent that it quickly became the fastest vessel afloat.

The Roman Empire did not favor alternative energy sources. Roman mills, which were primarily used for grinding grain, utilized animal or human power. As the empire expanded, the number of slaves in Roman society increased. This severely limited the incentive to find an alternative energy source and helped create an anti-technology mindset among Rome's leaders. In time, it became an accepted belief that mechanization would actually disrupt society because it would reduce the size of the labor force, and thousands of permanently unemployed workers would pose a constant threat of rioting and political unrest.

Impact

The fall of Rome and the onset of the European Middle Ages accelerated the search for alternative energy sources. Europe actually experienced a short Industrial Revolution in the thirteenth century. This occurred because technology helped bring about a revolution in energy. The windmill was the first successful attempt to harness the power of inanimate energy. Over time, improvements were made that allowed the top of the mill to be rotated in the direction of the wind, so as to take full advantage of this new source of energy. The rotating windmill became a powerful motor, creating the equivalent of 20 to 30 horsepower. This was a turning point in the production of power. It became the first reliable inanimate source of energy. It also helped to undermine the traditional classical view that all energy had to come from animals or humans. The first windmills were used for irrigation and grinding grain. Eventually they would power the medieval textile industry.

The first important "spin off" of the new development in wind power was in marine engineering. The construction of the caravel, which would be the primary vessel of European expansion, centered upon the harnessing of wind power through the invention of the lateen or triangular sail. Not unlike the concept of rotating the top of the windmill, the new rigging allowed sailors to rotate the sail in the direction of the wind. This new technological advancement had a profound economic effect on Europe. Captains no longer had to wait for the most favorable winds. This helped decrease the time of the voyages, which in turn reduced the cost. Productivity was also increased because more wind could be harnessed and thus the size of the ships could be increased; this had a positive impact on the amount of cargo that could be carried.

The first nation to take advantage of this new technology was Portugal. Prince Henry the Navigator (1394-1460) created a research institute at Sagres, where he brought in experts in cartography, marine engineering, and geography to study and research new methods of navigation. His experts combined the new wind energy of the of the triangular sail with the power and accuracy of newly developed ship cannons to become the dominate power on the high seas by the 1440s. The orientation toward wind power focused some of Henry's experts on studying the major wind patterns of the Atlantic Ocean. This knowledge allowed Portuguese sea captains to explore the African coast in greater detail and, at the same time, shorten their voyages to an even greater extent because they knew where to pick up the prevailing westerly winds that would propel them quickly back to Portugal. The financial impact of wind energy on fifteenth-century Portugal was impressive. The greater time spent moving up and down the African coast increased the nation's wealth from trade in slaves and gold. This dominance eventually spread to the Indian Ocean trading complex. In little over half a century the Portuguese took control of all the vital ports, harbors, and choke points in this extensive trading network. Without the advances in wind power the Portuguese would have been unable to acquire this transoceanic dominance.

The windmill also brought about significant changes in European economic and intellectual life. The effectiveness of the windmill increased a second time with the adoption of a horizontal axis. This allowed the energy created by the mill to be directed to the production of many important products. For the first time in history, machines were used to mass-produce paper, and inexpensive paper increased the flow of information throughout Western Europe. This medieval paper industry helped create the groundwork for the concept of mass production. It also initiated a passion for mechanization. Extensive advances in productivity showed that machines could increase the standard of living for all people, and this helped to create a mechanical worldview. Eventually the idea that the universe could be described as a large machine would come to dominate the Western mind.

Natural philosophers began to accept that this mechanized universe was controlled by certain laws of nature. These laws could be discovered through the exercise of human reason, and the knowledge found within them could be applied to the benefit of the human community. This created the perception that all of the problems facing humanity could be solved and that, in fact, knowledge was power. This concept, in turn, contributed to the idea of progress, the belief that life in all its dimensions would become more favorable with each passing generation. People would no longer have to suffer from disease, starvation, nor all the other problems that traditionally plagued humankind. These problems could be overcome by properly studying their characteristics and their interaction with the environment.

As time went on, the success of the windmill would set the example for future advancement in developing sources of inanimate energy. This pursuit of energy would coincide with continued developments in mechanization and mass production. By the eighteenth century, Europe would begin history's second great increase in material productivity, the Industrial Revolution.

RICHARD D. FITZGERALD

Further Reading

Cipolla, Carlo M. *Guns, Sails, and Empires: Technological Innovation and the Early Phases of European Expansion, 1400-1700.* New York: Pantheon Books, 1965.

Duby, Georges. *Rural Economy and Country Life in the Medieval West.* Translated by Cynthia Postan. Philadelphia: University of Pennsylvania Press, 1998.

Powers, Eileen. *Medieval People.* New York: Harper Perennial, 1997.

The Influence of Water Mills on Medieval Society

Overview

There were many sources of power used before the Industrial Revolution of the eighteenth century. The use of slave labor was the first source of large-scale power. This was followed by advances in animal power that were made possible by the invention of tools such as the horse collar. Even more significant was the success of medieval technology in harnessing water and wind power. The waterwheel is one of the oldest sources of power known to man. It was the first type of power harnessed by man that was not generated by animals or humans. When combined with the proper equipment to form a mill, waterwheels were used to grind grain, drive sawmills, power lathes, move pumps, forge bellows, make vegetable oils, and power textile mills. It served as the main source of power for medieval Europe and necessitated that most towns needed to exist near water to make use of this type of power source. It was estimated in *The Domesday Book* (a book based on William the I's survey of England in 1086) that there were nearly 6,000 water mills in England at that time, and many sources believe that this number more than doubled in the next two hundred years. The water mill served as a primary power supply until the advent of the steam engine during the Industrial Revolution.

From a modern perspective, the operating principles of the water-powered mill are quite simple. To generate energy, water is directed to a wheel and propels it in a circular motion. The spinning wheel transfers the power to a drive shaft that can be used to move many pieces of equipment. These were originally used to turn millstones and grind grain. Later, this generated power was harnessed to drive other types of tools.

Background

There is evidence that water power has been used since at least 300 B.C. in Egypt. It is possible that that this technology may have been adapted from cultures such as the Persians or the Chinese. The earliest known examples of water mills used past examples of water power to utilize wheels that were flat on the water and attached directly to the drive shaft in a horizontal design. When the wheel turned, so did the drive shaft. Because this type of setup was inefficient, waterwheels with a vertical design were soon being manufactured. These types required different engineering because there was a need for gears and cogs to transfer the power to the mills. There were two types of vertical waterwheels put into use at this time. The undershot wheel rests directly in the stream and depends upon the force of the water to push the wheel. Therefore, without a constant level and flow of water, the wheel cannot generate much force, and it is useless in times of low water flow. The overshot model is much more efficient and depends much less on the amount and force of water because it uses the force of gravity to help drive the wheel. Water is channeled to the wheel via a flume or pipe and is dropped directly on the paddle of the wheel. The wheel spins and drives the shaft allowing the power to be harnessed as the user sees fit. As technology increased towards the later Middle Ages, milling operations became more and more complex.

The earliest form of grinding grain between two stones was adapted for use in a water mill. Grain was pounded between two millstones until it became meal. The bottom millstone was fixed while the top millstone that was powered by the waterwheel could be separated to control how coarse the meal turned out. Both stones were corrugated so that the grinding motion of the top stone would then crush the meal to a desired consistency. Additional wheat to be ground could be added to the mill through an opening in the top stone. The meal was then sifted through sieves to obtain flour.

A group of individuals who took full advantage of the water mill technology during the Middle Ages were the Cistercian monks. This monastic order was founded in the year 1098, just after the waterwheel had revolutionized western Europe. Early in the twelfth century, St. Bernard (1090-1153) took over the order and attempted to gain social freedom by utilizing water mills to provide financial independence. Within the next 50 years, the Cistercians had reached the cutting edge of water-power and agricultural technology. Monasteries were built on artificially manufactured canals that ran throughout the complex. This source of running water provided power for activities such as milling, woodcutting, forging metals, and making olive oil. It was also a source of fresh water for daily needs and fulfilled the needs for sewage disposal. The Cistercian monasteries were great examples of organized factories that proved to be important in commerce of that time.

Other sources of power that appeared during the Middle Ages were the windmill and the tidal mill. The windmill appeared before the end of the twelfth century. While not efficient due to a dependence on the amount of the prevailing wind for power, windmills could grind grain and perform other tasks similar to water mills. As the technology advanced, more efficient windmills were developed. These enabled power to be utilized in areas that were far from sources of water, provided there was a reliable amount of wind. The tidal mill, which appeared around the same time, attempted to use the power of the changing tide to provide energy for the mill. While their use did not appear to be widespread, the tidal mill more than likely had a significant favorable impact on the local populations that used them.

Impact

Water mills helped to change the way of life in Medieval Europe, and affected all levels of society from each individual to entire countries. Certainly water mills had an immediate and direct impact on the people who operated them. This positive influence would have been primarily in the saving of time and money. People could do a larger amount of work in a shorter amount of time and for lower costs with a water-powered mill. While not usually considered to be part of the Industrial Revolution, the mill was a precursor to that era. The price of human labor was quite expensive, so allowing a mill to do the majority of work was very cost effective. One person could now do the same job as many with the help of the power generated by the waterwheel. It does not seem, however, that many people used this technology to increase their leisure time. Rather, it seems that this technological advancement was used to increase greatly the manufacture of certain goods and materials for sale and profit.

The mill often served to shift the industrial organization and power from urban centers to more rural areas closer to water sources. Thus towns became more powerful, often at the expense of cities. One good example of this was the application of water power to the industrial process known as fulling. Fulling was the process of shrinking and thickening cloth. Prior to its use in water mills in the thirteenth century, fulling was carried out by individuals stomping on the cloth by foot or beating it with a bat. This was obviously a very time-consuming and labor-intensive process. The fulling mill allowed the work to be done by wooden hammers powered by water. Now, only one man was needed to ensure that the cloth moved properly through the machinery. This process revolutionized the industry and initiated reform. The majority of work was now centered in rural areas instead of urban centers.

The effect that this mechanization had on the establishment of national markets cannot be overlooked. Now that goods were produced at a faster rate, with greater quantity, and at less expense, new economic frontiers could be explored. Large national markets were established to find outlets for the increased availability of goods. Water mills diminished much of the human labor costs by providing power for grinding grains and other goods; tanning hides; pressing vegetables for oil; sawing wood; forging metals; polishing armor; pulverizing rock; operating blast-furnace bellows; and crushing mash for beer. The water mill served as the major source of power prior to the invention of the steam engine. Its technology was constantly being upgraded, and new uses were found for the generated power.

These advances in technology that led to the improvement of the water mill were eventually applied in other fields. As an example, the switch from the horizontal to vertical waterwheel required gears to be used on the drive train to transfer the power. This mechanized process became quite complex with successive improvements and these ideas were later adapted on a smaller scale to make clocks and other similar mechanical devices.

Water mills also served to change the balance of power, both locally and nationally. On a local level, whoever operated and controlled the mill had the most power. With a working mill, the town could prosper from the increase in trade. The increased output of goods intensified the demand for raw materials, which was largely met by local merchants. As revenues increased, the town could afford greater protection and thereby was safer. These same ideas could be applied on a larger scale for a countrywide level. The use of water mills enabled countries such as England to open new markets and significantly benefit from this commerce. The water mill had a significant influence on medieval society and left its mark at many levels.

JAMES J. HOFFMANN

Further Reading

Derry, T. K., and T. I. Williams. *A Short History of Technology: From the Earliest Times to A.D. 1900*. Oxford: Oxford University Press, 1993.

Gies, F., and J. Gies. *Cathedral, Forge, and Waterwheel: Technology and Invention in the Middle Ages*. London: HarperTrade, 1995.

Holt, R. *The Mills of Medieval England*. London: Longman, 1988.

The Ancestral Puebloans and the Cliff Palaces at Mesa Verde

Overview

Eight hundred years ago in the American Southwest, a group of indigenous peoples almost literally carved out a home in rock walls of the mesas and canyons. The Anasazi Indians built villages in seemingly inaccessible alcoves of cliff walls. The imposing sandstone structures in the villages themselves are perhaps the work of the most advanced pre-Columbian culture in North America.

The Anasazi had abandoned the spectacular cliff houses 200 years before the first European explorers, who gave Mesa Verde (Spanish for "green table") its name, would visit the region. The area was largely unknown to modern scholars until the American push for western expansion in the mid-nineteenth century. A group of cowboys rediscovered the cliff dwelling villages of Mesa Verde in the 1880s. Almost immediately, the described "cliff palaces" fascinated archaeologists, but despite a century of research, excavation, and discovery, relatively little is known about the Anasazi who built them.

Background

The first inhabitants of Mesa Verde arrived in the region about 550. At first primarily nomadic hunter-gatherer bands, they lived in pithouses on the mesa tops or in shallow caves and rock shelters. These kinds of rock shelters were common in the early inhabitation periods in other regions of North America. Over the next few centuries, the groups became more sedentary and developed horticultural skills.

The environs that surrounded Mesa Verde were diverse, but far from lush. In the heart of plateau and canyon lands, the indigenous peoples of Mesa Verde had to be constantly concerned with the variable local water supply. The semi-arid climate limited building materials. Despite these possible obstacles, the Anasazi built successful agricultural settlements, primarily raising maize. By the seventh century, agriculture was the main means of sustenance for the people of Mesa Verde. As their agricultural practices grew more sophisticated, so did their social patterns. The Anasazi began to cluster greater groups of houses into small villages. They built row houses above ground, using local building materials such as wood and mud.

Sometime before 1000, dwellings had advanced from the mud row and pit houses to larger adobe structures. The Adobe structures were often multistoried and had dozens of rooms. This advancement allowed for the village structure to expand to include even more families. Additional families meant that farming practices became even more collective and systematized. This was the predominant village physical and social structure until emergence of Mesa Verde's classical period around 1100.

In the classical period, the character of Anasazi village structure altered dramatically. The villages became more enclosed and the expertise of Anasazi masonry transformed the village from architecturally functional, to aesthetically interesting. Straight-coursed walls and round towers adorned structures that were now being soundly constructed with double-coursed, carefully quarried sandstone. However, the Anasazi did not inhabit these stone villages atop the mesas for more than a few generations.

A shift in population patterns, for which archaeologists have yet to fully account, again altered the architectural design of Anasazi communities. The villages on the plateaus were largely abandoned, and people returned to the rock shelters in the cliff walls. Some scholars believe this move was for defense or religious reasons; others theorize that villages returned to the rock shelters because of the abundance of water running off of and seeping through the rock in these areas. South-facing alcoves were possibly sought to provide shelter from the elements and heat in the winter. Regardless of the reasons, the Anasazi reconstructed their villages on inset ledges of the mesa cliffs. These are the expertly constructed stone "cliff dwellings" for which the Mesa Verde region is famed.

The cliff dwellings were constructed out of single courses of stone designed to fit the alcoves in which they were built. Though there were several examples of poorer masonry than the village dwellings on the mesa top, the organization and scale of the cliff villages far surpassed those of its predecessor. There were separate structures for living, cooking, storage, and ceremonial uses. Many of the rooms had plastered walls that were intricately painted with bright pigments. Several series of wooden ladders and rope bridges, as well as natural tunnels, provided access to the mesa top or valley below. The village was kept amazingly clean and free of debris. Refuse (including the dead) was dumped down the cliff slope. In the remnants of these refuse piles, archaeologists have learned the most about the Anasazi.

A little over a century later, by 1300, these elaborate cliff dwellings and most of the Mesa Verde area were abandoned. There is evidence at that time of drought, so a possible explanation of the sudden abandonment of the cliff villages was massive crop failure and subsequent starvation.

Impact

Though the common name for the indigenous peoples of Mesa Verde is the Anasazi, meaning "ancient ancestors" in Navajo, many Native Americans and anthropologists now refer to them as the ancestral Puebloan. This change in name reflects an ongoing debate over the which modern peoples are most closely related to the inhabitants of Mesa Verde.

Obviously, without the aid of written documents, what we do know about the ancestral Puebloan was garnered from archaeological research. Archaeologists and anthropologists have no way of knowing exactly how their society was structured. Relying on comparative models from modern Pueblo peoples in the American Southwest, scholars can approximate how ancestral Puebloan society, religious traditions, and political institutions might have paralleled. Most likely, extended families lived together in one home. Members of such clans worked the same agricultural plots and perhaps even had their own family ceremonial rooms, or kivas. Modern Pueblo society is matrilineal (descent is determined through female ancestry) and most anthropologists believe the ancestral Puebloans were also, although it is likely that only men participated in the political aspects of society.

Technologically, the ancestral Puebloans were a lithic, or stone tool, culture. No evidence of metalworking or metal tools have been found among the classical period remains at Mesa Verde. Like most other indigenous cultures, the ancestral Puebloans used stone projectile points and knives. Though their tool assemblage is fairly typical, the ancestral Puebloans were expert basket-makers. This craft was a signature of the region until the introduction of ceramic pottery in the middle periods sent basket-making into decline. Pottery was easier to make, durable, and simplified cooking, thus it swiftly gained popularity. Before the abandonment of the region, several ceramic styles distinctive to Mesa Verde emerged, most featuring painted designs.

Like many other indigenous groups, the ancestral Puebloans established an elaborate trade network. They traded agricultural surplus with local neighbors, as well as specialty craft items such as baskets, leather goods, stone tools, and textiles. These specialty items were also used as trade goods on a far grander scale between communities that were geographically much further away. Remains of seashells, turquoise, and cotton—all indicative of trade from village to village, over hundreds of miles—have been excavated at Mesa Verde.

However unremarkable the daily material cultural of the peoples of classical Mesa Verde may seem, the architectural achievements of the last 200 years of their habitation in the area are outstanding in scale, technique, and appearance. The largest site at Mesa Verde, "Cliff Palace" has 217 rooms, several storage spaces, and 23 kivas. The population of this village alone could have ranged from 200 to nearly 1000. Several smaller cliff dwelling villages are found in the area, all well-planned and organized communities similar in structure to the village at Cliff Palace. Architecture and civic planning was perhaps the greatest legacy of the ancestral Puebloans' short golden age at Mesa Verde.

In 1906, Mesa Verde was declared a National Park. The park occupies nearly 80 square miles of plateau land above the Mancos and Montezuma Valleys and contains hundreds of individual archaeological sites which span the 750 years of settled occupation in the region. In 1978, it was added to the list of World Heritage Sties, granting further protection to its abundant cultural resources. It is the only U.S. National Park created solely to protect man-made works.

ADRIENNE WILMOTH LERNER

Further Reading

Roberts, David. *In Search of the Old Ones: Exploring the Anasazi World of the Southwest.* New York: Simon & Schuster Trade, 1995.

The Khmer Capital at Angkor

Overview

The famous ruins at Angkor Wat in northwestern Cambodia are remnants of the ancient capital of the Khmer Empire, which at its height ruled much of Southeast Asia. The Angkor complex, which covered approximately 77 square miles (199 sq km), was the cultural center of the empire from the ninth through the fifteenth centuries.

Background

About 2,000 years ago, the people of Southeast Asia lived in simple settlements along the coast and in valleys suitable for farming. Their way of life involved growing rice and root crops, and raising pigs and water buffalo. Their religion was animistic; that is, they believed that spirits were associated with land, trees, rivers, mountains, and other natural objects. They also practiced ancestor worship.

Over the next few hundred years, Indian merchants began seeking sea routes for trade with China. Southeast Asia became a convenient way station. Groups of Indian settlers established themselves near the ports, and brought new religious and cultural ideas to the Khmers in Cambodia as well as to other people of the region.

The Indian contributions to Southeast Asian civilizations during this period included the Sanskrit writing system, astronomy, laws, literature, and the idea of a centralized government with a powerful king. They also introduced both Hinduism and Buddhism. The two religions were combined to some degree in their Southeast Asian manifestations, and superimposed upon the animistic beliefs that are still common in the region today.

In the eighth century, Cambodia consisted of a number of principalities, which were consolidated into the Khmer Empire by Jayavarman II. (The suffix -varman was always added to the names of Khmer kings, and means protector.) Beginning in the year 790, Jayavarman II began a series of military campaigns to extend his territory, moving his capital several times. In 802, he proclaimed himself universal king and ruled from an area near the present-day Roluos until his death in 850.

Yasovarman I, who reigned from 889 to 900, moved the capital to Yasodharapura, which would eventually grow to include the temple complex known as Angkor. Except for a brief period during which a usurper ruled from a rival capital, Angkor was to be the political and cultural center of the Khmer Empire for the next 500 years.

Impact

In about 1113, Suyavarman II seized the Khmer throne by murdering his great-uncle and ruled until 1150, when he may have been murdered in turn. Suyavarman II built the great temple at Angkor Wat, dedicated to the Hindu god Vishnu, who was thought to be a protector of the world and would bring back moral order. Khmer temples were designed in accordance with a consistent geometric plan and aligned with the east-west axis. They had a central tower in an open courtyard, surrounded by a high wall with one or more entry gates. Some temples elaborated upon these essential elements, and might include long galleries, pavilions, bathing pools, and additional towers as well. Angkor Wat is an enormous structure with five large towers symbolizing the mountains where the Hindu deities were said to dwell.

Besides temples, another favorite construction project of the Khmer kings were huge reservoirs, or barays, each covering up to 20 square miles (52 sq. km) and accompanied by a network of canals and moats. The barays served as a large-scale irrigation system, to store rainwater and divert the water of the Great Lake, or Tonle Sap, as it retreated after the monsoon season. The water was channeled to rice fields, allowing three crops each year. The food surplus encouraged rampant population growth. Angkor is believed to have housed one million people in the twelfth century, when only 30,000 lived in Paris. The reservoirs may also have provided for urban needs, including transportation, drinking water, and bathing.

In 1177, the Champa state in Vietnam launched a surprise attack by sea, sailed up the Mekong River to the Khmer capital, and set fire to it. They ruled Cambodia for the next four years. Jayavarman VII headed a successful rebellion to regain the capital in 1181, and became king at the age of 55. He reigned for about 40 years, until he died in his 90s.

Jayavarman VII enjoyed a vigorous and productive old age, and the Khmer Empire reaped the fruits of his labors, although some of his subjects rebelled against his free-spending ways. His construction program included a walled city north of Angkor Wat, called Angkor Thom, Great City. He also built large numbers of monuments and guest houses. However, many of these deteriorated faster than earlier buildings like Angkor Wat, because of the poor quality of the building stone available by Jayavarman VII's time. He was especially interested in the transportation infrastructure, and built more roads and bridges than all the previous Khmer kings put together.

In 1190, Jayavarman VII took revenge upon the Champa, imprisoning the Champa king and annexing his territory. Although the Khmer ruler was a devout Buddhist mystic, he was also a skilled military leader, and greatly expanded the borders of his empire. At its height, it included parts of modern Thailand, Vietnam, Laos, Burma and much of the Malay Peninsula as well as Cambodia.

Some of the later temples at Angkor included images from Mahayana Buddhism. This type of Buddhism peaked during the reign of Jayavarman VII and is now centered in the Himalayas. It uses the Sanskrit language and stresses the veneration of images of Bodhisattvas, or "Enlightenment Beings", who had declined to enter Nirvana in order to return to Earth and ease the suffering of humanity. Bodhisattva images appeared in several forms at Angkor, along with Hindu deities.

In the thirteenth century, a short-lived Hindu religious revival led to the destruction of the Buddhist images at Angkor, but Theraveda Buddhism, a conservative form of the religion originating in Ceylon and expressed in the Pali language, was soon declared Cambodia's state religion. This school of Buddhism is now practiced throughout Southeast Asia. Angkor's last stone temple was built in about 1290. The Khmers maintained their capital at Angkor until 1432, and then began shifting southward in response to Thai invasions. Eventually they settled in Phnom Penh, which was less vulnerable to attacks from the north and better situated for maritime trade. It remains the Cambodian capital today.

Despite the loss of its status as the imperial capital in the fifteenth century, the ancient city was not immediately abandoned. Monks remained to serve at Angkor Wat. The royal court returned a few times for brief periods. But the empire itself was crumbling as its overextended central control began to weaken. Historians speculate that environmental factors such as depleted forests or drought may have contributed to its end. By the 1700s, when European traders and missionaries began traveling to Cambodia, they described Angkor as mysterious ruins hidden in the jungle, but as well known to local people as Rome was in the West.

Many of the stone temples and other buildings at Angkor have collapsed. The wooden palaces and houses of the ancient Khmer cities rotted away centuries ago. In poverty-stricken Cambodia, looting is a constant threat to the remaining architectural treasures. Parts of the Angkor site are inaccessible because of land mines, grim reminders of twentieth century hostilities. Although the stone structures that survived into modern times were largely undamaged by the decades of conflict, their caretakers were not as lucky. Many were executed by the Khmer Rouge during the civil war, which began in the 1970s and resulted in the deaths of about one million Cambodians.

In 1991, the United Nations Educational, Scientific and Cultural Organization (UNESCO) began assisting the new Royal Cambodian Government in establishing international efforts to preserve the remains of Angkor. The next year, Angkor was added to UNESCO's World Heritage List, which recognized it as one of humanity's most important cultural sites.

SHERRI CHASIN CALVO

Further Reading

Chandler, David. *A History of Cambodia.* Boulder, CO: Westview Press, 1983.

Dagens, Bruno. *Angkor: Heart of an Asian Empire.* New York: Harry N. Abrams, 1995.

Mabbett, Ian, and David Chandler. *The Khmers.* Oxford: Blackwell, 1995.

MacDonald, Malcolm. *Angkor and the Khmers.* London: Oxford University Press, 1987.

The Medieval Castle

Overview

Walled fortifications began with the founding of the first cities in the ancient world. Their design remained unaltered for almost four thousand years. Not until the late tenth century, when the first castles were built in western Europe, did a substantive change occur in the construction of fortifications. The evolution of the castle coincided with the emergence of a new political system called feudalism. For several centuries, castles played a crucial role in European history. However, by the end of the thirteenth century they had lost their military, political, and social significance and were being abandoned.

Background

In the ancient world cities were often fortified, especially if they were vulnerable to attack by outside forces. The purpose of these defensive walls was to protect the public; they were group strongholds. Over time, attackers developed techniques to penetrate these fortifications, ranging from scaling the walls with ladders, to tunneling under them, or using rams to batter them down. Machines were invented to assist in assaults: towers that could be rolled against the walls and artillery such as catapults to hurl missiles over them. Defenders, however, found ways to counter each of these methods of attack.

With the decline and fall of the Roman Empire, urban society collapsed in western Europe; people deserted cities for rural communities. Trade with the eastern Mediterranean world withered, leaving governments in the West with no money to spend maintaining expensive fortifications around the decaying urban centers. As monarchies such as the Frankish kingdoms grew weaker, outside invaders began to penetrate into the heart of western Europe.

Islamic forces known as the Saracens took Spain and pushed far into France before being turned back at the Battle of Poitiers (732 or 733), but remained a threat to Italy and southern France. The Magyars attacked from eastern Europe, carrying out their incursions from the late ninth century to the 950s. The third group of invaders, the Vikings or Norsemen, were the most dangerous because their shallow-draft ships allowed them to sail far up the major rivers of Europe, leaving few areas safe from their murderous raids, which culminated in the tenth century. Since governments were powerless to deal with these invasions, a new political and military system called feudalism evolved to meet these threats.

Under feudalism, monarchs gave much of their land to their strongest nobles as fiefs (estates) in return for the nobles' commitments to provide warriors on horseback (knights) to combat the invaders. To obtain the services of these knights, the nobles (now vassals of the king) would give up some of the land they had received to lower-ranking nobles, who promised to supply some of the knights needed. This second tier of nobles would repeat the process with weaker nobles. When raids occurred, the king would call on his vassals for knights, who would in turn call on their vassals. Groups of several dozen heavily armored men who fought on horseback were quickly assembled and the raiders, who fought on foot, were turned back.

This system worked, but the monarchs had paid a heavy price. They had lost control of much of their land to their vassals. In most of Europe, the political history of the later Middle Ages focused on the monarchs' successful attempts to regain control of their kingdoms from their nobles. Since the key military weapon from the tenth to the thirteenth centuries was the armored knight on horseback, this struggle hinged on the ability of the nobles to protect both their vassals and horses. The castle was developed to perform that function.

Impact

Feudalism and castles first developed in the Normandy region of northwest France. To protect themselves from sudden raids by the Vikings, the nobles needed strongholds to which they could retreat until they gathered their vassals for battle. These strongholds were the first castles. In 1066, the castle and feudal system were forcibly introduced into England in the Norman Conquest. Soon, feudalism and castles were established all over Europe. As the outside threats faded, nobles fought power struggles with each other and with monarchs anxious to restore central authority.

A castle was the fortified home of a member of the feudal nobility. Unlike earlier large-scale fortifications, its purpose was not to protect a large urban population but rather a noble, his family, and his retainers. It was designed to be defended by a small group of soldiers. Once danger had passed, the noble would venture forth and reestablish control over his fief. The medieval castle provided both a base for its garrison to dominate the local countryside and protection for that garrison from a superior invading force. It was the economic, administrative, and legal center of local control. Strong nobles who held huge fiefs or fiefs in widely separated areas thus needed more than one castle.

The earliest castles were of motte and bailey construction. The motte was a steep cone of earth surrounded by a deep ditch, formed when dirt was taken to build the mound. It was flat at the top, where a wooden tower, surrounded by a timber palisade, served as the noble's home and stronghold. The bailey was a large area around the motte which was also protected by a ditch and wooden palisade. Within the bailey were stables, barns, workshops, and other buildings. If an enemy penetrated the bailey, its defenders fled to the motte. Built of timber and earth, these early castles were relatively easy to construct using unskilled labor.

Since there was only room on the motte for very small towers (with deplorable sanitary conditions), castle construction quickly evolved. By the early twelfth century, stone began replacing timber in the towers and walls. Because skilled masons were needed for this work, only the wealthier, higher-ranking nobles could now afford to build castles. As nobles built larger dwellings, the motte was too small for their foundations. It disappeared and the defensive emphasis shifted to strengthening the tower (called the keep or donjon) and the outer walls of the bailey. Ditches surrounded these outer walls and wherever possible were filled with water for further protection. Access to the castle was controlled by a drawbridge which spanned the ditch or moat. By the late twelfth century, the "concentric" castle evolved, with a strong outer wall to keep sappers (tunnelers who removed the foundations of a wall causing it to collapse) and catapults away from the main fortification. This was a higher, stronger inner wall that surrounded the bailey and a massive keep. The Crusaders' Crac des Chevaliers and Richard I's Château Gaillard at Les Andelys in France were the most famous concentric castles.

Where possible, castles were built to take advantage of the surrounding terrain. Those on rugged hills or ridges were harder to assault than those on flat land. A river or swamp providing water for a moat often helped decide where a castle was built. But a source of drinking water was always the determining factor of a castle's site. A garrison could store enough food to withstand a siege (one Crusader castle held a five-year supply of grain), but it had to be located where springs or wells provided a constant supply of fresh water.

The medieval castle was overcrowded, unsanitary, and an unpleasant place to live. It was cold, drafty, and dark; fireplaces were located in the walls of the keep and provided little heat. Their ventilation was so poor that the rooms were filled with discoloring and unhealthy smoke. For defensive reasons there were few windows, merely narrow slits in the wall so that arrows could be fired at attackers. Bathrooms consisted of privies set in an outer wall or in a room built out over the ditch or moat. Their smell was barely tolerable.

The castle's purpose, however, was not comfort but defense. Forcing a castle garrison to surrender was difficult. The easiest method would seem to be surrounding it and starving the defenders into submission. In practice this rarely happened because the besiegers ran the risk of relieving forces arriving to rescue the defenders. Besides, the besiegers had to feed themselves and often ran out of food before the garrison did. Consequently, direct assaults on castles were common. A siege technique that was occasionally successful was to undermine the foundations of the walls by sapping. The defenders would try to intercept these tunnels by sinking counter-shafts, which was quite difficult to do. The main reason sapping was not more frequently used was that skilled miners were needed to carry out the tunneling, which was very dangerous. It was also very time-consuming.

Direct assault on the walls and gate was a more common method of attack. Covered battering rams and large movable towers were rolled against the walls or gate. To counter these threats, great rounded towers pierced with narrow slits were built along the walls allowing the defenders to fire arrows at those attacking the walls. Crenellations (open spaces on parapets) were built along the top of the walls to provide protection for the defenders. By the thirteenth century, the base of the walls was thickened to form a sloping incline that was difficult to scale or batter through. Machicolations (holes) were made in the roofs of gateways through which arrows, stones, and boiling pitch or water could be thrown down on attackers. Wooden shields were extended out from the tops of the walls for the same purpose. Portcullises (iron grates) fitted in stone grooves could be lowered to increase the protection of the gates.

Because of these difficulties, attackers relied heavily on siege "engines," artillery using tension, torsion, or counterpoise to hurl missiles (usually stones) into the castle or against its walls. The trebuchet and other siege engines were cumbersome but could hurl 300-pound rocks well over one hundred yards. The defenders countered by placing their own machines on the castle's walls and towers where the greater height gave them greater distance.

Although the defenders seemed to have the advantage, by the fourteenth century the castle became obsolete. It is commonly assumed that the development of gunpowder and cannon caused this decline, but early cannons were too weak and difficult to transport to play a role in siege warfare. It was the growing power of the monarchs that made castles obsolete. As urban society revived, monarchs made alliances with the towns, agreeing to protect them from the nobles in return for taxes. With this money, monarchs hired thousands of foot soldiers and armed them with pikes and long bows. Groups of a few dozen mounted knights were no match for these armies and monarchs were able to restore central authority; feudalism collapsed.

There was now no military or political role for the castle. Nor could many nobles afford the growing expense of building and maintaining them. We will never know how many castles were built since the majority were of the timber motte and bailey type, which have vanished leaving no trace. In England alone, at least half the 1,500 castles built since 1066 were deserted by 1300. In those that were still inhabited, the nobles spent their money seeking comfort, not protection. The castle was evolving into the manor house; in France, "château" no longer meant feudal castle, but rather a large country house.

ROBERT HENDRICK

Further Reading

Anderson, William. *Castles of Europe: From Charlemagne to the Renaissance*. London: Elek Books, 1970.

Bradbury, Jim. *The Medieval Siege*. Woodbridge, UK: Boydell Press, 1992.

Burke, John. *Life in the Castle in Medieval England*. London: Batsford, 1978.

Fedden, Robin and John Thomson. *Crusader Castles*. London: John Murray, 1957.

Kennedy, Hugh. *Crusader Castles*. Cambridge: Cambridge University Press, 1994.

O'Neil, Bryan. *Castles and Cannon: A Study of Early Artillery Fortifications in England* . Westport, Connecticut: Greenwood Press, 1975.

Platt, Colin. *The Castle in Medieval England and Wales*. New York: Scribner's, 1982.

Pounds, Norman J. G. *The Medieval Castle in England and Wales: A Social and Political History* . Cambridge: Cambridge University Press, 1990.

Toy, Sidney. *Castles: A Short History of Fortifications from 1600 B.C. to A.D. 1600*. London: William Heinemann, 1939.

The Gothic Cathedral: Height, Light, and Color

Overview

The Gothic cathedral was one of the most awe-inspiring achievements of medieval technology. Architects and engineers built churches from skeletal stone ribs composed of pointed arches, ribbed vaults, and flying buttresses to create soaring vertical interiors, colorful windows, and an environment celebrating the mystery and sacred nature of light. Based on empirical technology, the medieval cathedral provided the Middle Ages with an impressive house of worship, a community center, a symbol of religious and civic pride, and a constant reminder of the power and presence of God and the church.

Background

The growing impact and power of the Christian church in western Europe after the fall of Rome in 400 influenced church architecture. In Mediterranean Europe where sunny skies and hot summer days mandated buildings with small window space and thick walls, the Romanesque style dominated church architecture. However, in the northern and western regions of the continent, cloudy days and less intense summer heat were common so designers developed a style that attempted to maximize interior light and uninterrupted interior heights. Architects sought a style that would provide larger windows to illuminate the buildings' interiors. Because a cathedral nave flooded with light would have a dramatic effect on the faithful, vast window space became a necessary characteristic of the Gothic style and responded to one of the goals of a growing and dominant religion in the medieval era.

The Crusades also affected the development of the Gothic style. Crusaders returning from the Holy Land brought with them many relics, and church fathers wanted to display these holy objects prominently. Devout Christians often undertook several pilgrimages in a lifetime; because hordes of pilgrims paid homage to these relics the numbers of worshipers entering those churches increased intensifying the need for a greater amount of interior light and space.

The use of light as a factor in worship and in understanding the mystical paralleled another chief goal of the medieval cathedral builder: the pursuit of greater and greater interior heights. At a time when religion dominated everyday life and when the faithful spent an average of three days a week at a worship service, church leaders sought an architectural style which created a sense of awe, a sense of the majesty and power of God for anyone who entered the church. Waging a constant battle against gravity, master masons, who both designed and built these cathedrals, wanted to create as much uninterrupted vertical space as possible in their stone structures. These soaring heights provided a dramatic interior which served to reinforce the power of the church.

Medieval master masons used three architectural devices to create the Gothic style: the pointed arch, the ribbed vault, and the flying buttress. The pointed arch, a style that diffused to the West from the Arabic world, permitted the use of slender columns and high, large open archways. These stone arches were essential in the resultant stone bays that provided the basic support system for a Gothic cathedral freeing the area between arches from supporting the building. For the church's interior, these "curtain walls" added to the delicacy, openness, light and verticality of the space. The curtain walls on the building's exterior were filled with glass, often stained or colored glass, conveying some Biblical or other sacred tales.

The use of ribbed vaults for cathedral ceilings complemented the pointed arch as an architectural element. By carrying the theme of slender stone members from the floor through the ceiling, ribbed vaults reinforced the sense of height and lightness in the building. In a visual and structural sense, these vaults connected several stone columns throughout the building, emphasizing the interconnected stone elements which produced a skeletal frame that was both visually dramatic and structurally elegant.

The flying buttress completed the trio of unique Gothic design elements. In essence, this kind of buttress, typically used on the exterior of a church, supplemented the structural strength of the building by transferring the weight of the roof away from the walls onto these exterior elements surrounding the edifice. Often added as a means of addressing a problem of cracking walls in an existing building, these buttresses were incorporated so artfully into the exterior design of the cathedral that they became a hallmark of the Gothic style. By freeing the walls from supporting much of the weight of the cathedral roof, the flying buttress allowed medieval architects to pursue their goal of reaching ever greater interior heights.

The combination of these new architectural elements, which defined the Gothic style, along with the Church's interest in increased interior light, space, and height, resulted in a new technology heavily influenced by religion. Religion's goals provided the impetus for a daring empirical technology; at the same time, technological methods allowed the church to achieve an innovative awe-inspiring space within a new architectural style.

Impact

The Abbot Suger of St.-Denis near Paris first promoted the Gothic style in medieval France. As the leading French cleric of his time, Suger headed the mother church of St.-Denis with its strong ties to the French crown. When he sought to transform that church into an impressive center for pilgrimages and royal worship, he turned to the emerging Gothic style. Gothic elements would allow him to create a building with soaring heights, with curtain walls to fill with stories and lessons in glass, and with a display of light used to represent mystery and divinity. For Suger, the Gothic style created a transcendental aura, a theology of light and he hailed it as "[the]ecclesiastical architecture for the Medieval world." Suger's architectural preferences spread throughout France so effectively that the country became home to the most impressive and successful Gothic cathedrals. His notion that architecture could serve as theology appealed to the Church with its great influence over a mass of illiterate believers. The Gothic cathedral became a huge edifice of stories, signs, and symbols filled with church teachings and lessons for any who passed by or entered these churches. For many people of the Middle Ages, the cathedral became the poor man's Bible.

The cathedral itself was a citadel of symbols. The orientation of the building usually positioned the altar facing east toward the Holy Land with the floor plan in the shape of a cross. Exteriors contained sculptural elements representing both sacred and secular themes. A depiction of the Last Judgment often adorned the west portal so all who entered were reminded of their ultimate fate. Usually, the west portal also consisted of three entryways to mirror the doctrine of the Trinity. Interiors contained rose and other stained glass windows with the same mix of the sacred and the secular scenes present on the exterior. Rose windows themselves served as representations of infinity, unity, perfection, and the central role of Christ and the Virgin Mary in the life of the Church. The interplay of geometry and light in rose windows and the special qualities of changing color tones and glowing window glass in all of the stained glass windows created a visual experience with mystical and magical qualities that transported a viewer into a world far different from his or her mundane medieval surroundings. Sculptures within, along with paintings, tapestries, and geometric patterns in columns and walls, added to the teaching environment; inside a cathedral one could not escape being exposed to lessons or stories. Add to these the awe one felt by the great interior heights and the cathedral's impact was overwhelming, reinforcing the church's power and influence in the medieval world.

In addition to its role as a center of church lessons, the cathedral served as a source of community pride. Often the largest structure in a city or town, the church served as community center, theater, concert hall, circus ring, and meeting place. The cathedral at Amiens in northern France, for example, could house the entire population of the city. Often sited on the highest point in a city or in the city center, the cathedral dominated the cityscape. With its soaring towers and spires it could be seen for miles around and became a symbol of a city much as skyscrapers or tall monuments define cities in modern society. Because the cathedral was a source of civic as well as religious pride, cities vied with each other to build the largest or the tallest churches. As a multi-purpose structure, the cathedral served as much more than a house of worship.

Anyone who visits an extant Gothic cathedral today quickly understands the impact it had on medieval life, religion, and technology. Just as religion dominated the era, the cathedrals themselves dominated, and continue to dominate, much of the landscape of western Europe leaving no question regarding the major force in people's lives.

For example, Gothic cathedrals commanded the physical landscape with interior and exterior heights not matched until the late nineteenth century. External central cathedral towers rising as high as 450 feet (137 m) and uninterrupted interior space of 130-160 feet (40-49 m) from floor to ceiling overwhelm modern visitors much as they did medieval worshipers centuries ago.

Because Christianity reigned over every aspect of medieval society, the sacred and the secular became intertwined so that a cathedral played, and continues to play, both ecclesiastical and civic roles. With so much interior space, it remains the center for many special occasions as well as regular church activities.

Likewise, the cathedral as a marvel of an empirical technology, using relatively simple tools and skilled craftsmen aided by a large labor force, remains an impressive example of the interaction of technology and religion. That linkage has had an impact so strong in the Western world that the Gothic style has become synonymous with church architecture. The neo-Gothic style appears in many churches, and even skyscrapers, built in the nineteenth and twentieth centuries.

Standing today as reminders of a historical era, the Gothic cathedrals provide insights into the power of religion, the achievements of technology, and the role of civic pride and responsibility. Their impact has endured over the centuries and continues to inspire awe in both the sacred and the secular worlds just as they did when these magnificent stone structures were first built in the Western world several centuries ago.

H. J. EISENMAN

Further Reading

Courtenay, Lynn T. *The Engineering of Gothic Cathedrals.* Brookfield, VT: Ashgate Publishing, 1997.

Favier, Jean. *The World of Chartres.* NY: Harry N. Abrams Incorporated, 1990.

Gimpel, Jean. *The Cathedral Builders.* NY: Grove Press, 1983.

Johnson, Paul. *British Cathedrals.* NY: William Morrow & Company, 1980.

Morris, Richard. *Cathedrals and Abbeys of England and Wales: The Building Church, 600-1540.* NY: W.W. Norton & Company, 1979.

von Simson, Otto. *The Gothic Cathedral: Origins of Gothic Architecture and the Medieval Concept of Order.* Princeton, NJ: Princeton University Press, 1974.

Swaan, Wim. *The Gothic Cathedral.* NY: Parklane, 1981.

Medieval Feudalism and the Metal Stirrup

Overview

The European feudal system was intimately linked with the creation of a military upper class. Historians widely credit the development of a relatively stable feudal system as one of the main factors behind the rise of the modern nation-state in European history. The feudal system that rose to prominence in medieval Europe effectively grew out of and displaced the older social organization, principally Germanic in origin, which revolved around small groups of warriors. Changing attitudes toward land ownership and fealty, or obedience, to a lord necessitated the creation of a new social system.

These changes, however, may be linked to new advances in military technology. Because the feudal system empowered a military aristocracy, changes in warfare that accompanied the new social system must be taken into consideration. One of the main technological innovations that had a lasting impact on feudalism was the metal stirrup, which allowed for the dominance of cavalry in the feudal system and enabled high-intensity shock combat.

Background

The feudal system grew and developed as the European nobility gained greater and greater amounts of land. Furthermore, feudalism grew from changing attitudes toward the concepts of property and inheritance. The feudal system is based on the notion of tenure. This means that a holder of a piece of land has a right to that land, a right that is granted by another party. Mere possession of land, however, does not constitute tenure: possession in and of itself does not account for the ownership of land.

In the older Germanic tribal organization, the lord, known as the ring-giver, maintained his leadership role through the distribution of gifts to his followers. In the feudal system, on the other hand, the lord operated as a landlord. The landlord maintained control over the military organization through his granting of tenure. Land formed the basis of military organization. The king granted plots of land in exchange for the service of heavily-armed cavalry troopers. These knights, then, owed their allegiance directly to the king. At the same time, knights were grounded, linked to specific plots of land in a manner that clearly differentiated feudalism from the tribal military society that preceded it.

However, because knights were responsible for military service, and were subject to up to 60 days of military service a year, the knights needed to sublet their lots in order to build up their armed forces and to ensure that the land was managed effectively. In this situation, social hierarchy was based solely on land distribution. This, combined with the arrival of the stirrup in Europe by the eighth century, helped to merge military distinctions and class distinctions. It may be claimed, then, that this unique intermingling of military and class concerns resulted from the technological advances that elevated the importance of the heavily-armored mounted warrior in medieval Europe.

Impact

The introduction of the stirrup into Europe is indicative of drastic changes in warfare styles throughout Asia and the Middle East. The metal stirrup was invented by the Chinese, and scholars presume that it resulted from the influx of Indian culture into China during an extended period of Buddhist missionary activity around the fourth and fifth centuries. Prior to this, archaeologists are unable to find evidence of actual stirrups. Furthermore, literature from the period makes no mention of the device. The stirrup as we know it is not mentioned by the Chinese until the early fifth century. However, by A.D. 477, the stirrup was in common use.

Scholars contend that only stirrups made of metal were capable of transforming warfare practices. Archaeologists have found evidence of saddle straps, or of loose leather surgicles used as mounting aids, from as early as A.D. 300 in India and the Middle East. However, these were not particularly effective in battle. These leather straps frequently broke, and were unable to support the weight of a heavily-armed fighter. In fact, there are many accounts of warriors injured or killed by their own weapons after slipping off their mounts. But severe hand-to-hand combat required warriors to carry many weapons with them in case they were disarmed or one of their weapons happened to break. A stable mounted fighting force was an effective fighting force.

Indeed, scholars have found manuscripts that document initial response to the stirrup in the Arab world. A ninth-century Middle Eastern account describes the reaction of the seventh-century warrior 'Ali (600?-661). According to this account, "when he placed his foot in the *rikab,* he said, 'In the name of God' three times." Different sources record the same reaction, hinting that the stirrup was considered a gift from the heavens. These ninth-century manuscripts indicate that within 30 years of 'Ali's discovery, the stirrup, or rikab, became a standard feature in the Middle East. Also, nearly three centuries after 'Ali's reaction, these accounts provide detailed descriptions of the early stirrup.

The metal stirrup was so significant because it provided lateral support to the rider. But its importance was also tied to earlier developments such as the leather saddle, with its well-defined pommel and cantle. When combined, these devices effectively linked the horse and the rider. Such a combination enabled this fighting unit to deliver blows guided and amplified by animal energy. The warrior was able to inflict a large amount of damage. Such a change engendered mounted shock combat, a military development that shifted the focus from the axe-wielding infantry man to the heavily-armored, lance-carrying knight.

But only metal stirrups enabled this transformation. They provided the perfect complement to the leather saddle, and both of these devices were especially suited for the heavy horse, the ancestor of the draught horse and the medieval destrier. These horses were desired for combat because they could support heavily-armored knights and still move effectively.

This shift in focus was most significant in Carolingian France. Archaeological evidence indicates that, once the stirrup reached Europe in the early eighth century, there was a drastic shift from infantry combat to heavily-armored mounted combat. Frankish weapons underwent a rapid transformation at this time. Before the stirrup, the freeman with his battle-axe served as the mainstay of a kingdom's military. In France this was particularly true. The battle-axe used by Frankish troops was even named the francisca. This battle-axe, and the barbed javelin, another infantry weapon, disappeared in the eighth century. The weapons that replaced them hint at the supremacy of the cavalry. A spear with a heavy base and spurs below the blade became a standard weapon; the wide base forced the horse to absorb the impact of a blow and the spurs prevented the spear from becoming too deeply embedded in bone and tissue. Prior to the stirrup, horse-carried warriors were frequently jerked from their mounts because their spears had penetrated too deeply into their victims. A spear with a cross-piece, such as the Carolingian wing-spear, indicated that the medieval Franks were the first to realize the possibilities offered by the metal stirrup.

Use of the stirrup alone, however, did not cause social transformation or military victory. As the Frankish archaeological record reveals, a device such as the stirrup had to be used in conjunction with other technologies in order to create a total transformation in warfare. The Anglo-Saxons, for instance, who were crushed by the Normans in 1066, did not utilize the stirrup to its full potential.

The Frankish society that dominated medieval Europe utilized this new type of warfare; along with this came the social structure called feudalism. In essence, shock combat was effective, but also very expensive. Horses were extremely costly, and armor and weapons were just as difficult to secure. In the barter-economy of the eighth century, the military equipment for a single knight cost as much as 20 oxen. Indeed, in 761, a man named Isanhard acquired a horse and a sword through the sale of his ancestral lands and a slave. Skilled craftsmen were few and far between, but armor was necessary to secure a place in court and to possess political power.

The feudal aristocrat's rank was contingent on his role as a warrior. In order to maintain his equipment, he had to further sublet his own land and, in that way, secure the services of others. In the feudal system, the entitlement to wealth was linked to social responsibility. All freemen were required to bear arms when necessary. Freedom and service accompanied one another. This was the combination that powered the Germanic tribes and the Frankish infantry. But this direct link between freedom and service was complicated by the need for high-maintenance cavalry. In feudalism, true freedom was linked with economic position. The freeman was no longer distinguished from lower classes by certain rights or privileges. The feudal system erected additional levels of stratification that eventually produced a gulf between the warrior aristocracy and the greater mass that constituted the peasantry.

Military service, then, became a matter of class. Those unable to fight on horseback suffered socially as well. As the system developed, these distinctions became hereditary. The possession of land and the ability to secure more land was restricted to a select group of families who passed their estates down to their offspring and relatives.

Such a gap between rich and poor was, perhaps, unavoidable. The feudal system differentiated among people merely on the basis of who could and could not afford to purchase and support an armored horse. As the example of Isanhard reveals, only the extremely wealthy could afford to do this. Such economic restrictions could even limit the size of a king's cavalry, his strongest fighting force.

Charlemagne (742-814) attempted to alleviate this problem through a system that, had it been effective, might have limited the degree of social division that characterized feudalism. Charlemagne attempted to increase the size of his cavalry by forcing less prosperous freemen to pool their resources. Thus, a group of freemen would have been able to provide a single horseman. However, the political instability of the time prevented the full implementation of such a system. Furthermore, the rights and responsibilities of freemen in such an agreement were not clearly delineated.

While the feudal system developed in response to numerous impulses and influences, it was, above all, a military system. As such, the history of military technology provides intriguing glimpses into the motivations that accompanied the formation of this political system. These connections between material developments and the social fabric broaden our understanding of a distant time.

DEAN SWINFORD

Further Reading

Brooke, Christopher. *From Alfred to Henry III: 871-1272.* New York: Norton, 1961.

Contamine, Philippe. *War in the Middle Ages.* Translated by Michael Jones. New York: Basil Blackwell, 1984.

Edge, David, and John M. Paddock. *Arms and Armor of the Medieval Knight.* New York: Crescent Books, 1988.

Heath, Ian. *Armies of the Dark Ages: 600-1066.* Sussex: Wargames Research Group, 1980.

Jones, Archer. *The Art of War in the Western World.* Urbana: University of Illinois Press, 1987.

White, Lynn. *Medieval Technology and Social Change.* Oxford: Clarendon Press, 1962.

The Great Musical Machine: Origins of the Pipe Organ

Overview

The early history of European music is well entwined with the history of Christianity. At the very center of their mutual development stands the pipe organ. The organ and the music written for it reached a pinnacle of importance during the seventeenth century, but one must look to developments during the Middle Ages to understand how the organ came to be a part of the structure—literally—of the major Christian churches, and to appreciate its extraordinary mechanical complexity. The pipe organ was both the most important musical instrument and, along with the clock, the most complicated machine of the late Middle Ages and Renaissance.

Background

The very name "organ" reveals the dual place of that instrument in the history of music and the history of technology. The term "organon" was first used by Plato (427?-347 B.C.) and Aristotle (384-322 B.C.) to denote any kind of tool; only later did it come to refer specifically to the well-engineered assembly of pipes and bellows that make up the musical instrument known in English as the organ. The invention of the organ in antiquity is credited to Ctesibius, an Alexandrian engineer of the second century B.C. This instrument, and all the organs that followed, was characterized by four basic technological elements: 1) something to pressurize air, such as a lever- or pulley-operated pump; 2) a vessel in which to store air; 3) a mechanism such as a keyboard to control air flow; and 4) a series of different sized pipes across which the air can be directed to produce musical tones. Ctesibius's machine was praised in classical accounts for its impressive use of hydraulic principles, rather than for its musical qualities. However, two centuries after his invention, many references to organs and organ playing began to appear. Organs were a common feature in Roman life, providing music for the various spectacles of theatre, circuses, banquets, and other public events. While Ctesibius's organ and other early Roman instruments used water to maintain air pressure in the pipes, air bellows became more common than water systems sometime around the second century A.D. Some fragments of Roman organs have been recovered by archaeologists, but most of our evidence about them comes from textual descriptions and numerous artistic depictions.

There is no evidence of the existence of organs in western Europe from the fifth to the eighth centuries. The Byzantine Empire centered at Constantinople, however, continued the secular use of Greco-Roman musical instruments, including the organ. The organ was re-introduced to the West in 757 when a Byzantine leader sent an organ as a diplomatic gift to Pepin, father of the great king Charlemagne (742-814). This organ, with an elaborate system of pipes, stops, and bellows, was celebrated as an engineering marvel, and was used for public rather than religious ceremony. Although the historical record is incomplete, it seems that by sometime in the ninth century, organs had become a common element in Western European musical culture.

Most of these early medieval organs were used in strictly secular settings. Small, simple organs that could be carried, known as "portative" organs, appear in many illustrations from medieval manuscripts—the portative organ was often used as a symbol in such illustrations to represent music quite generally. But these organs, which featured a small number of pipes and a hand-operated bellows, could play only a single melodic line and had a limited range of volume and pitch. They disappeared by the sixteenth century, replaced by more versatile instruments for performing ensembles and accompanying secular singers.

Larger organs, called "positive" organs, were still moveable but significantly more complicated in their construction, featuring multiple rows of pipes and a keyboard. They required two players, one to operate the bellows alone, and could perform polyphonic music. The bellows systems were quite complex and used an array of weights to control air pressure. While some positive organs were located in churches (instead of or in addition to a larger stationary organ), like the portatives they were used primarily for the performance of secular music. Existing music manuscripts suggest that by the fourteenth century, there was a thriving tradition of chamber music written for such keyboards.

How and when organs first came to be used in and accepted by the church remains one of the major mysteries in the history of Western music. The early Christians were opposed to the use of any musical instrument in church, and they took a particularly dim view of the organ because of its association with extravagance and luxury. Yet, by the thirteenth century, evidence shows that organs were being built for many European churches and cathedrals. Their first function may have been similar to that of bells, calling the faithful to worship but not used in the mass itself. Historians can only speculate about the construction, distribution, and function of organs between 750 and 1250. It is likely that the population growth, economic well-being, and expansion of craft skills that accompanied the rise of Charlemagne helped make the organ popular among churchmen as well as laymen. Monasteries in these centuries became important centers of culture and craftwork, and organs may well have flourished along with the growing musical tradition and technological skill in these religious communities. Despite the limitations of historical evidence about the earliest church organs, bits of surviving accounts suggest that during this period ample attention was beginning to be paid to the problems of organ construction, particularly theories of pipe design, and that large stationary organs had been built for churches in various places throughout Europe.

Impact

By 1450 the organ had assumed a prominent place in liturgical music. Organs evolved alongside musical notation, both serving to fix notes and the relations between them for the first time. Detailed technological accounts of organ design and construction—and a few surviving examples—indicate a sophisticated level of craftsmanship and engineering. Different countries and regions developed characteristic organ styles: some had multiple keyboards while others had only one, some had pedals and others did not. The mechanisms used to stop pipes varied considerably, as did the materials used throughout the organ. These organs differed in their sound and capabilities as well as their construction, but in general as organs became more complex, they were capable of producing a great volume as well as a wide variety of tone colors. These massive stationary organs built in Europe's largest cathedrals and abbey churches were imposing physically and aurally. These organs produced the loudest—and the lowest—sounds known to man, and their performance surely added profundity and drama to the religious service.

The repertoire of organ music is the largest and oldest found for any instrument. Almost all of this music was intended for performance in a church, either solo or to accompany choral or congregational singing. By the seventeenth century the organ reached its final form, and found its way to the center of the European musical culture. The most important organ composer was Johann Sebastian Bach (1685-1750). Bach wrote extensively for the organ and was the first composer to fully exploit the organ's capacity for polyphony. The organ clearly inspired Bach, and his masterful compositions for organ strongly influenced the later evolution of all of Western music. After Bach's death in 1750 organ building entered a slow decline. While the organ retained its place in ecclesiastical music, it ceased to develop technologically and composers—increasingly employed in secular rather than religious posts—turned their interest to other instruments.

The nineteenth and twentieth centuries saw the installation of organs not only in churches, but in numerous other public venues. In a fascinating convergence of old and new technology, organs were used in early-twentieth-century cinemas to accompany silent films. Special theater organs were designed that could make colorful sounds not welcomed or needed in more conservative church organ compositions. The organ itself evolved in other ways. Most notably, electric substitutes for mechanical organs were popular for decades, both in smaller churches and in many American homes. By the end of the twentieth century, these electric organs were themselves being replaced by more sophisticated electronic synthesizers.

Organs in the Middle Ages and Renaissance were marvels of technological complexity, employing the highest craft and engineering knowledge to produce arrays of pipes, stops, keys, and bellows that could turn air into the most startling and awe-inspiring music. No cathedral was sufficiently majestic without a grand organ at its heart. Organs enriched the musical world of the church, and were carried without ecclesiastical concern into the new Protestant churches after the Reformation. The traditional pipe organ remains the loudest, largest, most mechanically complicated musical instrument, and its sound is both unmistakable and unforgettable.

LOREN BUTLER FEFFER

Further Reading

Baker, David. *The Organ: A Brief Guide to Its Construction, History, Usage, and Music.* Buckinghamshire, UK: Shire, 1993.

Hopkins, Edward J. *The Organ, Its History and Construction.* Amsterdam: F.A.M. Knuf, 1972.

Niland, Austin. *Introduction to the Organ.* London: Faber, 1968.

Sumner, William. *The Organ: Its Evolution, Principles of Construction, and Use.* New York: St. Martin's, 1973.

Williams, Peter. *A New History of the Organ from the Greeks to the Present Day.* Bloomington: Indiana University Press, 1980.

Williams, Peter. *The Organ in Western Culture, 750-1250.* New York: Cambridge University Press, 1993.

The Technology of the Incas and Aztecs

Overview

When Spanish conquistadors arrived in the Americas in the 1500s, among the native civilizations they encountered were two great empires. The Aztec Empire covered much of central Mexico, and had its capital at Tenochtitlan, the site of modern Mexico City. The Incas, from their capital at Cuzco, ruled a territory that stretched 4,000 miles along the western coast of South America and up into the Andean highlands. These civilizations never developed the wheel or used animals for hauling, and the Incas had no system of writing. Nevertheless, they built great cities with highly developed religious, political and economic structures, and were accomplished in the arts, creating fine jewelry, textiles and pottery.

Background

The Aztecs were part of a highly developed cultural tradition in Mesoamerica, today's Mexico and Central America. Among the peoples of the region were the Olmecs, whose civilization flourished as early as 1200 B.C., the Teotihuacan people, who built the greatest ancient city in the Americas, the Toltecs, and the Mayans. Common features of Mesoamerican culture included pyramids and temples in which human sacrifice was practiced, polytheism, a calendar, hieroglyphic writing, large commercial markets, and a ball game laden with religious symbolism.

The Aztecs began as a nomadic tribe, until they settled in a swampy area of Mexico and began building their city of Tenochtitlan in the fourteenth century. The Aztec Empire grew by conquest, and the Aztecs prospered by demanding tribute from the subjugated peoples. Captured enemy warriors supplied many of their human sacrifice victims, although women and children were sacrificed as well.

In South America, the Incas also built upon the accomplishments of their predecessors and their neighbors. These included the Nazca, Moche, Huari, Chimu and Tiahuanaco peoples. Complex societies were formed in the Andes and the coastal valleys beginning about 1800 B.C. Much of their culture was assimilated and became the foundation for the Inca civilization in the mid-1400s A.D.

Impact

Farming was very important for the civilizations of the Americas. Both the Aztecs and the Incas were excellent farmers, despite having no animals suitable for pulling plows or carrying heavy loads. Llamas were native to the Andes, but they could only carry small loads. In Mesoamerica, there were no pack animals at all. There were no wheeled carts, or even wheelbarrows. Although wheeled toys and decorations have been found at Mesoamerican sites, the wheel was never put to practical use. Human labor was marshaled to do all the agricultural work required to feed the population. The main tool was the wooden digging stick, used for turning the soil and planting seed.

Without animals, the farmers of the Americas found other ways to increase their productivity. The Aztecs built up plots of land called *chinampas* in the middle of marshy lakes by piling up layers of aquatic vegetation and rich mud from the lake bottom, along with animal and human manure. The result was an extremely fertile soil that, coupled with the warm climate of the region, could support up to seven harvests per year. Around the edges of the chinampas they planted willow trees. The extensive root systems of the willows helped to keep the soil from washing away. In the center they grew crops such as corn, beans, squash, tomatoes and avocado, flowers, and medicinal herbs. Corn was a staple of their diet, and it was the Aztecs who first introduced it to Europeans.

The Incas farmed the highlands, where special care had to be taken to prevent soil erosion on the hillsides. They practiced terraced farming, carving flat plots out of the hillside in a stairstep pattern. This greatly increased the amount of land that was available for cultivation, and helped prevent the soil from running off due to wind and rain. They also employed sophisticated irrigation methods. Using these techniques, Andean farmers cultivated potatoes, another important New World contribution to the European diet. Corn was an important crop in this region as well as in Mesoamerica.

In addition to farming, the Incas and Aztecs depended on hunting and fishing for their food supply. Their weapons included blowguns, bows and arrows, spears flung with a spear-thrower for greater distance, and slings made of braided yarn. The hunter held both ends of the sling, with a stone supported in a cradle at its center, and whirled it around his head. The stone was ejected by releasing one end of the sling. These weapons were surprisingly accurate and could be used at long range, both for hunting and in battle. Warriors also fought with wooden clubs and swords or spears edged with obsidian blades.

As fishermen, Incas and Aztecs employed a variety of techniques including angling, nets and harpoons. The bag-shaped nets of the Aztecs, woven from agave fibers, were not much different than some of the nets still used in Mexico today. Aztec canoes, used for fishing and transportation, were made from hollowed-out tree trunks. In Inca territory, in the Andes and on the South American coast, fewer trees were available, so the canoes were made from bundles of reeds woven together.

Both the Aztecs and the Incas were great builders of cities, despite the lack of wheeled carts to haul materials. Burdens that could be managed by a single man were carried in large baskets that were supported on the back and steadied by a strap across the forehead. Scholars believe that sleds, levers, or ropes must have been used to move heavier loads.

Tenochtitlan, the Aztec capital, impressed even the conquistadors. It sat in the middle of Lake Texcoco, connected to the mainland by three elevated stone causeways. Wooden drawbridges could be raised to allow boats to pass. There were also canals, both within the city and for long-distance transportation. Tenochtitlan was much larger than any European city of its time, and had wide, straight streets, stone aqueducts to bring fresh water from springs in the nearby hills, and a large, well-organized marketplace. Because of the swampy ground, the buildings sat on wooden pilings, a construction technique later adopted by the Spaniards.

The city was centered on a large pyramidal temple, the site of the human sacrifices. Around it were palaces and a ball court. The ball game, called *ulama*, was played with a rubber ball that could be propelled only using the hips. It was restricted to noblemen, and represented the battle between day and night. It was also intended as an offering for a good harvest. Like captured enemies, losing ballplayers were often sacrificed to the gods.

Aztec homes were built of adobe around a courtyard and religious shrine, and furnished with reed mats and low tables. The kitchen was equipped with a hearth fire and jars or bins for foods preserved by salting or drying in the sun. There were also grindstones for making corn flour. The flour was then cooked into a porridge called *atole* or made into *tortillas* that were cooked on a flat stone griddle. Tortillas are still central to the cuisine of the region.

The homes had adjacent bathhouses heated by a fireplace and used for taking steam baths. Water was thrown onto the hot walls to make the steam. Bathing was not only considered necessary for personal cleanliness; it also was part of religious purification rituals.

The Incas were known for their skill as stonemasons. Their buildings were constructed from huge stone blocks fitted so precisely that no mortar was needed to hold them together. Today their ruins withstand earthquakes that topple modern buildings. Yet this was accomplished with only stone hammers for cutting and wet sand for polishing. The Inca capital of Cuzco was built in the Andes with the mountains and the high walls of the fortress of Sacsahuaman for defense. In their palaces, the kings could enjoy stone baths into which the water from mountain springs was channeled.

THE AZTECS' TWO CALENDARS

Our current calendar is complicated. Seven days in a week, 12 months in a year, and each of those months is either 30 or 31 days long. Except, of course, for February, which is either 28 or 29 days in length, depending on the year. Now, think of having two calendars, one for religious purposes and one for non-religious matters. This is what the Aztecs appropriated from the Zapotec, their predecessors in Central America.

The Aztec religious calendar had 13 months of 20 days each. It formed the basis for religious ceremonies, for deciding "lucky" days based on the date of one's birth, and all other religious functions. The non-religious calendar had 365 days, divided into 18 months of 20 days each, plus an additional 5 days, which were considered very unlucky. Because of the differing lengths of the Aztec calendars, they were in synchrony only once every 52 years. Unfortunately for current scholars, the Aztecs would often refer to a date only by the name of the day, month, and the current year in the 52-year cycle. This has led to some degree of confusion for historians, who often have no way of telling which calendar cycle was referred to. Two dates are known for certain; the date that Cortez entered the Aztec capital city of Tenochtitlán (November 8, 1519) and the date of the surrender of Cuauhtémoc on August 13, 1521.

P. ANDREW KARAM

The famed Incan city of Machu Picchu was built shortly before the conquistadors arrived. However, its location was so remote that it was not discovered by outsiders until 1911. It had 143 stone buildings, of which about 80 were houses; the rest were dedicated to religious and ceremonial purposes. Incas also practiced human sacrifice in their temples, but less frequently than the Aztecs. The typical Inca house was a one-room structure made either of adobe or stone blocks, with a thatched roof and trapezoidal openings for doors and windows.

Complex civilizations like those of the Aztecs and Incas required the keeping of records. The Aztecs used hieroglyphs, or picture-writing, to represent objects and ideas in carvings, paintings, and long strips of paper called *codices*. Their counting system was based on units of 20, rather than the decimal system based on the number 10 that we use today. Their 365-day calendar consisted of 18 months that were 20 days long, plus five extra days. Astrology was important to their system of beliefs, and so the calendar was invested with religious meaning.

The Incas had no system of writing. Instead, they used bundles of cord called *quipus* to keep their numerical records. The quipu was made up of a horizontal cord with a series of strings suspended from it. The length of the cord, the color and position of the individual strings suspended from it, plus the type of knots upon them all meant something to Inca record-keepers. Quipus were used for census, taxation, and other administrative and commercial purposes.

Both cultures wove cloth using a simple back-strap loom that can still be seen in use by their remote descendants. The material being woven is stretched between two wooden poles. One pole is fixed to a tree or other support, and the other is fastened to a belt around the user's waist. Aztec cloth was generally made of plant fibers, such as cotton or fiber from the maguey cactus. Incas obtained wool from llamas and alpacas. Brightly colored garments and headdresses made of tropical bird feathers were reserved for special occasions and the nobility.

The pottery wheel was not known in either culture; nevertheless, the Incas and Aztecs were skilled at making highly decorated pottery and ceramics. The ability to craft beautiful jewelry and ritual objects from precious metals was developed thousands of years ago in the Andes, where gold was near the surface and could be obtained by panning the earth near rivers and streams. The knowledge spread to Mesoamerica in about 850 B.C. Intricate objects were molded using the "lost wax" method. The desired shape was delicately carved in beeswax, and then covered with clay to form a mold. Heated over a charcoal fire, the wax melted and ran out, and the clay shell was used as a mold for the molten metal. When the trinket was cooled, it was removed by breaking the clay. Precious stones were used for adornments and ceremonial objects, in combination with gold or alone. Turquoise and jade were particularly favored.

When the conquistadors arrived in the Americas in the sixteenth century, they were probably shocked by the practice of human sacrifice, but the lure of the gold and jewels they found led to atrocities of their own. Although some Spanish priests and laymen protested, native people who refused to give up their treasures were summarily massacred. Many others were forced to abandon the farms that supported them, and were enslaved. They were put to work mining more gold, which was shipped back to the royal court and ecclesiastical authorities in Spain. Some of it can still be seen in the gilt-covered interiors of churches there.

The Incas and Aztecs offered little resistance. In part this was because the conquistadors, although relatively few in number, had the advantages of horses, armor and guns. They also carried diseases that were new to the Americas, and took a fearsome toll. But another important reason these fierce warrior civilizations crumbled so quickly is that they believed from the beginning that they were doomed. The Aztec emperor Montezuma had been hearing rumors of strange and powerful men and perceiving omens of imminent disaster. When Cortes arrived among the Aztecs, they at first believed him to be the god Quetzalcoatl. One of the last Inca kings, Huayna Capac, heard from a soothsayer that both the royal line and his empire would be wiped out. The unfortunate oracle was promptly executed for being the bearer of bad news. Before long the cities of the Incas and Aztecs were destroyed, their rulers murdered, and Spain ruled much of the Americas.

SHERRI CHASIN CALVO

Further Reading

Boone, Elizabeth Hill. *The Aztec World.* Washington, DC: Smithsonian Institution, 1994.

Karen, Ruth. *Kingdom of the Sun: The Inca.* New York: Four Winds Press, 1975.

McIntyre, Loren. *The Incredible Incas and Their Timeless Land.* Washington, DC: National Geographic Society, 1975.

Stuart, Gene S. *The Mighty Aztecs.* Washington, DC: National Geographic Society, 1981.

Townsend, Richard. *The Aztecs.* London: Thames and Hudson, 1992.

Warburton, Lois. *Aztec Civilization.* San Diego: Lucent Books, 1995.

Medieval Trade Fairs and the Commercial Revolution

Overview

By A.D. 1200, Europe was in the process of changing from a medieval agricultural economy to one based upon interregional trade, which contributed to the growth of large urban centers. Many of these cities evolved from successful trade fairs established along busy trade routes. In turn, they engendered a commercial revolution that would eventually change medieval society.

Background

At its height, the Roman Empire extended from southwest Asia to the British Isles. Travel and communication were based upon an extensive network of roads that linked the four corners of the empire to the city of Rome. The Romans were the greatest military and civil engineers of the ancient world and many of their roads and bridges are still in operation today.

Over time, social, political, and economic problems weakened the empire, and by the end of Pax Romana, or "Roman Peace," in A.D. 180 it was in decline. Rome did not fall overnight, however. It took almost three centuries of incremental defeats and setbacks to finally bring the once mighty empire down. As problems increased, Roman authorities gradually withdrew from the border areas in order to reinforce Rome's defenses closer to home. When the army was pulled back, the protection of the far-flung empire went with it, and gradually it became impossible to maintain the established economic structure. Gangs of bandits roamed the countryside, attacking undefended farms and merchant caravans. Small, independent farmers and merchants found it impossible to continue their traditional methods of production. This brought about significant economic change, including the decline of both urban life and trade.

A large, interregional trading network under the Roman Empire was replaced by self-sufficient estates called manors. Every item that was required for survival was grown, raised, or constructed on these estates. The lord of the manor had complete control over the destiny of everyone on the estate. This system developed into the social, economic, and political model known as feudalism.

The Christian church was the major source of unity during this period of decentralized authority. The belief system of the church offered a guidepost to salvation. It was widely accepted by the population of Europe that if one followed the laws of the church, one could attain salvation. The monks and priests also provided an essential service for the monarch of each country. Since churchmen were the best educated members of early medieval society, they were essential to the day-to-day operation of the government. In time these men created an extensive and important bureaucracy in the governments of western Europe. When Pope Urban II called for a holy war, or crusade, to free the Holy Land from Muslim domination, the vast majority of European leaders answered the call. As a military operation the Crusades were a complete disaster, but their economic and cultural impact would prove to be extensive. In the area of economics, the Crusades reintroduced spices, especially pepper, silk, and perfume, back into European society. Of these three products, spices would prove to be the most significant because of its use for both preserving and flavoring food.

This early influx of goods was accompanied by an agricultural revolution based upon the application of new theories and inventions. Two of these new inventions, the horseshoe and the steel plow, allowed the farmers of western Europe to place thousands of acres of rich, fertile land under cultivation. The heavy steel plow had the ability to dig deeply into the rich soil, turning it over for planting. The horseshoe helped create a new, reliable source of power because it allowed horses to move safely through the fields without fear of damaging their hoofs on stones, rocks, and roots. Farmers also adopted the three-field system. Under this model one field would be planted with grain, the second with crops such as peas and beans, and the third would be left fallow. The peas and beans had a threefold impact on the system: the crops added to the diets of both the peasants and their livestock, and the roots and stems of these plants, when plowed under, increased the fertility of the field. The combination of increased agricultural output, population pressure, and the reintroduction of products because of the Crusades set the stage for the revitalization of trade in western Europe.

Impact

By the thirteenth century the great increase in trade and manufacturing lead to a substantial increase in the urban population of Europe. This is most readily evident in the establishment of trading and manufacturing centers in northern Italy and northwest France. The introduction of sea trade, most importantly in the Baltic and North seas, led to the development of a northern European economic federation known as the Hanseatic League. Merchants would establish fairs along these trade routes. In turn, other businessmen would take advantage of these fairs and construct and establish inns, stables, and banking institutions to service the people working at the fairs. New cities sprang up as the result of this economic activity.

These new cities were unique to the European environment. Other civilizations and empires kept strict control over their urban centers. Harsh laws and heavy taxation severely limited the initiative of merchants living in these areas. This was not the case in Europe. The vast majority of the new cities enjoyed independent status. National leaders knew that it was to their advantage to allow a considerable amount of freedom to the inhabitants of these cities. Over time, Europe began to develop a proto-capitalistic society in which the market, not the nobility, directed the economy.

This new urban economic environment was based upon talent and initiative. Success was not wholly the result of an accident of birth, but flowed from the application of intelligence and hard work. This new reality began to peel away the structure of traditional medieval society. No longer did a bright, aggressive young man have to accept that his life would be controlled by his social status at birth (women, however, remained largely excluded from such economic self-determination). This new economic system stimulated both economic and social mobility. A new, vibrant middle class was created that developed skills to take advantage of this new market economy.

Historians refer to this change as the Commercial Revolution, and revolutionary it was. Political, economic, and social power no longer rested solely in the hands of the wealthy and powerful landlords. The engine of the new economy was the middle class. The logic of the old medieval structure was based upon the fact that the lord and his vassals created and maintained an environment in which the production of food, shelter, and clothing was assured. The stability of this economic system depended upon the power of the noble class. The new economic order, which was based upon the movement of goods, shifted the location of that activity from the countryside to the new urban areas. This reduced the power of the local landlord and increased the importance of the merchant class.

This shift in power would play an important role in the development of the early modern European nation-state. In time, alliances were formed between monarchs and merchants. The traditional feudal model, which involved numerous feudal lords, created a significant barrier to trade. Often merchants had to pay multiple taxes to move their goods from one location to another. Every time they moved into a new feudal domain, another tax would be levied against them. Merchants found it far better for business if they could pay one yearly fee to the king, rather than numerous feudal lords, and rely solely on the king's protection.

In turn, the king viewed this as an opportunity to decrease the power of the nobles. With the tax revenue from the merchants, the king could afford to develop and maintain a strong standing army. This armed force could be used to ensure the safe movement of goods along the trading routes of the nation. It also enabled the king to equip his force with modern weapons, such as the long bow and cannon. These new tools of warfare created a tactical imbalance between the forces of the king and those of his nobles. Traditionally, the military power of the feudal lord rested in the defensive structure of his castle and the fighting power of his knights. The cannon quickly reduced the defensive stature of the medieval castle. Most importantly, the long bow was able to penetrate the armor of the mounted knight, which completely negated his effectiveness. Though powerful and deadly, the effective use of the long bow was easily acquired. Any peasant could be trained in a short period of time to become very accurate with this weapon. This resulted in a gradual but steady shift of power from the feudal lords to the king.

The new economy also led to an increase in anti-Semitism. The new proto-capitalistic economy created an extensive banking industry, which posed a major theological problem for European society. The Christian church had always prohibited usury, the charging of interest on borrowed money. The early Church fathers regarded this practice as unethical, as it was thought that a loving Christian should never profit from someone else's misfortune. In this new economic reality, the borrowing of money became a part of everyday business life, but the Church, which had deep reservations about the new concept of profit, refused to change its teachings. As a result, many of the banking houses were controlled by members of the Jewish faith, whose religion did not prohibit usury. The negative stereotype of the Jewish moneylender was established at this time.

A new, vibrant Europe emerged from this new economic system. Medieval trade fairs and the cities they helped create established a political, social, and economic worldview based upon the belief that any individual (again, primarily men rather than women) had the right to shape his own destiny and that success would be forever determined by talent, initiative, and drive. The new political philosophy of democratic capitalism would reorder society in Europe. Political legitimacy would come to be based upon a government's ability to create an environment in which individual talent and initiative could thrive.

RICHARD D. FITZGERALD

Further Reading

Abu-Lughod, Janet. *Before European Hegemony: The World System A.D. 1250-1350.* Oxford: Oxford University Press, 1989.

Lopez, Robert S. *The Commercial Revolution of the Middle Ages.* New York: Cambridge University Press, 1976.

Sapori, Armando. *The Italian Merchant in the Middle Ages.* New York: Norton, 1970.

Biographical Sketches

Roger Bacon
c. 1214-1292
English Philosopher, Educational Reformer, and Franciscan Monk

Roger Bacon played a key role in the early stages of the movement which eventually led to the Scientific Revolution. Instead of relying on rational deductions from the statements of ancient authorities for truths about the natural world, he advocated that confirmation by observation or experiment using the methods of mathematics should be required. While not successful in having his ideas accepted during his lifetime, his writings brought attention to this new way of thinking. He is regarded as an important medieval proponent of experimental science.

Bacon was born into a prominent family. His education emphasized the classics, geometry, arithmetic, music, and astronomy. He received his baccalaureate at Oxford University (c. 1233) and his masters at Paris (c. 1241). He subsequently lectured (1240-1247) on newly translated Aristotelian texts, helping introduce Aristotelian thought to Europe. Returning to Oxford in 1247, he remained there as a scholar and teacher until 1257.

At Oxford, Bacon was inspired by Robert Grossteste (c. 1175-1253) and Adam of Marisco (a.k.a. Adam Marsh). Grossteste is regarded as one of the earliest influences in the development of modern scientific thought. He advocated developing comprehensive laws based on personal observations of nature, used mathematics to present and explicate these laws, and wrote on experimental subjects, especially optics. Marsh, a Franciscan theologian, differed from most of the churchmen of the day by advocating the importance of experiential, as well as philosophical, knowledge. Bacon enthusiastically took the ideas of these two men as his own and became an outspoken advocate of scientific ideas and their place in university education. He attacked the Scholastics (the churchmen who controlled the universities and who, in general, opposed these new ideas) as ignorant conservatives.

In 1257, perhaps because of illness and/or financial problems, Bacon became a Franciscan monk. His religious superiors did not appreciate his unorthodox views and his outspoken contempt for authority and attempted to silence him. In 1266, Bacon appealed to Pope Clement IV for financial support for his work, proposing an encyclopedia that would interrelate all knowledge and use science to confirm the Christian faith. The Pope thought that Bacon had already completed the work and asked to see a copy, commanding Bacon not to reveal his interest to anyone. Bacon set to work, and during 1267-1268 wrote, in secret, three volumes: *Opus maius, Opus minus,* and *Opus tertium,* in which, among other things, he proposed a complete reform of education in which the sciences, includ-

Roger Bacon. *(Archive Photos. Reproduced with permission.)*

ing observation and exact measurement, would play a major part. Unfortunately, Bacon's hopes evaporated when Clement died in 1268.

Bacon subsequently began three other encyclopedias: *Communia naturalium, Communia mathematica,* and *Compendium philosophiae*. His blunt criticism of contemporary philosophers and theologians combined with what were considered to be heretical ideas involving alchemy and the mysticism of Joachim of Fiore, led to his imprisonment by the Franciscans in 1277. He spent the next 14 years imprisoned in a Paris convent. Upon his release he returned to Oxford and died soon afterward, a defeated and largely misunderstood man.

Bacon made some scientific contributions. He was the first European to describe a process for making gunpowder. He proposed motorized vehicles for land, air, and water, as well as eyeglasses and the telescope. But many of Bacon's ideas were not original. He performed few experiments himself; he was more an advocate for the scientific method than a practitioner. But his writings could not be ignored, and they, especially their general attitude toward the new scientific method of acquiring knowledge, played an important role in the development of the ideas that eventually led to the Renaissance and the Scientific Revolution.

J. WILLIAM MONCRIEF

Bénézet

c. 1165-c. 1184

French Engineer

Bénézet, sometimes known as Benedictus, remains as enigmatic a figure as any in the history of engineering. It is not clear if he received any formal training as a builder, and indeed he claimed he had received his knowledge through divine inspiration. Only one thing about Bénézet's story is clearly defined and that is the creation for which he is known: the Pont d'Avignon, first great medieval bridge and one of the first to be built in France since the fall of the Roman Empire seven centuries earlier.

Bénézet would later be canonized, with a feast day on April 14, but some listings of saints do not include him—no doubt because so many aspects of his story have a hint of legend about them. It is worth noting that the account of how he came to build the Pont d'Avignon is similar to the story of King David in the Old Testament. Like David, Bénézet was a shepherd boy from the hinterlands suddenly thrust into the spotlight, claiming special powers; as with David, these claims drew an understandably incredulous reaction from persons in authority; and again like David, he amazed all when he lived up to his claims, all to the greater glory of God.

The dates of Bénézet's life are unclear, as is his original profession. He is sometimes referred to as a priest, though more often he is described as a shepherd. The distinction is crucial, because if he were the latter, it is extremely unlikely that he had any education, and would almost certainly have been completely ignorant of reading, writing, and mathematics.

As for the city of Avignon, located in southeastern France near Marseilles, during the period from 1309 to 1417 it would serve as the seat for a series of popes and antipopes. But this was long after Bénézet's time, and in his era Avignon faced a powerful obstacle in its quest to be recognized as a great town. The city stood on the Rhône, which at that time served as the boundary between France and the Holy Roman Empire, and the current alongside Avignon was so strong that no bridge could span the river.

The problem of the Rhône bridge had discouraged even talented Roman builders in ancient times; but in the early 1170s, an untutored shepherd went to the bishop of Avignon with a plan he claimed to have received from God. Naturally, the prelate approached Bénézet's claims with skepticism. One legend maintains that Bénézet overcame the bishop's qualms by lifting a giant stone block and casting it into the river at the spot where, according to Bénézet, God had told him to build the bridge.

Regardless of the specifics, Bénézet won approval for the project and work began in 1177. Over the years that followed, the builders encountered numerous challenges—challenges that Bénézet is alleged to have overcome through miracles. He either died in 1184 or 1186, several years before the completion of the bridge in 1188, and was buried in a chapel on one of its piers.

The completed Pont d'Avignon, built of wood, would be destroyed in 1226, after which it was reconstructed in stone. It would undergo numerous renovations over the years that followed, and became immortalized in a French folk song, "Sur le pont d'Avignon." Today the bridge is a United Nations World Heritage Site.

JUDSON KNIGHT

Taddeo Gaddi

c. 1300-1366

Italian Architect and Painter

Floods in 1333 destroyed a bridge, built in 1177, that crossed the Arno River in Florence, and 12 years later, the city unveiled a new bridge, designed by the painter Taddeo Gaddi. The Ponte Vecchio was the first segmental arch bridge built in the West, and, as such, represents profound engineering achievement.

Gaddi, whose full name was Taddeo di Gaddo Gaddi, was born in Florence in about 1300. His father was the mosaicist Gaddi di Zanobi, and his godfather the painter Giotto (1276-1337). Eventually Gaddi went to work for Giotto, and lived with him for 24 years. Later Giorgio Vasari (1511-1574), who coincidentally designed a corridor for the Ponte Vecchio, would write in his famous *Lives of the Artists* that Gaddi "surpassed his master in color" and, in some works, "even in expression."

Though his most well-known paintings are in Florence, Gaddi's work can be found in other locations, including Pisa, Arezzo, and Pistoia. His earliest paintings, including *The Stigmatization of Saint Francis,* date from the 1320s and reflect the influence of his master Giotto; but by the 1330s, the most fruitful decade of his career as a painter, he had begun to show his own style in works such as the fresco cycle in the Baroncelli chapel of San Croce in Florence.

These and other works helped Gaddi amass an impressive fortune, and added to his growing reputation. He had meanwhile married Francesca Albizzi Ormanni, with whom he had three sons. Two of their sons, Agnolo and Giovanni, grew up to be painters themselves. Gaddi had also embarked on a second career as an architect, designing the Ponte Trinita, a bridge downriver from Florence that was destroyed in the 1500s.

As for the Ponte Vecchio ("old bridge" in Italian), work began in the mid-1330s. Gaddi's design called for fewer piers in the stream—just two—than did the traditional semicircular arch design handed down from the Romans, and this meant that the bridge created fewer obstructions to boat traffic, thus enabling a greater flow of commerce at the river port. It also permitted floodwaters to move more easily, which in turn reduced the threat of a great inundation. The Ponte Vecchio was designed as an inhabited urban bridge, with shops atop its three powerful masonry arches. In time it became a two-story roadway with the addition of Vasari's corridor, which connects the nearby Uffizi and Pitti palaces with other urban centers.

Gaddi continued work as a painter in the years following the completion of the Ponte Vecchio. Among the works from his later career are *The Virgin and Child Enthroned* (1355) and a fresco on the east wall of the San Croce refectory. Also exactly 600 years after the completion of the Ponte Vecchio, the bridge was one of the few spared by the Nazi army as it retreated through northern Italy in World War II.

JUDSON KNIGHT

I-Hsing

fl. 700s

Chinese Buddhist Monk, Mathematician, and Astronomer

I-Hsing is most known for his contributions to the development of a water clock with an escapement to control the speed and regularity of the clock's movements. This advancement allowed more accurate timekeeping correct to within 15 minutes a day.

Records on the life of I-Hsing, also known as I-Xing, are scarce and center on his association with the construction of a water clock in eighth century China. Early in the K'ai-Yuan reign of 713-741, the crown recruited Buddhist monk I-Hsing to work with administrative official Liang Ling-Tsan (Liang Ling-Zan) and lead the building of a bronze astronomical instrument, a water clock that would tell time and indicate the movements of the constellations.

Water clocks date back to around 3,000 B.C. in China. This type of clock relies on the constant drip of water from one vessel into another to provide a steady rate from which the passage of time could be determined. While the Chinese were developing water clocks, other countries were using such timekeeping methods as sundials or astronomical observation to tell time. The Chinese technique differed substantially in that it provided a method that did not depend on clear weather. Sun dials and astronomical observation required skies clear enough for the sun to cast a shadow on the dial or for the viewing of at least a portion of the night sky. Water clocks had no such requirements and continued operating regardless of weather.

Water clocks also allowed observers to maintain a constant unit of time. For many centuries—even after the development and refinement of water clocks—many societies used so-called temporal hours, time periods which varied from season to season and even day to day.

I-Hsing's mathematical and astronomical skills provided the theory for the water clock that he and Liang would oversee. Liang's engineering skills, on the other hand, gave form to I-Hsing's ideas. The final water clock included a celestial sphere, essentially a globe marked with representations of equatorial constellations as well as degrees of movement. As viewed from Earth, the stars appear to revolve around the planet, and the celestial sphere mimicked that movement. Outside the globe, the men added two rings: one to represent the sun's movement and one to follow the moon's travels.

To run the clock, I-Hsing and Liang employed a wheel that would turn as water steadily dropped from a tank onto the wheel. As this driving wheel turned the celestial sphere, sun ring, and moon ring moved in accordance with their current locations in the sky. In addition, the two designers added a bell-and-drum system to alert those within earshot of the current time. This intricate device took a place in the Wu Ch'êng Hall of the crown palace.

In 1092, Chang Ssu-Hsün followed in the footsteps of I-Hsing and Liang and created another great astronomical clock in China. Building upon the former designers' escapement, he constructed a 33-ft (10-m) tall water-clock tower. Water clocks continued to be used and refined in China, while European nations began to develop mechanical clocks. Eventually the accuracy of the mechanical clocks exceeded that of the water clocks. Water clocks began to disappear, and now are rarely found except in museums.

LESLIE A. MERTZ

Jayavarman VII

c. 1120-c. 1215

Khmer Emperor

Jayavarman VII was in his sixties when he took the throne of the Khmer or Angkor Empire, as Cambodia was known in ancient times, yet he reigned for three decades. During that time, he rebuilt the temple cities of Angkor Wat and Angkor Thom, and undertook the building of other temples, as well as hospitals and roads. He also expanded the boundaries of the Khmer Empire, which reached its greatest extent under his leadership.

An earlier ruler of the same name, Jayavarman II (r. c. 790-850), had founded the Khmer Empire and established Hinduism as its state religion, and some time after 900, the Khmers had carved the Hindu temple city Angkor Thom out of the jungle. Angkor Thom covered 5 square miles (12.8 square kilometers), and included a moat, high walls, temples, palaces, and a tower, all carved with detailed images of Hindu deities. Suryavarman II (r. 1113-50) began the building of Angkor Wat, the more famous—though actually the smaller—of the two temple cities. He also conquered a number of surrounding kingdoms, but after his death the empire went into a period of decline.

Jayavarman, whose reign harkened back to that of Suryavarman both in its building projects and in its territorial expansion, was the son of Dharanindravarman II (r. 1150-60), and the brother—or possibly the cousin—of Yasovarman II (r. 1160-66). His first wife, Jayarajadevi, was a devout Buddhist, and under her influence Jayavarman adopted that religion. After her death, he married her older sister, also a strong Buddhist.

The empire weakened under Yasovarman's rule, and in 1166 the rebel Tribhuvanadityavarman (r. 1166-77) seized power. Jayavarman chose to remain in the country, lying low and biding his time. Then in 1177 neighboring Champa (now part of Vietnam) invaded the Khmer Empire, ravaging the land and destroying much of Angkor

Wat and Angkor Thom. As a result, Jayavarman was able to gain Khmer support, leading a revolt of his own that ousted the Chams and led to his installation on the throne in 1181.

By this time Jayavarman was about 60 years old, and as though to make up for lost time, in the years that followed he undertook vigorous building programs and campaigns of expansion. Khmer builders under his orders constructed a Buddhist pyramid temple called the Bayon, which contained his mausoleum. They also built funerary temples dedicated to his parents, as well as numerous provincial temples. Just as Suryavarman had populated the earlier version of Angkor Wat with statues representing him as Vishnu, now temples throughout the empire bristled with representations of Jayavarman as the Buddha.

The boundaries of that empire expanded greatly under the reign of Jayavarman, who regained all the territory held by Suryavarman and went on to conquer additional areas in what is now Vietnam, Laos, Burma, and Malaysia. He also restored and expanded an earlier system of imperial highways. Along these roads, which radiated from the Bayon and palace buildings throughout Khmer lands, his workers built more than 100 rest houses, as well as some 100 hospitals.

Impressive as these undertakings were, however, the greatest of Jayavarman's building efforts took place at Angkor Wat and Angkor Thom. The versions of these that survive are largely the product of reconstruction, repair, and expansion undertaken by Jayavarman, under whose auspices the formerly Hindu temple of Angkor Wat became a monument to the Buddha—and to himself. A single tower at Angkor Thom might have as many as six dozen representations of his face, glowering down at the viewer from every possible angle.

Like the Gothic cathedrals built in France around the same time, Angkor Wat was a gigantic "sermon in stone." Carvings illustrated aspects of both the Buddhist and Hindu religions—Buddhism had risen out of Hinduism, much as Christianity and Islam did from Judaism—as well as the Khmer culture and daily life. This, however, was a world utterly foreign to the European mind. For one thing, Angkor Wat was not a place where the common people were invited to enter and worship, as they were at Notre Dame or Chartres; it was set aside purely for the royal house. Furthermore, one can only imagine a European priest's reaction to the many sculptures showing bare-chested beauties—yet this was an everyday sight in the humid jungles of Southeast Asia, where Khmer women wore wraparound skirts with nothing covering their breasts.

The pace of Jayavarman's building projects was extremely quick, and in some cases the workmanship shows this fact. It is likely that he felt a sense of urgency due to his advanced age. It is also possible that he suffered from leprosy, a dreaded disease involving gradual wasting of muscles, deformity, and paralysis that was relatively common until modern times.

When he died, Jayavarman left behind considerable physical evidence that he had once ruled a great and mighty empire—an empire that was doomed to be overtaken by outside invaders. Jayavarman's building projects and foreign conquests exhausted the energies and resources of his people, and in the years that followed his reign, Cambodia gradually went into terminal decline. By 1431 the Thais had completed their conquest of the Angkor Empire.

Despite his many efforts to preserve his memory, Jayavarman's name disappeared from later Cambodian histories, no doubt because the Thai conquerors did not want their Khmer subjects to know the former greatness of their land. Angkor Wat and Angkor Thom themselves, having been vacated by the Thais, succumbed to the surrounding jungles, their towers choked by vines and their inner courts the home of snakes and spiders. Only in the 1860s were these temple cities rediscovered—ironically, by another group of conquerors, the French colonists. Subsequent archaeological investigation revealed the record of Jayavarman and his highly advanced realm, and in later years he became a national hero.

His development of a Khmer welfare state gained Jayavarman admirers in modern times, but the Communists of the Khmer Rouge used his name as partial justification for a campaign of genocide following their seizure of power in 1975. Believing that the moats at Angkor Wat and Angkor Thom had been part of a massive irrigation system, the Khmer Rouge concluded that Jayavarman had turned his empire into a vast "rice factory," and sought to do the same through a vast network of slave-labor camps. In fact the Khmer Rouge, who killed more than 20% of the nation's people before losing power to Vietnamese invaders in 1978, were incorrect: Jayavarman and Suryavarman had not built the moats for practical purposes, but to symbolize aspects of Buddhist and Hindu cosmology.

JUDSON KNIGHT

Liang Ling-Tsan

fl. 700s

Chinese Engineer

Liang Ling-Tsan, also known as Liang Ling-Zan, made his reputation in eighth century China. He was half of a two-man team that led the construction of the first clock escapement in history.

An engineer, Liang was also a member of the Crown Prince's bodyguard. When officials of the K'ai-Yuan reign issued an edict for the construction of an astronomical instrument, they selected Liang and a Chinese monk named I-Hsing (I-Xing) to oversee the project. I-Hsing was also a noted astronomer and mathematician. The two men and a team of artisans and technicians designed a water clock with an escapement, or mechanism, to control the speed and regularity of the clock's movements, and thus allow more accurate timekeeping ability.

Water clocks dated back nearly 4,000 years before the Liang-I-Hsing project began. A very simple water clock might involve two vessels, one placed higher than the other. The empty, lower container would be placed so that it would collect drips issuing from a tiny hole in the higher container. When a day had passed (such as sunrise to sunrise), the lower vessel's water level would be marked, along with equal increments of that level. If the day was to be divided into 24 hours, which was becoming a more common time division in China, the lower vessel would be marked into 24 divisions. Once this water clock was constructed, an observer could determine the time by simply viewing the water level as it related to the hour marks on the lower vessel.

The water clocks, unlike sundials, allowed the ability to envision hours as stable entities with unchanging lengths. For hundreds of years, people from many cultures divided each day and each night into a set number of hours, regardless of the length of the day or night. In the summer, the longer days were divided into a set number of equal-length hours, and the shorter nights into a set number of equal-length hours. A daytime hour in the summer might last 70 minutes, whereas a nighttime hour might last only 50 minutes. As the days became shorter with the approach of autumn, a daytime hour would decrease in length and a nighttime hour would increase. With the water clocks, which relied on a near-constant drip rate rather than time of year and length of daylight, the opportunity arose for unchanging hour lengths.

Although water clocks can be simple devices, the water clock constructed by Liang and I-Hsing was anything but. They used water to turn a driving wheel that set the clock in motion. To regulate the clock's movements, they also employed what is credited as the first mechanical escapement. The resulting bronze clock presented a celestial map, gave the time and the location of the sun and moon, and depicted the movement of the equatorial constellations. The bronze astronomical instrument was a showpiece in the palace, rivaled in notoriety only some three centuries later when Chang Ssu-Hsün used many of the ideas of I-Hsing and Liang to construct a large and intricate water-clock tower.

LESLIE A. MERTZ

Offa

r. 757-796

Anglo-Saxon King

Ruler of Mercia, a kingdom in England, Offa left behind what is undoubtedly the third most well-known structure of pre-Norman Britain, after Stonehenge and Hadrian's Wall. Like the latter, Offa's Dyke is a line between one nation and its enemies. But whereas Hadrian intended his wall as a form of protection for Roman Britain—a purpose in which it failed miserably, as the fifth-century invasion of Offa's Anglo-Saxon ancestors made clear—Offa built his earthen dyke simply as a line of demarcation.

Offa's birth year is unknown, and his life prior to 757, when he became king of Mercia in southern England, is a mystery. Upon assuming the throne, he proceeded to bring southern England to the greatest degree of political unification and stability it had enjoyed since the Anglo-Saxon period began three centuries before.

When Offa's cousin Aethelbald (r. 716-757) was murdered, sparking a civil war, Offa quelled the rebellion with ruthless use of power. In the process, he seized control of the land and suppressed the smoldering remnants of insurrection both in Mercia and surrounding vassal kingdoms. The result of these efforts was the creation of a single state that ruled most of southern England.

As the first truly significant Anglo-Saxon king, Offa set out to establish diplomatic relations with the two most powerful forces in western Europe at the time: the Carolingian Empire, and the church. Offa and Charlemagne (742-

814) had several disagreements, but just before Offa's death in 796 they signed a commercial treaty. Perhaps even more remarkable was his relationship with the pope, who created a temporary archbishopric in Lichfield to offset the power of Canterbury's archbishop. The office of the archbishop was and is the highest office in the English church, though today that church is no longer affiliated with Rome. Because Canterbury was located in realms belonging to Kent, enemies of Mercia, Offa was willing to grant the pope greater authority over the English church in exchange for the creation of the new archbishopric.

Late in his reign, Offa called for the creation of an earthen wall to mark his kingdom's western border with Wales. This wall became Offa's Dyke, which runs for some 150 miles (240 km) from the Dee estuary in the north to the River Wye in the south. The builders used natural barriers wherever possible, but were still forced to construct 81 miles (130 km) of dyke—a length nearly 13 miles (20 km) greater than Hadrian's Wall.

Whereas Hadrian's Wall was made of stone and garrisoned with soldiers, Offa meant for his dyke simply to serve as a clear line between his realm and the "barbarians" to the west. Offa's Dyke certainly made for a formidable barrier: even today, it is as tall as 8.25 feet (2.5 m) in some places, and with the ditch beside it, is as wide as 65.5 feet (20 m). The wall, representing the work of thousands of men, runs perfectly straight for miles at a time, a testament to Anglo-Saxon engineering skills.

Another technological contribution of Offa's reign was his establishment of a new form of coinage. Coins minted by the Mercian kingdom bore the king's name and image, along with the name of the government minister who was responsible for ensuring the quality of the coins. This tradition continues even today on U.S. paper currency, which bears the signature of the treasury secretary, and the rules of coinage established by Offa prevailed in Britain for many centuries following his death.

JUDSON KNIGHT

Pi Sheng

fl. 1030s-1040s

Chinese Inventor

The inventor of the world's first movable-type printing press was not Johannes Gutenberg (c. 1395-1468), nor was the first such press built in Europe. In fact China was the home of the first printing press to use movable type, as opposed to printing from carved blocks—which was also invented in China. It was the blacksmith and alchemist Pi Sheng who developed movable type from baked clay, and he did so four centuries before Gutenberg.

The details of Pi Sheng's life are unknown; indeed, sources differ as to the date of his invention, which could have occurred anywhere between 1034 and 1048. It is much easier, however, to discern the historical and technological context in which he created his press.

As with printing, paper had first made its appearance in China. It is possible that the Chinese were making paper as early as 49 B.C., but its invention is usually credited to Tsai-lung (c. 48-118). Paper was the first of the four inventions—including printing, the magnetic compass, and gunpowder—regularly cited by historians as the greatest technological contributions of premodern China. The lag between its invention in China and its development in Europe was also longer for paper than for the other three: not until the fourteenth century did Europeans begin making paper.

Then there was block printing, whereby a printer carved out characters on a piece of wood. No single individual is credited with the invention of block printing: most likely it was the creation of seventh-century Buddhist monks who needed copies of sacred texts faster than they could produce them by hand. For ink, they used the black substance secreted by burning wood and oil in lamps. Later, when Westerners adopted this innovation as well, they incorrectly called it "India ink."

The world's first printed text was a Buddhist scroll, later discovered in Korea and probably printed in China between 704 and 751. Within a few centuries, block printing in China—particularly by Buddhist monks—had assumed massive proportions. Thus by 1000, the Buddhists had printed all their scriptures, an effort that required 130,000 wood blocks and took 12 years to complete.

Clearly it would constitute an improvement if, rather than carving out a block of wood every time he wanted to print something, a printer had at his disposal precast pieces of type. Then whenever he wanted to print a document, he could simply assemble the characters he needed. This was the achievement of Pi Sheng, who developed type made out of baked clay. He placed pieces of type in an iron frame lined with warm wax, then pressed down on them with a board until the surface was perfectly flat. After the wax cooled, he used the tray of letters to print pages.

Impressive though it was, and in spite of the fact that Chinese printers used it for a few centuries, Pi Sheng's invention never really caught on in China. The reason was that with some 30,000 characters in the Chinese language, it was actually faster to carve out a block than to sort through endless trays of pre-cast blocks. The movable-type press made much more sense in Europe, where alphabets had only a few dozen characters.

When Gutenberg developed his press in about 1450 he used metal, a far more efficient material than clay, for his type. This, too, had been pioneered in the East: in about 1390 the Korean emperor Tsai-Tung had ordered his printers to create type made out of bronze.

In modern printers' jargon, "pi type" refers to type that uses an irregular font. Apparently this term is a reference to the Chinese father of movable-type printing.

JUDSON KNIGHT

Richard of Wallingford

c. 1292-1336

English Scholar

Best known for the astronomical clock he constructed while serving as abbot of St. Albans in England, Richard of Wallingford was perhaps the first known clockmaker in history. A man of wide-ranging interests, Richard studied and recorded tides, wrote on arithmetic and trigonometry, designed astronomical instruments, and conducted studies of the heavens. At the same time he maintained one of the most powerful positions in the medieval English church.

Sometimes referred to as Richard Wallingford, the latter being the name of his hometown, Richard was the son of a blacksmith who died when he was 10 years old. After that time he came under the protection of William of Kirkeby, a prior at the Benedictine abbey of St. Albans. This institution figured heavily in Richard's life, and remained his home until William sent him to school at Oxford. He studied there for six years, gaining his B.A., and returned to St. Albans in 1314, when he was 23 years old. After three years he was ordained, then he returned to Oxford, where he remained until 1327.

It was at Oxford that Richard wrote most of his significant works, and he later commented that he felt some shame for his failure to concentrate on theological studies while at the institution. Instead, he focused on mathematics and astronomy, producing the *Quadripartitum* and *Tractus de sectore*, the first texts on trigonometry written in Latin. He also created an astronomical instrument he called the Albion, which he described in *Tractus Albionis*. The Albion was an equatorium, used for calculating planetary positions according to the system of epicycles established by Ptolemy (c. 100-170). In addition, Richard wrote works on horoscopes and astrology, as well as several ecclesiastical texts.

Soon after Richard earned a degree as bachelor of theology, he received word that the abbot of St. Albans, Hugh of Eversdon, had died. He was summoned to Avignon, and appointed to take Hugh's place. Not only was the role of abbot at St. Albans a politically powerful one in the England of those days, the St. Albans abbey was a technological center of no small significance. The abbey owned a horse-driven mill in which £100—a staggering sum for the times—had been invested, and it held an important role as a textile manufacturer.

While in Avignon, Richard contracted what was then called leprosy, though it was more likely some other disease. Syphilis, scrofula, and tuberculosis have all been offered as possibilities. Whatever the case, his health began to suffer in the last decade of his life, but this did not prevent him from undertaking his most important work, his astronomical clock.

Timekeeping devices of one sort or another had long been a feature of monasteries, where monks relied on their accuracy for dividing the day into various cycles of prayer. The clocks they used, however, were water clocks, which were susceptible to changes in temperature and to impurities in the water itself, which altered the flow of liquid. By 1271, Robertus Anglicus (fl. 1270s) was writing that clockmakers were looking for a way of using weights to drive a wheel on a steady velocity, making a single revolution a day. The first such mechanical clock, in England at least, appeared at Dunstable Priory in 1283; but Richard's clock, besides being the first whose inventor is known, was an improvement on its predecessor in many respects.

The clock he built at St. Albans showed not only the time, but the season, as well as the course of the Sun, Moon, and planets. In other words, it was a true astronomical clock. It had a wheel with 120 geared teeth at the bottom that drove all the other wheels, including one with 115 teeth that produced a revolution just 0.03 seconds longer than the sidereal day. There was also a wheel of 331 teeth, which rather than being circular was shaped in such a fashion that it replicated the Sun's equatorial velocity.

In his last years, Richard became increasingly infirm as a result of his "leprosy," and he died on May 23, 1336. The clock sat in the south transept at St. Albans until around 1546, when Henry VIII dissolved the monastery and the timepiece itself disappeared.

JUDSON KNIGHT

Alessandro di Spina
d. 1313
Italian Inventor

The identity of the man who invented reading glasses has long remained a subject of speculation, with opinions divided between Alessandro di Spina and Salvino degli Armati (d. 1317). Little is known about either, and it is likely that both contributed significantly to the development of spectacles; however, a slim majority of scholarly opinion seems to favor Spina.

Magnifying glasses, as well as lenses that used the sun's heat to create combustion, had been known since ancient times. The ancients, however, seem to have been unaware of refraction, or of the relationship between the shape of a lens and its magnifying qualities. Nor do they seem to have applied the concept of magnification to the creation of devices for aiding vision. Only in the eleventh century did Ibn al-Haitham (Alhazen; 965-1039) recognize the correlation between the curved surface of a semi-spherical lens and its powers of magnification.

Later, Robert Grosseteste (c. 1175-1253) became interested in experiments with magnifying lenses, and instilled this interest in his most famous pupil, Roger Bacon (1213-1292). The latter went on to conduct a number of experiments with mirrors and lenses, and suggested in his *Opus majus* (1268) that lenses properly shaped might have a corrective effect on persons with poor eyesight. Bacon himself did not carry his experiments very far, but it is likely his writing paved the way for the development of spectacles two decades later.

Of the two men who laid claim on the invention of eyeglasses, little is known. Armati developed his lenses in Florence between 1285 and 1299, whereas Spina's have been dated as early as 1282. Scholars are more certain about the location where spectacles made their first appearance: probably in Venice, and certainly in northern Italy. Soon the invention spread to the Netherlands as well.

In some sources, Spina is cited as a friend of Armati, and indeed Armati supporters have maintained that the latter actually made the first pair of glasses *for* Spina. It is quite possible the two men knew each other, though far from certain. As for Spina's profession, he is commonly cited as a Dominican monk. He died in Pisa in 1313.

JUDSON KNIGHT

Abbot Suger of St.-Denis
1081?-1151
French Architect, Politician, and Church Administrator

Suger, through his promotion of the redesign and reconstruction of the Abbey Church of St. Denis, near Paris, France, is regarded as the originator of gothic architecture. He also had great effect on the ideology of church decoration, was close advisor to two kings, and served for two years as regent of France.

The son of Helinand, a minor French knight, Suger was born north of Paris in the general vicinity of St.-Denis. When he was about ten, his family gave him as an oblate to the Benedictine Abbey of St.-Denis, the royal church of the patron saint of France. For the next ten years he lived in a tiny cell at L'Estrée while he completed his theological, liturgical, legal, and Latin education. Early in the twelfth century he studied at several schools along the Loire River, then returned to St.-Denis.

After 1106 Suger frequently represented the legal and political interests of St.-Denis at church councils, in Rome, and elsewhere. He gradually extended his connections with royal and ecclesiastical authorities, and became Abbot of St.-Denis in 1122. He was counselor to King Louis VI from 1124 to 1130, then minister to Louis VI from 1130 to 1137 and to Louis VII from 1137 to 1151. He ruled France as regent from 1147 to 1149 during the absence of Louis VII on the Second Crusade. Regarding his influence on French politics, Suger has been compared to Cardinal Richelieu in the court of Louis XIII.

Suger's life's project was to enlarge, beautify, and rebuild the Abbey Church of St.-Denis. This was accomplished through three major construction campaigns: the nave (west) from 1135 to 1140, the sanctuary and choir (east) from 1140 to 1144, and the exterior of the nave (north and south) from 1144 to 1150. Each phase was consecrated as it was completed. Wealthy bishops, nobles, royals, and merchants lavishly obeyed Suger's call for donations of money, gems, and artwork to furnish the new church.

Suger exemplified the bold, lively Catholicism of his time. He claimed that all ornamentation, when it reflected inward spirituality and faith, was in the service of God—and the more elaborate, expensive, and precious the ornamentation, the better. Influenced by stories he heard of the riches of the Church of Hagia Sophia in Istanbul, he developed what came to be the standard medieval arguments for church decoration. Throughout his career, Suger was often at odds with St. Bernard of Clairvaux, the zealous Cistercian monk who opposed most ostentation in churches. Suger was as strict a disciplinarian as Bernard, and followed the Benedictine rule methodically, but still claimed that the more extravagant the church and its fittings, the greater the gift to God, and the greater the glory of God.

Suger put into practice a philosophy of art that was prominent in both pagan and Christian neo-Platonic thought since the third century. Neo-Platonism is the modification of the philosophy of Plato (427-347 B.C.) that was developed mainly by the pagan Plotinus (205-270) and his followers. Plotinus believed that truth, being, goodness, beauty, and value were all one. He had a very high opinion of art, because the production and contemplation of objects of beauty could (under the right circumstances) direct a person toward the contemplation of beauty itself, and thus toward a mystical awareness of the divine. Christian neo-Platonists easily adapted this philosophy of art to Christian theology, claiming that religious art could be an effective means of conversion, devotion, and spiritual renewal.

Suger agreed with his contemporary, the artisan and philosopher of art, Theophilus Presbyter, that artists are intermediaries, like priests, uniquely situated, gifted, and commissioned to bring worshippers closer to God. The intermediationism of Suger and Theophilus, proclaiming a mutually beneficial relationship of worship and art, derives from the general twelfth-century Benedictine reception of sixth-century Patristic texts, such as those of Gregory the Great and Pseudo-Dionysius the Areopagite.

Among Suger's extant writings are *The Book of Abbot Suger of St.-Denis on What Was Done During his Administration, The Little Book of the Consecration of the Church at St.-Denis, The Life of Louis the Fat, The History of Louis VII,* and several letters. All are important source documents for understanding twelfth-century France.

ERIC V.D. LUFT

Suryavarman II
r. 1113-1150
Khmer Emperor

Ruler of the Khmer or Angkor Empire in what is today Cambodia, Suryavarman II spent much of his reign battling for control of Southeast Asia. Despite his preoccupation with war, however, he managed to direct the building of the world's largest religious structure, the temple city called Angkor Wat.

Suryavarman's birth year is unknown, and for all practical purposes his biography begins with his ascension to the Khmer throne in 1113. Long before, the Buddhist emperor Suryavarman I (d. 1050) had ruled a powerful, united realm. Despite his many wars to subdue neighboring lands, this earlier Suryavarman had found time to plan numerous public works projects, including irrigation systems, monasteries, and several temples. But in the years since, Cambodia had fallen into disunity, and by the time Suryavarman II assumed the throne, the turmoil had lasted for half a century.

Consolidating his power in part through foreign conquest, Suryavarman marched his troops to the east, west, and south. He did not venture northward: to that direction lay China, with which the Khmer Empire had not had diplomatic relations for more than two centuries. He re-established contact with China in 1116, and soon obtained for his nation a highly beneficial position as vassal to the much larger empire. This in turn gave him a free hand to deal with his neighbors, and during his reign Cambodia grew to include much of what is today Thailand, Burma, Laos, and Vietnam.

Whereas earlier monarchs were Buddhists, Suryavarman embraced Hinduism. Both religions came from India, though Buddhism had been transmitted through the Chinese, who had begun adopting the religion in the early centuries A.D. Hinduism, by contrast, had entered

Cambodia directly, brought by merchants who traded with the Khmers. Firmly entrenched in Hindu practices, Suryavarman had his own personal guru, a priest named Divakarapandita, and he resolved to build a temple to Vishnu.

That temple, begun early in his reign and still not completed at the time of his death, was Angkor Wat. In fact Angkor Wat is more like a city than a "mere" temple: the surrounding moat alone, an engineering feat in itself, is 2.5 miles (4 kilometers) long and 600 feet (180 meters) wide. Inside its walls is an enormous temple complex of towers guarding a central enclosure, an architectural symbol of Hindu beliefs concerning the outer and inner worlds. Thousands of statues and relief sculptures, depicting everything from lotus rosettes to asparas (heavenly nymphs) to prancing animals, decorate the inner courts. Others showed Suryavarman in a variety of guises both as king and as a god, the incarnation of Vishnu.

Conflict with the Dai Viet and Champa kingdoms in Vietnam occupied much of Suryavarman's attention from 1123 onward, and those conflicts would ultimately endanger Angkor Wat itself. Suryavarman himself died while still at war with Champa, and in the years that followed, the Chams swept into Khmer lands and ravaged much of Angkor Wat. Only with the ascension to power of medieval Cambodia's other great ruler, Jayavarman VII (c. 1120-1219), in 1181, were the Chams driven out and Angkor Wat restored.

JUDSON KNIGHT

Guido da Vigevano
fl. 1330s?
Italian Inventor and Physician

To judge from his writings and drawings, Guido da Vigevano (sometimes referred to as Guido Vigevano), was one of the most colorful figures of medieval technology. Along with Leonardo da Vinci (1452-1519), he is credited for developing the concept of a tank, and is also cited as the man who first conceived the idea of an automobile.

Most of what scholars know about Guido comes from his masterwork, *Texaurus regis Franciae,* which he presented to Philip VI of France (r. 1328-1350) in 1335. By that time the numbered crusades to the Holy Land had long since ended with the fall of Acre (now in Israel) to the Muslims in 1291, but "holy wars" of one form or another would continue until a few years after the Turks' capture of Constantinople in 1453. Certainly many Europeans still expected to be called to a crusade in the Holy Land at any time, and it was for this purpose that Guido wrote his text.

The composition of Guido's great opus reveals his dual roles as physician and inventor. Hence the first nine folios are devoted to the subject of health, and provide the king with information regarding the preservation of his physical well-being in far-off Palestine. The bulk of the manuscript, however—14 folios—concerns the subject of military technology.

Guido's concepts of warfare, as they emerge from the pages of his startling text, reveal his genius. Because wood was scarce in the Near East, he suggested that Philip not rely on the landscape to provide him with siege equipment; rather, he should use prefabricated materials that could be separated into relatively small parts and carried on horseback. Anticipating the highly mobile style of warfare practiced by armies today, Guido devoted considerable attention to the subject of assembling and disassembling equipment, and to the proper joints and construction that would make such activities viable. He offered designs for folding pontoon bridges and boats that could be rapidly assembled and presented new concepts in body armor.

Guido also included two designs for self-propelled wagons, forerunners of the automobile and (since they were armored machines of warfare) the tank. One would be driven by a crank, the other by a kind of highly sophisticated windmill-and-gear assembly. More than 150 years after Guido, Leonardo would create his own tank design, and no doubt the creator's artistry is one reason why his drawing is much more well-known: Guido, whose sketches lack perspective (a concept yet to be discovered at that time), was certainly no artist.

Nor were his far-fetched creations really adapted to the harsh pragmatism of late medieval Europe—though they might have been quite well-suited to the late nineteenth century, with its much more advanced technology. In any case, none of his machines got past the drawing board, because instead of going to the Holy Land, Philip in 1337 plunged France into a war with England. The conflict, which came to be known as the Hundred Years' War, would last until 1453, and by then Guido's enormously prescient designs would be all but forgotten.

JUDSON KNIGHT

Villard de Honnecourt

fl. 1220s?

French Architect

Villard de Honnecourt is known for a single portfolio consisting of 33 parchment leaves. These leaves contain drawings of French cathedrals and include, as Villard writes in the portfolio, sound advice on the techniques of masonry and on the devices of carpentry. This portfolio, due to its architectural illustrations, was extremely influential during the Gothic revival of the nineteenth century.

Very little is known of Villard de Honnecourt's life and career. It is more than likely that he was born in the village of Villard-sur-l'Escault, which is south of Cambrai in the Picardy region of France. Nothing is known of his training, schooling, or employment.

Villard's manuscript, which was apparently completed during the 1220s or 1230s, contains architectural drawings, depictions of church furnishings and mechanical devices, studies of human and animal figures, and geometrical figures, as well as illustrations and descriptions of masonry and carpentry techniques. However, there seems to be no clear theme that organizes these drawings. The manuscript appears to be either a collection of random sketches or a journal compiled from Villard's extensive travels.

While the reasons for his travels are not known, it is assumed that Villard traveled as far as the abbey of Pilis in Hungary, and visited the French cathedrals of Cambrai, Chartres, Lyon, Meaux, and Reims, as well as the cathedral of Lausanne in Switzerland.

During the French and English Gothic revival movements of the mid- nineteenth century, this eclectic portfolio was rediscovered and published. The architectural focus of the Gothic revival led to an undue focus on Villard's architectural drawings. As a result, art historians speculated that Villard was both an architect and a trained mason.

In fact, some scholars have gone so far as to attribute the designs of the cathedrals in Villard's portfolio to Villard himself. However, there is no record of Villard in any extant documents that detail the work of medieval artisans. Indeed, recent investigations have further cast doubt on Villard's role as a master craftsman or architect. For example, practical stereotomical formulas in the portfolio were often taken as evidence of Villard's training as a mason. However, in 1901 researchers discovered that these formulas were later additions inscribed by another hand. The breathless pronouncements of the nineteenth century that Villard "erected churches throughout the length and breadth of Christendom" have been replaced by the more moderate supposition that Villard was no more than an educated and inquisitive traveler who recorded details from his journeys.

His esteem as an accomplished craftsman is a result of the efforts of the caretakers of his manuscript. His architectural drawings vary considerably from the actual buildings upon which they are modeled. Details are added or deleted, and the overall compositional quality of these drawings indicates that Villard actually understood very little regarding the construction and design of medieval buildings.

Evidence suggests that, after Villard lost possession of the manuscript, several scholars attempted to repaginate its parchment leaves. In the fifteenth century, eight leaves were lost by someone named Marcel, who attempted several repagination schemes. By 1600 the portfolio belonged to the Felibien family. Later, it was moved to the Parisian monastery of Saint Germain-des-Pres. In 1795 it was added to the French national collections.

The nineteenth-century rediscovery of the portfolio led scholars to identify Villard as a master architect and the manuscript as an encyclopedia of the architectural knowledge of the Gothic period. However, these drawings are now regarded as important due to their antiquarian appeal. Thus, Villard is generally viewed as a well-traveled thirteenth-century French artist whose portfolio provides a glimpse into the interests of an era.

DEAN SWINFORD

William the Conqueror

c. 1027-1087

Norman-English Military Leader and King

Although he did not personally invent a single item or develop any specific piece of technology, William the Conqueror had more impact on the material culture of the English-speaking world than all but a handful of individuals. He is best known, of course, as the leader of the Norman invasion that in 1066 supplanted the Anglo-Saxon kings who had ruled England for some six centuries. The Norman invasion—the watershed event in all of English history—brought with it innovations in warfare, political organization, record-keeping, taxation, architecture, and most of all language that are still felt today.

William descended from a line of Vikings or "Northmen"—hence the name Normans—that had lived in northern France for about two centuries prior to his birth. The illegitimate son of Duke Robert I of Normandy (d. 1035) and a tanner's daughter named Herleve, William struggled for many years with the stigma surrounding his birth and his maternal family's low place in society. In 1035, Duke Robert went on a pilgrimage to Jerusalem, and before leaving convinced the nobles within the duchy of Normandy to recognize William as his legitimate heir. He died on the return trip, and in the years that followed young William faced enormous difficulty in establishing and maintaining control over Normandy. Knighted at 15, he survived a rebellion at age 19, and by his early twenties had emerged as a powerful leader. In 1052 or 1053 he married Matilda of Flanders, and they enjoyed a happy marriage that produced four sons and five or six daughters.

By the early 1060s, William turned his attention to England, on which the Normans had entertained designs ever since one of their own, Emma, married Ethelred the Unready (968?-1016) in 1002. The couple's son Edward the Confessor (1003-1066) became king in 1042, and when he died in early January 1066, many Normans took this as a sign that the time had come to place their claim on the throne of England. In this they were opposed by Harold (c. 1022-1066), leading member of the Godwinesons, the dominant family in Anglo-Saxon England.

Taking advantage of the fact that Harold was distracted by a conflict with Norway, William landed his army in southern England on September 28, and the next day took the town of Hastings. The English and Normans fought at Hastings on October 14, and though Harold's army put up a good fight, it was no match for the Normans. Harold himself died in battle, and William received the English crown on Christmas Day.

From a technological standpoint, the Norman conquest was interesting for several reasons. History records few notable land-sea invasions—certainly not on the scale of William's, which involved 400 large and 1,000 small craft—prior to 1066. As such it served as a prototype for another famous invasion, this time from England to Normandy, in 1944. The Norman conquest was also recorded for posterity in the era's equivalent of film, a 231- (70-m) ft-long scroll called the Bayeux Tapestry.

The latter shows the Normans' improvements in the technology of warfare, particularly larger, deeper saddles with stirrups. The superiority of Norman cavalry combined with their use of skilled archers helped them gain the advantage over the Anglo-Saxon infantry at Hastings, and further determined the direction of medieval military tactics. For the wounded, the Normans brought with them a new variety of transportation, a four-wheeled cart bearing a hammock strung between two poles called the hammock-wagon.

As he had done earlier in Normandy, William spent much of his time as king securing his power and faced a number of foes, including his son and half-brother. To secure his control, he reformed England's political organization, greatly strengthening the royal power and centralizing government while granting local earls fiefs that ensured their loyalty to the crown. As part of this process—and with an eye toward increasing the tax burden of the English people—in 1085 he ordered an intensive study of the nation's lands and properties, the *Domesday Book,* by far the most thorough census up to its time.

Meanwhile, the most lasting effects of the Norman invasion began to work their way into English culture. Norman architecture would prove highly influential on English buildings for centuries to come, but even more important was the Norman effect on the English language. The French-speaking Normans brought a whole new vocabulary to England, whose language was closely related to German. As a result, English today has an amazing array of words, some derived from the French and Latin, others from the German, and historians of the language cite 1066 as the dividing line between Old English and Middle English. A final emblem of William's lasting mark is the fact that, though his direct descendants ceased to rule in 1135, all English rulers to the present day can trace their ancestry back to him.

In spite of his greatness as a leader, William's latter years were sad ones. He lost Matilda in 1083 and grew so extraordinarily fat that on a military campaign in the summer of 1087 he injured his stomach on his pommel or saddlehorn. The wound led to an illness from which he would never recover, and he died on September 9, 1087. His body had become so bloated that the pallbearers had a hard time fitting it into the tomb and in the struggle to wedge it in, the corpse burst open. The smell of William's decomposing body filled the church, an inglorious end to an otherwise glorious career.

JUDSON KNIGHT

Biographical Mentions

Abu Ishaq al-Sahili
fl. c. 1325

Spanish Arab architect who was one of the most highly acclaimed builders of the medieval Islamic world. A native of Granada, Abu Ishaq was brought to West Africa by the Malian emperor Mansa Musa. The emperor apparently contracted Abu Ishaq's services during his celebrated *hajj* or pilgrimage to the Muslim holy city of Mecca, during which time Musa spent prodigious amounts of gold and established himself as the first sub-Saharan African ruler widely known throughout the Western world. Abu Ishaq returned with Musa to Mali in around 1325, and there was ordered to build the largest mosque in the region. Five centuries later, British traveler Henry Barth wrote of the mosque, "its stately appearance made a deep impression on my mind."

Anne of Bohemia
1366-1394

Bohemian queen who popularized the sidesaddle in England. The wife of the English king Richard II and daughter of Holy Roman Emperor Charles IV, Queen Anne was an arbiter of fashion during her time and, as such, was emulated by others. The sidesaddle, in addition to its practical style, was a form of liberation, as it allowed gentlewomen to ride in a way that accommodated their dress, which was quite bulky and elaborate at that time. Multilingual, well educated, and charming, Anne was also instrumental in introducing the works of Christian reformer John Wycliffe to her native Bohemia and was an inspiration to the poet Chaucer.

Salvino degli Armati
fl. c. 1285-1317

Italian inventor sometimes credited with the development of eyeglasses. A Florentine, Armati created his glasses between 1285 and 1299, some two decades after Roger Bacon (1213-1292) suggested in his *Opus majus* that properly shaped lenses might have a corrective effect on persons with poor eyesight. Though Armati certainly developed one or more pair of spectacles, credit is somewhat more often given to Alessando di Spina (d. 1317), a Dominican monk who may have created his own pair as early as 1282.

Banu Musa
fl. 800s

Banu Musa was the name of three brothers, Jafar, Ahmad, and Al-Hasan, all important ninth-century Arab mathematicians who continued and expanded the mathematics developed by the early Greeks. The three brothers received an excellent education in Baghdad, where they studied geometry, mathematics, and astronomy. Their scientific contributions included the concept of geometric proofs, as well as their accurate measurement of the length of a year (365 days and 6 hours long).

Ch'iao Wei-Yo
fl. 980s

Chinese engineer who in 984 built the world's first canal lock. Assigned to the enormous building project of the Grand Canal connecting the Yangtze and Yellow rivers, Ch'iao Wei-Yo devised a system that made use of a chamber enclosed by movable gates between two stretches of water differing in height. Water either entered through sluices to raise the boat, or was drained to lower the boat to the appropriate level. Despite this early invention of the canal lock in China, the invention gained little use there. Instead, it became much more widely implemented in the West, where it first appeared in 1396.

Chang Ssu-Hsün
fl. 900s

Chinese inventor who is best known for his invention (in 976) of the chain drive for mechanical clocks. The first chain drive, designed by Philon of Byzantium in about 250 B.C., was used in a catapult. Mechanical clocks did not appear in Europe until the early fourteenth century, and the first Western chain drive was not developed until 1770.

Charlemagne (also known as Charles I)
742-814

European ruler who conquered and united most of Europe, creating an entity reminiscent of the Roman Empire. He was proclaimed the first emperor of the Holy Roman Empire by Pope Leo III in 800. Charlemagne raised the level of cultural and intellectual life throughout Europe, founding schools and bringing together at his court intellectuals from throughout the Empire. His reign initiated the intellectual and political recovery of Europe from the Dark Ages.

Charles V
1338-1380

French king noted for his support of scholarship. Known as Charles le Sage (the Wise), he became regent in 1356 after the capture of his father, John II, by the English at Poitiers during the Hundred Years' War (1337-1453). In bailing out his father he ceded territory to the English, but later won it back. Charles reigned from 1364-80 and, though he raised taxes on his people, was credited as a wise and fair ruler. He opened one of the first important libraries in Europe, the National Library in Paris, in 1373, and was also a patron of art and literature. In 1380, however, he banned the study of alchemy in France, and made it a crime to possess alchemical instruments.

Jean de Chelles
fl. 1200s

French artisan who is credited with the design and construction of the most famous of the three rose windows that adorn Notre Dame Cathedral in Paris. This beautiful rose medallion stained-glass window, entitled "The Glorification of the Virgin" and attributed to de Chelles, was added to the north face of the massive cathedral in 1240-45. In addition to the rose window, de Chelles also added a bay and elaborate façade to create the north transept and began construction on a bay and south transept, which were finished under the direction of another artisan, Pierre de Montreuil.

Saint Cyril
827?-869

Greek librarian and missionary, also known as Constantine the Philosopher, who invented Cyrillic script. The youngest child of a Greek father and a Slavic mother, Cyril studied science, music, geography, and languages at the royal school of Magnaura in Constantinople, then part of the Byzantine Empire. Afterwards, he was appointed librarian of Saint Sofia and later professor of philosophy at Magnaura. Along with his brother, Methodius, Cyril helped create the Slavic alphabet, based upon Cyrillic script, which was used to translate Christian texts.

Edward I
1239-1307

English king known, among other things, for reforms, such as his standardization of measurements. During his reign (1272-1307), Edward greatly curtailed the power of feudal lords and in 1295 summoned England's first parliament. In 1305 he standardized the acre as a unit of land measurement. A participant in the Ninth Crusade (1270-72), Edward was among the monarchs that Bar Sauma, the Chinese-Turkish Nestorian monk, met during his trip to Europe in 1287-88. Edward spent a good deal of his reign suppressing revolts in Wales and Scotland—he was the English king depicted in the 1995 Academy Award-winning film *Braveheart*—and died on his way to suppress a revolt in Scotland under Robert the Bruce.

Guyot de Provins
fl. 1200s

French poet who provided one of the earliest written mentions of a magnetic compass. In his satire *La Bible* (c. 1205), Guyot described a compass used by sailors under the command of Holy Roman emperor Frederick I Barbarossa—who drowned in 1190 on his way to take part in the Third Crusade.

Hartmann von Aue
1170?-1220?

German poet who made the first Western reference to the magnetic compass. A Swabian knight as well as epic poet and minnesinger, Hartmann was firmly rooted in the courtly and Arthurian traditions of literature in the High Middle Ages. The work mentioning the compass dates from about 1200.

Henry IV
1366-1413

English king who, like his contemporary Charles V, outlawed the practice of alchemy in his realms. Henry, who spent much of his reign (1399-1413) in battles to consolidate his power, issued the edict in 1404. By that time several states in western Europe had prohibited alchemical practice. Rulers such as Henry and Charles apparently feared it as a challenge to their authority because, if alchemy really did work, the production of gold by a private individual would play havoc with state finances, much as counterfeiting would in a modern economy. Despite its modern status as a pseudo-science, alchemy was an important precursor of modern chemistry, perhaps to an even greater extent than astrology was to astronomy.

Htai Tjong
fl. 1400s

Korean king who in 1403 ordered the creation of 100,000 pieces of cast bronze type for printing. Movable-type printing had been in use in China since its creation by alchemist Pi Sheng in about 1045. Pi Sheng, however, had used clay type, and the Korean implementation of bronze type—more than 40 years before Johannes Gutenberg used a metal movable-type press in Europe—constituted a major improvement.

Abul Qasim ibn Firnas
d. 873

Arab Spanish inventor who devised and demonstrated a glider. A native of Cordoba, Ibn Firnas experimented with the manufacture of glass and developed a chain of rings to depict the motions of the stars and planets. Eventually, however, he found his greatest interest in flying and constructed a rudimentary flying machine, which much of Cordoba's population gathered to watch him demonstrate. Unfortunately, after flying just a short distance, the craft fell to earth, severely wounding Ibn Firnas, who died shortly afterward.

Jabir ibn Hayyan
721?-815?

Arab alchemist who pioneered the development of many chemical processes. He prepared steel and other metals, used manganese dioxide in glassmaking, and devised dying and tanning techniques. He also prepared hydrochloric, citric, and tartaric acids, as well as ammonium chloride and aqua regia. Jabir is best known for modifying the Greek doctrine of four elements, maintaining that they combine to form sulfur (idealized principle of combustibility) and mercury (idealized principle of metallic properties), from whence all metals are formed. Jabir believed that, in principle, it was possible to transmute one metal into another, an idea widely believed until the rise of the phlogiston theory in the late seventeenth century.

Al-Jazari
fl. 1200s

Arab inventor who is remembered for his design of five water-raising machines. The first two, powered by animals, used an open channel and a scoop. The third machine relied on water power and a series of gears to lift pots, which in turn raised the water. His fourth device was the first known machine to use a crank. The fifth machine was the most complex and utilized a cog wheel, pistons, and suction pipes.

Konrad Kyeser von Eichstadt
fl. early 1400s

German author of the *Bellifortis* (1405), also called the *Kyeser Codex,* in which he compiled everything that he knew about contemporary technology, primarily warfare. Interestingly, the work also contains one of the earliest known references to a "chastity belt," an iron device that Italian men are reported to have imposed upon women to assure their sexual abstinence while the men were away, usually at war.

Brunetto Latini
1230-1294

Italian encyclopedist who in his *Tesoro* (1260) made an early mention of the compass's use at sea, as well as superstitions surrounding the compass's seemingly magical powers. A prominent Florentine, Latini greatly influenced Italian poet Dante Alighieri. Though in the *Inferno* section of Dante's major work, *The Divine Comedy,* Latini is shown suffering, condemned because of his homosexuality, Dante's persona in the work addresses Latini with obvious respect.

Robert de Luzarches
1199?-?

French architect who is credited with the rebuilding of the cathedral of Notre Dame in Amiens, France, after the original was destroyed by fire in 1218. Luzarches's cathedral, begun around 1220, became the standard design throughout France and beyond, emulated for its successful counterbalance of weight and strength and its Gothic style. The Amiens Cathedral is the largest of the three great Gothic cathedrals built in France during the thirteenth century and remains the largest in France to this day.

Marcus Graecis
1100s

Byzantine scholar who provided the earliest written description of Greek fire. This incendiary, developed for military purposes, had existed for several centuries and gave Byzantine forces a decisive advantage in numerous engagements. Marcus, also known as Marcus Gracchus, gave the formula for Greek fire thus: "Take pure sulfur, Tarter [salt produced by the reaction of tartaric acid with a base], Sarcocolla [Persian gum], pitch, dissolved nitre, petroleum [available in surface deposits throughout the Middle East and nearby areas] and pine resin; boil these together, then saturate tow [linty cloth] with the result and set fire to it."

Menahem ben Saruq
c. 910-c. 970

Spanish Jewish poet and lexicographer who compiled the *Mahberet,* the first dictionary of Hebrew. Menahem, secretary to Hisdai ibn

Shaprut (c. 915-c. 975), helped spawn a golden age in Hebrew philology.

Filippo da Modena or Fioravante da Bologna
fl. 1430s

Italian engineers who built Italy's first canal lock, and indeed one of the first pound locks in Europe, at Milan. Bringing together two staunches or gates between the Via Arena and the Naviglio Grande, they managed to get the gates as close together as possible, and thus were able to reduce water loss—in other words, to "lock" the water in.

Tomasso da Modena
c. 1325-1379

Italian painter who was the first to depict eyeglasses. The painting was a 1352 portrait of Hugh of Provence, showing its subject holding a pair of glasses—which at that time, less than a century after the invention of spectacles, were regarded as a status symbol. Among Modena's other works are 40 panels that depict the monks of a Dominican chapter house reading, writing, and praying; an altarpiece commissioned by Charles IV of Bohemia; and a portrait of Cardinal Nicholas of Rouen.

Peter of Colechurch
1150-1205

English engineer who directed the building of the first London bridge. Construction on the bridge began in 1176 and continued past the time of his death 29 years later. The resulting stone structure was an impressive achievement, including 19 pointed arches and a drawbridge. Because its piers were so wide, the 900-ft (274.3-m) waterway beneath was reduced to less than a third of its original width. In later years, Londoners built houses and shops on the bridge that overhung its sides. The medieval London Bridge was replaced in 1831, more than six centuries after its construction. By contrast, the 1831 bridge lasted only until 1973, when it was moved to Lake Havasu in Arizona, and a new London Bridge was built.

Petrus Peregrinus
fl. 1200s

French scholar noted for his observations on the magnetic compass and for his work as a military engineer. In 1269 Petrus, also known as Pierre Pelerin de Maricourt, was involved in a siege by the French army against the city of Lucera in southern Italy. While occupied on projects such as making machines for slinging stones and fireballs against the city, he began to consider the concept of a perpetual-motion machine. In Petrus's conception, a wheel could be kept constantly in motion by use of a magnet. He explained this idea in *Epistola de magnete,* a letter on the subject written to a layman.

Theophilus Presbyter
fl. 1100s

German Benedictine monk and craftsman who, between 1110 and 1140, wrote *De diversis artibus,* an extensive three-volume work about the techniques of all known contemporary crafts, including stained glass, metalwork, ivory carving, and manuscript illumination. The Latin text, which makes the earliest known European reference to paper, was translated into English in 1961. Theophilus Presbyter was probably the pseudonym of Roger of Helmarshausen.

Berthold Schwarz
fl. 1300s

German monk and alchemist who was once credited with the discovery of gunpowder (c. 1313) in Europe, though Roger Bacon is now known to have discussed it earlier. A statue in the town of Freiberg claims that he invented gunpowder and firearms, and a copper engraving found in numerous books on explosives depicts Schwarz firing a charge of gunpowder with a flint of steel, marked with the date 1380. Schwarz has also been cited as the inventor of the cast bronze cannon in Europe. However, there is no clear evidence of his discoveries, or even of his life (no birth or baptismal record, tombstone, monastery roll, or other documents attest to his existence), and it is possible that "Black Berthold" is a myth.

Shih Tsung
fl. 900s

Chinese emperor who commissioned a cast-iron sculpture in 954 to commemorate his victory over the Tartars. The work, which weighs about 40 tons (60,963 kg), was cast from a massive piece of cast iron and became known as the Great Lion of Tsang-chou.

Ulman Stromer
fl. 1300s

German inventor who established a paper mill in Nuremberg, Germany, after seeing similar paper mills in Italy. Stromer's mill used water-powered hammers to beat the material, a method that the Chinese had already developed but was not brought to Europe until Muslims established paper mills in Spain. Mills similar to

Stromer's enabled the success of Gutenberg's printing press in the mid-fifteenth century and the cultural revolution that followed.

Mariano di Jacopo Taccola
1382-c. 1453

Italian inventor responsible for the keel-breaker, a security device for ships, and the trebuchet, a siege engine. As an early figure of the Renaissance, Taccola saw himself as one helping to restore the knowledge of the ancients, and he came to be known as the "Archimedes of Siena." His keel breaker, intended to discourage pirates from stealing ships, was designed to render a vessel useless by piercing its hull if anyone attempted to seize the craft without first disengaging the trigger mechanism. The trebuchet was a wood and iron catapult for pummeling an enemy city's walls with stones.

Tseng Jung-Liang (Zeng Gung-lyang)
fl. 1000s

Chinese author of *Wu Ching Tsung Yao* (1044), an encyclopedia in which early Chinese versions of gunpowder are mentioned. Tseng wrote that these mixtures, often containing petrochemicals and even garlic or honey, were useful in flamethrowing devices, fireworks, and rockets. Though the Chinese probably used firecrackers as early as the sixth century, these did not contain gunpowder. Tseng's manuscript supports the theory that the Chinese had indeed invented gunpowder by the eleventh century.

Heinrich de Vick
fl. c. 1379

German clockmaker who created the first mechanical clock for which a complete description exists. The first known public clock that struck the hours appeared in Milan in 1335, and others made their debut around Europe in the years that followed. De Vick's, built in 1379, was regulated by a balance, and used a verge or crown-wheel escapement.

Walter de Milemete
fl. 1300s

English scholar who created the first Western illustration of a firearm. The drawing, which depicts a small cannon for firing arrows, appeared in Walter's *De officiis regnum* (1326). Firearms had been used in the 1324 siege of the German town of Metz, and in 1326 a Florentine document mentioned a bronze gun capable of firing iron balls.

Cheng Yin
fl. 800s

Reputed Chinese author of a ninth-century work entitled "Classified Essentials of the Mysterious Tao of the True Origin of Things," which contains an important early description of a type of Chinese gunpowder. The author warns that the substance is dangerous, causing serious burns if mixed incorrectly, and that the experimenter should exercise extreme caution.

Bibliography of Primary Sources

Eichstadt, Konrad Kyeser von. *Bellifortis* (1405). Also called the *Kyeser Codex*, this work is a compilation of everything that von Eichstadt knew about contemporary technology, primarily warfare. Interestingly, the work also contains one of the earliest known references to a "chastity belt," an iron device that Italian men are reported to have imposed upon women to assure their sexual abstinence while the men were away, usually at war.

Neckam, Alexander. *De Utensilibus* (On Instruments) (1180). Earliest known mention of the compass in a European text. Neckam discussed the use of a magnetic compass by sailors for navigation.

Menahem ben Saruq. *Mahberet* (c. 950). The first dictionary of Hebrew, this work helped spawn a golden age in Hebrew philology.

Theophilus Presbyter. *De diversis artibus* (between 1110 and 1140). An extensive three-volume work about the techniques of all known contemporary crafts, including stained glass, metalwork, ivory carving, and manuscript illumination. The Latin text, which makes the earliest known European reference to paper, was translated into English in 1961. Theophilus Presbyter was probably the pseudonym of Roger of Helmarshausen.

Tseng Jung-Liang. *Wu Ching Tsung Yao* (1044). A Chinese military text that describes how soldiers magnetized iron shards by heating and then rapidly cooling them. They then floated the piece on water to indicate direction. The encyclopedia also mentions early Chinese versions of gunpowder. Tseng wrote that these mixtures, often containing petrochemicals and even garlic or honey, were useful in flamethrowing devices, fireworks, and rockets. Though the Chinese probably used firecrackers as early as the sixth century, these did not contain gunpowder. Tseng's manuscript supports the theory that the Chinese had indeed invented gunpowder by the eleventh century.

Villard de Honnecourt. Untitled. (c.1220s or 1230s). A single portfolio consisting of 33 parchment leaves. These leaves contain drawings of French cathedrals and include, as Villard writes in the portfolio, sound advice on the techniques of masonry and on the devices of carpentry. This portfolio, due to its architectural illustrations, was extremely influential during the Gothic revival of the nineteenth century.

JOSH LAUER

General Bibliography

Anderson, E. W. *Man the Navigator.* London: Priory Press, 1973.

Asimov, Isaac. *Adding a Dimension: Seventeen Essays on the History of Science.* Garden City, NY: Doubleday, 1964.

Basalla, George. *The Evolution of Technology.* New York: Cambridge University Press, 1988.

Benson, Don S. *Man and the Wheel.* London: Priory Press, 1973.

Boorstin, Daniel J. *The Discoverers.* New York: Random House, 1983.

Bowler, Peter J. *The Norton History of the Environmental Sciences.* New York: W. W. Norton, 1993.

Brock, W. H. *The Norton History of Chemistry.* New York: W. W. Norton, 1993.

Brooke, John Hedley. *Science and Religion: Some Historical Perspectives.* New York: Cambridge University Press, 1991.

Bruno, Leonard C. *Science and Technology Firsts.* Edited by Donna Olendorf, guest foreword by Daniel J. Boorstin. Detroit: Gale, 1997.

Bud, Robert and Deborah Jean Warner, editors. *Instruments of Science: An Historical Encyclopedia.* New York: Garland, 1998.

Butterfield, Herbert. *The Origins of Modern Science, 1300-1800.* New York: Macmillan, 1951.

Bynum, W. F., et al., editors. *Dictionary of the History of Science.* Princeton, NJ: Princeton University Press, 1981.

Campbell, Anna Montgomery. *The Black Death and Men of Learning.* New York: Columbia University Press, 1931.

Carnegie Library of Pittsburgh. *Science and Technology Desk Reference: 1,500 Frequently Asked or Difficult-to-Answer Questions.* Washington, D.C.: Gale, 1993.

Crombie, Alistair Cameron. *Medieval and Early Modern Science.* Garden City, NY: Doubleday, 1959.

Crone, G. R. *Man the Explorer.* London: Priory Press, 1973.

De Groot, Jean. *Aristotle and Philoponus on Light.* New York: Garland, 1991.

Ellis, Keith. *Man and Measurement.* London: Priory Press, 1973.

Good, Gregory A., editor. *Sciences of the Earth: An Encyclopedia of Events, People, and Phenomena.* New York: Garland, 1998.

Grant, Edward. *The Foundations of Modern Science in the Middle Ages: Their Religious, Institutional, and Intellectual Contexts.* New York: Cambridge University Press, 1996.

Grant, Edward. *Physical Science in the Middle Ages.* New York: Wiley, 1971.

Grattan-Guiness, Ivor. *The Norton History of the Mathematical Sciences: The Rainbow of Mathematics.* New York: W. W. Norton, 1998.

Gullberg, Jan. *Mathematics: From the Birth of Numbers.* Technical illustrations by Pär Gullberg. New York: W. W. Norton, 1997.

Hasan, Ahmad Yusuf, and Donald Routledge Hill. *Islamic Technology: An Illustrated History.* New York: Cambridge University Press, 1992.

Hellemans, Alexander and Bryan Bunch. *The Timetables of Science: A Chronology of the Most Important People and Events in the History of Science.* New York: Simon and Schuster, 1988.

Hellyer, Brian. *Man the Timekeeper.* London: Priory Press, 1974.

Holmes, Edward, and Christopher Maynard. *Great Men of Science.* Edited by Jennifer L. Justice. New York: Warwick Press, 1979.

Hooper, Nicholas, and Matthew Bennett. *The Cambridge Illustrated Atlas of Warfare: The Middle Ages, 768-1487.* New York: Cambridge University Press, 1995.

Hoskin, Michael. *The Cambridge Illustrated History of Astronomy.* New York: Cambridge University Press, 1997.

Kren, Claudia. *Medieval Science and Technology: A Selected, Annotated Bibliography.* New York: Garland, 1985.

Lankford, John, editor. *History of Astronomy: An Encyclopedia.* New York: Garland, 1997.

Lindberg, David C. *Theories of Vision from al-Kindi to Kepler.* Chicago: University of Chicago Press, 1976.

Lindberg, David C., editor. *Science in the Middle Ages.* Chicago: University of Chicago Press, 1978.

Multhauf, Robert P. *The Origins of Chemistry.* New York: F. Watts, 1967.

Porter, Roy. *The Cambridge Illustrated History of Medicine.* New York: Cambridge University Press, 1996.

Reeds, Karen. *Botany in Medieval and Renaissance Universities.* New York: Garland, 1991.

Sarton, George. *Introduction to the History of Science.* Huntington, NY: R. E. Krieger Publishing Company, 1975.

Smith, Roger. *The Norton History of the Human Sciences.* New York: W. W. Norton, 1997.

Stiffler, Lee Ann. *Science Rediscovered: A Daily Chronicle of Highlights in the History of Science.* Durham, NC: Carolina Academic Press, 1995.

Talbot, C. H. *Medicine in Medieval England.* London: Oldbourne, 1967.

Temkin, Owsei. *Galenism: Rise and Decline of a Medical Philosophy.* Ithaca, NY: Cornell University Press, 1973.

Travers, Bridget, editor. *The Gale Encyclopedia of Science.* Detroit: Gale, 1996.

Whitehead, Alfred North. *Science and the Modern World: Lowell Lectures, 1925.* New York: The Free Press, 1953.

World of Scientific Discovery. Detroit: Gale, 1994.

Young, M. J. L. et al., editors. *Religion, Learning, and Science in the Abbasid Period.* New York: Cambridge University Press, 1990.

Young, Robyn V., editor. *Notable Mathematicians: From Ancient Times to the Present.* Detroit: Gale, 1998.

Index

Numbers in bold refer to main biographical entries